Karl K. Sabelfeld

Random Fields and Stochastic Lagrangian Models

Karl K. Sabelfeld

Random Fields and Stochastic Lagrangian Models

Analysis and Applications in Turbulence and Porous Media

De Gruyter

Physics and Astronomy Classification Scheme 2010: 92.60.hk, AMS 65c05, 76f55.

ISBN 978-3-11-029664-8
e-ISBN 978-3-11-029681-5

Library of Congress Cataloging-in-Publication Data
A CIP catalog record for this book has been applied for at the Library of Congress.

Bibliographic information published by the Deutsche Nationalbibliothek
The Deutsche Nationalbibliothek lists this publication in the Deutsche Nationalbibliografie; detailed bibliographic data are available in the Internet at http://dnb.dnb.de.

Typesetting: PTP-Berlin Protago-TEX-Production GmbH, www.ptp-berlin.eu
Printing and binding: Hubert & Co. GmbH & Co. KG, Göttingen
♾ Printed on acid-free paper

Printed in Germany

www.degruyter.com

Preface

This book presents advanced stochastic models and simulation methods for random flows and transport of particles by turbulent velocity fields and flows in porous media. Two main classes of models are constructed: (1) turbulent flows are modeled as synthetic random fields which have certain statistics and features mimicking those of turbulent fluid in the regime of interest, and (2) the models are constructed in the form of stochastic differential equations for stochastic Lagrangian trajectories of particles carried by turbulent flows. In both these classes, we develop Random flight models for the trajectories of tracer particles in turbulence and in flows through porous media. The boundary value problems in stochastic formulation for high-dimensional PDEs present a powerful research instrument in many modern branches of science and technology, in particular, in the turbulence simulation, transport in porous media, random load analysis in mechanical systems, geodesy, composite materials, elastography for biological tissues, acoustic scattering from rough surfaces, defects in metals, X-ray diffraction analysis of epitaxial layers, dislocations in crystals, etc. Interesting example is related to the coagulation of particles carried in turbulent flows governed by Smoluchowski nonlinear systems of coagulation equations with random coefficients. We present detailed results of numerical simulations for these applied problems and discuss stochastic interpretations related to the physics of the relevant problems.

A considerable amount of material is devoted to the random field description and different stochastic simulation methods, mainly the stochastic spectral and Fourier-wavelet methods for homogeneous vector Gaussian random fields, and the Karhunen–Loève expansions for inhomogeneous random fields. Of special interest are the so-called partially homogeneous random fields which we use in the development of stochastic models for boundary value problems with random boundary conditions, in particular, for Stokes flows, which are presented in the last chapter.

The book is written for mathematicians, physicists, and engineers studying processes associated with probabilistic interpretation, researchers in applied and computational mathematics, in environmental and engineering sciences dealing with turbulent transport and flows in porous media, as well as nucleation, coagulation, and chemical reaction analysis under fluctuation conditions. It can be of interest for students and post-graduates studying numerical methods for solving stochastic boundary value problems of mathematical physics and dispersion of particles by turbulent flows and flows in porous media.

Acknowledgments. Support of the Russian Fund of Basic Research under grant 12-01-00635-à is kindly acknowledged.

Berlin, Novosibirsk, June 2012 *Karl K. Sabelfeld*

Contents

Chapter 1

Introduction

1.1 Why random fields?

Probabilistic approach and stochastic simulation are becoming more and more popular in all branches of science and technology, especially in problems where the data are randomly fluctuating, or they are highly irregular in a deterministic sense. As a rule, in such problems it is very difficult and expensive to carry out measurements to extract the desired data. Important examples we deal with in this book are the turbulent flow simulation [146], and construction of flows through porous media [34, 65]. The temporal and spatial scales of the input parameters in this class of problems vary enormously, and their behavior is very complicated, so that there is no chance of describing it deterministically. In the stochastic approach, the input parameters are considered as random fields, and one needs to know only a few of their functions, such as the mean and correlation tensor, whose behavior in time and space is much more regular, so that usually it is considerably easier to extract them through measurements.

In most applications, it is assumed that the random fields are Gaussian, or that they can be obtained by a functional transformation of Gaussian fields. Generally, it is very difficult to construct efficient simulation methods for inhomogeneous random fields, even if they are Gaussian. Therefore, the most developed methods deal with homogeneous or quasihomogeneous random fields where the characteristic scales of the variations of the means of the field are considerably larger than the correlation scale. There are highly intensive studies and literature concerned with the simulation of homogeneous random fields.

The most important class of simulation methods seeks to construct stochastic models based on spectral representations. We shall consider general real-valued Gaussian homogenous random fields $u(\mathbf{x})$ defined on the multidimensional Euclidean space $I\!R^d$. Under quite general conditions, a real-valued Gaussian homogenous random field $u(\mathbf{x})$ can be represented through a stochastic Fourier integral [146]:

$$u(\mathbf{x}) = \int_{I\!R^d} \mathrm{e}^{-2\pi \mathrm{i}\mathbf{k}\cdot\mathbf{x}} E^{1/2}(\mathbf{k}) \tilde{W}(\mathrm{d}\mathbf{k}), \tag{1.1}$$

where $\tilde{W}(\mathrm{d}\mathbf{k})$ is a complex-valued white noise random measure on $I\!R^d$, with $\tilde{W}(B) = \overline{\tilde{W}(-B)}$, $\langle \tilde{W}(B) \rangle = 0$, and $\langle \tilde{W}(B)\overline{\tilde{W}(B')} \rangle = \mu(B \cap B')$ for the Lebesgue measure μ and all Lebesgue-measurable sets B, B'. We use angle brackets $\langle \cdot \rangle$ to denote statistical (ensemble) averages. The spectral density $E(\mathbf{k})$ is a nonnegative even function representing the strength (energy) of the random field associated to the wavenumber $\mathbf{k}$, meaning the length scale $1/|\mathbf{k}|$ and direction $\mathbf{k}/|\mathbf{k}|$.

We mention several simulation methods based on the spectral representation: (i) the discrete spectral method (DSM) [226] which is simply a deterministic discrete approximation of the Fourier Stieltjes integral; (ii) the randomized spectral method (RSM) [102, 142, 191]) which is based on a randomized approximation of the same Fourier Stieltjes integral; (iii) the Fourier wavelet method (FWM) [49,50,104,118] is a different approximation of the Fourier Stieltjes integral based on reexpansion in a special family of orthogonal functions, and is obtained by an expansion of the Gaussian white noise in a wavelet basis.

Another class of methods includes methods which deal with the expansions in the physical space, in the relevant system of orthonormal functions: (i) methods based on expansions in the wavelet basis (WM) [231,264]; (ii) the Karhunen–Loève expansion method (K–L) [230,249] based on the expansions in eigenfunctions of the correlation operator – note that this also works for inhomogeneous random fields; (iii) the moving averages mMethod (MAM) [140], based on the representation of the random field in the form of a convolution of a deterministic function (more precisely, a Fourier transform of a square root of the spectral function) with the Gaussian white noise in the physical space. We mention the the fast Fourier transform spectral method (FFTSM) (e. g., see [39, 168]) which is a particular case of the discrete spectral method whose nodes are chosen as a diadic mesh to apply further the fast Fourier method. The matrix factorization method (MFM) [40, 224] and the circulant embedding method (CEM) [41] are based on the Holessky decomposition of the covariance matrix.

The methods listed above all have their advantages as well as their disadvantages. For example, DSM, RSM, and MAM are simple and convenient for implementation; they provide the possibility to calculate the values of the random field at some points on demand. But in multidimensional cases, DSM and MAM are less efficient. FFTSM is also simple for implementation, but it calculates the random field only on a diadic mesh and has therefore a disadvantage that the samples are periodic. Furthermore, FWM and WM models are efficient for simulating multiscale processes but they are difficult in implementation.

The K–L model is highly efficient but is not universal, since it is necessary to solve the eigenvalue problem for the correlation operator.

More details about the above mentioned methods can be found in [25, 47, 54, 103, 104, 165] where a comparative analysis of some methods is also given. In particular, in [25, 47], RSM and FWM are compared by analyzing a fractal random field with the spectral function $F(k) = k^{-\alpha}$ $(1 < \alpha < 3)$, where the calculated structure function was compared with the exact result. The main conclusion is that to construct the samples of a multiscale random field with a fixed desired accuracy, the cost of RSM is considerably lower than that of FWM if $\lg(l_{\max}/l_{\min}) \le 4$ where $l_{\min}$ and $l_{\max}$ are the minimal and maximal spatial scales of the random field, respectively. In [104] we have shown that a logarithmically uniform subdivision of the spectral space (we have introduced such a subdivision in [119]) when calculating two- and a few-point statistical characteristics of the fractal random field, the RSM is more efficient than FWM for all values of $l_{\max}/l_{\min}$. In particular, when calculating the structure function of a multiscale random field with $\alpha = -5/3$, $l_{\max}/l_{\min} = 10^{12}$ it was found that the cost of FWM was 12 times larger than that of RSM; results were obtained for 9 decades, with a fixed accuracy.

Up to now, we discussed the calculation of statistical characteristics by ensemble averaging over the samples constructed by the relevant method. In many practical problems (e. g., in underground hydrology) only data obtained through spatial averaging is at hand, for instance, statistical characteristics obtained by spatial averages, or over a family of Lagrangian trajectories generated in one fixed sample of the field (e. g., see [35, 65]). If the random field is ergodic, then the ensemble averages can be well approximated by the appropriate space averages. This is very important when a boundary value problem with random parameters is solved: then, in contrast to the ensemble averaging, we have to solve the problem only once and then make the relevant space averaging. In practical calculations, to increase the efficiency, it is sometimes reasonable to combine both the space and ensemble averaging, e. g., see [99, 120]. The same technique is used also in simulation of turbulent transport [63, 240].

And so we stress that good ergodic properties of the constructed random field model are very important and highly desired in practical problems. In [104] we studied the ergodic properties of RSM and FWM. Calculations of structure functions through ensemble and space averaging have shown that the ergodic properties of FWM are much better than those of RSM. Therefore, to obtain a good approximation through space averaging in RSM, it is necessary to take many thousands of harmonics per each decade. However, this conclusion was made only for random processes (i. e., random fields depending on one scalar variable).

Thus, as discussed above, random fields provide a useful mathematical framework for representing disordered heterogeneous media in theoretical and computational studies. The random fields may appear in a very simple form in the problem, and then the focus is only on a detailed and accurate simulation of the samples. For instance, in many optic problems there is a need to simulate a random surface with a given correlation function, while the reflected light intensity is then easily calculated as in integral over a certain angle region.

In more sophisticated models, random fields enter differential or integral equations in the form of a coefficient, a kernel, a right-hand side, a boundary condition, or even the boundary itself.

1.2 Some examples

Let us give some examples. First we mention turbulent transport, where the velocity field representing the turbulent flow is modeled as a random field $\vec{v}(\mathbf{x}, t)$ with statistics encoding important empirical features, and the temporal dynamics of the position $\vec{X}(t)$ and velocity $\vec{V}(t) = \frac{\mathrm{d}\vec{X}}{\mathrm{d}t}$ of immersed particles is then governed by equations involving this random field such as

$$m\,\mathrm{d}\vec{V}(t) = -\gamma\big(\vec{V}(t) - \vec{v}(\vec{X}(t),t)\big)\,\mathrm{d}t + \sqrt{2k_{\mathrm{B}}T\gamma}\,\mathrm{d}\mathbf{W}(t), \tag{1.2}$$

where m is particle mass, γ is its friction coefficient, k_{B} is Boltzmann's constant, T is the absolute temperature, and $\mathbf{W}(t)$ is a random Wiener process representing molecular collisions.

In more complicated stochastic models of turbulent flows, both the drift and the dispersion terms are constructed as some functions using data extracted from physical and statistical laws.

Let us consider a passive scalar dispersed by the turbulent velocity field. The passive scalar is assumed to follow the streamlines of the flow. We assume that the source of particles is quite arbitrary; for instance, it might be situated on the surface or in the space, or even at given points. Let us denote by $q(\mathbf{x},t)$ the spatial-temporal density distribution function of the source, i. e,, the number of emitted particles per unit volume in a unit time interval at the phase point $(\mathbf{x},t)$. Initially, the spatial density of particles is given by $q_0(\mathbf{x})$. The particles are transported by a 3D turbulent velocity field $\vec{u}(\mathbf{x},t) = (u_1(\mathbf{x},t), u_2(\mathbf{x},t), u_3(\mathbf{x},t))$. Let us denote by $\mathbf{X}(t;\mathbf{x}_0,t_0)$ and $\mathbf{V}(t;\mathbf{x}_0,t_0) = \vec{u}(\mathbf{X}(t;\mathbf{x}_0,t_0),t)$ the Lagrangian spatial coordinates and the velocity, respectively. Then, neglecting molecular diffusion, the instantaneous concentration $c(\mathbf{x},t)$ is governed by

$$\frac{\partial c(\mathbf{x},t)}{\partial t} + \sum_{i=1}^{3} u_i(\mathbf{x},t)\frac{\partial c}{\partial x_i} = q(\mathbf{x},t), \quad t > 0, \quad c(\mathbf{x},0) = q_0(\mathbf{x}). \tag{1.3}$$

The turbulent velocity field $\vec{u}(\mathbf{x},t)$ is considered to be an incompressible 3D random field. Accordingly, the concentration $c(\mathbf{x},t)$, satisfying equation (1.3) with random coefficients u_i, is a scalar random field, and we are interested in calculating the mean concentration $\langle c(\mathbf{x},t)\rangle$ and the mean fluxes $\langle u_i(\mathbf{x},t)c(\mathbf{x},t)\rangle$, $i = 1,2,3$. These functions can be evaluated by tracking Lagrangian trajectories, which are obtained in turn by solving the system of stochastic differential equations

$$\begin{aligned} d\mathbf{X}(t) &= \mathbf{V}(t)dt, \\ d\mathbf{V}(t) &= \mathbf{a}(t,\mathbf{X}(t),\mathbf{V}(t))dt + \sqrt{C_0\bar{\varepsilon}(\mathbf{X}(t),t)}\,d\mathbf{W}(t), \end{aligned}$$

where the function $\mathbf{a}$ is to be defined in each specific situation, C_0 is the universal Kolmogorov constant ($C_0 \approx 4 \div 6$), $\bar{\varepsilon}(\mathbf{x},t)$ is the mean dissipation rate of the kinetic energy of turbulence, and $\mathbf{W}(t)$ is a standard 3D Wiener process.

A second example concerns transport through porous media, such as groundwater aquifers, in which the hydraulic conductivity $\mathsf{K}(\mathbf{x})$ is modeled as a random field reflecting the empirical variability of the porous medium. The Darcy flow rate $\vec{q}(\mathbf{x})$ in response to pressure applied at the boundary is governed by the Darcy equation

$$\begin{aligned} \vec{q}(\mathbf{x}) &= -\mathsf{K}(\mathbf{x}) \operatorname{grad} \phi(\mathbf{x}), \\ \operatorname{div} \vec{q} &= 0, \end{aligned} \tag{1.4}$$

in which the random hydraulic conductivity function appears as a coefficient, and the applied pressure is represented in the boundary conditions for the internal pressure head ϕ. Our concern is with the computational simulation of random fields for applications such as these. Interesting insights into the dynamics of transport in disordered media can be achieved already through relatively simple random models for the velocity field, such a finite superposition of Fourier modes, with each amplitude independently evolving according to an Ornstein–Uhlenbeck process. Here efficient and accurate numerical simulations of the flow can be achieved through application of the well-developed literature on simulating stochastic ordinary differential equations.

In conventional deterministic numerical methods, boundary value problems for random PDEs are solved as follows. First, one constructs a synthesized sample of the input random parameter. Then the obtained deterministic equation is solved numerically, say, by the finite element method, and gives the solution in all points of the grid domain. These two steps are repeated many times, so that the obtained statistics are sufficient for calculation of the desired sufficiently accurate averages. This approach is used in stochastic finite element methods (e. g., see [2, 150, 230, 249]). Obviously, this technique is generally time consuming, and to solve problems of practical interest one needs supercomputers to extract sufficient statistical information.

In the Monte Carlo approach, the algorithms are designed so that the solution is calculated only in the desired set of points without constructing the solution in the whole domain (e. g., see [191, 203, 204]). To evaluate different statistical characteristics of random boundary value problems we use the double randomization technique (e. g., see [191]).

This approach is possible if the desired statistical characteristics (e. g., the mean or the correlation tensor) can be represented in the form of a double expectation over the input random parameters, and over the trajectories of a Markov process used in a stochastic estimator for solving the deterministic equation. The advantage of this method is that there is no need to solve the equation many times, hence, the cost of this method is drastically decreased compared to the stochastic finite element method. The well-known drawback of stochastic simulation methods should be mentioned: the error behaves like $\varepsilon \sim O(N^{-1/2})$, where N is the number of samples; hence, it is reasonable to apply the Monte Carlo methods if the desired accuracy is not too high. So, for realistic applied problems the typical Monte Carlo accuracy lies in the range of 0.1 % to several percent.

The basic idea behind double randomization can be explained by the following very simple example. Assume we have to evaluate an integral

$$J(x;\omega) = \int_D f(x, y; \omega) dy, \tag{1.5}$$

where $f(x, y; \omega)$ is a random function indexed through x, y, defined on a probability space, ω being the relevant random element. Obviously of interest are statistical characteristics of the random process $J(x; \omega)$, like the expectation $\langle J(x; \omega)\rangle$ and the covariance $\langle J(x_1; \omega) J(x_2; \omega)\rangle$.

In deterministic methods, to calculate the expectation $\langle J(x; \omega)\rangle$, one has first to construct a sample of the random function f, say, $f(x, y; \omega_1)$, and then calculate the integral $J(x; \omega_1)$ by one of the quadrature formulas. This is then repeated N times, N large enough to guarantee that the average over N samples provides a good approximation to $\langle J(x; \omega)\rangle$. Thus, in short, one must solve a deterministic problem (in this case, evaluation of the integral) N times, N being the number of samples.

Double randomization is based on the representation of the desired functional as a double expectation. Indeed, we choose an arbitrary probability density function $p(y)$, $y \in D$, arbitrary enough but so that $p(y) \neq 0$ for y, where $f(x, y; \omega) \neq 0$ for all x and ω. Then we can write

$$J(x; \omega) = E_p\big[f(x, \xi; \omega)/p(\xi)\big] \tag{1.6}$$

where E_p stands for the average over the random points ξ distributed in D according to the density p. Therefore,

$$\langle J(x; \omega)\rangle = \big\langle E_p\big[f(x, \xi; \omega)/p(\xi)\big]\big\rangle = E_{(\omega,p)}\big[f(x, \xi; \omega)/p(\xi)\big], \tag{1.7}$$

where $E_{(\omega,\xi)}$ stands for averaging over random elements (ω, ξ). This statement is exactly the Fubini theorem, and it shows that the desired result can be obtained by averaging over random samples of ω and ξ.

This approach works also when the deterministic problem is not simply an integral evaluation, but a PDE, or an integral equation, or nonlinear system of equations, e. g., like the Smoluchowski equations. The main challenging problem here is to transform the solution of the original problem to the evaluation of an expectation over relevant stochastic elements, including averaging in functional spaces. In our case the expectations are often constructed over Markov chains and Gaussian random fields. Note that in applied problems, we deal also with generalized random fields, but in Monte Carlo simulations they do not complicate the situation, but, quite the contrary, the stochastic simulation algorithms can efficiently use this feature.

Let us illustrate this by the following simple but important example when the input randomness enters the problem as a generalized 2D random field $u(x, \boldsymbol{\omega})$, $x \in [0, L]$, $\boldsymbol{\omega} = (\omega_1, \omega_2, \ldots, \omega_m)$, in the form:

$$u(x, \boldsymbol{\omega}) = \sum_{j=1}^{m} \delta(x - \omega_j), \tag{1.8}$$

where δ is the Dyrac delta function, and the random points $\omega_1, \omega_2, \ldots, \omega_m$ are distributed on $[0, L]$ with a density $p(\boldsymbol{\omega})$ independent of the spatial coordinate x. The random points $\omega_1, \omega_2, \dots, \omega_m$ may have a quite different distribution on $[0, L]$, i. e., they may be all independent of each other, they may form a Markov chain with a certain transition probability density, or they may be placed almost periodically, with small but correlated random shifts ω_j from a mean fixed step $\langle \delta\omega_j \rangle$: $\omega_{j+1} = \omega_j + \delta\omega_j$ where $\langle \delta\omega_j \rangle = \rho^{-1} = m/L$ is the mean density of the points on $[0, L]$. Thus the random field $u(x, \boldsymbol{\omega})$ is stationary with the mean $\langle u(x, y; \boldsymbol{\omega}) \rangle = \rho$, as can be found by direct calculation.

This kind of random process enters the boundary elastic displacements produced by the nets of misfit dislocations in crystals [87], in many problems related to renewal processes, in the analysis of statistic of neuronal spike trains (for instance, see [52, 68, 180]), etc.

In the analysis of dislocations in crystals, the goal is to evaluate the x-ray diffraction peak profiles from distributions of misfit and threading dislocations [87]. The x-ray scattering amplitude from a film of thickness d is given by an integral of the form

$$A(q_x, q_z) = \int_{-\infty}^{\infty} dx \int_0^d dz \, \exp\left\{ \mathrm{i}[q_x x + q_z z + V(x, z)] \right\}, \tag{1.9}$$

where

$$V(x, z) = \int K(x - x', z) u(x', \boldsymbol{\omega}) dx' \tag{1.10}$$

is the resulting displacement field due to all misfit dislocations, the kernel $K(x - x', z)$ is the Green function given explicitly, and the random field $u(x', \omega)$ is defined by (1.8). The scattered intensity $I(q_x, q_z) = \left\langle |A(q_x, q_z)|^2 \right\rangle$ can be directly calculated by the Monte Carlo double randomization method, as explained above, by a randomized evaluation of the integral representation:

$$I(q_x, q_z) = \int_{-\infty}^{\infty} dx \int_0^d \int_0^d dz_1 \, dz_2 \, e^{\mathrm{i}[q_x x + q_z (z_1 - z_2)]} \left\langle e^{\mathrm{i}[V(x_1, z_1) - V(x_2, z_2)]} \right\rangle. \tag{1.11}$$

Note that in the case where the random field u is Gaussian, it is possible to evaluate the expectation explicitly (see [87]).

Thus the boundary value problems with random coefficients, parameters, random source terms, stochastically distributed boundary functions, or even with randomly moving boundaries are used as a powerful instrument in modern science and technology. We mention here applied fields such as structural mechanics, composite materials [2], porous media and soils [34, 99, 196, 260], biological tissues [258], geodesy [182, 212], turbulence, [13, 103, 104, 146, 191], etc. In engineering-related stochastic boundary value problems, the common computational techniques include Monte Carlo methods, stochastic finite elements, finite difference, and spectral methods. Among these methods, the finite volume and boundary element techniques are the methods most adaptable to problems in solid and structural mechanics characterized with highly irregular and complex structures [2, 230, 249]. We mention also classical potential problems dealing with random boundary conditions and sources [31] where the Monte Carlo methods are very efficient (e. g., see [191, 206–208]).

The book is organized as follows. Chapter 2 presents different simulation methods for Gaussian random fields. Chapter 3 deals with the stochastic Lagrangian models of relative dispersion of a pair of fluid particles. In Chapter 4 a new version of the 2-particle relative turbulent dispersion model is developed. A combined Eulerian–Lagrangian model is presented in Chapter 5. In Chapter 6 we describe a stochastic Lagrangian model extended to the intermittent turbulence. Chapter 7 presents results of numerical experiments. A 1-particle stochastic Lagrangian model for a horizonatlly homogeneous turbulent flow is described in Chapter 8. Formulation of the footprint problem and the methods based on backward Lagrangian trajectories are presented in Chapter 9. Applications of the stochastic Lagrangian models to evaluate the particle transport in the boundary layer of the atmosphere are given in Chapter 10. Comparisons of different 2-particle models are described in Chapter 11. Algorithms for concentration and fluxes in turbulent flows by stochastic Lagrangian models are presented in Chapter 12. Application to the footprint problem for the case of an abrupt change of roughness is given in Chapter 13. A Lagrangian stochastic model for the transport in porous medium is presented in Chapters 14 and 15. Chapter 16 deals with the analysis of the coagulation of aerosol particles in intermittent turbulent flows. Finally, in Chapter 17 we give an example of a Stokes flow which is governed by Stokes equations with random boundary conditions.

1.3 Fundamental concepts

We introduce two main classes of stochastic simulation models presented in this book: an *Eulerian* class which is based on a random velocity field model defined on a fixed coordinate system. The velocity field is generated over a prescribed spatial domain, but by a direct stochastic construction rather than the much more expensive simulation of the nonlinear Navier–Stokes PDE's.

In the *Lagrangian* class the motion of fluid particles is stochastically modeled as a random trajectory $X(t)$ of any immersed particle representing, for example, a tracer, pollutant, or chemical reactant. It is computed using the local value of the stochastically constructed velocity. The random trajectory thus can be defined as a random process determined from the random ordinary differential equation $\frac{dX(t)}{dt} = u(t; X(t)$ where $u(\tau, r)$ is a random field, or, alternatively, it can be determined from a stochastic Ito-type differential equation $dX(t) = A\,dt + B\,dW(t)$.

These two classes of models use different mathematical apparatus: the Eulerian stochastic models focus on the simulation of spatial-temporal random fields with a desired statistical characteristics, while the stochastic Lagrangian models are based on the so-called Ito-type stochastic differential equations, known in physics as Langevin-type equations.

As mentioned in [103], the primary challenge in most Eulerian fluid Monte Carlo simulations is the generation of a synthetic random velocity $u(x;t)$ which has certain statistics and features mimicking those of a turbulent fluid in the regime of interest. Thus a fully developed turbulent flow at a sufficiently high Reynolds number should possess a wide inertial range of scales over which the statistics of the velocity field assume a self-similar fractal structure. A quantitative way to express this criterion is

$$\langle |u(x+r;t) - u(x;t)|^2 \rangle = S_v^I |r|^{2H} \quad \text{for } L_K \ll |r| \ll L_0,$$

where $\langle \cdot \rangle$ denotes a statistical average, $0 < H < 1$ is the Hurst exponent which takes the Kolmogorov value $H = 1/3$ for fully developed turbulence, L_K is the dissipation length scale and L_0 is the integral length scale, which define the extent of the inertial scaling range, and S_v^I is a (dimensional) scaling prefactor. Other desired properties in turbulence simulations are the incompressibility of the fluid and appropriate geometric symmetries such as isotropy.

1.3.1 Random functions in a broad sense

Let us start with an informal description of random functions. A random function, as well as a deterministic function, is defined by some dependence $\xi(\theta)$ which describes a mapping from the space of parameters $\theta \in \Theta$ to the space of values of $\xi(\theta)$, thus it can be real- or complex-valued, or generally vector-valued. But in the stochastic case, $\xi(\theta)$ is a random value for each fixed θ. Thus, a random function is a family of random variables $\xi(\theta) = \xi(\theta, \omega)$ defined on a probability space $(\Omega, \mathcal{F}, P)$ depending on a parameter $\theta \in \Theta$. The deterministic function $\xi(\theta, \omega_0)$ for a fixed value $\omega = \omega_0$ is called a sample function (or a sample trajectory). Here we recall that a finite set of random variables $\xi_1, \xi_2, \ldots \xi_n$ is fully defined by the mutual distribution function

$$F(x_1, x_2, \ldots, x_n) = P\{\xi_1 < x_1, \xi_2 < x_2, \ldots, \xi_n < x_n\}. \tag{1.12}$$

In the case of random functions, we have to characterize an infinite family of random variables. So we can say that the infinite family of random variables $\xi(\theta)$ is defined if statistical characteristics of any finite sets of random variables

$$\xi(\theta_1), \xi(\theta_2), \ldots \xi(\theta_n), \quad \theta_i \in \Theta, \quad i = 1, 2 \ldots, n; \quad n = 1, 2 \ldots\ , \tag{1.13}$$

i. e., any finite-dimensional distributions are defined, which means, the random function $\xi(\theta)$ is defined by its finite-dimensional distributions

$$F_{\theta_1, \theta_2, \ldots, \theta_n}(x_1, x_2, \ldots, x_n), \quad \theta_i \in \Theta; \quad i = 1, 2 \ldots, n; \quad n = 1, 2 \ldots\ , \tag{1.14}$$

and any function $F_{\theta_1, \theta_2, \ldots, \theta_n}(x_1, x_2, \ldots, x_n)$ is interpreted as a mutual distribution function of the set of random variables (1.13).

To make this interpretation correct, the family of distributions should satisfy some assumptions. These assumptions are quite natural:

$$F_{\theta_1,\theta_2,\dots,\theta_n,\theta_{n+1},\dots,\theta_{n+m}}(x_1, x_2, \dots, x_n, +\infty, \dots, +\infty)$$
$$= F_{\theta_1,\theta_2,\dots,\theta_n}(x_1, x_2, \dots, x_n), \quad (1.15)$$

$$F_{\theta_1,\theta_2,\dots,\theta_n}(x_1, x_2, \dots, x_n) = F_{\theta_1,\theta_2,\dots,\theta_n}(x_{i_1}, x_{i_2}, \dots, x_{i_n}), \quad (1.16)$$

where $i_1, i_2, \dots, i_n$ is any permutation of the indices $1, 2 \dots, n$.

Thus we are in a position to give the following definition.

Random function on a set of parameters $\theta \in \Theta$ having real values $\xi(\theta)$ is defined as a family of distributions (1.14) which satisfies the conditions (1.15) and 1.16). The functions $F_{\theta_1,\theta_2,\dots,\theta_n}(x_1, x_2, \dots, x_n)$ are called finite-dimensional distributions of the random function $\xi(\theta)$.

This definition is clear and simple, and is sufficient when we are interested in statistical characteristics for a finite set of the parameter's values. But it is not satisfactory when we are trying to characterize the function in its entirety, for all infinite values of the parameter. For instance, this definition cannot provide us with a graph of the random function. With this definition, we cannot even answer such an important question as whether the sample of the random function is continuous or differentiable. A different definition considers the random function as an element of an appropriate functional space and will be given later. In this section we deal with the definition of the random function in athewise sense given above by the family of distribution functions.

Generalization to vector random functions is obvious: a vector random function $\xi(\theta)$ is defined as a vector with n components, scalar random functions: $\xi(\theta) = (\xi_1(\theta), \xi_2(\theta), \dots \xi_n(\theta))$. The distribution function of this vector random function is the function of nm variables

$$F_{\theta_1,\theta_2,\dots,\theta_n}(x_{11}, x_{12}, \dots, x_{nm})$$
$$= P\{\xi_1(\theta_1) < x_{11}, \xi_1(\theta_2) < x_{12}, \dots, \xi_n(\theta_m) < x_{mn}\}.$$

The distribution functions can be represented through the probability density $f_{\theta_1,\dots,\theta_n}(x_1, \dots, x_n)$:

$$F_{\theta_1,\dots,\theta_n}(x_1, \dots, x_n) = \int_{-\infty}^{x_1} \cdots \int_{-\infty}^{x_n} f_{\theta_1,\dots,\theta_n}(y_1, \dots, y_n)\, dy_1 \dots y_n.$$

From this the following well-known property follows

$$f_{\theta_1,\dots,\theta_n}(x_1, \dots, x_n)$$
$$= \int_{-\infty}^{x_1} \cdots \int_{-\infty}^{x_n} f_{\theta_1,\dots,\theta_n,\theta_{n+1},\dots\theta_{n+m}}(x_1, \dots x_n, y_1, \dots, y_m)\, dy_1 \dots y_m.$$

A characteristic function of a finite-dimensional distribution is defined by

$$\varphi_{\theta_1,\ldots,\theta_n} = \mathbf{E}\exp\left\{\mathrm{i}\sum_{k=1}^{n} \xi(\theta_k)u_k\right\},$$

where $\mathbf{E}$ stands for the mathematical expectation, and $u_1, \ldots u_n$ are real numbers. If the density $f_{\theta_1,\ldots,\theta_n}$ exists, then

$$\varphi_{\theta_1,\ldots,\theta_n} = \int_{\boldsymbol{R}^n} \mathrm{e}^{\sum_{i=1}^{n} x_i u_i} f_{\theta_1,\ldots,\theta_n}(x_1,\ldots,x_n)\,dx_1\ldots dx_n,$$

i. e., the characteristic function is a Fourier transform of the probability density.

Moment functions of $\xi(\theta)$ are defined by

$$m_{j_1,\ldots,j_s}(\theta_1,\ldots,\theta_s) = \mathbf{E}[\xi(\theta_1)]^{j_1}\ldots[\xi(\theta_s)]^{j_s}, \quad j_k \ge 0, \quad (k = 1,2,\ldots,s)$$

if the expectation on the right-hand side exists for all $\theta_i \in \Theta$, $i = 1,2,\ldots,s$.

A random function $\xi(\theta)$ belongs to the class $L_p(\Theta)$, $(\xi \in L_p(\Theta))$ if $\mathbf{E}|\xi(\theta)|^p < \infty$ for all $\theta \in \Theta$. So if $(\xi \in L_p(\Theta)$, then all moments of the order $q \le p$ are finite.

If the characteristic functions of the finite-dimensional distributions are given, then the moments of any integer order can be obtained by taking derivatives. Indeed, if $(\xi \in L_p(\Theta)$, then

$$m_{j_1,\ldots,j_s}(\theta_1,\ldots,\theta_s) = (-1)^q \frac{\partial^q \varphi(u_1,\ldots,u_s)}{\partial u_1^{j_1}\ldots u_s^{j_s}}$$

for $q \le p$ where $q = j_1 + \cdots + j_s$.

Centered moments are defined by

$$\bar{m}_{j_1,\ldots,j_s}(\theta_1,\ldots,\theta_s) = \mathbf{E}\big[\xi(\theta_1) - m_1(\theta_1)\big]^{j_1}\ldots\big[\xi(\theta_s) - m_1(\theta_s)\big]^{j_s}.$$

Here $m_i(\theta)$ is the expectation of $\xi_i(\theta)$: $m_i(\theta) = \mathbf{E}\xi_i(\theta)$.

The *correlation function* $R(\theta_1,\theta_2)$ is defined by

$$R(\theta_1,\theta_2) = \mathbf{E}\big[\xi(\theta_1) - m(\theta_1)\big]\big[\xi(\theta_2) - m(\theta_2)\big].$$

The variance is the quantity $\sigma^2(\theta) = R(\theta,\theta)$, and the correlation coefficient is just the normalized correlation function:

$$r(\theta_1,\theta_2) = \frac{R(\theta_1,\theta_2)}{\sigma(\theta_1)\,\sigma(\theta_2)}.$$

If the random variables $\xi(\theta_1)$ and $\xi(\theta_2)$ are independent, then the correlation coefficient is zero. Note that the reverse is not true: zero correlation does not imply that the variables are independent. However, if the mutual 2D distribution of $\xi(\theta_1)$ and $\xi(\theta_2)$ is Gaussian, then the zero correlation implies that $\xi(\theta_1)$ and $\xi(\theta_2)$ are independent.

Generalization to complex-valued random functions is straightforward. A complex-valued random function is defined by $\zeta(\theta) = \xi(\theta) + i\eta(\theta)$ and can be considered as a 2-dimensional vector of real-valued random functions. For a complex-valued function $\zeta(\theta) \in L_p(\Theta)$ means that

$$\mathbf{E}|\zeta(\theta)|^p < \infty, \quad \theta \in \Theta, i.e., \quad \xi \in L_p(\Theta) \quad \text{and} \quad \eta \in L_p(\Theta).$$

The correlation function for a complex-valued random function is defined by

$$R(\theta_1, \theta_2) = \mathbf{E}\big[\xi(\theta_1) - m(\theta_1)\big]\overline{\big[\xi(\theta_2) - m(\theta_2)\big]},$$

where $\bar{\xi}$ stands for the complex conjugate of ξ.

The following properties of the correlation functions can be easily checked:

1. $R(\theta, \theta) \geq 0$, where the equality appears if and only if the random function is constant with probability one.
2. $R(\theta_1, \theta_2) = \overline{R(\theta_2, \theta_1)}$.
3. $|R(\theta_1, \theta_2)|^2 \leq R(\theta_1, \theta_1) R(\theta_2, \theta_2)$.
4. For any integer n, $\theta_1, \dots \theta_n$ and complex numbers $\lambda_1, \dots, \lambda_n$,

$$\sum_{j,k=1}^{n} R(\theta_j, \theta_k)\, \lambda_j \bar{\lambda}_k \geq 0.$$

Note that properties 1–3 follow from property 4.

For two random functions $\zeta_1(\theta)$ and $\zeta_2(\theta)$ (belonging to $L_p(\Theta)$) one defines the cross-correlation function

$$R_{\zeta_1\zeta_2}(\theta_1, \theta_2) = \mathbf{E}\big[\zeta_1(\theta_1) - \mathbf{E}\zeta_1(\theta_1)\big]\ \overline{\big[\zeta_2(\theta_2) - \mathbf{E}(\theta_2)\big]}$$

Extension to multidimensional complex-values random functions. Let $\zeta_1(\theta), \zeta_2(\theta), \dots, \zeta_r(\theta)$ be a set of random complex-valued functions. It is considered as a complex-valued random function $\zeta(\theta) = (\zeta_1(\theta), \zeta_2(\theta), \dots, \zeta_r(\theta))^T$, $\theta \in \Theta$. Here $(\cdot)^T$ stands for the transpose, so $\zeta(\theta) = (\zeta_1(\theta), \zeta_2(\theta), \dots, \zeta_r(\theta))^T$, $\theta \in \Theta$ is a column.

For two columns, $\xi = (\xi_1, \xi_2, \dots, \xi_r)^T$ and $\eta = (\eta_1, \eta_2, \dots, \eta_r)^T$, we define a matrix $\xi\eta^*$ by

$$\xi\eta^* = \begin{pmatrix} \xi_1\bar{\eta}_1 & \xi_1\bar{\eta}_2 & \dots & \xi_1\bar{\eta}_m \\ \xi_2\bar{\eta}_1 & \xi_2\bar{\eta}_2 & \dots & \xi_2\bar{\eta}_m \\ \dots & \dots & \dots & \dots \\ \xi_r\bar{\eta}_1 & \xi_r\bar{\eta}_2 & \dots & \xi_r\bar{\eta}_m \end{pmatrix}.$$

For a random vector function $\zeta(\theta) = (\zeta_1(\theta), \dots \zeta_1(\theta_r))^T$ we put

$$\mathbf{m}(\theta) = (m_1(\theta), \dots, m_r(\theta))^T = \mathbf{E}\zeta(\theta) = (\mathbf{E}\zeta_1(\theta), \dots, \mathbf{E}\zeta_r(\theta))^T,$$

$$R(\theta_1, \theta_2) = \Big(R_{ij}(\theta_1, \theta_2)\Big)_{i,j=1,\dots r} = \mathbf{E}\Big([\zeta(\theta_1) - \mathbf{m}(\theta_1)]\,[\zeta(\theta_2) - \mathbf{m}(\theta_2)]^*\Big)_{ij}$$
$$= \Big(\mathbf{E}[\zeta_i(\theta_1) - m_i(\theta_2)]\,\overline{[\zeta_j(\theta_2) - m_j(\theta_2)]}\Big)_{ij=1,\dots,r}.$$

The vector $\mathbf{m}(\theta)$ is an r-dimensional complex-valued vector function; it is the expectation, called also a mean of the random function $\zeta(\theta)$. The matrix $R(\theta_1, \theta_2)$ is called a correlation matrix.

1.3.2 Gaussian random vectors

There is an important class of random functions completely defined by their first two moments, the expectation and correlation function. They are called Gaussian random functions and by definition have a Gaussian form of the finite-dimensional distributions. So we recall here first the case of random vectors.

The *Gaussian* random vector $\xi = (\xi_1, \xi_2, \dots, \xi_n)^T$ is defined by its characteristic function

$$\varphi(u) = \mathbf{E}\,\mathrm{e}^{\mathrm{i}(\mathbf{u},\xi)} = \exp\left\{\mathrm{i}\,(\mathbf{m}, \mathbf{u}) - \frac{1}{2}(R\mathbf{u}, \mathbf{u})\right\}, \tag{1.17}$$

where $\mathbf{m} = (m_1, \dots, m_n)$, $\mathbf{u} = (u_1, \dots, u_n)$, R is a nonnegative definite real-valued symmetric matrix $R = (r_{ik})$, $\quad i, k = 1, 2, \dots, n$. Here we use the conventional definition of a scalar product so that $(\mathbf{m}, \mathbf{u}) = \sum_{k=1}^n m_k u_k$ and $(R\mathbf{u}, \mathbf{u}) = \sum_{j,k=1}^n r_{jk}\, u_j\, u_k$.

The following statement explains the role of the above definition.

The function

$$\psi(\mathbf{u}) = \exp\left\{\mathrm{i}(\mathbf{m}, \mathbf{u}) - \frac{1}{2}(R\mathbf{u}, \mathbf{u})\right\}$$

is a characteristic function of a random vector ξ if and only if the real-valued matrix R is non-negative definite and symmetric. The rank of the matrix R equals the dimension of a subspace where the distribution of the vector ξ is concentrated.

If r, the rank of the matrix R is less than n, then the random vector is concentrated in an r-dimensional hyperplane; hence it has no density. Such a distribution is called a singular Gaussian distribution. If $r = n$, the random vector ξ has the density

$$f(\mathbf{x}) = \frac{1}{\sqrt{\Delta(2\pi)^n}} \exp\left\{-\frac{1}{2}(R^{-1}(\mathbf{x} - \mathbf{m}), (\mathbf{x} - \mathbf{m}))\right\}, \tag{1.18}$$

where R^{-1} is the inverse of the matrix R, $\Delta = \det(R)$ is the determinant of R.

Let us give a series of statements which are well known from the probability theory, e. g., see [71].

1. In the expression (1.17), $\mathbf{m} = (m_1, \dots, m_n)^T$ is the vector of expectations, and R is the correlation function: $\mathbf{m} = \mathbf{E}\xi$, $r_{jk} = \mathbf{E}[(\xi_j - m_j)(\xi_k - m_k)]$.
2. If the correlation function of a Gaussian vector ξ is not singular, then there exists an n-dimensional probability density $f(\mathbf{x})$ which is defined by (1.18).
3. The mutual distribution of any group of components of a Gaussian vector is Gaussian.
4. If $\xi = (\xi_1, \dots, \xi_n)^T$ is a Gaussian vector, and random vectors $\xi' = (\xi_1, \dots, \xi_r)^T$, $\xi'' = (\xi_{r+1}, \dots, \xi_n)^T$ are noncorrelated, then ξ' and ξ'' are independent.
5. Gaussian distributions remain Gaussian under linear transformation.

Let us give some other properties of the Gaussian distributions which are useful in practice. Assume we are given two vectors with Gaussian distributions: $\xi = (\xi_1, \dots, \xi_n)^T$ and $\eta = (\eta_1, \dots, \eta_m)^T$. We are interested in the conditional distribution of the vector ξ, assuming that η is fixed. Without loss of generality we suppose that the correlation matrix R_{22} of the vector η is nonsingular. Indeed, if R_{22} is singular, it means some components of η are linearly dependent on the other components. Then, we exclude these components, and the dimension of η is decreased. So let $\mathbf{m}_1 = \mathbf{E}\xi$, $\mathbf{m}_2 = \mathbf{E}\eta$, and let R_{11} be the correlation matrix of the vector ξ, and R_{12} be the cross-correlation matrix of the vectors ξ and η: $R_{12} = \mathbf{E}(\xi - \mathbf{m}_1)\eta - \mathbf{m}_2)^*$. Let us introduce a vector $\tilde{\xi} = \mathbf{m}_1 + R_{12} R_{22}^{-1}(\eta - \mathbf{m}_2)$. Then the conditional distribution of the vector ξ, under the condition that η is fixed, is Gaussian with the conditional expectation $\mathbf{E}(\xi|\eta) = \tilde{\xi}$ and the conditional correlation matrix

$$\mathbf{E}\Big\{\big(\xi - \tilde{\xi}\big)\big(\xi - \tilde{\xi}\big)^* \big|\eta\Big\} = R_{11} - R_{12}R_{22}^{-1}R_{21}.$$

Notice the following important property: the matrix of the conditional correlations of the vector ξ, η fixed, is not random, and in particular, it does not depend on the value of η.

1.3.3 Gaussian random functions

A vector n-dimensional random function $\xi(\theta) = \{\xi_1(\theta), \dots \xi_n(\theta)\}$ is called a Gaussian random function if the mutual distribution function of all components of the random vectors $\xi(\theta_1), \dots, \xi(\theta_n)$ is Gaussian. The correlation matrix R of the mutual distribution of the vectors $\xi(\theta_1), \dots, \xi(\theta_n)$ has a dimension $sn \times sn$ and can be divided in square blocks of size $s \times s$ as follows:

$$R = \begin{pmatrix} R(\theta_1, \theta_1) & R(\theta_1, \theta_2) & \dots & R(\theta_1, \theta_n) \\ R(\theta_2, \theta_1) & R(\theta_2, \theta_2) & \dots & R(\theta_2, \theta_n) \\ \dots & \dots & \dots & \dots \\ R(\theta_n, \theta_1) & R(\theta_n, \theta_2) & \dots & R(\theta_n, \theta_n) \end{pmatrix},$$

where $R(\theta_1, \theta_2)$ is the correlation matrix of the function $\xi(\theta)$.

The reverse statement is true: for any real-valued vector function $m(\theta)$ and a non-negative definite symmetric matrix function $R(\theta_1, \theta_2)$ there exists an r-dimensional Gaussian random function for which $m(\theta)$ is the expectation, and $R(\theta_1, \theta_2)$ is the correlation matrix.

The Gaussian random functions play an extremely important role in many practical problems. This can be explained generally as follows. The real processes are usually affected by many random independent factors and the resulting superposition of these factors tends to a Gaussian distribution. This can be rigorously formulated as a limit theorem of normal correlations, which is a generalization of the well-known central limit theorem. Let us present this statement.

A sequence of random functions $\xi_n(\theta)$, $\theta \in \Theta$, $n = 1, \dots$ is said to be weakly convergent to a random function $\xi(\theta)$, $\theta \in \Theta$ if for any s the mutual distribution of the series of random variables $\{\theta_n(\theta_1), \dots, \xi(\theta_s)\}$ is weakly convergent, as $n \to \infty$, to the distribution of $\{\xi(\theta_1), \dots, \xi(\theta_s)\}$.

Theorem 1.1 (see [71]). *Assume we are given a family of sums of random functions*

$$\eta_n(\theta) = \sum_{k=1}^{m_n} \alpha_{nk}(\theta), \quad \theta \in \Theta, \quad n = 1, 2 \dots$$

and the following conditions are satisfied:

1. *For fixed n, the random variables* $\alpha_{n1}(\theta_1), \alpha_{n2}(\theta_2), \dots, \alpha_{nm}(\theta_m)$ *are all mutually independent for each* $\theta_1, \theta_2, \dots \theta_m$, *and have finite second moments such that*

$$\mathbf{E}\alpha_{nk}(\theta) = 0, \quad \mathbf{E}\alpha_{nk}^2(\theta) = b_{nk}^2(\theta).$$

2. *The correlation function* $R_n(\theta_1, \theta_2) = \mathbf{E}[\eta_n(\theta_1)\eta_n(\theta_2)]$, *converges to a limit:*

$$\lim_{n \to \infty} R_n(\theta_1, \theta_2) = R(\theta_1, \theta_2).$$

3. *The sums* $\eta_n(\theta) = \sum_{k=1}^{m_n} \alpha_{nk}(\theta)$ *satisfy, for each* θ, *the Lindeberg condition: for each* $\tau > 0$

$$\frac{1}{B_n^2} \sum_{k=1}^{m_n} \int_{|x| > \tau B_n} x^2 \, d\Pi_{nk}(\theta, x) \to 0,$$

where $\Pi_{nk}(\theta, x)$ *is the distribution function of the random variable* $\alpha_{nk}(\theta)$*, and*

$$B_n^2 = \sum_{k=1}^{m_n} b_{nk}^2(\theta) = R_n(\theta, \theta).$$

Then the random function $\eta_n(\theta)$ *is weakly convergent, as* $n \to \infty$*, to a Gaussian random function with zero expectation and correlation function* $R(\theta_1, \theta_2)$*.*

1.3.4 Random fields

Assume that our parameter θ is a point $\mathbf{x} \in I\!R^n$. Let $\zeta(\mathbf{x}) = (\zeta^1(\mathbf{x}), \ldots, \zeta^d(\mathbf{x}))^T$ be a vector random function possibly with complex values, defined for all $\mathbf{x} \in I\!R^n$. It is called a sl random field. Note that if $n = 1$, one uses the notion of *random process*. The random field $\zeta(\mathbf{x})$ is called homogeneous in a broad sense if

$$\mathbf{E}\zeta(\mathbf{x}) = \mathbf{m} = \text{const}, \quad \mathbf{E}(\zeta(\mathbf{x}) - \mathbf{m})(\zeta(\mathbf{x}) - \mathbf{m})^* = R(\mathbf{x} - \mathbf{y}),$$

where $R(\mathbf{x})$ is a continuous matrix function, the correlation of the homogeneous random field. The matrix function $R(\mathbf{x})$ is nonnegative definite. This means that for any d-dimensional complex vectors z_k, points $\mathbf{x}_k \in I\!R^m, \quad k = 1, \ldots, n$, and for any integer n

$$\sum_{k,j=1}^{n} \mathbf{z}_k^* R(\mathbf{x}_k - \mathbf{x}_j)\mathbf{z}_j \geq 0.$$

A random field $\zeta(\mathbf{x})$, $\mathbf{x} \in I\!R^m$ is called meansquare continuous (m.s.c.) if from $\mathbf{x}^{(n)} \to \mathbf{x}$ it follows that $\mathbf{E}|\xi(\mathbf{x}^{(n)}) - \xi(\mathbf{x})|^2 \to 0$ as $n \to \infty$.

The correlation functions and nonnegative definite functions are related by the following theorem.

Theorem 1.2 (Bochner–Khinchin theorem). *A matrix function* $R(\boldsymbol{x}), \boldsymbol{x} \in I\!R^d$ *is a correlation function of a homogeneous, m.s.c. random field if and only if it can be represented in the form*

$$R(\boldsymbol{x}) = \int_{\boldsymbol{R}^d} e^{\mathrm{i}(x, bfu)}\, F(d\mathbf{u}), \tag{1.19}$$

where $F(A)$ *is a matrix-valued complex countable-additive function defined on Borel sets in* $I\!R^m$ *such that* $\mathbf{z}^* F(A)\mathbf{z} \geq 0$ *for any complex vector* $\mathbf{z}$ *and any Borel set* $A \subset I\!R^d$*, and the trace* $Sp\ F(I\!R^m) < \infty$*.*

A random field $\zeta(\mathbf{x})$ is called *isotropic* if it is homogeneous and it correlation function $R(\mathbf{x})$ depends on the length $|\mathbf{x}|$. Thus for an isotropic random field $R(\mathbf{x}) = R(\rho)$

1.3.5 Stochastic measures and integrals

In this section we introduce integrals with respect to stochastic measures, known as stochastic integrals.

Let (Ω, F, P) be a probability space, $\mathbf{E}f = \int dP$ is the expectation, $L_2(\Omega) = L_2(\Omega, F, P)$ is a class of random variables ζ with finite second moment. Let X be a set, and $\mathcal{K}$ be a semi-ring of the subsets of X. Assume that each $\Delta \in \mathcal{K}$ is related to a complex-valued random variable $\zeta(\Delta)$ which satisfies the following conditions:

1. $\zeta(\Delta) \in L_2(\Omega), \quad \zeta(\emptyset) = 0,$
2. $\zeta(\Delta_1 \bigcup)\Delta_2) = \zeta(\Delta_1) + \zeta(\Delta_2)(\mathrm{mod}(P), \ \text{ if } \Delta_1 \bigcap \Delta_2 = \emptyset,$
3. $\mathbf{E}\zeta(\Delta_1)\overline{\zeta(\Delta_2)} = m(\Delta_1 \bigcap \Delta_2),$

where $m(\Delta)$ is a function defined on the sets of $\mathcal{K}$.

A family of random variables $\{\zeta(\Delta),\ \Delta \in \mathcal{K}\}$ satisfying the conditions 1–3 is called an elementary orthogonal stochastic measure, and $m(\Delta)$ is its structure measure. The orthogonality property of the stochastic measure is expressed by condition 3: if $\Delta_1 \bigcap \Delta_2) = \emptyset$, then $\mathbf{E}\zeta(\Delta_1)\overline{\zeta(\Delta_2)} = 0$. From the definition of $m(\Delta)$ follows that it is nonnegative:

$$m(\Delta) = \mathbf{E}|\zeta(\Delta)|^2 \geq 0, \quad m(\emptyset) = 0$$

and additive, i. e., if $\Delta_1 \bigcap \Delta_2) = 0$, then

$$\begin{aligned} m\Big(\Delta_1 \bigcup \Delta_2\Big) &= \mathbf{E}|\zeta(\Delta_1) + \zeta(\Delta_2)|^2 \\ &= m(\Delta_1) + m(\Delta_2) + 2m\Big(\Delta_1 \bigcap \Delta_2\Big) \\ &= m(\Delta_1) + m(\Delta_2). \end{aligned}$$

Now, a class $L_0\{\mathcal{K}\|$ of *simple functions* is introduced:

$$f(x) = \sum_{r=1}^{n} c_r \chi_{\Delta_r}(x), \quad \Delta_r \in \mathcal{K}, \quad r = 1, 2, \ldots n, \tag{1.20}$$

where n is arbitrary, and $\chi_A(x)$ is the indicator of the set A, $c_1, \ldots c_r$ are complex numbers.

The *stochastic integral* of a simple function $f(x) \in L_0(\mathcal{K})$ with respect to a stochastic measure $\zeta(\Delta)$ is defined by the formula

$$\eta = \int f(x)\zeta(dx) = \sum_{r=1}^{n} c_r \zeta(\Delta_r). \tag{1.21}$$

For any two functions $f(x), g(x) \in L_0(\mathcal{K})$ the following equality holds:

$$\mathbf{E}\Big(\int f(x)\zeta(dx) \int \overline{g(x)\zeta(dx)}\Big) = \int f(x)\bar{g}(x)\, m(dx). \tag{1.22}$$

Assume that m satisfies the semi-additive condition and hence can be prolonged to a complete measure $\{X, \mathcal{B}, m\}$. Then $L_0\{\mathcal{K}\}$ is a linear subset of the Hilbert space $L_2(m) = L_2\{\{X, \mathcal{B}, m\}$. Denote by $L_2\{\mathcal{K}\}$ the close of $L_0(\mathcal{K})$ in $L_2(m)$. Now we introduce a linear span $L_0\{\zeta\}$ of the family of random variables $\zeta(\Delta), \Delta \in \mathcal{K}\}$ i. e., a set of random variables which can be represented in the form (1.21); the space $L_2(\zeta)$ is defined as a closure of $L_0(\zeta)$ in the Hilbert space of random variables $L_2(\Omega, F, P)$. Notice that the relation (1.21) establishes an isometrical mapping $\eta = \psi(f)$ between $L_0(\mathcal{K})$ and $L_0(\zeta)$. This mapping can be prolonged to an isometry between $L_2\{\mathcal{K}\}$ and $L_2\{\zeta\}$. If $\eta = \psi(f)$, $f \in L_2\{\mathcal{K}\}$, then we define $\eta = \psi(f) = \int f(x)\zeta(dx)$. The random variable η is then called a stochastic integral of f with respect to the measure ζ. From this follows the following.

The following statement is true (see [71].

Theorem 1.3.

(a) *For a simple function* (1.20) *the stochastic integral is defined by the formula* (1.21).

(b) *For any f and g from $L_2\{m\} = L_2\{X, \mathcal{B}, m\}$, the equality* (1.22) *holds.*

(c) $\int \big[\alpha f(x) + \beta g(x)\big]\zeta(dx) = \alpha \int f(x)\zeta(dx) + \beta \int g(x)\zeta(dx)$.

(d) *For an arbitrary sequence of functions $f^{(n)}(x) \in L_2\{X, \mathcal{B}, m\}$ such that*

$$\int |f(x) - f^{(n)}(x)|^2 \, m(dx) \to 0,$$

the following relation is true:

$$\int f(x)\,\zeta(dx) = \underset{n\to\infty}{\text{l.i.m.}} \int f^{(n)}(x)\,\zeta(dx).$$

Here l.i.m. *means a limit in mean square sense, i. e.,* $\underset{n\to\infty}{\text{l.i.m.}}\, \xi_n = \xi$ *implies* $\lim_{n\to\infty} \mathbf{E}|\xi_n - \xi|^2 = 0$.

The existence of a sequence of simple functions approximating an arbitrary function $f(x0 \in L_2\{X, \mathcal{B}, m\}$ follows from the general theorems of the measure theory. Thus the stochastic integral can be considered as a mean square limit of the relevant integral sums.

Let us denote by $\mathcal{B}_0$ the class of all subsets $A \in \mathcal{B}$ with $m(A) < \infty$, and define a random function of sets $\tilde{\zeta}(A)$ by

$$\tilde{\zeta}(A) = \int \chi_A(x)\zeta(dx) = \int_A \zeta(dx). \tag{1.23}$$

We list the following properties of this function:

1. $\tilde{\zeta}(A)$ is defined on the class of sets $\mathcal{B}_0$;
2. if $A_n \in \mathcal{B}_0$, $n = 0, 1, \ldots, A_0 = \bigcup_{n=1}^{\infty} A_n, q \quad A_k \bigcap A_r = \emptyset$ for $k \neq r, k > 0$, then $\tilde{\zeta}(A_0) = \sum_{n=1}^{\infty} \tilde{\zeta}(A_n)$ in mean square sense;
3. $\mathbf{E}\tilde{\zeta}(A)\tilde{\zeta}(B) = m(A \bigcap B), \quad A, B \in \mathcal{B}_0$;
4. $\tilde{\zeta}(\Delta) = \zeta(\Delta) \quad$ for $\Delta \in \mathcal{K}$.

An *orthogonal stochastic measure* is defined as a random set function $\tilde{\zeta}$ satisfying the above conditions 1–4.

Note that property 4 means that $\tilde{\zeta}$ is a prolongation of the elementary stochastic measure $\zeta(\Delta)$. Thus the following statement is true [71].

Theorem 1.4. *If the structure measure of the elementary stochastic measure $\zeta(\Delta)$. is semi-additive, then $\zeta(\Delta)$ can be prolonged to a stochastic measure $\tilde{\zeta}(\Delta)$.*

Note that

$$\int f(x)\zeta(dx) = \int f(x)\tilde{\zeta}(dx)$$

since $L_2\{\zeta\} = L_2\{\tilde{\zeta}\}$.

1.3.6 Integral representation of random functions

Using the results of the previous section we can represent random functions via stochastic integrals. Let Θ be an arbitrary parameter set, and (Ω, F, P) a probability space. Assume first that a p-dimensional random vector function $\xi(\theta)$, $\theta \in \Theta)$ can be written in the form

$$\xi(\theta) = \int g(\theta, x)\zeta(dx), \tag{1.24}$$

where $\zeta()$ is a stochastic measure on a measurable space $(X, \mathcal{B})$ with its values in C^p and a structure measure $m()$, and $g(\theta, x)$ is a scalar function such that for any $\theta \in \Theta$

$$g(\theta, x) \in L_2(m_0) = L_2(X, \mathcal{B}, m_0), \quad m_0(\Delta) = \operatorname{Sp} m(\Delta) = \sum_{k=1}^{p} m_{kk}(\Delta).$$

The correlation function reads

$$B(\theta_1, \theta_2) = \mathbf{E}\xi(\theta_1)\xi^*(\theta_2) = \int g(\theta_1, x)\overline{g(\theta_2, x)}\, m(dx). \tag{1.25}$$

We recall that $(X, \mathcal{B}, m_0)$ is a space with a complete measure, $L_2(m_0) = L_2(X, \mathcal{B}, m_0)$ is a Hilbert space of $\mathcal{b}$-measurable complex-valued functions which are square m_0-integrable:

$$L_2(m_0) = \left\{ f : X \to C : \int |f(x)|^2 m_0(dx) < \infty \right\}.$$

By $L_2\{g\}$ we denote the closure in $L_2(m_0)$ of a linear span generated by the family of functions $\{g(\theta, x),\ \theta \in \Theta\}$. Then, $L_2\{g\}$ is a linear closed subspace of $L_2(m_0)$.

If $L_2\{g\} = L_2(m_0)$, then the family of functions $\{g(\theta, x),\ \theta \in \Theta\}$ is called complete in $L_2(m_0) = L_2(X, \mathcal{B}, m_0)$.

Let $\{g(\theta),\ \theta \in \Theta\}$ be a Hilbert random function with its values in C^p, and $L_0\{\xi\}$ is a set of all random vectors

$$\eta = \sum_{k=1}^{n} c_k \xi(\theta_k), \quad n = 1, 2, \ldots, \quad \theta_k \in \Theta,$$

where c_k are arbitrary complex numbers, and $L_2\{\xi\}$ is a closure of $L_0\{\xi\}$ in the sense of mean square convergence of random vectors.

A family of random vectors $\{\eta_\alpha, \alpha \in A\}$, $\eta_\alpha \in L_2(\Omega)$ is called subordinate to a random function $\xi(\theta),\ \theta \in \Theta\}$ if $\eta_\alpha \in L_2(\xi),\ \alpha \in A$.

For random functions whose correlation function can be represented in the form (1.25) the following theorem can be formulated, see [71].

Theorem 1.5. *Assume that the correlation matrix of a random function* $\{\xi(\theta),\ \theta \in \Theta\}$ *can be written in the form* (1.25) *where* m *is a positive definite matrix measure on* $(X, \mathcal{B})$, $g(\theta, x) \in L_2(m_0)$, $\theta \in \Theta$ $(m_0 = Spm)$, *and the family* $\{g(\theta, x), \theta \in \Theta\}$ *is complete in* $L_2(X, \mathcal{B}, m_0)$. *Then* $\xi(\theta)$ *can be represented in the form* (1.24) *where* $\{\zeta(B), B \in \mathcal{B}\}$ *is a stochastic orthogonal measure which is subordinate to the random function* $\{\xi(\theta),\ \theta \in \Theta\}$ *with a structure measure* $m()$, *and the equality* (1.24) *holds with probability one for each* θ.

Let $\xi(\mathbf{x}) = (\xi^1(\mathbf{x}), \ldots, \xi^d(\mathbf{x}))$, $\mathbf{x} \in \mathbb{R}^m$ is centered mean square continuous homogeneous (in a broad sense) random field with real- or complex values. Let $R(\mathbf{x})$ be the correlation function of this field. By the Bochner–Khinchin theorem, this correlation function obeys the representation (1.19). This representation is a particular case of (1.25) in which $X = \Theta = \mathbb{R}^m$, $g(\theta, \mathbf{x}) = \exp\{\mathrm{i}(\mathbf{x}, \theta)\}$. Since the family $\{\exp\{\mathrm{i}(\mathbf{x}, \theta)\},\ \theta \in \mathbb{R}^m\}$ is complete in $L_2(m_0) = L_2(X, \mathcal{b}, m_0)$ where m_0 is an arbitrary bounded measure on the sigma-algebra $\mathcal{B}^m$ of $\mathbb{R}^m$, we come to the following corollary of the previous theorem which is known as the *spectral theorem* (see [71]).

Theorem 1.6. *Any centered mean square continuous homogeneous vector random field* $\xi(\theta), \theta \in \mathbb{R}^m$ *can be represented in the form*

$$\xi(\theta) = \int e^{\mathrm{i}(x,\theta)}\, \zeta(d\mathbf{x}), \quad \theta \in \mathbb{R}^m,$$

where $\zeta(A)$, $A \in \mathcal{B}^m$ *is a vector orthogonal measure on* $\mathcal{B}^m$ *which is subordinate to* $\xi(0)$. *Between* $L_2\{\xi\}$ *and* $L_2\{F\}$, $F_0 = SpF$ *there is an isometry providing*

(a) $\xi(\theta) \in L_2\{\xi\} \leftrightarrow e^{\mathrm{i}(x,\theta)} \in L_2\{F_0\}$;

(b) *if* $\eta_i \leftrightarrow g_i(\boldsymbol{x}), \eta_i \in L_2\{\xi\}, g_i \in L_2\{F_0\}$, $i = 1, 2$, *then*

$$\eta_i = \int g_i(\boldsymbol{x})\,\zeta(d\boldsymbol{x}), \quad \boldsymbol{E}\eta_1\eta_2^* = \int g_1(\boldsymbol{x})\overline{g_2(\boldsymbol{x})}\,F(d\boldsymbol{x}).$$

1.3.7 Random trajectories

A random trajectory can be defined as follows. A random trajectory is defined as a solution to an ordinary differential equation with a random field in the right hand side:

$$\frac{dX_i(t,\omega)}{dt} = f_i(X(t,\omega),t,\omega), \quad t \geq t_0,\ \omega \in \Omega,\ i = 1,\ldots,n, \tag{1.26}$$

where $f(x,t) = f(x,t,\omega) = (f_1(x,t,\omega),\ldots,f_n(x,t,\omega))$, $x \in I\!R^n$ is a vector random field, $X(t,\omega) = (X_1(t,\omega),\ldots,X_1(t,\omega))$. The samples of the random field $f(x,t)$ are assumed to be smooth enough in the sense that for each sample the classical solution of the deterministic equation (1.26) exists. An important difference to the classical deterministic ordinary differential equation is only in that in the considered case the right-hand side depends on a parameter $\omega \in \Omega$ where Ω is a probabilistic space. So in contrast to the stochastic differential equations of the Ito type, the solution is well defined in the classical sense, and there is no need to develop a special calculus for studying (1.26) . However it does not imply that all the question about the solutions to (1.26) can be answered by adapting the relevant results of the classical theory of differential equations. For example, the existence of the solution of an initial value problem for (1.26) is not equivalent to the existence of the solution for all (or almost all) samples (i. e., all values of ω), one needs the existence of the solutions as random processes on one common interval (not depending on ω). Let us give an illustrating example. Assume we solve the following equation:

$$\frac{dX}{dt} = \xi\,X^2, \quad X \in I\!R^1,\ t \in [0,\infty),$$

where $\xi = \xi(\omega)$ is a standard Gaussian random variable. The solution of this equation with the prescribed initial value $X(0) = x_0$ can be written explicitly as

$$X(t,\omega) = \begin{cases} 0, & \text{if } x_0 = 0, 0 \leq t < \infty \\ \left(\frac{1}{x_0} - \xi(\omega)t\right)^{-1}, & \text{if } |x_0| \neq 0, 0 \leq t < \delta. \end{cases}$$

Here

$$\delta = \delta(x_0,\omega) = \begin{cases} \frac{1}{\xi\,x_0}, & \text{if } x_0\xi > 0 \\ \infty, & \text{if } x_0\xi < 0 \end{cases}$$

is the explosion time instant, i. e., $\lim_{t\to\delta} |X(t) = \infty$. This means, that for any $x_0 \neq 0$, the solution may explode arbitrarily quickly, with positive probability, i. e., for any $\varepsilon > 0$, we get $P(\delta < \varepsilon) > 0$. This implies that the problem has no sample solutions.

Another issue is the existence of statistical characteristics of the solution to (1.26). For example in the framework of the conventional theory of differential equations it is not possible to answer the question of whether or not the first moment $\mathbf{E}|X(t)|$ is finite, i. e., when $\int_\Omega |X(t,\omega)\, P(d\omega) < \infty$. Nevertheless, many results in stochastic differential equations of the type (1.26) were obtained (e. g., see [9, 37, 228]) via an extension of the relevant results of the classical differential equation theory. For example, for the existence of a sample solution to (1.26) on a finite interval, say, $I = [0,1]$, it is necessary to prove the existence of measurable random processes $\xi_1(t,\omega), \xi_2(t,\omega)$, $(t,\omega) \in I \times \omega$, such that $P\{\xi_i(t)\,dt < \infty\} = 1$, $i = 1,2$, and almost all samples of the right-hand side $f(x,t) = f(x,t,\omega)$ should satisfy the linear growth condition

$$|f(x,t)| \leq \xi_1(t) + \xi_2(t)\,|x| \quad \text{for all } x \in I\!R^n .$$

This result follows from the classical theory of differential equations (e. g., see [75]).

1.3.8 Stochastic differential, Ito integrals

As already mentioned above, the motion of fluid elements and aerosol particles are often described by Langevin stochastic differential equations, known in the theory of stochastic processes as Ito stochastic differential equations. We have given the definition of stochastic integrals with respect to general stochastic measure, and so we could just refer to a particular case of Wiener measure to define the Ito integrals. Let us however begin with considerations which are closer to physics.

1.3.9 Brownian motion

Let us again consider the ordinary differential equation (1.26), but now the randomness is specifically entering this equation as an additive noise:

$$\frac{dX}{dt} = b(t, X_t) + \sigma(t, X_t) \cdot \text{"random noise"}, \tag{1.27}$$

where b and σ are some given deterministic functions. So intuitively, the solution X_t is a random process with some distribution and correlations caused by the input noise W_t = "random noise". Again from physical intuition, we might assume that W_t has the following properties: (i) W_{t_1} and W_{t_2} are independent if $t_1 \neq t_2$; (ii) W_t is a stationary process, i. e., the joint distributions of $\{W_{t_1+t}, \dots, W_{t_k+t}\}$ do not depend on t; (iii) $\mathbf{E}W_t = 0$ for all t.

However, it turns out there does not exist any "reasonable" stochastic process satisfying (i) and (ii): such a process cannot have continuous trajectories (e. g., see [70]). If we assume $\mathbf{E}|W_t^2| = 1$, then the function $(t, \omega) \to W_t(\omega)$ cannot even be measurable, with respect to the σ-algebra $\mathcal{B} \times \mathcal{F}$ where $\mathcal{B}$ is the Borel σ-algebra on $[0, \infty)$ (e. g., see [158]). So the functions W_t belong to other class we considered above, generalized stochastic functions, constructed as a probability measure in the space of tempered distributions on $[0, \infty)$, and not as a probability measure on the much smaller space $\mathbb{R}^{0,\infty}$, like an ordinary stochastic process.

We will avoid this construction and turn to a description of stochastic differential equations which uses random processes with independent increments. Let us start with the physical process of Brownian motion.

Let us consider a particle which is moving on a line: the particle starts from the origin $\xi(0) = 0$, and during a small time increment Δt it makes a jump to the left with probability $1/2$, and to the right with probability $1/2$. So during time t the particle makes $n = t/\delta t$ jumps. We put

$$x_i = \begin{cases} \Delta x & \text{with probability } 1/2, \\ -\Delta x & \text{with probability } 1/2. \end{cases}$$

The total displacement of the particle reads as $\xi(t) = \sum_{i=1}^{n} x_i$. Assuming that x_i are mutually independent and equally distributed, the variance (dispersion) can be found as

$$D\xi(t) = D\sum_{i=1}^{n} x_i = n\, D\, x_i = n(\Delta x)^2 = \frac{t}{\Delta t}\,(\Delta x)^2.$$

Let us consider two arbitrary times s and t, where $s < t$. Then, the first and second summands in $\xi(t) = [\xi(t) - \xi(s)] + \xi(s)$ are independent, and in addition, the displacement during the time interval (s, t) depends obviously on $t - s$, since physically the character of the motion is not changing in the time, hence, $\xi(t) - \xi(s)$ and $\xi(t - s)$ are equally distributed. Therefore,

$$D\xi(t) = D[\xi(t) - \xi(s)] + D\xi(s) = D\xi(t - s) + D\xi(s),$$

which implies that $D\xi(t)$ is a linear function of t: $D\xi(t) = t\sigma^2$, where σ^2 is a constant called a diffusion coefficient. Thus we have

$$D\xi(t) = t\sigma^2 = \frac{t}{\Delta t}\,(\Delta x)^2 = n\, Dx_i,$$

and due to the Gaussian distribution of x_i

$$P\left(\frac{\xi(t)}{\sigma\sqrt{t}} < x\right) = P\left(\frac{1}{\sqrt{nDx_i}}\sum_{i=1}^{n} < x\right).$$

Taking a limit as $n \to \infty$, we conclude by the central limit theorem that

$$P\left(\frac{1}{\sqrt{nDx_i}}\sum_{i=1}^{n} < x\right) \to \frac{1}{\sqrt{2\pi}}\int_{-\infty}^{x} e^{-z^2/2}\,dz,$$

and hence

$$P\left(\frac{\xi(t)}{\sigma\sqrt{t}} < x\right) = \frac{1}{\sqrt{2\pi}} \int_{-\infty}^{x} e^{-z^2/2}\, dz,$$

which implies that $\xi(t)$ is a Gaussian random variable with zero mean and variance $\sigma^2 t$.

Since the increment $\xi(t-s) - \xi(s)$ is distributed as $\xi(s)$ for all $s \geq 0$, we conclude that $\xi(t-s) - \xi(s)$ has a Gaussian distribution with zero mean and variance $\sigma^2 t$, $s \geq 0$. In this context, we call a stochastic process $W(t)$ *Brownian motion* if $W(0)$ and $W(t-s) - W(s)$ are independent and have a Gaussian distribution with zero mean and variance $\sigma^2 t$ for all $s \geq 0$.

We are now in a position to give an interpretation of the stochastic differential equation of the Ito type we started with in (1.27). So let us take the time subdivision $0 = t_0 < t_1 < \cdots < t_m = t$ and write a discrete version of (1.27):

$$X_{k+1} - X_k = b(t_k, X_k)\, \Delta t_k + \sigma(t_k, X_k)\, W_k \Delta_k, \tag{1.28}$$

where $X_j = X(t_j)$, $W_k = W_{t_k}$, $\Delta_k = t_{k+1} - t_k$. Now we look at the value $W_k \Delta t_k$ as an increment of a random process $\{V_t\}_{t \geq 0}$, $\Delta V_k = V_{k+1} - V_k$, and assume, in accordance with the assumptions (i)–(iii) made above for W_t, that V_t have stationary independent increments with zero mean. It can be shown that the only such process with continuous samples is the Brownian motion B_t. Thus we put $V_t = B_t$ and obtain from (1.28)

$$X_k = X_0 + \sum_{j=0}^{k-1} b(t_j, X_j)\Delta t_j + \sum_{j=0}^{k-1} \sigma(t_j, X_j)\Delta B_j. \tag{1.29}$$

Now we can take a limit, as $\Delta t_j \to 0$, in the mean squared sense we introduced above in the general considerations, and we arrive at

$$X(t) = X(0) + \int_0^t b(s, X_s)\, ds + \int_0^t \sigma(s, X_s)\, dB_s. \tag{1.30}$$

Thus the solution is expressed through the stochastic integral $\int_0^t \sigma(s, X_s)\, dB_s$ where the stochastic measure dB_s is generated by the Brownian motion. The construction of this integral can be carried out following the general construction of stochastic integrals given above. Alternatively, it can be done following a direct scheme using the Wiener process begun at the origin. So suppose $0 \leq S < T$ and $f(t, \omega)$ are given. As in the general approach presented above, first the stochastic integral for a simple class of functions f is defined, and then it is extended using some approximating procedure. So first assume that the function f has the form

$$\varphi(t, \omega) = \sum_{j \geq 0} e_j(\omega)\, \chi_{[j \cdot 2^{-n}, (j+1)2^{-n}]}(t), \tag{1.31}$$

where ξ_I is the indicator of the interval I, and n is a natural number, and e_j are some functions to be defined properly. For functions φ the stochastic integral is defined by

$$\int_S^T \varphi(t,\omega)\, dB_t(\omega) = \sum_{j\geq 0} e_j(\omega)\left[B_{t_{j+1}} - B_{t_j}\right](\omega), \tag{1.32}$$

where

$$t_k = t_k^{(n)} = \begin{cases} k\cdot 2^{-n} & \text{if } S \leq k\cdot 2^{-n} \leq T, \\ S & \text{if } k\cdot 2^{-n} < S, \\ T & \text{if } k\cdot 2^{-n} > T. \end{cases}$$

Then the standard extension procedure is carried out similarly to what we presented above (for more on this see [158]).

1.3.10 Multidimensional diffusion and Fokker–Planck equation

In this book, we will often work with multidimensional diffusions governed by systems of stochastic differential equations. The relevant probability density function satisfies the multidimensional Fokker–Planck equation. So let us consider first a 1-dimensional stochastic differential equation

$$dX(t) = b(X_t,t)dt + \sigma(X_t,t)\, dB_t \tag{1.33}$$

for a random trajectory starting at $X(0) = x_0$, with drift $b(X_t,t)$ and diffusion coefficient $\sigma^2(X_t,t)$. The probability density $f(x,t) = \langle\delta(x - X(t,x_0))\rangle$ satisfies the Fokker–Planck equation

$$\frac{\partial f(x,t)}{\partial t} = -\frac{\partial}{\partial x}\left[b(x,t)f(x,t)\right] + \frac{1}{2}\frac{\partial^2}{\partial x^2}\left[\sigma^2(X_t,t)\, f(x,t)\right]. \tag{1.34}$$

In many dimensions, we deal with $\mathbf{X}_t$, an n-dimensional random vector, and $\mathbf{W}_t$, an n-dimensional standard Wiener process. The multidimensional stochastic differential equation reads

$$d\mathbf{X}_t = \mathbf{b}(\mathbf{X}_t,t)\, dt + \sigma(\mathbf{X}_t)\, d\mathbf{W}_t \qquad \mathbf{X}_t(0) = \mathbf{x}_0, \tag{1.35}$$

where $\mathbf{b}(\mathbf{X}_t,t) = (b_1,\dots,b_n)$ is the drift vector, and $\sigma(\mathbf{x},t)$ is a positive-definite matrix.

The probability density function $f(\mathbf{x},t)$ for the random vector $\mathbf{X}_t$ satisfies the Fokker–Planck equation which is the following parabolic equation

$$\frac{\partial f(\mathbf{x},t)}{\partial t} = -\sum_{j=1}^{n}\frac{\partial}{\partial x_j}\left[b_j(\mathbf{x},t)f(\mathbf{x},t)\right] + \frac{1}{2}\sum_{i=1}^{n}\sum_{j=1}^{n}\frac{\partial^2}{\partial x_i^2\,\partial x_j^2}\left[\sigma^2(\mathbf{x},t)f(\mathbf{x},t)\right], \tag{1.36}$$

where $\sigma^2(\mathbf{x},t)$ is the diffusion tensor, a matrix with the entries

$$\sigma_{ij}^2(\mathbf{x},t) = \frac{1}{2}\sum_{k=1}^{n}\sigma_{ij}(\mathbf{x},t)\sigma_{jk}(\mathbf{x},t).$$

1.3.11 Central limit theorem and convergence of a Poisson process to a Gaussian process

Let us start with the central limit theorem (CLT).

Assume that we have a sequence of zero mean, i.i.d. (*arbitrarily distributed*!) random variables X_i, $i = 1, \ldots$, $P_X(x)$ being the probability distribution of X_i, and σ^2, the variance of X_i. Let

$$Z = \frac{X_1 + \cdots X_n}{\sqrt{n}}.$$

The CLT says that if $\sigma^2 < \infty$, then Z converges (in distribution) to the Gaussian distribution $N(0, \sigma^2)$.

The proof is extremely simple. We show that the characteristic function $\varphi_Z(t) = \langle e^{\mathrm{i}tZ} \rangle$ converges to the characteristic function of the Gaussian variable $N(0, \sigma^2)$, i. e., to $\exp[-\sigma^2 t^2/2]$.

Indeed, for any distribution $P_X(x)$ we have for its characteristic function

$$\varphi_X(t) = \int e^{itx}\, dP_X(x) = 1 - \frac{1}{2}\sigma^2 t^2 + \cdots. \tag{1.37}$$

But, since X_i are all mutually independent, we can write, using the well-known property of characteristics functions $\varphi_X(t\sigma) = \varphi_{\sigma X}(t)$, that

$$\varphi_Z(t) = \left[\varphi_{X_1}\Big(\frac{t}{\sqrt{n}}\Big)\right]^n \approx \Big(1 - \frac{1}{2}\frac{\sigma^2 t^2}{n}\Big)^n \to \exp\,[-\sigma^2 t^2/2]. \tag{1.38}$$

Notice that the omitted terms in (1.37) give a $n^{-3/2}$ contribution to (1.38), hence it vanishes, as $n \to \infty$.

Remark 1.1. The assumption that the means of X_i are equal to zero is by no means necessary. Just notice that $\varphi_{X+m}(t) = e^{imt}\varphi_X(t)$ where $m = \langle X \rangle$.

Now let us prove that a Poisson distribution with a mean m (for simplicity we assume that m is an integer number), having the discrete probability distribution $P_i = \frac{m^i}{i!}\, e^{-m}$, and the characteristic function

$$\varphi(t) = \exp\,[m(e^{\mathrm{i}t} - 1)], \tag{1.39}$$

converges to a Gaussian distribution $N(m, m)$ if m is sufficiently large.

Clearly, from CLT we could immediately conclude this result if we could represent the Poissonian random variable as a sum of i.i.d variables. But this can indeed be done, since the random variable having a Poisson distribution with the mean m can be represented as a sum of m i.i.d. random numbers whose means are equal to 1. This follows immediately from the structure of the characteristic function (1.39). So, by the CLT this sum converges to $N(m, m)$; however, m should be large enough. In practice, starting from $m = 10$ the approximation is already adequate, but we will show below that the asymmetry of the Poissonian law is of the order of $m^{-1/2}$, and so, to have a good symmetry property, m should be chosen appropriately large!

Final remark: we could of course directly prove that the characteristic function of the Poisson distribution with the mean m converges to the characteristic function of the Gaussian law $N(m, m)$ as m goes to infinity.

Indeed, let us first rewrite the characteristic function of the Poisson distribution as follows:

$$\varphi(t) = e^{m(e^{it}-1)} = e^{imt} \exp[-mt^2/2 + imt^3/6 + \cdots].$$

So it remains to prove that the second exponent tends, as m increases, to the characteristic function of the Gaussian random variable $m^{1/2}\xi$, where ξ is a standard Gaussian random variable from $N(0, 1)$. Let us make the change of variable $t = t'/\sqrt{m}$. Using the property $\varphi_{\sqrt{m}\xi}(\frac{t'}{\sqrt{m}}) = \varphi_\xi(t')$, we come to the final conclusion looking at the transformation after substituting $t = t'/\sqrt{m}$ that

$$\exp[-mt^2/2 + imt^3/6 + \cdots] \quad \text{transforms to} \quad \exp\left[-\frac{t'^2}{2} + O\left(\frac{t'^3}{m^{1/2}}\right)\right],$$

since the function in the right-hand side converges, for any fixed t', to the characteristic function of $\xi \sim N(0, 1)$, as $m \to \infty$.

One useful remark from the last equality is that the asymmetry of the Poissonian distribution has the order of $m^{-1/2}$.

Now we pose a question: Is the above result concerning the convergence of a Poisson distribution to a Gaussian distribution also true for random processes? To be specific we choose an interpretation of the random points as dislocations in crystals which cause elastic displacements.

Let $\xi_1, \dots \xi_n, \dots$ be the random positions of dislocations on a line, sampled according the Poisson distribution, and let $v(x, \xi_j)$ be the elastic displacement at the point x caused by the dislocation positioned at ξ_j. Then, the total displacement is the superposition

$$V(x) = \sum_j v(x, \xi_j), \tag{1.40}$$

which may be written as a stochastic integral with respect to Poissonian measure:

$$V(x) = \int_{x_0}^{x} v(x, y)\, dN(y) \tag{1.41}$$

Here $N(x)$ is a Poisson random process which is characterized as follows:

(a) $N(x)$ is a random function with independent increments, taking integer values, and is almost certainly nondecreasing.

(b) The probability that the number of dislocations belonging to any finite interval be infinite is zero.

(c) The distribution of these dislocation positions is homogeneous in x.

(d) m is the expectation of the number of dislocations belonging to any unit interval.

(e) The probability that the number of dislocations belonging to any interval (x, x'), $x < x'$, be equal to n is equal to

$$e^{-m(x'-x)} \frac{[m(x'-x)]^n}{n!}.$$

The logarithm $\psi(w)$ of the characteristic function of $V(x)$ is equal to

$$\psi(w) = m \int_{-\infty}^{\infty} \left[e^{\mathrm{i}w\, v(y)} - 1\right] dy.$$

Now we can study the limit as $m \to \infty$. The conditions of convergence of V to the Gaussian process can be formulated as follows: there exist a number a and a positive number b such that

$$\lim_{m\to\infty} \left[\psi(w/b) - \mathrm{i}aw/b\right] = -w^2/2.$$

Indeed, let us suppose that $v(x)$ is integrable and v^2 is integrable. Then,

$$\psi'(w) = \mathrm{i}m \int_{-\infty}^{\infty} v(y)\, e^{\mathrm{i}w\, v(y)} dy, \quad \psi'(0) = \mathrm{i}m \int_{-\infty}^{\infty} v(y)\, dy$$

and

$$\psi''(w) = -m \int_{-\infty}^{\infty} v^2(y) e^{\mathrm{i}w\, v(y)} dy, \quad \psi''(0) = -m \int_{-\infty}^{\infty} v^2(y)\, dy.$$

Let

$$\mu_1 = \int_{-\infty}^{\infty} v(y)\, dy, \quad \mu_2 = \int_{-\infty}^{\infty} v^2(y)\, dy.$$

Then,

$$\psi(w) = \mathrm{i}m\mu_1 w - \frac{m}{2}\mu_2 w^2 + w^2 O(w).$$

where $O(w) \to 0$ as $w \to 0$. Thus the constants a and b exist and can be taken as $a = m\mu_1$, and $b = \sqrt{m\mu_2}$.

If our function v depends only on the difference $x - x'$, the random function V tends to a homogeneous Gaussian random process, as $m \to \infty$, which is uniquely defined by its correlation function

$$B(h) = \int_{-\infty}^{\infty} v(y) v(y+h)\, dy.$$

Chapter 2

Stochastic simulation of vector Gaussian random fields

Randomized spectral models (RSM) and randomized Fourier wavelet models (FWM) for simulation of homogeneous Gaussian random fields based on spectral representations and plane wave decomposition of random fields are presented following our paper [118]. Extensions of FWM to vector random processes are constructed. Convergence of the constructed Fourier wavelet models (in the sense of finite-dimensional distributions) under some general conditions on the spectral tensor is given. A comparative analysis of RSM and FWM is made by calculating Eulerian statistical characteristics of a 3D isotropic incompressible random field through an ensemble and space averaging.

2.1 Introduction

In the present chapter we deal with an analogous comparative analysis of RSM and FWM for 3-dimensional random fields which we developed mainly in [118]. Since the cost of both methods is proportional to the number of decades of the simulated random field, $\lg(l_{\max}/l_{\min})$, it is not necessary to handle multiscale random fields. In this chapter we simulate a 3D isotropic incompressible random field whose longitudinal correlation function is exponential. In the comparative analysis we have also included a stratified randomized Fourier wavelet model (SRFWM), a new version of FWM which we introduce in this chapter. We also generalized the FWM to vector random fields (we are aware only of papers dealing with FWM for scalar processes and isotropic fields; see [25,47–51,103]). The convergence analysis in different functional spaces for FWM was given in our paper [117].

The chapter is organized as follows. In Section 2.2, we present the basic formulas for the discrete expansions of random fields (starting from its spectral representation) in orthonormal system of functions. In Section 2.3, we give the basics of the Fourier wavelet expansions, as well as the wavelet expansions, to stress the difference between these two representations. The randomization of spectral expansions is described in Section 2.4. In Section 2.5, the Fourier wavelet model for random processes is presented, with a study of the error when a finite number of terms is taken in the relevant series expansion. Here we compare the model with the original method published in [47]. In Section 2.6, we present an approach known as a plane wave decomposition of random fields, which suggests a general technique of random field simulation through samples of stochastic processes. We describe different versions (deterministic, randomized, and stratified randomized) for choosing the nodes in the stochastic integral representation through the plane wave decomposition. In Section 2.7, we present the simulation results for a 3D isotropic incompressible random field carried out by different versions of RSM and FWM. We compare the Eulerian longitudinal and transversal correlation functions. All these statistical characteristics were calculated both by ensemble and space averaging. The main results are summarized in the Conclusion, and some technical details are included in Appendices A and B of Section 2.9.

2.2 Discrete expansions related to the spectral representations of Gaussian random fields

2.2.1 Spectral representations

We deal in this chapter with real-valued homogeneous Gaussian l-dimensional vector random fields $\vec{u}(\mathbf{x}) = (u_1(\mathbf{x}), \ldots, u_l(\mathbf{x}))^T$, $\mathbf{x} \in I\!R^d$ with a given correlation tensor $B(\vec{r})$ with entries

$$B_{ij}(\vec{r}) = \langle u_i(\mathbf{x}+\vec{r})\, u_j(\mathbf{x})\rangle, \quad i,j = 1,\ldots l, \tag{2.1}$$

or with the corresponding spectral tensor F:

$$F_{ij}(\mathbf{k}) = \int_{\boldsymbol{R}^d} e^{-\mathrm{i}\,2\pi\,\mathbf{k}\cdot\vec{r}} B_{ij}(\vec{r})\, d\vec{r}, \quad B_{ij}(\vec{r}) = \int_{\boldsymbol{R}^d} e^{\mathrm{i}\,2\pi\,\vec{r}\cdot\mathbf{k}} F_{ij}(\mathbf{k})\, d\mathbf{k}, \quad i,j = 1,\ldots l. \tag{2.2}$$

We will assume that the condition $\int_{\boldsymbol{R}^d} |B_{jj}(\vec{r})|\, d\vec{r} < \infty$ is satisfied, which ensures that the spectral functions F_{ij} are uniformly continuous with respect to $\mathbf{k}$. Here B_{jj} is the trace of B, i. e., we use here and in the following the summation convention under repeated indices. Note that a weaker assumption that B is squared integrable guarantees only the existence of the spectral tensor in the space L_2.

Let $Q(\mathbf{k})$ be an $l \times n$-matrix defined by

$$Q(\mathbf{k})Q^*(\mathbf{k}) = F(\mathbf{k}), \quad Q(-\mathbf{k}) = \bar{Q}(\mathbf{k}). \tag{2.3}$$

Here the star stands for the complex conjugate transpose, which is equivalent to taking two operations: the transpose ^{T}, and the complex conjugation of each entry. Then the spectral representation of the random field is written as follows (e. g., see [70, 146, 259]):

$$\vec{u}(\mathbf{x}) = \int_{\boldsymbol{R}^d} e^{\mathrm{i}\,2\pi\,\mathbf{k}\mathbf{x}}\, Q(\mathbf{k})\, \mathbf{Z}(d\mathbf{k}), \tag{2.4}$$

where the column-vector $\mathbf{Z} = (Z_1, \ldots Z_n)^T$ is a complex-valued homogeneous n-dimensional white noise on $I\!R^d$ with a unite variance and zero mean:

$$\langle \mathbf{Z}(d\mathbf{k})\rangle = 0, \quad \langle Z_i(d\mathbf{k}_1)\, \bar{Z}_j(d\mathbf{k}_2)\rangle = \delta_{ij}\, \delta(\mathbf{k}_1 - \mathbf{k}_2)\, d\mathbf{k}_1\, d\mathbf{k}_2, \tag{2.5}$$

satisfying the condition $\mathbf{Z}(-d\mathbf{k}) = \bar{\mathbf{Z}}(d\mathbf{k})$. Note that the last condition can be satisfied as follows. Let $\mathbf{Z} = (\mathbf{Z}_R + \mathrm{i}\mathbf{Z}_I)/\sqrt{2}$ where $\mathbf{Z}_R$ and $\mathbf{Z}_I$ are independent homogeneous real-valued white n-dimensional vector noises with a unity variance and zero mean defined on the half space $[0,\infty) \times \mathbb{R}^{d-1}$. Now, we prolong the measures $\mathbf{Z}_R$ and $\mathbf{Z}_I$ on $(-\infty, 0) \times \mathbb{R}^{d-1}$ by $\mathbf{Z}_R(d\mathbf{k}) = \mathbf{Z}_R(-d\mathbf{k})$, and $\mathbf{Z}_I(d\mathbf{k}) = -\mathbf{Z}_I(-d\mathbf{k})$. All these conditions ensure that the random field (2.4) has real-valued components.

There is a different spectral representation which we will use further. Let us introduce a complex random field

$$\mathbf{w}(\mathbf{x}) = \vec{u}_R(\mathbf{x}) + \mathrm{i}\,\vec{u}_I(\mathbf{x}) = \int_{\mathbb{R}^d} e^{\mathrm{i}\,2\pi\,\mathbf{k}\mathbf{x}} Q(\mathbf{k})\,\mathbf{W}(d\mathbf{k}). \tag{2.6}$$

Here $\mathbf{W}(d\mathbf{k}) = \mathbf{W}_R(d\mathbf{k}) + \mathrm{i}\mathbf{W}_I(d\mathbf{k})$ is an n-dimensional complex white noise where the real and imagine parts are two independent homogeneous white noises on the whole space $\mathbb{R}^d$ with unit variance and zero mean. It is not difficult to see that $\vec{u}_R(\mathbf{x})$ and $\vec{u}_I(\mathbf{x})$ are independent Gaussian random fields both having the spectral tensor $F(\mathbf{k})$.

2.2.2 Series expansions

In this section, we deal with the expansions

$$\vec{u}(\mathbf{x}) = \sum_{\alpha \in \mathcal{A}} G_\alpha(\mathbf{x})\,\xi_\alpha, \tag{2.7}$$

where $G_\alpha(\mathbf{x})$ is a system of deterministic functions (or possibly matrices), ξ_α is a family of random variables (possibly vectors), $\mathcal{A}$ is a countable (finite or not) index set.

Our purpose is to construct the system G_α and the family ξ_α so that the random field (2.7) has the desired spectral tensor (2.2).

2.2.3 Expansion with an even complex orthonormal system

Let us choose the system of scalar functions $\varphi_\alpha(\mathbf{k})$ as a set of generally complex valued even functions ($\varphi_\alpha(-\mathbf{k}) = \bar{\varphi}_\alpha(\mathbf{k})$) which are orthonormal and complete in $L_2(\mathbb{R}^d)$ equipped with the scalar product $(f,g) = \int_{\mathbb{R}^d} f(\mathbf{k})\bar{g}(\mathbf{k})\,d\mathbf{k}$:

$$(\varphi_\alpha, \varphi_\beta) = \int_{\mathbb{R}^d} \varphi_\alpha(\mathbf{k})\bar{\varphi}_\beta(\mathbf{k})\,d\mathbf{k} = \delta_{\alpha\beta}, \quad \alpha, \beta \in \mathcal{A}, \tag{2.8}$$

where $\delta_{\alpha\beta}$ is the Kronecker symbol. Since $B_{jj}(0) = \int F_{jj}(\mathbf{k})\,d\mathbf{k} < \infty$, and $QQ^* = F$, we conclude that all the entries of the matrix Q belong to $L_2(\mathbb{R}^d)$. Then, we expand

$e^{i 2\pi \mathbf{k}\cdot\mathbf{x}} Q(\mathbf{k})$, as a function of $\mathbf{k}$, in the system of orthonormal functions $\varphi_\alpha(\mathbf{k})$:

$$e^{i 2\pi \mathbf{k}\cdot\mathbf{x}} Q(\mathbf{k}) = \sum_{\alpha\in\mathcal{A}} G_\alpha(\mathbf{x})\, \varphi_\alpha(\mathbf{k}), \quad G_\alpha(\mathbf{x}) = \int_{\mathbb{R}^d} e^{i 2\pi \mathbf{k}\mathbf{x}} Q(\mathbf{k}) \bar{\varphi}_\alpha(\mathbf{k})\, d\mathbf{k}. \tag{2.9}$$

We now substitute this representation in (2.4), and obtain the expansion (2.7) with

$$\boldsymbol{\xi}_\alpha = \int_{\mathbb{R}^d} \varphi_\alpha(\mathbf{k})\, \mathbf{Z}(d\mathbf{k}), \quad \alpha \in \mathcal{A}. \tag{2.10}$$

Note that by the construction and the assumptions made, the random vectors $\boldsymbol{\xi}_\alpha$ and the functions $G_\alpha(\mathbf{x})$ are all real valued. Notice also that $\boldsymbol{\xi}_\alpha$ are mutually independent standard Gaussian random vectors since

$$\langle \boldsymbol{\xi}_\alpha\, \boldsymbol{\xi}_\beta^* \rangle = \mathrm{I} \int_{\mathbb{R}^d} \varphi_\alpha(\mathbf{k}) \bar{\varphi}_\beta(\mathbf{k})\, d\mathbf{k} = \mathrm{I}\, \delta_{\alpha\beta},$$

where I is a $n \times n$ identity matrix.

Thus we have constructed an expansion of type (2.7) with independent Gaussian random vectors.

2.2.4 Expansion with a real orthonormal system.

Let us construct another expansion starting from a set of orthonormal scalar real valued functions $g_\alpha(\mathbf{x})$ complete in $L_2(\mathbb{R}^d)$:

$$(g_\alpha, g_\beta) = \int_{\mathbb{R}^d} g_\alpha(\mathbf{x})\, g_\beta(\mathbf{x})\, d\mathbf{x} = \delta_{\alpha\beta}.$$

Then,

$$\vec{u}(\mathbf{x}) = \sum_{\alpha\in\mathcal{A}} g_\alpha(\mathbf{x}) \boldsymbol{\xi}_\alpha, \tag{2.11}$$

where the family of random column-vectors of dimension l is defined by

$$\boldsymbol{\xi}_\alpha = \int \vec{u}(\mathbf{x}) g_\alpha(\mathbf{x})\, d\mathbf{x}. \tag{2.12}$$

Substitution of the representation (2.2) in (2.12) yields

$$\boldsymbol{\xi}_\alpha = \int_{\mathbb{R}^d} \hat{g}_\alpha(-\mathbf{k})\, Q(\mathbf{k})\, \mathbf{Z}(d\mathbf{k}), \quad \hat{g}_\alpha(\mathbf{k}) = \int_{\mathbb{R}^d} e^{-i 2\pi \mathbf{k}\mathbf{x}} g_\alpha(\mathbf{x})\, d\mathbf{x}. \tag{2.13}$$

Note that $\boldsymbol{\xi}_\alpha$ are real-valued Gaussian random column vectors of dimension l; these vectors are generally correlated, in contrast to the family (2.10).

2.2.5 Complex valued orthogonal expansions

A different expansion can be obtained if we expand $e^{\mathrm{i}\,2\pi\,\mathbf{k}\cdot\mathbf{x}}Q(\mathbf{k})$ in (2.6) in the form of (2.9), where φ_α is a set of arbitrary (possibly not even) orthonormal complex-valued functions complete in $L_2(I\!R^d)$, while $G_\alpha(\mathbf{x})$ and the family $\boldsymbol{\xi}_\alpha$ are defined in (2.9) and (2.10), respectively. This gives

$$\vec{u}_R(\mathbf{x}) = \sum_\alpha \left(G_\alpha^{(R)}(\mathbf{x})\,\boldsymbol{\xi}_\alpha^{(R)} - G_\alpha^{(I)}(\mathbf{x})\,\boldsymbol{\xi}_\alpha^{(I)} \right),$$
$$\vec{u}_I(\mathbf{x}) = \sum_\alpha \left(G_\alpha^{(R)}(\mathbf{x})\,\boldsymbol{\xi}_\alpha^{(I)} - G_\alpha^{(I)}(\mathbf{x})\,\boldsymbol{\xi}_\alpha^{(R)} \right), \tag{2.14}$$

where

$$G_\alpha(\mathbf{x}) = G_\alpha^{(R)}(\mathbf{x}) + \mathrm{i}\,G_\alpha^{(I)}(\mathbf{x}), \quad \boldsymbol{\xi}_\alpha = \boldsymbol{\xi}_\alpha^{(R)} + \mathrm{i}\,\boldsymbol{\xi}_\alpha^{(I)}.$$

Analogously, we can generalize the expansion (2.11) to the general complex-valued case. Let $g_\alpha(\mathbf{x})$ be an arbitrary complete system of complex-valued orthonormal functions in $L_2(I\!R^d)$, and let $\boldsymbol{\xi}_\alpha$ be a family of random column vectors defined in (2.13). Then,

$$\vec{u}_R(\mathbf{x}) = \sum_\alpha \left(g_\alpha^{(R)}(\mathbf{x})\,\boldsymbol{\xi}_\alpha^{(R)} - g_\alpha^{(I)}(\mathbf{x})\,\boldsymbol{\xi}_\alpha^{(I)} \right),$$
$$\vec{u}_I(\mathbf{x}) = \sum_\alpha \left(g_\alpha^{(R)}(\mathbf{x})\,\boldsymbol{\xi}_\alpha^{(I)} + g_\alpha^{(I)}(\mathbf{x})\,\boldsymbol{\xi}_\alpha^{(R)} \right), \tag{2.15}$$

where

$$g_\alpha(\mathbf{x}) = g_\alpha^{(R)}(\mathbf{x}) + \mathrm{i}\,g_\alpha^{(I)}(\mathbf{x}), \quad \boldsymbol{\xi}_\alpha = \boldsymbol{\xi}_\alpha^{(R)} + \mathrm{i}\,\boldsymbol{\xi}_\alpha^{(I)}.$$

2.3 Wavelet expansions

In this section we consider an important example of expansions based on even orthogonal systems described in the first subsection of Section 2.2. For simplicity we present here the 1-dimensional case, $d = 1$. Our goal is to construct the expansion of a homogeneous Gaussian vector random process $\vec{u}(x) = (u_1(x), \ldots, u_l(x))^T$, $x \in I\!R$ with a given spectral tensor $F(k)$. Again, we assume that $F = Q\,Q^*$, where $Q(k)$ is $l \times n$-dimensional matrix satisfying the condition $Q(-k) = \bar{Q}(k)$.

The orthonormal system of functions φ_α is constructed as follows. Let us introduce the notations: $\phi(x)$ and $\psi(x)$, $x \in I\!R$ are orthonormal scaling and wavelet functions, respectively (e. g., see [27]), and

$$\phi_{mj}(x) = 2^{m/2}\phi(2^m\,x - j), \quad \psi_{mj}(x) = 2^{m/2}\psi(2^m\,x - j), \tag{2.16}$$

where $m, j = \dots, -2, -1, 0, 1, 2, \dots$. It is known (e. g., see [27, 78]) that the system of functions

$$\{\phi_{m_0 j}\}_{j=-\infty}^{\infty}, \quad \left\{\{\psi_{mj}\}_{j=-\infty}^{\infty}, \quad m \geq m_0\right\} \tag{2.17}$$

is, for an arbitrary fixed integer m_0, a complete set of orthonormal functions in $L_2(I\!R)$, and moreover, by Parseval equality, the relevant Fourier transforms of these functions

$$\{\hat{\phi}_{m_0 j}\}_{j=-\infty}^{\infty}, \quad \left\{\{\hat{\psi}_{mj}\}_{j=-\infty}^{\infty}, \quad m \geq m_0\right\} \tag{2.18}$$

compose also a complete set of orthonormal functions in $L_2(I\!R)$.

2.3.1 Fourier wavelet expansions

Thus we choose, as the family φ_α, the system of functions (2.18).

From (2.9) and (2.7) we find that

$$\vec{u}(x) = \sum_{j=-\infty}^{\infty} G_{m_0 j}^{(\phi)}(x)\, \boldsymbol{\xi}_j + \sum_{m=m_0}^{\infty} \sum_{j=-\infty}^{\infty} G_{mj}^{(\psi)}(x) \boldsymbol{\xi}_{mj}, \tag{2.19}$$

where $\boldsymbol{\xi}_j$, $\boldsymbol{\xi}_{mj}$ is a family of mutually independent standard real-valued Gaussian random vectors of dimension n, and $G_{mj}^{(\phi)}(x)$, $G_{mj}^{(\psi)}(x)$ are $l \times n$-dimensional matrices defined by

$$G_{mj}^{(\phi)}(x) = \int_{-\infty}^{\infty} e^{\mathrm{i}\, 2\pi\, kx}\, Q(k)\, \bar{\hat{\phi}}_{mj}(k)\, dk, \quad G_{mj}^{(\psi)}(x) = \int_{-\infty}^{\infty} e^{\mathrm{i}\, 2\pi\, kx}\, Q(k)\, \bar{\hat{\psi}}_{mj}(k)\, dk.$$

It is clear that

$$\begin{aligned} \hat{\phi}_{mj}(k) &= 2^{-m/2}\, e^{-\mathrm{i}\, 2\pi\, k\, j\, 2^{-m}}\, \hat{\phi}(2^{-m}\, k), \\ \hat{\psi}_{mj}(k) &= 2^{-m/2}\, e^{-\mathrm{i}\, 2\pi\, k\, j\, 2^{-m}}\, \hat{\psi}(2^{-m}\, k). \end{aligned} \tag{2.20}$$

Now we can define the analog of G_α by substituting $\bar{\varphi}_\alpha(k)$ in (2.9) with these expressions, which yields

$$\begin{aligned} G_{mj}^{(\phi)}(x) &= \int_{-\infty}^{\infty} e^{\mathrm{i}\, 2\pi\, kx}\, Q(k)\, \hat{\phi}_{mj}(-k)\, dk \\ &= \int_{-\infty}^{\infty} e^{-\mathrm{i}\, 2\pi\, kx}\, \bar{Q}(k)\, \hat{\phi}_{mj}(k)\, dk \\ &= \int_{-\infty}^{\infty} e^{-\mathrm{i}\, 2\pi\, kx}\, \bar{Q}(k)\, 2^{-m/2}\, e^{-\mathrm{i}\, 2\pi\, k\, j\, 2^{-m}}\, \hat{\phi}(2^{-m}\, k)\, dk \\ &= \int_{-\infty}^{\infty} e^{-\mathrm{i}\, 2\pi\, k'(2^m x + j)}\, 2^{m/2} \bar{Q}(2^m k') \hat{\phi}(k')\, dk'. \end{aligned} \tag{2.21}$$

Analogously,

$$G_{mj}^{(\psi)}(x) = \int_{-\infty}^{\infty} e^{-\mathrm{i}\,2\pi\, k'(2^m x+j)}\, 2^{m/2} \bar{Q}(2^m k')\hat{\psi}(k')\, dk'. \tag{2.22}$$

For convenience, we define

$$\begin{aligned} \mathcal{F}_m^{(\phi)}(y) &= \int_{-\infty}^{\infty} e^{-\mathrm{i}\,2\pi\, ky}\, 2^{m/2} \bar{Q}(2^m k)\hat{\phi}(k)\, dk, \\ \mathcal{F}_m^{(\psi)}(y) &= \int_{-\infty}^{\infty} e^{-\mathrm{i}\,2\pi\, ky}\, 2^{m/2} \bar{Q}(2^m k)\hat{\psi}(k)\, dk, \end{aligned} \tag{2.23}$$

hence

$$G_{m_0 j}^{(\phi)}(x) = \mathcal{F}_{m_0}^{(\phi)}(2^{m_0} x + j), \quad G_{mj}^{(\psi)}(x) = \mathcal{F}_m^{(\psi)}(2^m x + j), \tag{2.24}$$

and finally,

$$\vec{u}(x) = \sum_{j=-\infty}^{\infty} \mathcal{F}_{m_0}^{(\phi)}(2^{m_0} x + j)\, \boldsymbol{\xi}_j + \sum_{m=m_0}^{\infty} \sum_{j=-\infty}^{\infty} \mathcal{F}_m^{(\psi)}(2^m x + j)\, \boldsymbol{\xi}_{mj}. \tag{2.25}$$

2.3.2 Wavelet expansion

In this section we deal with the expansion of the type (2.11) and use the scaling and wavelet functions (2.16) as the orthonormal system G_α. Thus the expansion reads

$$\vec{u}(x) = \sum_{j=-\infty}^{\infty} \phi_{m_0 j}(x)\, \boldsymbol{\xi}_{m_0 j}^{(\phi)} + \sum_{m=m_0}^{\infty} \sum_{j=-\infty}^{\infty} \psi_{mj}(x) \boldsymbol{\xi}_{mj}^{(\psi)}, \tag{2.26}$$

where $\boldsymbol{\xi}_{m_0 j}^{(\phi)}$ and $\boldsymbol{\xi}_{mj}^{(\psi)}$ is a set of Gaussian column-vectors of dimension n defined as follows:

$$\begin{aligned} \boldsymbol{\xi}_{mj}^{(\phi)} &= \int_{-\infty}^{\infty} Q(k)\, \hat{\phi}_{mj}(-k)\, \mathbf{Z}(dk) \\ &= \int_{-\infty}^{\infty} Q(k)\, 2^{-m/2}\, e^{\mathrm{i}\,2\pi\, k\, j\, 2^{-m}}\, \bar{\hat{\phi}}(2^{-m} k)\, \mathbf{Z}(dk), \end{aligned} \tag{2.27}$$

and analogously,

$$\boldsymbol{\xi}_{mj}^{(\psi)} = \int_{-\infty}^{\infty} Q(k)\, 2^{-m/2}\, e^{\mathrm{i}\,2\pi\, k\, j\, 2^{-m}}\, \bar{\hat{\psi}}(2^{-m} k)\, \mathbf{Z}(dk). \tag{2.28}$$

Note that in contrast to the Fourier wavelet expansions, here the random vectors $\xi^{(\phi)}_{mj}$ and $\xi^{(\psi)}_{mj}$ are correlated, and in particular, in calculations one needs the following correlations:

$$\langle \xi^{(\phi)}_{mj_1} \, (\xi^{(\phi)}_{mj_2})^T \rangle = 2^{-m} \int_{-\infty}^{\infty} e^{\mathrm{i}\, 2\pi\, k(j_1-j_2)} |\hat{\phi}(2^{-m}k)|^2 \, F(k)\, dk, \tag{2.29}$$

$$\langle \xi^{(\phi)}_{mj_1} \, (\xi^{(\psi)}_{mj_2})^T \rangle = 2^{-m} \int_{-\infty}^{\infty} e^{\mathrm{i}\, 2\pi\, k(j_1-j_2)} \Big(\bar{\hat{\phi}}(2^{-m}k)\hat{\psi}(2^{-m}k)\Big) \, F(k)\, dk, \tag{2.30}$$

$$\langle \xi^{(\psi)}_{mj_1} \, (\xi^{(\psi)}_{mj_2})^T \rangle = 2^{-m} \int_{-\infty}^{\infty} e^{\mathrm{i}\, 2\pi\, k(j_1-j_2)} |\hat{\psi}(2^{-m}k)|^2 \, F(k)\, dk. \tag{2.31}$$

It is worth noting that in the expansion (2.26), a biorthogonal system of wavelet functions can be used (see [231]).

2.3.3 Moving averages

Let $F(\mathbf{k})$ be an $l \times l$-dimensional spectral tensor of a random field, and $Q(\mathbf{k})$ be an $l \times n$-dimensional complex-valued matrix satisfying (2.3). Since obviously $B_{jj}(0) = \int F_{jj}(\mathbf{k})\, d\mathbf{k} < \infty$, we see by (2.3) that the entries of the matrix are square integrable, i. e., $Q(\mathbf{k}) \in L_2(I\!R^d)$. Therefore the $l \times n$-dimensional matrix

$$G(\mathbf{x}) = \int_{\mathbf{R}^d} e^{\mathrm{i}\, 2\pi\, \mathbf{kx}}\, Q(\mathbf{k})\, d\mathbf{k} \tag{2.32}$$

is well defined, with real-valued square integrable entries.

We now define an l-dimensional homogeneous Gaussian random field

$$\vec{u}(\mathbf{x}) = \int_{\mathbf{R}^d} G(\mathbf{x}-\mathbf{y})\, \mathbf{W}(d\mathbf{y}), \tag{2.33}$$

where $\mathbf{W}(d\mathbf{y}) = (W_1(d\mathbf{y}), \dots, W_n(d\mathbf{y}))^T$, and $W_i(d\mathbf{y}), i = 1, \dots, n$ are real-valued independent homogeneous Gaussian white noises in $I\!R^d$ with united variance and zero mean. Then the random field (2.33) has the spectral tensor $F(\mathbf{k})$. Indeed,

$$\begin{aligned}\Big\langle \vec{u}(\mathbf{x}+\vec{r})\, \vec{u}^T(\mathbf{x}) \Big\rangle &= \Big\langle \int_{\mathbf{R}^d} G(\mathbf{x}+\vec{r}-\mathbf{y})\, \mathbf{W}(d\mathbf{y}) \Big(\int_{\mathbf{R}^d} G(\mathbf{x}-\mathbf{y})\, \mathbf{W}(d\mathbf{y}) \Big)^T \Big\rangle \\ &= \int_{\mathbf{R}^d} G(\mathbf{x}+\vec{r}-\mathbf{y})\, G^T(\mathbf{x}-\mathbf{y})\, d\mathbf{y} = \int_{\mathbf{R}^d} G(\mathbf{z}+\vec{r})\, G^T(\mathbf{z})\, d\mathbf{z}.\end{aligned}$$

Taking the Fourier transformation of the right-hand side we get

$$\int_{\mathbb{R}^d} e^{-\mathrm{i}\,2\pi\,\mathbf{k}\vec{r}}\,d\vec{r}\int_{\mathbb{R}^d} G(\mathbf{z}+\vec{r})\,G^T(\mathbf{z})\,d\mathbf{z}$$

$$= \int_{\mathbb{R}^d} e^{-\mathrm{i}\,2\pi\,\mathbf{k}(\vec{r}+\mathbf{z})}\,G(\vec{r}+\mathbf{z})\,d(\vec{r}+\mathbf{z})\int_{\mathbb{R}^d} e^{\mathrm{i}\,2\pi\,\mathbf{k}\mathbf{z}},G^T(\mathbf{z})\,d\mathbf{z}$$

$$= \int_{\mathbb{R}^d} e^{-\mathrm{i}\,2\pi\,\mathbf{k}\mathbf{x}}\,G^T(\mathbf{x})\,d\mathbf{x}\overline{\left(\int_{\mathbb{R}^d} e^{-\mathrm{i}\,2\pi\,\mathbf{k}\mathbf{z}}G^T(\mathbf{z})\,d\mathbf{z}\right)} = Q(\mathbf{k})Q^*(\mathbf{k}) = F(\mathbf{k}).$$

2.4 Randomized spectral models

A straightforward evaluation of the stochastic integral (2.4) is based on the Riemann sums calculation with fixed cells (see, e. g., [168, 226]). The integral is approximated by a finite sum

$$\vec{u}(\mathbf{x}) \approx \sum_{i=1}^{n}\left[\cos(2\pi\mathbf{k}_i\cdot\mathbf{x})\boldsymbol{\xi}_i + \cos(2\pi\mathbf{k}_i\cdot\mathbf{x})\,\boldsymbol{\eta}_i\right].$$

where $\mathbf{k}_i$ are deterministic nodes in the Fourier space, $\boldsymbol{\xi}_i$ and $\boldsymbol{\eta}_i$ are Gaussian random vectors with zero mean and relevant covariance. Efficient calculation of the above sum is usually carried out by the fast Fourier transform, which assumes that the nodes are chosen uniformly. It should be mentioned that this scheme suffers from an artificial periodicity in the scale of $1/\Delta k$ where Δk is the integration step in the Fourier space.

In randomized models, the nodes are chosen at random, with an appropriate probability distribution so that the model has the desired correlation structure. We mention by passing that this model is free from the above-mentioned artificial periodicity.

2.4.1 Randomized spectral models defined through stochastic integrals

Let us consider an l-dimensional random field $\vec{u}(\mathbf{x})$, $\mathbf{x}\in\mathbb{R}^d$ defined by the stochastic integral [70]:

$$\vec{u}(\mathbf{x}) = \int_{\mathbb{R}^d} \mathcal{H}(\mathbf{x},\mathbf{k})\,\mathbf{W}(d\mathbf{k}), \tag{2.34}$$

where (1) $\mathcal{H} : \mathbb{R}^d \times \mathbb{R}^{d_1} \to \mathcal{C}^{l\times n}$ is a matrix such that $\mathcal{H}(\mathbf{x},\cdot) \in L_2(\mathbb{R}^{d_1})$ for each $\mathbf{x}\in\mathbb{R}^d$; (2) $\mathbf{W}(\cdot) = \mathbf{W}_R(\cdot) + \mathrm{i}\,\mathbf{W}_I(\cdot)$, where $\mathbf{W}_R(\cdot)$ and $\mathbf{W}_I(\cdot)$ are two independent n-dimensional homogeneous Gaussian white noises on $\mathbb{R}^{d_1}$ with unit variance. Here $\mathcal{C}$ is the set of complex numbers.

Let us describe the randomized evaluation of the stochastic integral (2.34). Let $p : \mathbb{R}^{d_1} \to [0,\infty)$ be a probability density on $\mathbb{R}^{d_1}$: $\int p(\mathbf{k})\,d\mathbf{k} = 1$, and let $\mathbf{k}_1,\ldots,\mathbf{k}_{n_0}$ be independent equally distributed random points in $I\!R^{d_1}$ with density $p(\mathbf{k})$. Assume that $\boldsymbol{\zeta}_1, \dots, \boldsymbol{\zeta}_{n_0}$ is a family of mutually independent standard Gaussian complex random vectors of dimension n (i. e., $\boldsymbol{\zeta}_j = \boldsymbol{\xi}_j + \mathrm{i}\,\boldsymbol{\eta}_j$ with independent, n-dimensional real-valued standard Gaussian random vectors $\boldsymbol{\xi}_i$ and $\boldsymbol{\eta}_i$). Then the random field

$$\vec{u}_{n_0}(\mathbf{x}) = \frac{1}{\sqrt{n_0}} \sum_{j=1}^{n_0} \frac{1}{\sqrt{p(\mathbf{k}_j)}} \mathcal{H}(\mathbf{x}, \mathbf{k}_j) \boldsymbol{\zeta}_j \tag{2.35}$$

has the same correlation tensor as $\vec{u}(\mathbf{x})$, provided that $p(\mathbf{k})$ satisfies the condition

$$p(\mathbf{k}) > 0, \quad \text{if} \quad \exists \mathbf{x} \in I\!R^d : \mathcal{H}(\mathbf{x}, \mathbf{k}) \neq 0.$$

In practical calculations, to guarantee an equal presentation of different spectral regions, one uses a stratified randomization technique (e. g., see [142, 191]). Let us describe it briefly. Let $\{\Delta_i\}_{i=1}^N$ be a subdivision of the spectral space $I\!R^{d_1}$: $I\!R^{d_1} = \cup_{i=1}^N \Delta_i$, and $\Delta_i \cap \Delta_j = \emptyset$ if $i \neq j$. This generates the representation of the random field $\vec{u}(\mathbf{x})$ as a sum of independent random fields:

$$\vec{u}(\mathbf{x}) = \sum_{i=1}^N \vec{u}_i(\mathbf{x}), \quad \vec{u}_i(\mathbf{x}) = \int_{\Delta_i} \mathcal{H}(\mathbf{x}, \mathbf{k})\, \mathbf{W}(d\mathbf{k}).$$

Let $p_i : \Delta_i \to [0, \infty)$ $(i = 1, \dots, N)$ be a probability density on Δ_i: $\int_{\Delta_i} p(\mathbf{k})\, d\mathbf{k} = 1$ satisfying the condition $p_i(\mathbf{k}) > 0$, for $\mathbf{k} \in \Delta_i$ if $\exists \mathbf{x} \in I\!R^d$: $\mathcal{H}(\mathbf{x}, \mathbf{k}) \neq 0$. Then using the randomized representation for $\vec{u}_i(\mathbf{x})$ we get a stratified randomization model for $\vec{u}(\mathbf{x})$:

$$\vec{u}_{N,n_0}(\mathbf{x}) = \sum_{i=1}^N \frac{1}{\sqrt{n_0}} \sum_{j=1}^{n_0} \frac{1}{\sqrt{p_i(\mathbf{k}_{ij})}} \mathcal{H}(\mathbf{x}, \mathbf{k}_{ij}) \boldsymbol{\zeta}_{ij}, \tag{2.36}$$

where $\{\mathbf{k}_{ij}\}_{j=1}^{n_0} \subset \Delta_i$, $i = 1, \dots, N$ are mutually independent random points such that for fixed i the random points $\mathbf{k}_{ij}$, $j = 1, \dots$ are all distributed with the same density $p_i(\mathbf{k})$, and $\boldsymbol{\zeta}_{ij}$, $i = 1, \dots, N$; $j = 1, \dots, n_0$ are mutually independent, and independent of $\{\mathbf{k}_{ij}\}_{j=1}^{n_0}$ $i = 1, \dots, N$ family of n-dimensional complex valued standard Gaussian random variables.

By the construction, for any N and n_0, the random field $\vec{u}_{N,n_0}(\mathbf{x})$ has the same correlation tensor as that of $\vec{u}(\mathbf{x})$. As the central limit theorem says, by increasing n_0 (N fixed) the field $\vec{u}_{N,n_0}(\mathbf{x})$ is convergent to a Gaussian random field (e. g., see [18, 106, 107]). So the stratified randomization model $\vec{u}_{N,n_0}(\mathbf{x})$ can be considered as an approximation to $\vec{u}(\mathbf{x})$.

2.4.2 Stratified RSM for homogeneous random fields

Let now $\vec{u}(\mathbf{x})$, $\mathbf{x} \in I\!R^d$ be a homogeneous l-dimensional Gaussian random field with the spectral tensor $F(\mathbf{k})$, $\mathbf{k} \in I\!R^d$, and let $Q(\mathbf{k})$ be an $l \times n$-dimensional matrix satisfying the condition $Q(\mathbf{k})Q^*(\mathbf{k}) = F(\mathbf{k})$, $\mathbf{k} \in I\!R^d$. Then due to the spectral representation (2.6), the real and imaginary parts, $\vec{u}_R(\mathbf{x})$ and $\vec{u}_I(\mathbf{x})$ of the complex-valued field $\mathbf{w}(\mathbf{x})$ are two independent copies of the random field $\vec{u}(\mathbf{x})$. The field $\mathbf{w}(\mathbf{x})$ can be obviously represented (in view of (2.6)), as a stochastic integral (2.34) with the kernel $\mathcal{H}(\mathbf{x}, \mathbf{k}) = e^{i2\pi \mathbf{x}\cdot\mathbf{k}} Q(\mathbf{k})$. Therefore, the relevant stratified randomization model has the form

$$\mathbf{w}_{N,n_0}(\mathbf{x}) = \sum_{i=1}^{N} \frac{1}{\sqrt{n_0}} \sum_{j=1}^{n_0} \frac{e^{i2\pi \mathbf{x}\cdot\mathbf{k}_{ij}}}{\sqrt{p_i(\mathbf{k}_{ij})}} Q(\mathbf{k}_{ij}) \zeta_{ij} \tag{2.37}$$

under conditions on the densities p_i written as $p_i(\mathbf{k}) > 0$ for $\mathbf{k} \in \Delta_i$ if $\max_{mr} |q_{mr}(\mathbf{k})| > 0$ (where $Q = (q_{mr})$). Taking the real part of (2.37) we get the stratified randomization model of the original real-valued random field $\vec{u}(\mathbf{x})$:

$$\begin{aligned} \vec{u}_{N,n_0}(\mathbf{x}) = \sum_{i=1}^{N} \frac{1}{\sqrt{n_0}} \sum_{j=1}^{n_0} \frac{1}{\sqrt{p_i(\mathbf{k}_{ij})}} \Big\{ \Big[Q'(\mathbf{k}_{ij}) \cos\theta_{ij} - Q''(\mathbf{k}_{ij}) \sin\theta_{ij} \Big] \boldsymbol{\xi}_{ij} \\ + \Big[Q''(\mathbf{k}_{ij}) \cos\theta_{ij} + Q'(\mathbf{k}_{ij}) \sin\theta_{ij} \Big] \boldsymbol{\eta}_{ij} \Big\}. \end{aligned} \tag{2.38}$$

Here $\theta_{ij} = 2\pi \mathbf{k}_{ij} \cdot \mathbf{x}$, $\{\mathbf{k}_{ij}\}$, $i = 1, \ldots, N; j = 1, \ldots, n_0$ are the same as in (2.36), and $\boldsymbol{\xi}_{ij}$, $\boldsymbol{\eta}_{ij}$, $i = 1, \ldots, N; j = 1, \ldots, n_0$ is a family of mutually independent and independent of the set $\{\mathbf{k}_{ij}\}$ n-dimensional real-valued standard Gaussian random variables.

2.5 Fourier wavelet models

In the numerical implementation of (2.25) we have to: (i) choose the scaling and wavelet functions, (ii) evaluate the coefficients (2.23), and (iii) find a reasonable choice of the cut-off parameters m_0 and m_1 in the approximations

$$\sum_{j=-\infty}^{\infty} \mathcal{F}_{m_0}^{(\phi)}(2^{m_0}x + j)\, \xi_j \simeq \sum_{j=-b_0+\lfloor 2^{m_0}x \rfloor}^{b_0+\lfloor 2^{m_0}x \rfloor} \mathcal{F}_{m_0}^{(\phi)}(2^{m_0}x + j)\, \xi_j, \tag{2.39}$$

$$\sum_{m=m_0}^{\infty} \sum_{j=-\infty}^{\infty} \mathcal{F}_{m}^{(\psi)}(2^{m}x + j)\, \xi_{mj} \simeq \sum_{m=m_0}^{m_1} \sum_{j=-b_1+\lfloor 2^{m}x \rfloor}^{b_1+\lfloor 2^{m}x \rfloor} \mathcal{F}_{m}^{(\psi)}(2^{m}x + j)\, \xi_{mj}, \tag{2.40}$$

where $\lfloor a \rfloor$ stands for the integer part of a, and $b_0 = b_0(m_0)$, $b_1 = b_1(m)$ are such integers that supports of the functions $\mathcal{F}_{m_0}^{(\phi)}$ and $\mathcal{F}_m^{(\psi)}$ belong essentially to the intervals $[-b_0, b_0]$ and $[-b_1, b_1]$, respectively. We will deal here with a scalar random process $u(x)$, $x \in I\!R$. Extensions to vector random processes is straightforward.

2.5.1 Meyer wavelet functions

The Meyer wavelet functions $\phi(x)$ and $\psi(x)$ are defined by their Fourier transforms (e. g., see [36]):

$$\phi(x) = \int_{-\infty}^{\infty} e^{\mathrm{i}\,2\pi\,kx} \hat{\phi}(k)\,dk, \quad \psi(x) = \int_{-\infty}^{\infty} e^{\mathrm{i}\,2\pi\,kx} \hat{\psi}(k)\,dk, \tag{2.41}$$

where

$$\hat{\phi}(k) = \begin{cases} 1 & |k| \le 1/3, \\ \cos[\frac{\pi}{2}\nu(3|k|-1)], & 1/3 \le |k| \le 2/3 \\ 0, & \text{otherwise,} \end{cases} \tag{2.42}$$

$$\hat{\psi}(k) = \begin{cases} e^{-\mathrm{i}\pi\,k}\sin[\frac{\pi}{2}\nu(3|k|-1)], & 1/3 \le |k| \le 2/3 \\ e^{-\mathrm{i}\pi\,k}\cos[\frac{\pi}{2}\nu(\frac{3}{2}\,|k|-1)], & 2/3 \le |k| \le 4/3 \\ 0, & \text{otherwise.} \end{cases} \tag{2.43}$$

Here $\nu(x)$ is a smooth function satisfying the following conditions: $\nu(x) \equiv 0$ for $x \le 0$, $\nu(x) \equiv 1$ for $x \ge 1$, and $\nu(x) + \nu(1-x) = 1$ for $0 < x < 1$. As an example of such a function, we consider a function $\nu(x) = \nu_p(x)$ depending on a positive parameter p (see [47]):

$$\nu_p(x) = \frac{4^{p-1}}{p}\left\{[x-x_0]_+^p + [x-x_p]_+^p + 2\sum_{j=1}^{p-1}(-1)^j\,[x-x_j]_+^p\right\},$$

where $x_j = (1/2)[\cos(((p-j)/p)\pi) + 1]$, and $[a]_+ = \max(a, 0)$. The function ν_p is $p-1$ times continuously differentiable, therefore, choosing p sufficiently large, we can make the functions $\hat{\phi}$ and $\hat{\psi}$ smooth enough.

2.5.2 Evaluation of the coefficients $\mathcal{F}_m^{(\phi)}$ and $\mathcal{F}_m^{(\psi)}$

Here we give some technical details on the calculation of the functions (2.23) which in our case read

$$f_m(\xi) = \int_{-4/3}^{4/3} e^{-2\pi \mathrm{i}\,k\,\xi}\,g(k)\,dk, \tag{2.44}$$

where $g(k) = 2^{m/2} F^{1/2}(2^m k)\hat{\phi}(k)$ and $g(k) = 2^{m/2} F^{1/2}(2^m k)\hat{\psi}(k)$ for $\mathcal{F}_m^{(\phi)}$ and $\mathcal{F}_m^{(\psi)}$, respectively.

We calculate this function on the grid of points $\xi_j = -\frac{N}{2}\Delta\xi + (j-1)\Delta\xi$, $j = 1, \ldots, N$, where N is an even number and $\Delta\xi \geq 0$ is the grid step. In order to evaluate the truncated sums appearing in the Fourier wavelet representation (2.22), we must choose $N\Delta\xi/2 \geq b$.

We approximate the integral (2.44) by a Riemann sum:

$$f_m(\xi_j) = \int_{-a}^{a} e^{-2\pi \mathrm{i} k \xi} g(k)\, dk \simeq \sum_{l=1}^{N} \Delta k\, e^{-2\pi \mathrm{i} k_l \xi_j} g(k_l), \tag{2.45}$$

where

$$k_l = -a + (l - 1/2)\Delta k, \quad l = 1, \ldots, N; \quad \Delta k = \frac{2a}{N}.$$

We use the same number of points $N = 2^r$ (where r is some positive integer) to discretize the integral as we have done in representing $f_m(\xi)$ in physical space, so that we can apply the discrete fast Fourier transform. We also clearly need the cut-off in the integral in (2.45) to satisfy $a > 4/3$ (with $g(k)$ set to zero whenever evaluated for $|k| > 4/3$). Finally, the use of the fast Fourier transform requires that the steps in physical and wavenumber space be related through $\Delta\xi\Delta k = 1/N$. Indeed, simple transformations then yield

$$\begin{aligned} \xi_j k_l &= \left[-\frac{N}{2}\Delta\xi + (j-1)\Delta\xi\right][-a + (l-1/2)\Delta k] \\ &= \frac{N-1}{4} - \frac{j-1}{2}\left(1 - \frac{1}{N}\right) - \frac{l-1}{2} + \frac{(j-1)(l-1)}{N}, \end{aligned} \tag{2.46}$$

hence

$$f_m(\xi_j) \simeq \exp\left\{\pi \mathrm{i} (j-1)\left(1 - \frac{1}{N}\right)\right\} \sum_{l=1}^{N} G_l \exp\left\{-2\pi \mathrm{i} \frac{(j-1)(l-1)}{N}\right\}, \tag{2.47}$$

where

$$G_l = \Delta k\, g(k_l) \exp\left\{-2\pi \mathrm{i}\left[\frac{N-1}{4} - \frac{l-1}{2}\right]\right\},$$

which is in the form of a discrete Fourier transform.

The constraints imposed on the discretization of the integral (2.45) to obtain an expression amenable to fast Fourier transform imply the following sequence of choosing parameters. First a bandwidth value b is chosen according to the desired accuracy in the Fourier wavelet representation (2.22). Then a spatial resolution $\Delta\xi$ for $f_m(\xi)$ is selected, either according to the grid spacing h on a prespecified set of evaluation points or such that $f_m(\xi)$ can be calculated accurately enough by interpolation from the computed values. (In any event, we must have $\Delta\xi < 3/8$.) Next, a binary power $N = 2^r$ is chosen large enough so that $2b/N \leq \Delta\xi$. Then we set $a = \frac{1}{2\Delta\xi}$, and discretize the integral (2.45) with step size $\Delta k = 2a/N = 1/(N\Delta\xi)$.

2.5.3 Cut-off parameters

In practice, we use the approximation to $u(x)$:

$$u^{(FW)}(x) = \sum_{j=-b_0+\lfloor 2^{m_0}x\rfloor}^{b_0+\lfloor 2^{m_0}x\rfloor} \mathcal{F}_{m_0}^{(\phi)}(2^{m_0}x + j)\,\xi_j + \sum_{m=m_0}^{m_1} \sum_{j=-b_1+\lfloor 2^{m}x\rfloor}^{b_1+\lfloor 2^{m}x\rfloor} \mathcal{F}_{m}^{(\psi)}(2^{m}x + j)\,\xi_{mj}. \tag{2.48}$$

Recall that the parameters $b_0 = b_0(m_0)$ and $b_1 = b_1(m)$ are chosen from the criterion that the supports of the functions $\mathcal{F}_{m_0}^{(\phi)}$ and $\mathcal{F}_{m}^{(\psi)}$ belong essentially to the intervals $[-b_0, b_0]$ and $[-b_1, b_1]$, respectively.

Let us first suggest general arguments to the choice of parameters m_0 and m_1. Theoretically, m_0 can be arbitrarily, and for a fixed m_1, it would be reasonable to take the value of m_0 large enough, since the cost per sample is proportional to $m_1 - m_0 + 1$. However practical calculations show that an important criterion for the choice of m_0 is that the value 2^{m_0} is comparable with k_0, a characteristic wave number scale. This characteristic scale could be defined for instance as a value k_0, for which the integrals $\int_0^{k_0} F(k)\,dk$ and $\int_{k_0}^{\infty} F(k)\,dk$ are compared in the order of magnitude. It turns out that for values of m_0 for which 2^{m_0} is essentially larger than k_0, the dependence of $b_0 = b_0(m_0)$ on m_0 is exponential (see below), hence the cost of simulation is then increasing. When choosing the parameter m_1 one should remember that in the approximation of $u(x)$ by (2.48) the contributions of random harmonics with wave numbers from $k \in I\!R : |k| \geq 2^{m_1}$ are not taken into account. Therefore, to accurately simulate the random field in the interval $(l_{\min}, l_{\max})$, the value of m_1 should be taken so that $2^{m_1} > 1/l_{\min}$.

Thus, suppose these arguments have helped us to choose some values of m_0 and m_1. The next question then which arises is naturally the quantitative criterion of the quality of approximation (2.48). It is clear that for Gaussian processes, the main criterion is to have good accuracy in evaluation of the correlation function. Therefore, the correlation function of $u^{(FW)}$ should well approximate the true correlation function $B(r) = \langle u(x + r)u(x)\rangle$ in an interval $(l_{\min}, l_{\max})$ depending on the problem to be solved. In what follows we assume that this interval is given. Thus with the fixed values of parameters m_0 and m_1, the error of our approximation for the random field (2.48) is defined by

$$\varepsilon(m_0, m_1) = \sup_{l_{\min}\leq r\leq l_{\max}} \left|B(r) - B^{(FW)}(r)\right|, \tag{2.49}$$

where $B^{(FW)}(r) = \langle u^{(FW)}(x + r)u^{(FW)}(x)\rangle$. The random numbers ξ_j and ξ_{mj} in

(2.48) are mutually independent, and hence

$$B^{(FW)}(r) = \sum_{j:|j|\vee|j-\lfloor 2^{m_0}x\rfloor|\le b_0} \mathcal{F}^{(\phi)}_{m_0}(2^{m_0}r + j)\,\mathcal{F}^{(\phi)}_{m_0}(j) + \sum_{m=m_0}^{m_1} \sum_{j:|j|\vee|j-\lfloor 2^{m}x\rfloor|\le b_0} \mathcal{F}^{(\psi)}_{m}(2^{m}r + j)\,\mathcal{F}^{(\psi)}_{m}(j); \quad (2.50)$$

here $a \vee b = \max\{a, b\}$. The cost to calculate the value of the random field $u^{(FW)}$ in one point is proportional to

$$T(m_0, m_1) = b_0(m_0) + \sum_{m=m_0}^{m_1} b_1(m). \quad (2.51)$$

Proposition 2.1. *Assume that a spectral function $F(k)$ satisfies the condition (B.4) for some $s > 0$, and the function Q belongs to the Nikolskiĭ–Besov space $B^{\rho}_{1\infty}(I\!R)$, $\rho > 1/2$. Then the following estimation is true:*

$$\sup_{x\in\boldsymbol{R}} \sup_{r\in\boldsymbol{R}} |B(r) - \langle u^{(FW)}(x+r)u^{(FW)}(x)\rangle| \le \frac{C_s}{4^{m_1 s}} + \frac{C'_{m_0\rho}}{b_0^{2\rho-1}(m_0)} + \sum_{m=m_0}^{m_1} \frac{C''_{m\rho}}{b_1^{2\rho-1}(m)}, \quad (2.52)$$

with the relevant constants C_s, $C'_{m_0\rho}$ and $C''_{m\rho}$ depending on F, $\hat{\phi}$, and $\hat{\psi}$.

The proof is given in Appendix B of this chapter.

This estimation provides a justification for the construction of Fourier wavelet approximations. Indeed, (2.52) shows that the right-hand side of this estimation can be made arbitrarily small by a successive choice of the parameters. For example, we could first take some value of m_0, then choose $b_0(m_0)$ so that the second term in the right-hand side is small enough. Then choose m_1 so that the first term is small, and finally, choose $b_1(m), m = m_0, \ldots, m_1$ so that the last sum in (2.52) is small.

Unfortunately this approach cannot be practically used since it is difficult to calculate the coefficients C_s, $C'_{m_0 l}$ and C''_{ml}. In addition, the estimation (2.52) may be crude, and so the described choice of parameters may be not the best one. Therefore in practice, we recommend using a direct estimation of $\sup_{r\in\boldsymbol{R}} |B(r) - B^{(FW)}(r)|$ starting from (2.50).

2.5.4 Choice of parameters

In this section we discuss some aspects of the practical choice of the parameters of our model. First we study the influence of the parameters m_0 and b_0 on the accuracy of approximation. So let us consider the behavior of $\mathcal{F}^{(\phi)}_m(y)$ under the change of m. By

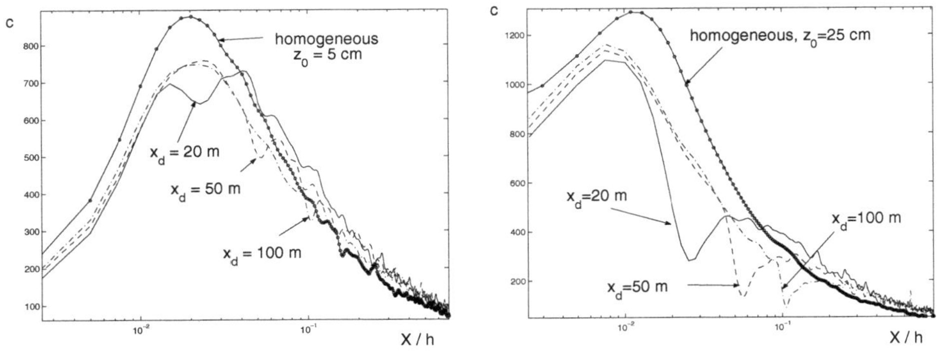

Figure 2.1. $\mathcal{F}_m^{(\phi)}(y)$ as a function of y, for different values of m. Left panel: $F(k) = \frac{2}{1+(2\pi k)^2}$; right panel: $F(k) = e^{-(\pi k)^2}$.

the definitions (2.23) and (2.42) of the functions $\mathcal{F}^{(\phi)}$ and $\hat{\phi}$, it follows that for large negative values of m

$$\begin{aligned}\mathcal{F}_m^{(\phi)}(y) &= \int_{-2/3}^{2/3} e^{-\mathrm{i}\,2\pi\, ky}\, 2^{m/2} Q(2^m k)\hat{\phi}(k)\, dk \\ &\approx \int_{-2/3}^{2/3} e^{-\mathrm{i}\,2\pi\, ky}\, 2^{m/2} Q(0)\hat{\phi}(k)\, dk = 2^{m/2} Q(0)\phi(-y). \qquad (2.53)\end{aligned}$$

provided that $Q(k)$ is continuous at the point zero. It implies that for such values of m, the function $\mathcal{F}^{(\phi)}(y)$ behaves analogously to the scaling function $\phi(y)$ (which decreases as $|y|^{-p}$ as $|y| \to \infty$, where p is a parameter characterizing the smoothness of the function $\hat{\phi}$ such that $\hat{\phi} \in B_{1\infty}^{p}(I\!R)$; see Appendix B, Lemma 2.2.

For simplicity, and for the sake of a clear presentation, let us assume that $F(k) = 0$ if $|k| \geq k_*$, for same $k_* > 0$. Then for positive numbers m such that $2^m/3 \geq k_*$

$$\begin{aligned}\mathcal{F}_m^{(\phi)}(y) &= \int_{-1/3}^{1/3} e^{-\mathrm{i}\,2\pi\, ky}\, 2^{m/2} Q(2^m k)\, dk \\ &\quad + \int_{1/3\leq|k|\leq 2/3} e^{-\mathrm{i}\,2\pi\, ky}\, 2^{m/2} Q(2^m k)\hat{\phi}(k)\, dk \\ &= \int_{-\infty}^{\infty} e^{-\mathrm{i}\,2\pi\, k 2^{-m} y}\, 2^{-m/2} Q(k)\, dk = 2^{-m/2} G(2^{-m} y). \qquad (2.54)\end{aligned}$$

Hence with the increase of m, the effective support of the function $\mathcal{F}_m^{(\phi)}$ is expanded exponentially. Experience tells us that such a broadening may happen in a more general case when the support of the spectral function is not compact. To illutrate this, we show in Figure 2.1 the plots of the functions $\mathcal{F}_m^{(\phi)}(y)$ for different values of m, for $F(k) = 2/(1 + (2\pi k)^2)$ (see the left panel) and $F(k) = e^{-(\pi k)^2}$ (right panel).

So this suggests that the parameter m_0 should be chosen so that at $m = m_0$, the exponential broadening of the support of $\mathcal{F}_m^{(\phi)}$ has not yet happened. This implies the condition $2^{m_0} \leq k_0$, where k_0 is a characteristic wave number scale. For these values of m_0 the width of the support of $\mathcal{F}_{m_0}^{(\phi)}$ is practically independent of m_0 (recall that for $m \to -\infty$ the support width is defined by the function ϕ, in view of (2.53)). As a reasonable approximation we can put $b_0 = 10$.

Let us analyse the behavior of $\mathcal{F}_m^{(\psi)}(y)$ for different values of m. The main difference between this function and the function $\mathcal{F}_m^{(\phi)}$ is that $\mathcal{F}_m^{(\psi)}$ is determined by the values of $Q(k)$ for $2^m/3 \leq |k| \leq 2^{(m+2)}/3$; while $\mathcal{F}_m^{(\phi)}$ depends on the values of $Q(k)$, $|k| \leq 2^{(m+1)}/3$.

Let us see what the influence of this difference is. Take an example of a spectrum $F(k)$, which for $|k| \geq k_0$ has the form

$$F(k) = C_F\, |k|^{-\alpha}, \quad |k| \geq k_0 \tag{2.55}$$

for some $k_0 > 0$ and $\alpha > 1$. Then, for $2^m/3 \geq k_0$, we conclude by the definitions (2.23) and (2.43) of the functions $\mathcal{F}^{(\psi)}$ and $\hat{\psi}$, respectively that for $m \geq \log_2(3k_0)$

$$\begin{aligned} \mathcal{F}_m^{(\psi)}(y) &= \int_{1/3 \leq |k| \leq 4/3} e^{-\mathrm{i}\,2\pi\, ky}\, 2^{m/2} Q(2^m k) \hat{\psi}(k)\, dk \\ &= 2^{-\frac{m}{2}(\alpha-1)} C_F^{1/2} A_\alpha(y), \end{aligned} \tag{2.56}$$

where

$$A_\alpha(y) = \int_{1/3 \leq |k| \leq 4/3} e^{-\mathrm{i}\,2\pi\, ky}\, |k|^{-\frac{\alpha}{2}} \hat{\psi}(k)\, dk.$$

It follows from this equality that the support of $\mathcal{F}_m^{(\psi)}$ does not broaden as m increases. Our experience shows that this property holds for many different spectral functions $F(k)$. We illustrate this in Figure 2.2, where $\mathcal{F}_m^{(\psi)}(y)$ for $F(k) = 2/(1 + (2\pi k)^2)$ (the left panel) and $F(k) = e^{-(\pi k)^2}$ (right panel) are presented. It is seen that for both spectral functions, $\mathcal{F}_m^{(\psi)}(y)$ are practically defined on $-10 \leq y \leq 10$. So we put $b_1 = 10$ for any m.

Thus when the values of b_0 and b_1 are fixed, we can study the dependence of the Fourier wavelet model on m_0 and m_1. As an example, let us consider the case $F(k) = 2/(1 + (2\pi k)^2)$ with the relevant correlation function $B(r) = \exp(-r)$. Since $k_0 \simeq 1$, we can put $m_0 = 0$ (recall that 2^{m_0} should be of the order k_0). In the first two rows of Table 2.1 the dependence of the error (2.49) on m_1 is given, for the fixed value $m_0 = 0$, and $l_{\min}$ and $l_{\max}$ are chosen as $l_{\min} = 0$ and $l_{\max} = 5$.

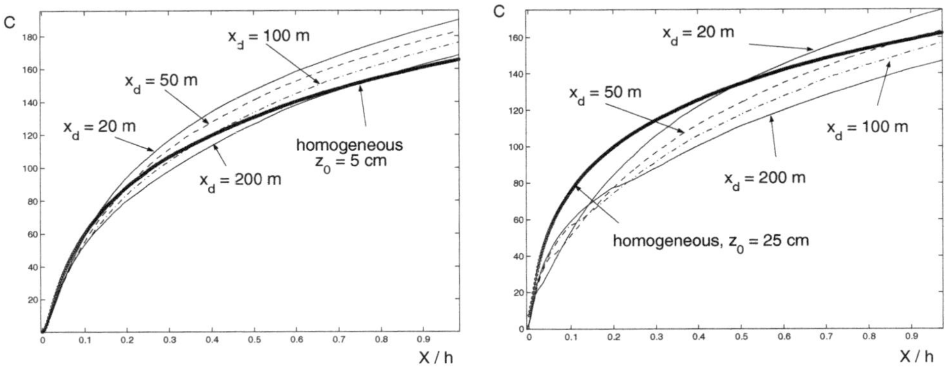

Figure 2.2. $\mathcal{F}_m^{(\psi)}(y)$ as a function of y, for different values of m. Left panel: $F(k) = \frac{2}{1+(2\pi k)^2}$; right panel: $F(k) = e^{-(\pi k)^2}$.

Table 2.1. The errors of the Fourier-wavelet models, for different values of m_0 and m_1. $F(k) = 1/(1 + (2\pi k)^2)$, $l_{\min} = 0$, $l_{\max} = 5$.

m_1	1	2	3	4	5	6	7
$\varepsilon(0, m_1)$	0.0455	0.0233	0.0121	0.0068	0.0041	0.0029	0.0024
m_0	−3	−4	−5	−6	−7	−8	−9
$\tilde{\varepsilon}(m_0, 3)$	0.2714	0.1464	0.0789	0.0476	0.0303	0.0217	0.0124

It can be seen that the error rapidly decreases with m_1, and for $m_1 = 3$ it is about 0.0121.

In this example, we have compared our Fourier wavelet model (2.48) with the Fourier wavelet model [47] which can be obtained by ignoring the first sum in (2.48).

Note that the choice of m_0 in the model [47] is more difficult.

In the last two rows of Table 2.1 we show the error $\tilde{\varepsilon}(m_0, m_1)$ of the model [47], for different values of m_0, for fixed $m_1 = 3$. The cost of calculation of $u^{(FW)}$ is proportional to $m_1 - m_0$ (provided that b_0 and b_1 do not depend on m), so the results of Table 2.1 show that in this example the cost of the method [47] is more than twice that of the model (2.48).

In all the above calculations we have fixed the parameter values as $b_0 = b_1 = 10$. Let us now examine the dependence of the error of the model (2.48) on the parameters b_0 and b_1.

For the case of $F(k) = 1/(1 + (2\pi k)^2)$ we have reached the accuracy $\varepsilon(0, 6) = 0.01$ with $b_0 = b_1 = 5$, and $\varepsilon(0, 6) = 0.0029$ was obtained with $b_0 = b_1 = 6$. This shows that the values $b_0 = b_1 = 10$ we recommended are in this case slightly overestimated.

We have also studied the error $\varepsilon(m_0, 6)$ for the model (2.48) for different values of m_0. We have fixed $b_1 = 10$, and the parameter $b_0 = b_0(m_0)$ was varying depending on m_0 in such a way that $\varepsilon(m_0, 6) \simeq 0.01$.

The computations give $b_0(0) = 5$, $b_0(1) = 7$, and then, $b_0(m_0)$ is exponentially increasing as $b_0(m_0 + 1) = 7 \cdot 2^{m_0}$, $m_0 \geq 0$. Hence, the cost of evaluation of the field (2.48) in one point is proportional to $T(m_0, 6) \simeq 7 \cdot 2^{m_0 - 1} + 10\,(m_1 - m_0 + 1)$ (see (2.51)). If we choose b_0 to ensure $\varepsilon(m_0, 6) \simeq 0.003$, then $b_0(0) = 6$, $b_0(1) = 9$, and $b_0(m_0 + 1) = 9 \cdot 2^{m_0}$, $m_0 \geq 0$. This shows that the cost exponentially increases with m_0.

Note that for vector processes $\vec{u}(x) = (u_1(x), \ldots, u_l(t))^T$, $x \in I\!R$ $(l > 1)$ the FW model is constructed analogously to the scalar case, as we obtained (2.48) from (2.25) by taking a finite number of terms. The following generalization of Proposition 2.1 can be given.

Proposition 2.2. *Assume that a spectral tensor $F(k)$ satisfies the condition*

$$\int_{-\infty}^{\infty} spF(k)(1 + |k|^2)^s\, dk < \infty \quad (spF = F_{jj})$$

for some $s > 0$, and the entries of the tensor Q defined by (2.3) belong to the Nikolskiĭ–Besov space $B^{\rho}_{1\infty}(I\!R)$, $\rho > 1/2$. Then the following estimation is true:

$$\sup_{x \in \boldsymbol{R}} \sup_{r \in \boldsymbol{R}} \max_{i,j=1,\ldots,l} |B_{ij}(r) - \langle u_i^{(FW)}(x + r) u_j^{(FW)}(x)|$$
$$\leq \frac{C_s}{4^{m_1 s}} + \frac{C'_{m_0\rho}}{b_0^{2\rho-1}(m_0)} + \sum_{m=m_0}^{m_1} \frac{C''_{m\rho}}{b_1^{2\rho-1}(m)},$$

with the relevant constants C_s, $C'_{m_0\rho}$ and $C''_{m\rho}$ depending on F, $\hat{\phi}$, and $\hat{\psi}$.

2.6 Fourier wavelet models of homogeneous random fields based on randomization of plane wave decomposition

2.6.1 Plane wave decomposition of homogeneous random fields

Our presentation of Fourier wavelet expansions was given for the 1-dimensional case, i. e., $\vec{u}(x)$, $x \in I\!R$. Direct generalizations for the high-dimensional case are possible on the basis of relevant multidimensional scaling and wavelet functions; see, e. g., [231, 264]. However, with the dimension the complexity of this kind of models increases drastically. Therefore, it is reasonable to try to construct high-dimensional models via 1-dimensional ones.

Here we present the plane waves decomposition method for simulating random fields using samples of relevant random processes. This approach is especially efficient for isotropic random fields; see, e. g., [47, 136].

Denote by S_{d-1} a unit sphere in $\mathbb{R}^d$, and let A_d be its surface area, $d\boldsymbol{\Omega}$ the area element.

Let us introduce a function $B : \mathbb{R} \times S_{d-1} \to \mathbb{R}^{n\times n}$ such that $B(\cdot;\boldsymbol{\Omega})$ is a correlation tensor for almost all $\boldsymbol{\Omega} \in S_{d-1}$. We introduce also a n-dimensional Gaussian random field $\mathbf{Z}(t;A)$, $t \in \mathbb{R}$, and A is a measurable subset of S_{d-1} which is defined by

$$\langle \mathbf{Z}(t;A)\rangle = 0; \quad \mathbf{Z}(t;A_1\cup A_2) = \mathbf{Z}(t;A_1) + \mathbf{Z}(t;A_2) \quad \text{for} \quad A_1\cap A_2 = \emptyset \tag{2.57}$$

$$\langle \mathbf{Z}(t_1;A_1)\mathbf{Z}^T(t_2;A_2)\rangle = \int_{A_1\cap A_2} B(t_1-t_2;\boldsymbol{\Omega})\,d\boldsymbol{\Omega} \tag{2.58}$$

for each $t_1, t_2 \in \mathbb{R}$ and measurable subsets A_1, A_2, A from S_{d-1}.

The block matrix $\mathcal{B}$ with the entries

$$B_{ij} = \int_{A_i\cap A_j} B(t_i-t_j;\boldsymbol{\Omega}), \quad i,j = 1,\dots,N \tag{2.59}$$

is positive definite and symmetric for all positive integers N and values $t_1,\dots t_N$ in $\mathbb{R}$, and measurable subsets $A_1,\dots,A_N$ in S_{d-1}. The proof of this statement will be presented in Appendix A of this chapter.

Thus the Gaussian random field $\mathbf{Z}(t;A)$ is well defined. Note that $\mathbf{Z}(t;\cdot)$ is a Gaussian white noise measure for fixed t.

Let $H_i(\boldsymbol{\Omega})$, $i = 1,2$ be complex-valued $l\times n$-matrices, depending on $\boldsymbol{\Omega} \in S_{d-1}$ satisfying the condition $\int \|H_i\|^2\,d\boldsymbol{\Omega} < \infty$. We will use the following property:

$$\Big\langle \int_{S_{d-1}} H_1(\boldsymbol{\Omega})\,\mathbf{Z}(t_1;d\Omega)\Big(\int_{S_{d-1}} H_2(\boldsymbol{\Omega})\,\mathbf{Z}(t_2;d\Omega)\Big)^*\Big\rangle = \int_{S_{d-1}} H_1(\boldsymbol{\Omega})B(t_1-t_2;\boldsymbol{\Omega})H_2^*(\boldsymbol{\Omega})\,d\boldsymbol{\Omega}. \tag{2.60}$$

Given the correlation tensor in τ, $B(\tau;\boldsymbol{\Omega})$, the spectral tensor in κ, $F_v(\kappa;\boldsymbol{\Omega})$ is defined by

$$F_v(\kappa;\boldsymbol{\Omega}) = \int_{-\infty}^{\infty} e^{-\mathrm{i}2\pi\kappa\tau} B(\tau;\boldsymbol{\Omega})\,d\tau. \tag{2.61}$$

Let $H(\boldsymbol{\Omega})$ be a $l\times n$ matrix defined on S_{d-1}. We define $\tilde{\vec{u}}(\mathbf{x})$ by

$$\tilde{\vec{u}}(\mathbf{x}) = \int_{S_{d-1}} H(\boldsymbol{\Omega})\mathbf{Z}(\mathbf{x}\cdot\boldsymbol{\Omega};d\boldsymbol{\Omega}). \tag{2.62}$$

Here the stochastic integral is taken over the white noise measure $\mathbf{Z}(t;A)$ (e.g., see [70]). We show now that

$$\begin{aligned}&\langle\tilde{\vec{u}}(\mathbf{x}+\vec{r})\tilde{\vec{u}}^T(\mathbf{x})\rangle\\&=\int_{\mathbb{R}^d} e^{\mathrm{i}\,2\pi\,\vec{r}\mathbf{k}}\frac{1}{k^{d-1}}\{H(\hat{\mathbf{k}})F_v(k,\hat{\mathbf{k}})H^*(\hat{\mathbf{k}})+H(-\hat{\mathbf{k}})F_v(-k,-\hat{\mathbf{k}})H^*(-\hat{\mathbf{k}})\}\,d\mathbf{k},\end{aligned}\tag{2.63}$$

where $k=|\mathbf{k}|$, $\hat{\mathbf{k}}=\mathbf{k}/k$. This implies that the spectral tensor $\tilde{F}(\mathbf{k})$ of the random field $\tilde{\vec{u}}(\mathbf{x})$ is given by

$$\tilde{F}(\mathbf{k})=\frac{1}{k^{d-1}}\{H(\hat{\mathbf{k}})F_v(k,\hat{\mathbf{k}})H^*(\hat{\mathbf{k}})+H(-\hat{\mathbf{k}})F_v(-k,-\hat{\mathbf{k}})H^*(-\hat{\mathbf{k}})\}.\tag{2.64}$$

By the definition, we get from (2.60)

$$\begin{aligned}&\left\langle\tilde{\vec{u}}(\mathbf{x}+\vec{r})\tilde{\vec{u}}^T(\mathbf{x})\right\rangle\\&=\left\langle\int_{S_{d-1}}H(\mathbf{\Omega})\,\mathbf{Z}((\mathbf{x}+\vec{r})\cdot\mathbf{\Omega};d\Omega)\Big(\int_{S_{d-1}}H(\mathbf{\Omega})\,\mathbf{Z}(\mathbf{x}\cdot\mathbf{\Omega};d\Omega)\Big)^*\right\rangle\\&=\int_{S_{d-1}}H(\mathbf{\Omega})B(\vec{r}\cdot\mathbf{\Omega};\mathbf{\Omega})H^*(\mathbf{\Omega})\,d\mathbf{\Omega}\\&=\int_{S_{d-1}}d\mathbf{\Omega}\int_{-\infty}^{\infty}e^{\mathrm{i}\,2\pi\,\kappa\vec{r}\cdot\mathbf{\Omega}}H(\mathbf{\Omega})F_v(\kappa;\mathbf{\Omega})\,H^*(\mathbf{\Omega})\,d\kappa\\&=\int_{S_{d-1}}d\mathbf{\Omega}\int_{0}^{\infty}e^{\mathrm{i}\,2\pi\,\kappa\vec{r}\cdot\mathbf{\Omega}}H(\mathbf{\Omega})F_v(\kappa;\mathbf{\Omega})\,H^*(\mathbf{\Omega})\,d\kappa\\&\quad+\int_{S_{d-1}}d\mathbf{\Omega}\int_{0}^{\infty}e^{-\mathrm{i}\,2\pi\,\kappa\vec{r}\cdot\mathbf{\Omega}}H(\mathbf{\Omega})F_v(-\kappa;\mathbf{\Omega})\,H^*(\mathbf{\Omega})\,d\kappa\\&=\int_{S_{d-1}}d\mathbf{\Omega}\int_{0}^{\infty}e^{\mathrm{i}\,2\pi\,\kappa\vec{r}\cdot\mathbf{\Omega}}\\&\quad\times\left\{H(\mathbf{\Omega})F_v(\kappa;\mathbf{\Omega})\,H^*(\mathbf{\Omega})+H(-\mathbf{\Omega})F_v(-\kappa;-\mathbf{\Omega})\,H^*(-\mathbf{\Omega})\right\}d\kappa\\&=\int_{\mathbb{R}^d}e^{\mathrm{i}\,2\pi\,\vec{r}\mathbf{k}}\frac{1}{k^{d-1}}\{H(\hat{\mathbf{k}})F_v(k,\hat{\mathbf{k}})H^*(\hat{\mathbf{k}})+H(-\hat{\mathbf{k}})F_v(-k,-\hat{\mathbf{k}})H^*(-\hat{\mathbf{k}})\}\,d\mathbf{k}.\end{aligned}\tag{2.65}$$

Let $\vec{u}(\mathbf{x})$, $\mathbf{x}\in\mathbb{R}^d$ be a zero mean vector homogeneous Gaussian random field of dimension l defined by its spectral tensor $F(\mathbf{k})$. We are in position now to construct the random field $\vec{u}(\mathbf{x})$ as a superposition of the plane waves. Indeed, let

$$F_v(\kappa;\mathbf{\Omega})=\frac{|\kappa|^{d-1}}{2}\,F(\kappa\,\mathbf{\Omega}),\quad H(\mathbf{\Omega})=\mathrm{I},\tag{2.66}$$

hence it follows from (2.63) that the random field

$$\vec{u}(\mathbf{x}) = \int_{S_{d-1}} \mathbf{Z}(\mathbf{x}\cdot\mathbf{\Omega};\, d\mathbf{\Omega}) \tag{2.67}$$

has the desired spectral tensor $F(\mathbf{k})$.

Thus, we give an integral representation of homogeneous Gaussian random fields through a superposition of plane waves $\mathbf{Z}(\mathbf{x}\mathbf{\Omega}; d\mathbf{\Omega})$ (see (2.62) and (2.67)). A numerical model can be constructed through an approximation of the integral over an unit sphere by a finite sum. Here one can use both deterministic and stochastic approaches.

2.6.2 Decomposition with fixed nodes

Let $\Delta\mathbf{\Omega}_i$, $i = 1, \dots, N_s$ be a finite partition of the unit sphere S_{d-1}: $S_{d-1} = \cup_{i=1}^{N_s} \Delta\mathbf{\Omega}_i$, and $\Delta\mathbf{\Omega}_i \cap \Delta\mathbf{\Omega}_j = \emptyset$ when $i \neq j$. Choosing nodes $\mathbf{\Omega}_i \in \Delta\mathbf{\Omega}_i$, $i = 1, \dots, N_s$, the following deterministic approximation to (2.62) can be constructed:

$$\tilde{\vec{u}}(\mathbf{x}) = \int_{S_{d-1}} H(\mathbf{\Omega})\mathbf{Z}(\mathbf{x}\cdot\mathbf{\Omega}; d\mathbf{\Omega}) \simeq \sum_{i=1}^{N_s} H(\mathbf{\Omega}_i)\mathbf{Z}(\mathbf{x}\cdot\mathbf{\Omega}_i; \Delta\mathbf{\Omega}_i), \tag{2.68}$$

where by definition (see (2.57) and (2.58)) $\mathbf{Z}(t; \Delta\mathbf{\Omega}_i)$, $t \in I\!R$, $i = 1, \dots, N_s$ are mutually independent zero mean stationary Gaussian processes with the following correlation structure:

$$\begin{aligned}\langle \mathbf{Z}(t+\tau; \Delta\mathbf{\Omega}_i)(\mathbf{Z}(t; \Delta\mathbf{\Omega}_i))^T \rangle &= \int_{\Delta\mathbf{\Omega}_i} B(\tau; \mathbf{\Omega})\, d\mathbf{\Omega} \\ &= \int_{\Delta\mathbf{\Omega}_i} \int_{-\infty}^{\infty} e^{\mathrm{i}2\pi\kappa\tau} F_v(\kappa; \mathbf{\Omega}) d\kappa d\mathbf{\Omega}.\end{aligned}$$

From this it follows that the process $\mathbf{Z}(t; \Delta\mathbf{\Omega}_i)$, $t \in I\!R$ has a spectral tensor given by $\int_{\Delta\mathbf{\Omega}_i} F_v(\kappa; \mathbf{\Omega}) d\mathbf{\Omega} \simeq |\Delta\mathbf{\Omega}_i| F_v(\kappa; \mathbf{\Omega}_i)$, where $|\Delta\mathbf{\Omega}_i| = \int_{\Delta\mathbf{\Omega}_i} d\mathbf{\Omega}$. Hence the process $\mathbf{Z}(t; \Delta\mathbf{\Omega}_i)$ can be approximated by $|\Delta\mathbf{\Omega}_i|^{1/2} \vec{v}^{(i)}(t)$ where $\vec{v}^{(i)}(t) = (v_1^{(i)}(t), \dots, v_n^{(i)}(t))^T$ is a random process with the spectral tensor $F_v(\kappa; \mathbf{\Omega}_i)$. From this we get by (2.68) the following approximation to the random field $\tilde{\vec{u}}(\mathbf{x})$, $\mathbf{x} \in I\!R^d$:

$$\tilde{\vec{u}}_{N_s}^{det}(\mathbf{x}) = \sum_{i=1}^{N_s} |\Delta\mathbf{\Omega}_i|^{1/2}\, H(\mathbf{\Omega}_i)\vec{v}^{(i)}(\mathbf{x}\cdot\mathbf{\Omega}_i). \tag{2.69}$$

By the construction it is clear that the smaller the subregions $\Delta\mathbf{\Omega}_i$, $i = 1, \dots, N_s$ the better $\tilde{\vec{u}}_{N_s}^{det}(\mathbf{x})$ approximates the random field with the spectral tensor (2.64). In particular the model

$$\vec{u}_{N_s}^{det}(\mathbf{x}) = \sum_{i=1}^{N_s} |\Delta\mathbf{\Omega}_i|^{1/2}\, \vec{v}^{(i)}(\mathbf{x}\cdot\mathbf{\Omega}_i), \tag{2.70}$$

where $\vec{v}^{(i)}(t) = (v_1^{(i)}(t), \dots, v_l^{(i)}(t))^T, i = 1, \dots, N_s$ are mutually independent zero mean Gaussian processes with spectral tensor $F^{(i)}(\kappa) = \frac{|\kappa|^{d-1}}{2} F(\kappa \boldsymbol{\Omega}_i)$, is an approximation of the random field $\vec{u}(\mathbf{x}) = (u_1(\mathbf{x}), \dots, u_l(\mathbf{x}))^T$, $\mathbf{x} \in I\!R^d$ having the spectral tensor $F(\mathbf{k})$.

Let us consider the case when the spectral tensor $F(\mathbf{k})$ has the form

$$F(\mathbf{k}) = H(\hat{\mathbf{k}}) H^*(\hat{\mathbf{k}}) \frac{2E(k)}{A_d k^{d-1}}, \tag{2.71}$$

where $H(\boldsymbol{\Omega})$, $\boldsymbol{\Omega} \in S_{d-1}$ is an $l \times n$-matrix satisfying the condition

$$H(-\boldsymbol{\Omega}) H^*(-\boldsymbol{\Omega}) = H(\boldsymbol{\Omega}) H^*(\boldsymbol{\Omega}), \tag{2.72}$$

and $E(k)$ is a scalar nonnegative even function, A_d is the area of the unit sphere in $I\!R^d$.

Thus the approximation to the random field $\vec{u}(\mathbf{x}) = (u_1(\mathbf{x}), \dots, u_l(\mathbf{x}))^T$, $\mathbf{x} \in I\!R^d$ with the spectrum (2.71) can be constructed by the formula

$$\vec{u}_{N_s}^{det}(\mathbf{x}) = \sum_{i=1}^{N_s} \left(\frac{|\Delta \boldsymbol{\Omega}_i|}{A_d} \right)^{1/2} H(\boldsymbol{\Omega}_i) \vec{v}^{(i)}(\mathbf{x} \cdot \boldsymbol{\Omega}_i) \tag{2.73}$$

where $\vec{v}^{(i)}(t) = (v_1^{(i)}(t), \dots, v_n^{(i)}(t))^T, i = 1, \dots, N_s$ are mutually independent zero mean stationary Gaussian random processes with independent components each having the same spectral function $F_v(\kappa) = E(\kappa)$:

$$\langle v_{j_1}^{(i_1)}(t + \tau) v_{j_2}^{(i_2)}(t) \rangle = \delta_{i_1 i_2} \delta_{j_1 j_2} \int_{-\infty}^{\infty} e^{\mathrm{i} 2\pi \kappa \tau} E(\kappa) \, d\kappa.$$

Let us give an example of subdivision on a 3D sphere, $S_2 = \cup_{i=1}^{N_s} \Delta \boldsymbol{\Omega}_i$. So, to choose a subdivision of the unite sphere in $I\!R^3$, we work in the spherical coordinates (θ, ϕ), $0 \le \theta \le \pi$, $0 \le \phi \le 2\pi$. First we fix an integer parameter $n_\theta \ge 1$ which defines the step $\Delta\theta = \pi / n_\theta$ and the relevant altitude mesh $\theta_j = (j - 1/2)\Delta\theta$, $j = 1, \dots, n_\theta$. For each fixed j, $(1 \le j \le n_\theta)$, we construct the latitude mesh $\phi_{jr} = (r - 1/2)\Delta\phi_j$, $r = 1, \dots, n_\phi^{(j)}$, where $n_\phi^{(j)} = \lfloor 2\pi \sin(\theta_j)/\Delta\theta \rfloor$, $\Delta\phi_j = 2\pi / n_\phi^{(j)}$. Finally, a 2-index subdivision $\{\{\Delta\boldsymbol{\Omega}_{j,r}\}_{r=1}^{n_\phi^{(j)}}, \ j = 1, \dots, n_\theta\}$ of the unit sphere S_2,

$$\Delta\boldsymbol{\Omega}_{j,r} = \left\{ (\theta, \phi) : \theta_j - \frac{\Delta\theta}{2} \le \theta < \theta_j + \frac{\Delta\theta}{2}; \quad \phi_{jr} - \frac{\Delta\phi_j}{2} \le \phi < \phi_{jr} + \frac{\Delta\phi_j}{2} \right\}, \tag{2.74}$$

is constructed, which consists of $N_s = N_s(n_\theta) = \sum_{j=1}^{n_\theta} n_\phi^{(j)}$ surface elements. In our calculations, we used the values of n_θ and N_s as given in Table 2.2. We choose

Table 2.2. Dependence of N_s on n_θ.

n_θ	4	6	8	10	16	30
N_s	20	44	78	124	320	1132

the nodes as follows:

$$\mathbf{\Omega}_{j,r} = (\sin(\theta_j)\cos(\phi_{jr}),\ \sin(\theta_j)\sin(\phi_{jr}),\ \cos(\theta_j)),$$
$$j = 1,\dots,n_\theta,\ r = 1,\dots,n_\phi^{(j)}. \quad (2.75)$$

Thus the 3D FWM based on the plane wave decomposition with deterministic nodes is

$$\vec{u}^{\,det}_{n_\theta}(\mathrm{x}) = \sum_{j=1}^{n_\theta}\sum_{r=1}^{n_\phi^{(j)}} \left(\frac{|\Delta\mathbf{\Omega}_{j,r}|}{4\pi}\right)^{1/2} H(\mathbf{\Omega}_{j,r})\,\vec{v}^{(jr)}(\mathrm{x}\cdot\mathbf{\Omega}_{j,r}), \quad (2.76)$$

where

$$|\Delta\mathbf{\Omega}_{j,r}| = \int_{\Delta\mathbf{\Omega}_{j,r}} d\mathbf{\Omega} = 4\pi\Delta\phi_j\,[\cos(\theta_j - \Delta\theta/2) - \cos(\theta_j + \Delta\theta/2)].$$

2.6.3 Decomposition with randomly distributed nodes

A randomized model of the field $\tilde{\vec{u}}(\mathrm{x})$, $\mathrm{x} \in I\!R^d$ defined by the stochastic integral (2.62) with the spectral tensor (2.64) can be represented in the form

$$\tilde{\vec{u}}^{\,rnd}_{N_s}(\mathrm{x}) = \frac{A_d^{1/2}}{N_s^{1/2}} \sum_{i=1}^{N_s} H(\boldsymbol{\omega}_i)\vec{v}^{(i)}(\mathrm{x}\cdot\boldsymbol{\omega}_i), \quad (2.77)$$

where $\boldsymbol{\omega}_i$, $i = 1,\dots,N_s$ is a family of independent unit isotropic vectors in $I\!R^d$, $\vec{v}^{(i)}(t) = (v_1^{(i)}(t),\dots,v_n^{(i)}(t))^T$, $i = 1,\dots,N_s$ is a set of mutually independent and stochastically independent of $\boldsymbol{\omega}_i$, $i = 1,\dots,N_s$, zero mean, stationary Gaussian random processes with the spectral tensor $F_v(\kappa,\boldsymbol{\omega}_i)$.

In the general case of anisotropic random fields, the following generalization of the model (2.77) can be given:

$$\tilde{\vec{u}}^{\,rnd}_{N_s}(\mathrm{x}) = \frac{1}{N_s^{1/2}} \sum_{i=1}^{N_s} \frac{1}{p^{1/2}(\boldsymbol{\omega}_i)} H(\boldsymbol{\omega}_i)\vec{v}^{(i)}(\mathrm{x}\cdot\boldsymbol{\omega}_i), \quad (2.78)$$

where $p(\mathbf{\Omega})$, $\mathbf{\Omega} \in S_{d-1}$ is a probability density function defined on a unit sphere S_{d-1}, and the random points $\boldsymbol{\omega}_i$, $i = 1,\dots,N_s$ are independently sampled on S_{d-1} according to the probability density $p(\mathbf{\Omega})$; the family $\vec{v}^{(i)}(t)$ is constructed the same as in (2.77).

Proposition 2.3. *Suppose that the density $p(\mathbf{\Omega})$ satisfies the condition*

$$p(\mathbf{\Omega}) > 0 \quad if \quad \int_{-\infty}^{\infty} F_v(\kappa; \mathbf{\Omega})\, d\kappa > 0. \tag{2.79}$$

Then for any N_s the function $\tilde{\vec{u}}_{N_s}^{rnd}(\mathrm{x})$ defined by (2.78) is a zero mean homogeneous random field (generally non-Gaussian) with the spectral tensor (2.64):

$$\langle \tilde{\vec{u}}_{N_s}^{rnd}(\mathrm{x} + \vec{r})(\tilde{\vec{u}}_{N_s}^{rnd}(\mathrm{x}))^T \rangle = \int_{\boldsymbol{R}^d} e^{\mathrm{i}2\pi\vec{r}\cdot\boldsymbol{k}} \tilde{F}(\boldsymbol{k})\, d\boldsymbol{k}. \tag{2.80}$$

Proof. The terms in the sum of (2.78) are independent; also, the process $\vec{v}^{(i)}(t)$ is independent of $\boldsymbol{\omega}_i$, and hence

$$\begin{aligned}\langle \tilde{\vec{u}}_{N_s}^{rnd}(\mathrm{x} + \vec{r})(\tilde{\vec{u}}_{N_s}^{rnd}(\mathrm{x}))^T \rangle &= \langle \tilde{\vec{u}}_1^{rnd}(\mathrm{x} + \vec{r})(\tilde{\vec{u}}_1^{rnd}(\mathrm{x}))^T \rangle \\ &= \langle H(\boldsymbol{\omega})\langle \vec{v}_{\boldsymbol{\omega}}((\mathrm{x} + \vec{r})\boldsymbol{\omega})\vec{v}_{\boldsymbol{\omega}}^T(\mathrm{x}\boldsymbol{\omega})|\boldsymbol{\omega}\rangle H^*(\boldsymbol{\omega})\rangle,\end{aligned}$$

where $\boldsymbol{\omega}$ is a random point on S_{d-1} having the density $p(\mathbf{\Omega})$. The random process $\vec{v}_{\boldsymbol{\omega}}(t)$ is stochastically independent of $\boldsymbol{\omega}$, having a spectral tensor $F_v(\kappa; \boldsymbol{\omega})$. Here $\langle \cdot|\boldsymbol{\omega}\rangle$ means a conditional averaging under a fixed $\boldsymbol{\omega}$. By this construction,

$$\langle \vec{v}_{\boldsymbol{\omega}}((\mathrm{x} + \vec{r})\boldsymbol{\omega})\vec{v}_{\boldsymbol{\omega}}^T(\mathrm{x}\boldsymbol{\omega})|\boldsymbol{\omega}\rangle = \int_{-\infty}^{\infty} e^{\mathrm{i}2\pi\kappa\vec{r}\cdot\boldsymbol{\omega}} F_v(\kappa; \boldsymbol{\omega})\, d\kappa.$$

Therefore,

$$\begin{aligned}&\langle H(\boldsymbol{\omega})\langle \vec{v}_{\boldsymbol{\omega}}((\mathrm{x} + \vec{r})\boldsymbol{\omega})\vec{v}_{\boldsymbol{\omega}}^T(\mathrm{x}\boldsymbol{\omega})|\boldsymbol{\omega}\rangle H^*(\boldsymbol{\omega})\rangle \\ &\qquad = \int_{S_{d-1}} d\mathbf{\Omega} \int_{-\infty}^{\infty} e^{\mathrm{i}\,2\pi\,\kappa\vec{r}\cdot\mathbf{\Omega}} H(\mathbf{\Omega}) F_v(\kappa; \mathbf{\Omega})\, H^*(\mathbf{\Omega})\, d\kappa.\end{aligned}$$

Then we can proceed as we have done after the second row in (2.65); this leads us to the desired relation (2.80). □

Assume that the spectral tensor of the random field to be simulated has a representation (2.71) where the tensor H satisfies (2.72). Then, it is reasonable to choose $p(\mathbf{\Omega}) \equiv 1/A_d$, which means the points on $\boldsymbol{\omega}_i$ are distributed uniformly; hence the simulation formula can be written as follows:

$$\vec{u}_{N_s}^{rnd}(\mathrm{x}) = \frac{1}{N_s^{1/2}} \sum_{i=1}^{N_s} H(\boldsymbol{\omega}_i)\vec{v}^{(i)}(\mathrm{x} \cdot \boldsymbol{\omega}_i), \tag{2.81}$$

where $\vec{v}^{(i)}(t) = (v_1^{(i)}(t), \ldots, v_m^{(i)}(t))^T, i = 1, \ldots, N_s$ are mutually independent zero mean stationary Gaussian random processes with independent components having the same spectral function $E(\kappa)$.

Decomposition with stratified randomly distributed nodes

It is possible to construct a stratified randomization model by choosing a subdivision $\{\Delta\boldsymbol{\Omega}_i\}_{i=1}^{N_s}$. Then in each $\Delta\boldsymbol{\Omega}_i$ a random point $\boldsymbol{\omega}_i \in \Delta\boldsymbol{\Omega}_i$ is sampled. For simplicity, we may assume that it is uniformly distributed in $\Delta\boldsymbol{\Omega}_i$. So under the condition (2.71) we can use, along with the models (2.73) and (2.81) the following model with stratified randomly distributed nodes:

$$\vec{u}_{N_s}^{srm}(\mathbf{x}) = \sum_{i=1}^{N_s} \left(\frac{|\Delta\boldsymbol{\Omega}_i|}{A_d}\right)^{1/2} H(\boldsymbol{\omega}_i)\vec{v}^{(i)}(\mathbf{x}\cdot\boldsymbol{\omega}_i). \tag{2.82}$$

A hybrid stratified randomized FWM based on the plane wave decomposition (2.82) and subdivision sets (2.74) in 3D case has the form

$$\vec{u}_{n_\theta}^{srm}(\mathbf{x}) = \sum_{j=1}^{n_\theta}\sum_{r=1}^{n_\phi^{(j)}} \left(\frac{|\Delta\boldsymbol{\Omega}_{j,r}|}{4\pi}\right)^{1/2} H(\boldsymbol{\omega}_{j,r})\,\vec{v}^{(jr)}(\mathbf{x}\cdot\boldsymbol{\omega}_{jr}), \tag{2.83}$$

where

$$\begin{aligned}
\boldsymbol{\omega}_{jr} &= (\sin(\tilde{\theta}_j)\cos(\tilde{\phi}_{jr}),\ \sin(\tilde{\theta}_j)\sin(\tilde{\phi}_{jr}),\ \cos(\tilde{\theta}_i j)),\\
\tilde{\theta}_j &= \arccos\big((1-\gamma_j)\cos(\theta_j - \Delta\theta/2) + \gamma_j\cos(\theta_j + \Delta\theta/2)\big),\\
\tilde{\phi}_{jr} &= \phi_{jr} + (\gamma_{jr} - 1/2)\Delta\phi_j,
\end{aligned}$$

and γ_j, γ_{jr}, ($j = 1,\dots,n_\theta$; $r = 1,\dots,n_\phi^{(j)}$) are mutually independent random numbers uniformly distributed on $[0,1]$.

2.6.4 Some examples

Isotropic vector fields

A homogeneous d-dimensional vector random field $\vec{u}(x)$, $\mathbf{x} \in I\!R^d$ is called isotropic if the random field $\vec{v}(\mathbf{x}) = U^T\,\vec{u}(U\mathbf{x})$ has the same finite-dimensional distributions as those of the random field $\vec{u}(\mathbf{x})$ for any rotation U with transpose U^T (e. g., see [146]). It is known that the general structure of the spectral tensor of an isotropic random field is the following [146]:

$$F_{ij}(\mathbf{k}) = \frac{2}{A_d\,k^{d-1}}\left\{E_1(k)\Big(\delta_{ij} - \frac{k_i\,k_j}{k^2}\Big) + E_2(k)\,\frac{k_i\,k_j}{k^2}\right\}, \tag{2.84}$$

where $k = |\mathbf{k}|$, A_d is the area of the unit sphere in $I\!R^d$, E_1 and E_2 are nonnegative scalar even functions.

The tensor (2.84) can be represented in the form

$$F(\mathbf{k}) = \frac{2E_1(k)}{A_d k^{d-1}} H_1(\hat{\mathbf{k}}) H_1^*(\hat{\mathbf{k}}) + \frac{2E_2(k)}{A_d k^{d-1}} H_2(\hat{\mathbf{k}}) H_2^*(\hat{\mathbf{k}}), \tag{2.85}$$

where $H_1 = (h_{ij}^{(1)})_{i,j=1}^d$ and $H_2 = (h_1^{(2)}, \dots, h_d^{(2)})$ are $d \times d$ and $1 \times d$ matrix functions defined on the unite sphere S_{d-1} by

$$h_{ij}^{(1)}(\mathbf{\Omega}) = \delta_{ij} - \Omega_i \Omega_j; \quad h_j^{(2)}(\mathbf{\Omega}) = \Omega_j,$$
$$i, j = 1, \dots, d, \quad \mathbf{\Omega} = (\Omega_1, \dots, \Omega_d) \in S_{d-1}.$$

The tensors $H_i, i = 1, 2$ obviously satisfy (2.72).

Thus using the decomposition with fixed nodes (2.73), the isotropic random field $\vec{u}(\mathbf{x})$ can be approximated by

$$\vec{u}_{N_s}^{det}(\mathbf{x}) = \sum_{i=1}^{N_s} \left(\frac{|\Delta \mathbf{\Omega}_i|}{A_d} \right)^{1/2} \left(H_1(\mathbf{\Omega}_i) \vec{v}_1^{(i)}(\mathbf{x} \cdot \mathbf{\Omega}_i) + H_2(\mathbf{\Omega}_i) \vec{v}_2^{(i)}(\mathbf{x} \cdot \mathbf{\Omega}_i) \right), \tag{2.86}$$

where $\vec{v}_1^{(i)}$, and $\vec{v}_2^{(i)}$ $i = 1, \dots, N_s$ are mutually independent zero mean, stationary d-dimensional, Gaussian random processes with independent components. Each component of $\vec{v}_1^{(i)}(t)$ has the spectral function $E_1(\kappa)$, and each component of $\vec{v}_2^{(i)}(t)$ has the spectral function $E_2(\kappa)$. Let $v^{(i)}(t) = H_2(\mathbf{\Omega}_i)\vec{v}_2^{(i)}(t) = \mathbf{\Omega}_i \cdot \vec{v}_2^{(i)}(t)$. It is not difficult to check that the spectral function of the scalar process $v^{(i)}(t)$, $t \in I\!R$ is $E_2(\kappa)$. Thus we can present the formula (2.86) in a form convenient for simulation:

$$\vec{u}_{N_s}^{det}(\mathbf{x}) = \sum_{i=1}^{N_s} \left(\frac{|\Delta \mathbf{\Omega}_i|}{A_d} \right)^{1/2} \left(H_1(\mathbf{\Omega}_i) \vec{v}_1^{(i)}(\mathbf{x} \cdot \mathbf{\Omega}_i) + v^{(i)}(\mathbf{x} \cdot \mathbf{\Omega}_i) \right), \tag{2.87}$$

where $v^{(i)}(t)$, $i = 1, \dots, N_s$ is a set of scalar random processes which are independent of each other and of $\vec{v}_1^{(i)}(t)$, $i = 1, \dots, N_s$ with the spectral function $E_2(\kappa)$. Analogously, using the decomposition with randomly distributed nodes (2.77), a d-dimensional isotropic random field with a spectral density (2.84)) can be simulated by

$$\vec{u}_{N_s}^{rnd}(\mathbf{x}) = \frac{1}{N_s^{1/2}} \sum_{i=1}^{N_s} \left(H_1(\boldsymbol{\omega}_i) \vec{v}_1^{(i)}(\mathbf{x} \cdot \boldsymbol{\omega}_i) + v^{(i)}(\mathbf{x} \cdot \boldsymbol{\omega}_i) \right), \tag{2.88}$$

where $\boldsymbol{\omega}_i$, $i = 1, \dots, N_s$ is a family of independent unit isotropic vectors in $I\!R^d$; $\vec{v}_1^{(i)}$ and $v^{(i)}$ $i = 1, \dots, N_s$ are mutually independent and independent of $\boldsymbol{\omega}_i$, $i = 1, \dots, N_s$, zero mean, stationary d-dimensional scalar Gaussian random processes with independent components. Components of $\vec{v}_1^{(i)}(t)$ are independent of each other and have the same spectral function $E_1(\kappa)$ while the random process $v^{(i)}(t)$ has the spectral function $E_2(\kappa)$. Note that the representation (2.88) is given in [136].

Remark 2.1. Note that the simulation formulas (2.87) and (2.88) can be rewritten in the 3D case in different forms which involve vector products, as is done in [102, 191]. Indeed, the tensor $\Delta(\boldsymbol{\omega})$ with entries $\Delta_{ij}(\boldsymbol{\omega}) = \delta_{ij} - \omega_i \omega_j$, $\boldsymbol{\omega} \in S_2$ can be represented as $\Delta(\boldsymbol{\omega}) = \mathcal{H}(\boldsymbol{\omega})\mathcal{H}^T(\boldsymbol{\omega})$ where $\mathcal{H}(\boldsymbol{\omega})$ is 3×3 tensor with entries $h_{ii}(\boldsymbol{\omega}) = 0$, $i = 1, 2, 3$, and

$$h_{12}(\boldsymbol{\omega}) = -h_{21}(\boldsymbol{\omega}) = -\omega_3,\ h_{13}(\boldsymbol{\omega}) = -h_{31}(\boldsymbol{\omega}) \\ = \omega_2,\ h_{23}(\boldsymbol{\omega}) = -h_{32}(\boldsymbol{\omega}) = -\omega_1.$$

Taking into account $\mathcal{H}(\boldsymbol{\omega})\,\vec{v} = \boldsymbol{\omega} \times \vec{v}$, we rewrite the representations (2.87) and (2.88) in the form

$$\vec{u}_{N_s}^{\,det}(\mathbf{x}) = \sum_{i=1}^{N_s} \left(\frac{|\Delta\boldsymbol{\Omega}_i|}{4\pi}\right)^{1/2} \Big(\boldsymbol{\Omega}_i \times \vec{v}_1^{(i)}(\mathbf{x}\cdot\boldsymbol{\Omega}_i) + v^{(i)}(\mathbf{x}\cdot\boldsymbol{\Omega}_i)\Big), \tag{2.89}$$

$$\vec{u}_{N_s}^{\,rnd}(\mathbf{x}) = \frac{1}{N_s^{1/2}} \sum_{i=1}^{N_s} \Big(\boldsymbol{\omega}_i \times \vec{v}_1^{(i)}(\mathbf{x}\cdot\boldsymbol{\omega}_i) + v^{(i)}(\mathbf{x}\cdot\boldsymbol{\omega}_i)\Big). \tag{2.90}$$

2.6.5 Flow in a porous media in the first order approximation

For saturated porous media flows, under time independent conditions, the so-called Darcy's velocity, or specific discharge, $\mathbf{q}$ is determined by the Darcy law:

$$\mathbf{q}(\mathbf{x}) = \theta(\mathbf{x})\vec{u}(\mathbf{x}) = -K(\mathbf{x})\nabla\big(\phi(\mathbf{x})\big), \tag{2.91}$$

where $\vec{u}$ is the pore velocity, θ, the porosity, ϕ, the hydraulic potential $\phi = z + pg/\rho$, ρ, the fluid density, and K is the hydraulic conductivity. The functions θ and K are key parameters of the flow. Experimental measurements show a highly heterogeneous behavior of K in space with the following remarkable property: when considering K as a random field, its distribution is well approximated by the log-normal law (e. g., see [34, 65]). Therefore, in models, the hydraulic log-conductivity $Y = \ln K$ is commonly considered to be a statistically homogeneous Gaussian random field . We denote $Y' = Y - <Y>$, $C_Y(\vec{r}) = \langle Y'(\mathbf{x}+\vec{r})Y'(\mathbf{x})\rangle$ the auto-correlation function of the hydraulic log-conductivity.

The porosity θ is also often considered in some models as a random field. However its variability is generally much smaller than that of K, and it is usually assumed to be constant.

In the framework of the method of small perturbations, when the variation of the hydraulic log-conductivity is small, the spectral tensor $F(\mathbf{k})$ of the velocity field $\vec{u}(\mathbf{x})$ in a 3D porous medium within a first order approximation can be represented by the form (see, e. g., [65])

$$F(\mathbf{k}) = \Delta(\hat{\mathbf{k}})\mathbf{U}(\Delta(\hat{\mathbf{k}})\mathbf{U})^T\, F_Y(\hat{\mathbf{k}}), \tag{2.92}$$

where $\mathbf{U} = (U_1, U_2, U_3)^T = \langle \vec{u} \rangle$ is the mean velocity,

$$F_Y(\hat{\mathbf{k}}) = \int_{\boldsymbol{R}^3} e^{-\mathrm{i}\, 2\pi\, \mathbf{k}\cdot\vec{r}}\, C_Y(\vec{r}) d\vec{r}$$

is the spectral function of the hydraulic log-conductivity, and $\Delta(\hat{\mathbf{k}})$ is a 3×3 tensor defined in the previous subsection. Assume that the hydraulic log-conductivity field $Y = \ln K$ is isotropic which implies that $F_Y(\mathbf{k}) = F_Y(k)$. Introducing $E(k) = 2\pi k^2 F_Y(k)$ and taking into account

$$F(\mathbf{k}) = \Delta(\hat{\mathbf{k}})\mathbf{U}(\Delta(\hat{\mathbf{k}})\mathbf{U})^T \frac{2E(k)}{4\pi k^2}$$

we get the following representation with fixed nodes for the velocity field $\vec{u}(\mathbf{x})$:

$$\vec{u}_{N_s}^{det}(\mathbf{x}) = \mathbf{U} + \sum_{i=1}^{N_s} \left(\frac{|\Delta \boldsymbol{\Omega}_i|}{A_d} \right)^{1/2} \Delta(\boldsymbol{\Omega}_i)\, \mathbf{U}\, v^{(i)}(\mathbf{x} \cdot \boldsymbol{\Omega}_i), \tag{2.93}$$

where $v^{(i)}(t)$, $i = 1, ..., N_s$ are mutually independent zero mean, stationary Gaussian scalar processes with the spectral function $E(k)$. Analogously, using a decomposition with randomly distributed nodes (e. g., see (2.77)), the fluctuations of a 3-dimensional random field with the spectral density (2.92) can be simulated by

$$\vec{u}_{N_s}^{rnd}(\mathbf{x}) = \mathbf{U} + \frac{1}{N_s^{1/2}} \sum_{i=1}^{N_s} \Delta(\boldsymbol{\omega}_i)\, \mathbf{U}\, v^{(i)}(\mathbf{x} \cdot \boldsymbol{\omega}_i), \tag{2.94}$$

where $\boldsymbol{\omega}_i$, $i = 1, \ldots, N_s$ is a family of independent unit isotropic 3D vectors , $v^{(i)}(t)$, $i = 1, \ldots, N_s$ is a set of mutually independent and independent of $\boldsymbol{\omega}_i$, $i = 1, \ldots, N_s$, zero mean, stationary Gaussian scalar processes with the spectral function $E(k) = 2\pi k^2 F_Y(k)$.

2.6.6 Fourier wavelet models of Gaussian random fields

We have presented above an approximation of a homogeneous Gaussian random field obtained by a superposition of plane waves. This approach makes it possible to simulate a random field $\vec{u}(\mathbf{x})$ through simulation of random processes $\vec{v}^{(i)}(t)$, $i = 1, ..., N_s$. If in the simulation of these random processes we use the Fourier wavelet method presented in Section 2.5, we obtain a Fourier wavelet model of Gaussian random fields based on the plane wave decomposition. If the random field is isotropic, or if its spectral tensor can be represented in the form (2.71) where the tensor H satisfies the condition (2.72), then the random processes $\vec{v}^{(i)}(t)$, $i = 1, ..., N_s$ all have the same spectrum, which simplifies the construction of the simulation formulas.

In the general case when the random field is anisotropic, the general simulation formula (2.70) is used. In this case the random processes $\vec{v}^{(i)}(t)$, $i = 1, ..., N_s$ may generally have different spectral functions. This complicates the simulations when precalculating the functions $\mathcal{F}_m^{(\phi)}$ and $\mathcal{F}_m^{(\psi)}$ which will also depend on the index i: $\mathcal{F}_{m,i}^{(\phi)}$ and $\mathcal{F}_{m,i}^{(\psi)}$, $i = 1, ..., N_s$. To this point we make the following remark. Assume that the random field $\vec{u}(\mathbf{x})$ is anisotropic, but can be transformed to an isotropic field by a linear transformation of $\mathbf{x}$. More precisely, assume that there exists a nonsingular matrix A such that the random filed $\vec{u}'(\mathbf{x}') = \vec{u}(A\mathbf{x}')$ is isotropic. Then to simulate the random field $\vec{u}$ we first simulate an isotropic random field $\vec{u}'(\mathbf{x}')$, and then use the representation $\vec{u}(\mathbf{x}) = \vec{u}'(A^{-1}\mathbf{x})$.

2.7 Comparison of Fourier wavelet and randomized spectral models

As mentioned in the introduction to this chapter, our experiments in the 1-dimensional case ($d = 1$) have shown that the randomized models are more efficient for evaluating ensemble averages than the FW models, or compatible in efficiencies, if multipoint statistics should be calculated (see [104]). The situation is opposite for spatial averaging.

In this section we compare the Fourier wavelet model (FWM) and the randomized spectral model (RSM) in the 3-dimensional case ($d = 3$). We simulate a 3-dimensional zero mean Gaussian isotropic incompressible random field $\vec{u}(\mathbf{x})$ with a spectral tensor

$$F_{ij}(\mathbf{k}) = \frac{2E(k)}{4\pi k^2}\Big(\delta_{ij} - \frac{k_i k_j}{k^2}\Big), \quad i,j = 1,2,3; \quad E(k) = \frac{8(2\pi k)^4}{\big(1 + (2\pi k)^2\big)^3}. \qquad (2.95)$$

It is known that the correlation tensor of isotropic incompressible fields is defined by $B_{LL}(r)$, the longitudinal, and $B_{NN}(r)$, transversal correlation functions (e. g., see [146]). In our case these functions can be represented explicitly (e. g., see [146]):

$$\begin{aligned} B_{LL}(r) &= \langle u_1(x+r,0,0)u_1(x,0,0)\rangle = e^{-r}, \\ B_{NN}(r) &= \langle u_2(x+r,0,0)u_2(x,0,0)\rangle = e^{-r}(1 - r/2). \end{aligned} \qquad (2.96)$$

The comparison of RSM and FWM is made by calculation of the correlation functions (2.96).

2.7.1 Some technical details of RSM

As a basis, we take the model (2.38) for the simulation of isotropic random field $\vec{u}(\mathbf{x})$. The spectral space is divided in $N = 3$ subsets $\Delta_i = \{\mathbf{k} : a_i \le |\mathbf{k}| < b_i\}$, $i = 1,2,3$

with $a_1 = 0$, $b_1 = a_2 = 0.34$, $b_2 = a_3 = 0.8$, $b_3 = \infty$. In each subset Δ_i, n_0 points $\mathbf{k}_{i,j}$, $j = 1, \dots, n_0$ are sampled with the same probability density:

$$p_i(\mathbf{k}) = \frac{1}{4\pi k^2} \frac{C_i}{(1 + (2\pi k)^2)}, \quad \mathbf{k} \in \Delta_i; \quad C_i = 1 / \int_{a_i}^{b_i} \frac{dk}{(1 + (2\pi k)^2)}.$$

Note that the spectral tensor (2.95) can be written in the form (2.3) with $Q(\mathbf{k}) = f^{1/2}(\mathbf{k}) \mathcal{H}(\hat{\mathbf{k}})$:

$$F(\mathbf{k}) = f^{1/2}(\mathbf{k})(\mathcal{H}\hat{\mathbf{k}}) \left(f^{1/2}(\mathbf{k}) \mathcal{H}(\hat{\mathbf{k}}) \right)^T; \quad f(\mathbf{k}) = 16\pi \frac{(2\pi k)^2}{(1 + (2\pi k)^2)^3},$$

and the 3×3-dimensional antisymmetric matrix $\mathcal{H}(\mathbf{\Omega})$, $\mathbf{\Omega} \in S_2$ is defined in the previous section. For a 3-dimensional vector $\boldsymbol{\xi}$ $\mathcal{H}(\hat{\mathbf{k}})\boldsymbol{\xi} = \hat{\mathbf{k}} \times \boldsymbol{\xi}$ ($\mathbf{a} \times \mathbf{b}$ is a vector product of vectors $\mathbf{a}$ and $\mathbf{b}$), hence in view of (2.38) we come to the simulation formula

$$\begin{aligned} \vec{u}_{N,n_0}(\mathbf{x}) = \sum_{i=1}^{N} \frac{1}{\sqrt{n_0}} \sum_{j=1}^{n_0} \left(\frac{f(\mathbf{k}_{ij})}{p_i(\mathbf{k}_{ij})} \right)^{1/2} \\ \times \left\{ (\boldsymbol{\omega}_{ij} \times \boldsymbol{\xi}_{ij}) \cos(\theta_{ij}) + (\boldsymbol{\omega}_{ij} \times \boldsymbol{\eta}_{ij}) \sin(\theta_{ij}) \right\}, \end{aligned} \tag{2.97}$$

where $\mathbf{k}_{ij} = k_{ij}\boldsymbol{\omega}_{ij}$, $\boldsymbol{\omega}_{ij}$, $i = 1, \dots, N$; $j = 1, \dots, n_0$ are mutually independent 3-dimensional unit isotropic vectors, and k_{ij} $i = 1, \dots, N$; $j = 1, \dots, n_0$ are random points (independent of each other and of $\boldsymbol{\omega}_{ij}$) distributed in $[a_i, b_i]$ with the density $p_i(k) = C_i/(1 + (2\pi k)^2)$. Here $\theta_{ij} = 2\pi k_{ij}\boldsymbol{\omega}_{ij} \cdot \mathbf{x}$, and $\boldsymbol{\xi}_{ij}$, $\boldsymbol{\eta}_{ij}$, $i = 1, \dots, N$; $j = 1, \dots, n_0$ are mutually independent and independent of the family $\mathbf{k}_{ij}$ standard 3-dimensional Gaussian random variables (with zero mean and unity covariance matrix). The points $\mathbf{k}_{ij}$ in (2.97) are sampled isotropically, which does not imply that the samples of $\vec{u}_{N,n_0}(\mathbf{x})$ are isotropic in space, especially when n_0 are relatively small. To improve the isotropic property, it is reasonable to use a stratified sampling over angles using the subdivision (2.74). Namely, we constructed the subdivision of the space $I\!R^3 = \sum_{i,j,r} \Delta_{ijr}$, where

$$\Delta_{ijr} = \{\mathbf{k} \in I\!R^3 : a_i \le |\mathbf{k}| < b_i; \hat{\mathbf{k}} = \mathbf{k}/|\mathbf{k}| \in \Delta\mathbf{\Omega}_{j,r}\},$$

where $i = 1, \dots, N$; $j = 1, \dots, n_\theta$; $r = 1, \dots, n_\phi^{(j)}$. In each element Δ_{ijr} we sample one point $\mathbf{k}_{ijr} = k_{ijr}\boldsymbol{\omega}_{ijr}$, where k_{ijr} is chosen in $[a_i, b_i)$ with the density $p_i(k) = C_i/(1 + (2\pi k)^2)$, while the random direction $\boldsymbol{\omega}_{ijr}$ is sampled uniformly in $\Delta\mathbf{\Omega}_{j,r}$:

$$\boldsymbol{\omega}_{ijr} = (\sin(\tilde{\theta}_{ij})\cos(\tilde{\phi}_{ijr}), \sin(\tilde{\theta}_{ij})\sin(\tilde{\phi}_{ijr}), \cos(\tilde{\theta}_{ij})),$$

where $\tilde{\theta}_{ij} = \arccos\big((1 - \gamma_{ij})\cos(\theta_j - \Delta\theta/2) + \gamma_{ij}\cos(\theta_j + \Delta\theta/2)\big)$, $\tilde{\phi}_{ijr} = \phi_{jr} + (\gamma_{ijr} - 1/2)\Delta\phi_j$. Here γ_{ij}, γ_{ijr} ($i = 1, \dots, N$; $j = 1, \dots, n_\theta$; $r = 1, \dots, n_\phi^{(j)}$) are mutually independent random numbers uniformly distributed in [0, 1]. The random point $\mathbf{k}_{ijr}$ is distributed with the density

$$p_{ijr}(\mathbf{k}) = \frac{1}{|\Delta\boldsymbol{\Omega}_{jr}|}\frac{C_i}{|\mathbf{k}|^2(1+(2\pi|\mathbf{k}|)^2)}.$$

Thus the RSM for an isotropic Gaussian incompressible random field with the spectral tensor (2.95) constructed from the described expansion has the form

$$\vec{u}_{N,n_\theta}(\mathbf{x}) = \sum_{i=1}^{N}\sum_{j=1}^{n_\theta}\sum_{r=1}^{n_\phi^{(j)}} \frac{\sqrt{f(\mathbf{k}_{ijr})}}{\sqrt{p_{ijr}(\mathbf{k}_{ijr})}} \times \left\{(\boldsymbol{\omega}_{ijr}\times\boldsymbol{\xi}_{ijr})\cos(\theta_{ijr}) + (\boldsymbol{\omega}_{ijr}\times\boldsymbol{\eta}_{ijr})\sin(\theta_{ijr})\right\}, \qquad (2.98)$$

where $\theta_{ijr} = 2\pi k_{ijr}\boldsymbol{\omega}_{ijr}\cdot\mathbf{x}$, and $\boldsymbol{\xi}_{ijr}$, $\boldsymbol{\eta}_{ijr}$, $i = 1,\dots,N$; $j = 1,\dots,n_\theta$; $r = 1,\dots,n_\phi^{(j)}$ are mutually independent and independent of the family $\mathbf{k}_{ijr}$, standard 3-dimensional Gaussian random variables (with zero mean and unity covariance matrix).

2.7.2 Some technical details of FWM

Simulation formula for a $3D$ isotropic incompressible random field $\vec{u}(\mathbf{x})$ with the spectral tensor (2.95) is constructed through the plane wave decomposition, without stratification, using the model (2.81), or with stratification, on the basis of (2.73) and (2.82), where the random processes $\vec{v}^{(i)}(t)$, $i = 1,\dots,N_s$ are constructed by FWM (2.48). In our case, the 3D FWM based on the plane wave decomposition without a stratification has the form

$$\vec{u}^{rnd}_{N_s}(\mathbf{x}) = \frac{1}{N_s^{1/2}}\sum_{i=1}^{N_s}\boldsymbol{\omega}_i\times\vec{v}^{(i)}(\boldsymbol{\omega}_i\cdot\mathbf{x}), \qquad (2.99)$$

where $\boldsymbol{\omega}_i$, $i = 1,\dots,N_s$ are mutually independent 3–dimensional unit isotropic vectors, and $\vec{v}^{(i)}(t)$, $i = 1,\dots,N_s$ are mutually independent 3-dimensional stationary Gaussian processes with independent identically distributed components. Each component of the process $\vec{v}^{(i)}(t)$ has the spectral function $F(k) = E(k)$.

In the case of stratified models, the subdivision of the unit sphere appeared in (2.73) and (2.82) is constructed as described in the previous section.

Thus the 3D FWM based on the plane wave decomposition with deterministic nodes (i. e., with (2.75)) is

$$\vec{u}^{det}_{n_\theta}(\mathbf{x}) = \sum_{j=1}^{n_\theta}\sum_{r=1}^{n_\phi^{(j)}}\left(\frac{|\Delta\boldsymbol{\Omega}_{j,r}|}{4\pi}\right)^{1/2}\boldsymbol{\Omega}_{j,r}\times\vec{v}^{(jr)}(\mathbf{x}\cdot\boldsymbol{\Omega}_{j,r}), \qquad (2.100)$$

where

$$\begin{aligned}\mathbf{\Omega}_{j,r} &= (\sin(\theta_j)\cos(\phi_{jr}),\ \sin(\theta_j)\sin(\phi_{jr}),\ \cos(\theta_j)),\\ |\Delta\mathbf{\Omega}_{j,r}| &= \int_{\Delta\mathbf{\Omega}_{j,r}} d\mathbf{\Omega} = 4\pi\Delta\phi_j\,[\cos(\theta_j-\Delta\theta/2)-\cos(\theta_j+\Delta\theta/2)].\end{aligned}$$

A hybrid stratified randomized FWM based on the plane wave decomposition (2.82) has the form

$$\vec{u}^{\,srm}_{n_\theta}(\mathbf{x}) = \sum_{j=1}^{n_\theta}\sum_{r=1}^{n_\phi^{(j)}}\left(\frac{|\Delta\mathbf{\Omega}_{j,r}|}{4\pi}\right)^{1/2}\boldsymbol{\omega}_{j,r}\times\vec{v}^{(jr)}(\mathbf{x}\cdot\boldsymbol{\omega}_{jr}), \tag{2.101}$$

where

$$\begin{aligned}\boldsymbol{\omega}_{jr} &= (\sin(\tilde{\theta}_j)\cos(\tilde{\phi}_{jr}),\ \sin(\tilde{\theta}_j)\sin(\tilde{\phi}_{jr}),\ \cos(\tilde{\theta}_i j)),\\ \tilde{\theta}_j &= \arccos\left((1-\gamma_j)\cos(\theta_j-\Delta\theta/2)+\gamma_j\cos(\theta_j+\Delta\theta/2)\right),\\ \tilde{\phi}_{jr} &= \phi_{jr}+(\gamma_{jr}-1/2)\Delta\phi_j,\end{aligned}$$

and γ_j, γ_{jr}, ($j = 1,\dots,n_\theta$; $r = 1,\dots,n_\phi^{(j)}$) are mutually independent random numbers uniformly distributed on $[0,1]$. The stationary random processes $\vec{v}^{(jr)}(t)$ ($j = 1,\dots,n_\theta$; $r = 1,\dots,n_\phi^{(j)}$), appeared in (2.100) and (2.101) are mutually independent 3-dimensional stationary Gaussian processes with independent identically distributed components. Each component of the process $\vec{v}^{(jr)}(t)$ has the spectral function $F(k) = E(k)$ (see (2.95)) and is simulated by FWM (2.48). Since all three components of all these random processes have the same spectral function, in the simulation formulas one and the same set of functions $\mathcal{F}^{(\phi)}_{m_0}$ and $\mathcal{F}^{(\psi)}_m$, $m = m_0,\dots,m_1$ is used. Parameters of the Fourier wavelet model (2.48) are taken as $b_0 = b_1 = 10; m_0 = 0, m_1 = 6$.

Now we give some details related to the simulation formulas (2.99)–(2.101). Assume we have to simulate a random field $\vec{u}^{\,det}_{n_\theta}(\mathbf{x})$ (or $\vec{u}^{\,srm}_{n_\theta}(\mathbf{x})$) in a bounded domain $D \subset I\!R^3$. We choose segments $[a_{jr}, b_{jr}]$ ($j = 1,\dots,n_\theta$; $r = 1,\dots,n_\phi^{(j)}$) so that $\mathbf{x}\cdot\mathbf{\Omega}_{j,r} \in [a_{jr}, b_{jr}]$ $\forall \mathbf{x} \in D$. Then we simulate by the FW model (2.48) the processes $\vec{v}^{(jr)}(t)$, $j = 1,\dots,n_\theta$; $r = 1,\dots,n_\phi^{(j)}$ on a sufficiently fine grid on $[a_{jr}, b_{jr}]$. The quantities $\vec{v}^{(ij)}(\mathbf{x}\cdot\mathbf{\Omega}_{i,j})$ are evaluated through a linear interpolation of the values $\vec{v}^{(ij)}$ obtained on this fine grid. The same can be done for the model (2.99).

The Eulerian correlation functions ($B_{LL}(r)$, $B_{NN}(r)$, defined above are calculated through ensemble and spatial averaging, by FWM and RSM.

2.7.3 Ensemble averaging

In Figure 2.3 we plot the functions $B_{LL}(r)$ and $B_{NN}(r)$ for a Gaussian isotropic incompressible random field with the spectral tensor (2.95). The number of MC samples here and in all ensemble averagings was taken as 16,000.

2.7.4 Space averaging

In this section we deal with the space averaging aiming at evaluating the Eulerian correlation functions on the basis of the ergodic property.

Let us discuss this in more detail.

Assume we need to evaluate the Eulerian correlation function $B_{LL}(r)$ using only one sample of the random field $\vec{u}(\mathbf{x})$. Since $\vec{u}(\mathbf{x})$ is homogeneous, the random process $\xi(x;r) = u_1(x+r,0,0)u_1(x,0,0)$, $x \in I\!R$ (r is fixed) is stationary, and by definition, $\langle\xi(x;r)\rangle = B_{LL}(r)$. Assuming that this process is ergodic (this is true under quite general conditions; e. g., see [259]) we can evaluate $\langle\xi(x;r)\rangle$ by a spatial averaging. In particular, $\langle\xi(x;r)\rangle \simeq \frac{1}{n_x}\sum_{i=1}^{n_x}\xi(x_i)$, where $x_i, i = 1,\dots,n_x$ is a set of points in $I\!R$. It is reasonable to choose the points so that the minimal distance between the points is larger than the characteristic correlation length of our random process $\xi(x;r)$. These arguments can be extended to the case of points $\mathbf{x}_i \in I\!R^3$ which leads to the relations

$$B_{LL}(r) \simeq \frac{1}{n_p}\sum_{i_x=-n_x}^{n_x}\sum_{i_y=-n_y}^{n_y}\sum_{i_z=-n_z}^{n_z} u_1(i_xL+r, i_yL, i_zL)u_1(i_xL, i_yL, i_zL),$$

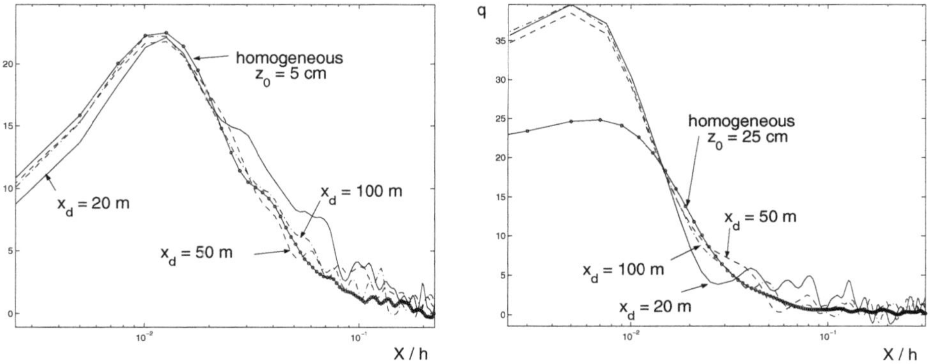

Figure 2.3. The longitudinal (B_{LL}) and transversal (B_{NN}) correlation functions calculated by RSM (model (2.97) with $n_0 = 25$, left panel), and FWM (model (2.99) with $N_s = 25$, right panel) models through ensemble averaging. Calculations: bold solid line; Explicit result (2.96): solid line. The number of MC samples in both cases was 16,000.

$$B_{NN}(r) \simeq \frac{1}{n_p} \sum_{i_x=-n_x}^{n_x} \sum_{i_y=-n_y}^{n_y} \sum_{i_z=-n_z}^{n_z} u_2(i_x L + r, i_y L, i_z L) u_2(i_x L, i_y L, i_z L),$$

where $L > 0$ is the grid size, $n_p = (2n_x + 1)(2n_y + 1)(2n_z + 1)$; in calculations we have taken $n_x = n_y = n_z = 7$ (hence the number of points $n_p = 3375$), and $L = 5$. In Figures 2.4 and 2.5 the correlation functions $B_{LL}(r)$ and $B_{NN}(r)$ evaluated by RSM (2.97) for different values of n_0 are presented. These results show that to reach the accuracy compared to that of the ensemble averaging calculations, (see Figure 2.3, left panel) even $n_0 = 1200$ cannot be considered as satisfied. Close results were obtained by the model (2.98) for $n_\theta \leq 30$ (i. e., $N_s \leq 1132$).

In Figure 2.6 we plot the functions $B_{LL}(r)$ and $B_{NN}(r)$ calculated by the use of models (2.99) (left panel) and (2.101) (right panel). The model (2.100) with $n_\theta = 10$ gave approximately the same results. The amount of statistics is $n_p = 15^3 = 3375$, so we see that the results are in a good agreement with the results of Figure 2.3 obtained through the ensemble averaging (note that the statistics in Figure 2.6 is about 4 times less than that of Figure 2.3).

Thus we conclude that to calculate the Eulerian correlation functions $B_{LL}(r)$ and $B_{NN}(r)$ by the space averaging with one sample of the field $\vec{u}(\mathbf{x})$, the number of harmonics N_s in RSMs (2.97), (2.98) should be taken to be around several thousand, to attain the same accuracy as FWMs provide with several of hundreds of harmonics. (in both methods, statistics was 3375).

2.8 Conclusions

Simulation methods of homogeneous Gaussian random fields based on randomized spectral representations and Fourier wavelet decomposition were presented. Extensions of Fourier wavelet method (FWM) from scalar processes and isotropic fields to general vector fields were suggested. Convergence of the constructed Fourier wavelet models (in the sense of finite-dimensional distributions) under some general conditions on the spectral tensor were given. A comparative analysis of RSM and FWM was made by calculating the Eulerian statistical characteristics of a 3D isotropic incompressible random field through an ensemble and space averaging.

The comparative analysis can be summarized as follows.

- In the case of ensemble averaging, in contrast to the 1-dimensional case, the complexity of the randomized spectral model (RSM) and 3D FWM are more or less compatible, because in FWM based on the plane-wave decomposition, the 1-dimensional processes are precalculated on a fine grid in advance, so that there is no need in the sophisticated management for random numbers choice. The best results are obtained when the number of 1-dimensional processes is about 10. Nevertheless, RSM seems to be preferable in this case, since its implementation is much easier.

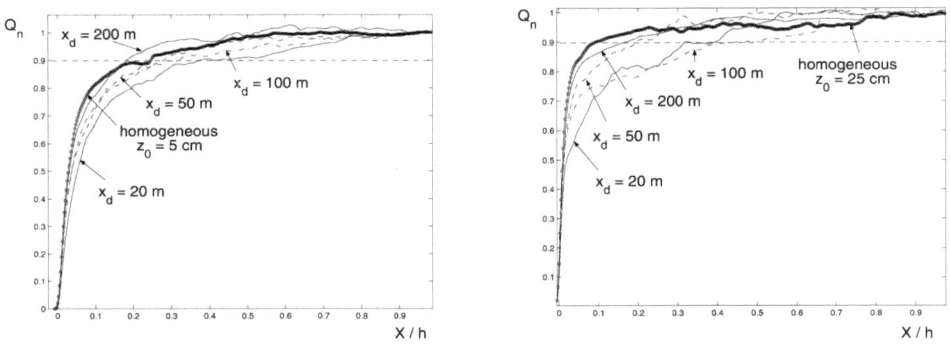

Figure 2.4. Eulerian longitudinal and transversal correlation functions calculated by RSM (2.97) with $n0 = 100$ (left panel) and $n0 = 400$ (right panel) obtained by space averaging. Bold solid line; calculations; solid line: explicit result.

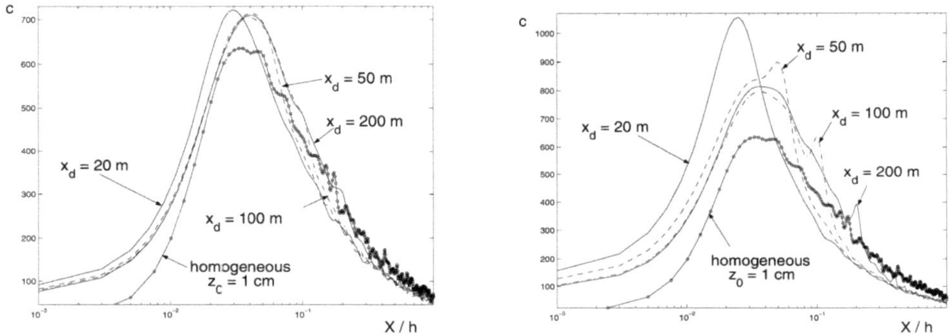

Figure 2.5. Eulerian longitudinal and transversal correlation functions calculated by RSM (2.97) with $n0 = 800$ (left panel) and $n0 = 1200$ (right panel) obtained by space averaging. Bold solid line: calculations; solid line: explicit result.

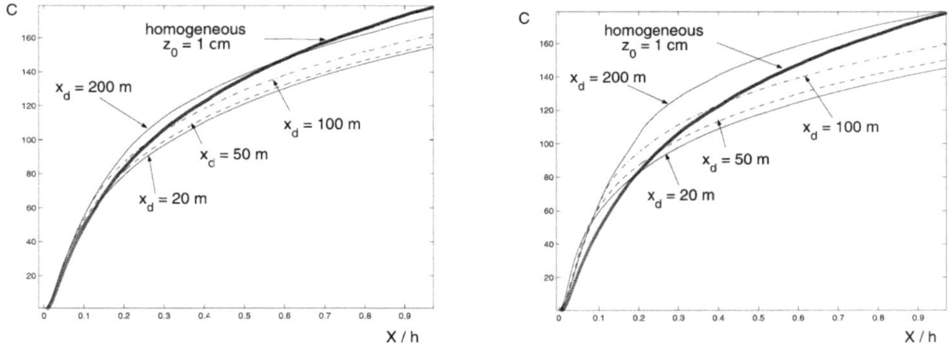

Figure 2.6. Eulerian longitudinal and transversal correlation functions calculated by FWM (2.99) with $Ns = 200$ (left panel) and (2.101) with $n_\theta = 10$ (right panel) obtained by space averaging. Bold solid line: calculations; solid line; explicit result.

- For space averaging, when evaluating Eulerian characteristics using one sample of the simulated random field, FWM shows higher efficiency because its samples have better ergodic properties. To get stable results by RSM, one needs a huge number (several thousands) of harmonics.

2.9 Appendices

2.9.1 Appendix A. Positive definiteness of the matrix $\mathcal{B}$

Here we prove that the matrix $\mathcal{B}$ with entries (2.59) is positive definite, i. e., for any N-dimensional vector-rows $\mathbf{c}_1, \dots, \mathbf{c}_N \in I\!R^m$

$$\mathbf{c}_i b_{ij} \mathbf{c}_j^T = \int_{A_i \cap A_j} \mathbf{c}_i B(t_i - t_j; \boldsymbol{\Omega}) \mathbf{c}_j^T \, d\boldsymbol{\Omega} \geq 0 \tag{A.1}$$

is true for all integers $N > 0$ and values $t_1, \dots, t_N$ in $I\!R$, and measurable subsets $A_1, \dots, A_N$ in S_{d-1}.

Let $\{\mathcal{D}_\alpha\}_{\alpha=1}^M$ be a subdivision of the set $A = \cup_{i=1}^N A_i$ (i. e., $\mathcal{D}_\alpha \cap \mathcal{D}_\beta = \emptyset$ if $\alpha \neq \beta$, and $A = \cup_{\alpha=1}^M \mathcal{D}_\alpha$) such that it is also a subdivision of A_i for any $i = 1, \dots, N$: $A_i = \cup_{\alpha_j=1}^{N_i} \mathcal{D}_{\alpha_j}$.

Let $\vec{u}_\alpha(t)$, $1 \leq \alpha \leq M$ be a family of n-dimensional vector columns, mutually independent zero mean Gaussian random processes with the correlation tensor $\int_{\mathcal{D}_\alpha} B(t, \boldsymbol{\Omega})\, d\boldsymbol{\Omega}$. Let us construct, for each $i = 1, \dots, N$, a zero mean Gaussian random process $\vec{v}_i(t)$ by putting $\vec{v}_i(t) = \sum_{j=1}^{N_i} \vec{u}_{\alpha_j}(t)$. Then obviously

$$\langle \vec{v}_i(t) \vec{v}_j^T(t) \rangle = \int_{A_i \cap A_j} B(t_i - t_j; \boldsymbol{\Omega})\, d\boldsymbol{\Omega}.$$

Introducing a random variable $\xi = \mathbf{c}_i \vec{v}_i(t_i)$ (here we recall about the summation convention) yields

$$\langle \xi^2 \rangle = \langle \mathbf{c}_i \vec{v}_i \vec{v}_j^T \mathbf{c}_j^T \rangle = \mathbf{c}_i b_{ij} \mathbf{c}_j^T.$$

which ensures (A.1).

2.9.2 Appendix B. Proof of Proposition 2.1

From the definition (2.48) we have for the random field $u^{(FW)}$

$$u(x) = u^{(FW)}(x) + \delta v_{m_0}^{(1)}(x) + \sum_{m=m_0}^{m_1} \delta v_m^{(2)}(x) + \delta v_{m_1}^{(3)}(x), \tag{B.1}$$

where

$$\delta v^{(1)}_{m_0}(x) = \sum_{j:|j-\lfloor 2^{m_0}x\rfloor|>b_0} \mathcal{F}^{(\phi)}_{m_0}(2^{m_0}x+j)\,\xi_j,$$

$$\delta v^{(2)}_{m}(x) = \sum_{j:|j-\lfloor 2^{m}x\rfloor|>b_1} \mathcal{F}^{(\psi)}_{m}(2^{m}x+j)\,\xi_{mj},$$

$$\delta v^{(3)}_{m_1}(x) = \sum_{m=m_1+1}^{\infty}\sum_{j=-\infty}^{\infty} \mathcal{F}^{(\psi)}_{m}(2^{m}x+j)\,\xi_{mj}.$$

The random variables ξ_j, ξ_{mj} are mutually independent, hence the terms in the right-hand side of (B.1) are also independent; therefore

$$\begin{aligned}\langle u(x+r)u(x)\rangle &- \langle u^{(FW)}(x+r)u^{(FW)}(x)\rangle \\ &= \langle \delta v^{(1)}_{m_0}(x+r)\delta v^{(1)}_{m_0}(x)\rangle + \sum_{m=m_0}^{m_1}\langle \delta v^{(2)}_{m}(x+r)\delta v^{(2)}_{m}(x)\rangle \\ &\qquad + \langle \delta v^{(3)}_{m_1}(x+r)\delta v^{(3)}_{m_1}(x)\rangle. \end{aligned} \tag{B.2}$$

Let us estimate the terms in the right hand side. First, we have for the last term

$$|\langle \delta v^{(3)}_{m_1}(x+r)\delta v^{(3)}_{m_1}(x)\rangle| \le \Big(\langle(\delta v^{(3)}_{m_1}(x+r))^2\rangle\,\langle(\delta v^{(3)}_{m_1}(x))^2\rangle\Big)^{1/2}. \tag{B.3}$$

Lemma 2.1. *Assume that for some $s > 0$*

$$I_s = \int_{-\infty}^{\infty} F(k)(1+|k|^2)^s\,dk < \infty. \tag{B.4}$$

Then a constant B_s depending only on s can be found such that

$$\langle(\delta v^{(3)}_{m_1}(x))^2\rangle \le \frac{B_s^2}{4^{(m_1+1)s}}\int_{-\infty}^{\infty} F(k)(1+|k|^2)^s\,dk\,. \tag{B.5}$$

Proof. Note that

$$\langle(\delta v^{(3)}_{m_1}(x))^2\rangle = \sum_{m=m_1+1}^{\infty}\sum_{j=-\infty}^{\infty} |\mathcal{F}^{(\psi)}_{m}(2^{m}x+j)|^2. \tag{B.6}$$

Further, by (2.24)

$$\begin{aligned}\mathcal{F}^{(\psi)}_{m}(2^{m}x+j) = G^{(\psi)}_{mj}(x) &= \int_{-\infty}^{\infty} e^{i2\pi kx}Q(k)\bar{\hat{\psi}}_{mj}(k)\,dk \\ &= \int_{-\infty}^{\infty} G(x+y)\psi_{mj}(y)dy,\end{aligned}$$

where $G(x) = \int_{-\infty}^{\infty} e^{i2\pi kx} Q(k)\,dk$. Consequently for fixed x, the quantities $\mathcal{F}_m^{(\psi)}(2^m x + j)$ $(m, j = \dots, -1, 0, 1, \dots)$ are Fourier coefficients in the expansion of $G(x + \cdot)$ with respect to the orthonormal system ψ_{mj} $(m, j = \dots, -1, 0, 1, \dots)$. Notice that the condition (B.4) can be formulated equivalently that the function G belongs to the space of Bessel potentials $H_2^s(\mathbb{R})$, since $G \in H_2^s(\mathbb{R})$ means that the function $Q(k)(1 + |k|^2)^{s/2}$ is from $L_2(\mathbb{R})$) and $||G||_{H_2^s} = I_s^{1/2}$ (for more details see [243]). Further we use the fact that $H_2^s(\mathbb{R})$ coincides with the Besov space $B_{22}^s(\mathbb{R})$, and the norms in these spaces are equivalent (see [243]). On the other hand, it is known (see [141]), that under quite general assumptions on the wavelet function $\psi(x)$ (Meyer's wavelet function satisfies this condition if the function $\nu(x)$ is smooth enough) the norm $||f||_{B_{pq}^s}$ in Besov's space $B_{pq}^s(\mathbb{R})$, $1 \le p, q \le \infty$, $s > 0$ can be equivalently defined through the wavelet coefficients $\beta_{mj} = \int f(x)\psi_{mj}(x)dx$ by

$$||f||_{B_{pq}^s} = ||f||_{L_p} + \left\{ \sum_{m=-\infty}^{\infty} \left(\sum_{j=-\infty}^{\infty} (|\beta_{mj}| 2^{m(s+\frac{1}{2}-\frac{1}{p})})^p \right)^{q/p} \right\}^{1/q}.$$

Since $G \in B_{22}^s$, we conclude

$$\sum_{m=-\infty}^{\infty} \sum_{j=-\infty}^{\infty} |\mathcal{F}_m^{(\psi)}(2^m x + j)|^2 4^{ms} \le B_s^2 ||G(x + \cdot)||_{H_2^s}^2$$

for some B_s depending only on s. From this we get by $||G(x+\cdot)||_{H_2^s}^2 = ||G||_{H_2^s}^2 = I_s$ that

$$\begin{aligned}
&\sum_{m=m_1+1}^{\infty} \sum_{j=-\infty}^{\infty} |\mathcal{F}_m^{(\psi)}(2^m x + j)|^2 \\
&\quad \le \sum_{m=m_1+1}^{\infty} \sum_{j=-\infty}^{\infty} |\mathcal{F}_m^{(\psi)}(2^m x + j)|^2 \frac{4^{ms}}{4^{(m_1+1)s}} \\
&\quad \le \frac{B_s^2}{4^{(m_1+1)s}} I_s.
\end{aligned}$$

This completes the proof of Lemma 2.1. □

This implies due to (B.3) and (B.5) that the last term in the right-hand side of (B.2) can be estimated as follows:

$$|\langle \delta v_{m_1}^{(3)}(x + r)\delta v_{m_1}^{(3)}(x)\rangle| \le \frac{B_s^2}{4^{(m_1+1)s}} \int_{-\infty}^{\infty} F(k)(1 + |k|^2)^s\,dk. \tag{B.7}$$

Now we turn to the estimation of the first two terms in the right-hand side of (B.2). We will use some results from the theory of Nikolskiĭ–Besov spaces (e. g., see [151, 243]).

A triple (r, j, l) is called admissible if $j \in I\!N$, $l \in I\!N_0$ and $j > r - l$. Here $I\!N_0 = \{0, 1, 2, \ldots\}$ and $I\!N = \{1, 2, \ldots\}$. Let us denote by $\Delta_h^{(j)} g$ the j-th difference of g:

$$\Delta_h g(\cdot) = g(\cdot + h) - g(\cdot), \ldots, \Delta_h^{(j)} g(\cdot) = \Delta_h \Delta_h^{(j-1)} g(\cdot).$$

For $1 \leq p, q \leq \infty$, $r > 0$ the space $B_{pq}^r(I\!R)$ consists of all functions f such that the norm

$$||f||_{B_{pq}^r} = ||f||_{L_p} + ||f||_{b_{pq}^r},$$

where

$$||f||_{b_{pq}^r} = \left(\int_{-1}^{1} \left(\frac{||\Delta_h^{(j)} f^{(l)}||_{L_p}}{|h|^{r-l}} \right)^q \frac{dh}{|h|} \right)^{\frac{1}{q}}, \quad 1 \leq q < \infty, \tag{B.8}$$

$$||f||_{b_{p\infty}^r} = \sup_{0<|h|\leq 1} |h|^{l-r} \, ||\Delta_h^{(j)} f^{(l)}||_{L_p} \tag{B.9}$$

makes sense and is finite for some admissible triple (r, j, l). The ambiguity in the choice of triple (r, j, l) is not essential: different admissible triples correspond to equivalent norms.

We will exploit the following imbeddings (e. g., see [151, 243]):

$$B_{pq}^r(I\!R) \hookrightarrow B_{p_1 q}^{\rho}(I\!R), \quad \rho = r\left(1 - \frac{1}{r}\left(\frac{1}{p} - \frac{1}{p_1}\right)\right), \quad 1 \leq p < p_1 < \infty, \tag{B.10}$$

$$B_{p\infty}^{r+\varepsilon}(I\!R) \hookrightarrow B_{pq}^r(I\!R) \hookrightarrow B_{pq_1}^r(I\!R), \quad 1 \leq q < q_1 \leq \infty, \quad \varepsilon > 0, \tag{B.11}$$

$$B_{p1}^l(I\!R) \hookrightarrow W_p^l(I\!R) \hookrightarrow B_{p\infty}^l(I\!R), \quad l \in I\!N_0, \quad 1 \leq p < \infty, \tag{B.12}$$

$$B_{pp}^l(I\!R) \hookrightarrow W_p^l(I\!R) \hookrightarrow B_{pp}^l(I\!R), \quad l \in I\!N_0, \quad 1 \leq p < \infty, \tag{B.13}$$

where $W_p^l(I\!R)$ is the Sobolev space, and $X \hookrightarrow Y$ for seminormed spaces X and Y means that $X \subset Y$ and there exists a constant $c > 0$ such that the inequality $||x||_Y \leq c||x||_X$ is fulfilled.

Lemma 2.2. *Let $\hat{f} \in B_{1\infty}^r(I\!R)$, $r > 1/2$. Then f is uniformly continuous, $f \in L_2(I\!R)$ and there exists a positive constant C_r depending only on r such that for all $x \in I\!R$*

$$|x|^r \, |f(x)| \leq C_r ||\hat{f}||_{B_{1\infty}^r}. \tag{B.14}$$

Proof. The uniform continuity of f follows from the fact that $\hat{f} \in L_1(I\!R)$ since $B_{1\infty}^r(I\!R) \subset L_1(I\!R)$. Then, for a positive $\varepsilon \in (0, r - 1/2)$ in view of (B.10) and (B.11)

$$B_{1\infty}^r(I\!R) \hookrightarrow B_{2\infty}^{r-1/2}(I\!R) \hookrightarrow B_{22}^{r-1/2-\varepsilon}(I\!R). \tag{B.15}$$

From this we get by (B.13) that $\hat{f} \in L_2(I\!R)$, hence $f \in L_2(I\!R)$. Let $l = \lfloor r \rfloor$ be the integer part of r. From

$$\Delta_h^{(l+2)} \hat{f}(k) = \int_{-\infty}^{\infty} e^{-i 2\pi kx}(e^{-i 2\pi hx} - 1)^{l+2} f(x)\, dx \tag{B.16}$$

and $f \in L_2(I\!R)$ it follows by the inverse Fourier transform

$$f(x)(e^{-i 2\pi hx} - 1)^{l+2} = \int_{-\infty}^{\infty} e^{i 2\pi kx} \Delta_h^{(l+2)} \hat{f}(k)\, dk. \tag{B.17}$$

Taking the absolut values and dividing this equation by $|h|^r$ we then take the supremum over $h \in I\!R$. This yields

$$|x|^r\, |f(x)| C_r' \leq \sup_{h \in \mathbb{R}} |h|^{-r}\, ||\Delta_h^{(l+2)} \hat{f}||_{L_1}, \tag{B.18}$$

where

$$C_r' = \sup_{t \in \mathbb{R}} \left\{ \frac{|e^{-i 2\pi t} - 1|^{l+2}}{|t|^r} \right\}. \tag{B.19}$$

Since the triple $(r, l + 2, 0)$ is admissible, and $\sup_{h \in \mathbb{R}} |h|^{-r}\, ||\Delta_h^{(l+2)} \hat{f}||_{L_1} \leq C_r'' ||\hat{f}||_{B^r_{1\infty}}$ for some C_r'' depending only on r, the proof of Lemma 2.2 is complete. □

To get an estimation for $\langle \delta v_{m_0}^{(1)}(x + r) \delta v_{m_0}^{(1)}(x) \rangle$ we estimate the quantity

$$\langle (\delta v_{m_0}^{(1)}(x))^2 \rangle = \sum_{j : |j - \lfloor 2^{m_0} x \rfloor| > b_0} |\mathcal{F}_{m_0}^{(\phi)}(2^{m_0} x + j)|^2. \tag{B.20}$$

We first estimate each term of this sum. The function $\mathcal{F}_m^{(\phi)}(y)$ has the Fourier transform $2^{m/2} Q(2^m k) \bar{\hat{\phi}}(k)$ (see (2.23)).

Hence if $\hat{\phi}$ is smooth enough and if the function $Q(k)$ belongs to the Nikolskiĭ–Besov space $B^\rho_{1\infty}(I\!R)$, $\rho > 1/2$, then $2^{m/2} Q(2^m k) \bar{\hat{\phi}}(k) \in B^\rho_{1\infty}(I\!R)$. Then from (B.14) we get

$$|\mathcal{F}_m^{(\phi)}(y)| \leq \frac{C_\rho}{|y|^\rho} ||2^{m/2} Q(2^m \cdot) \bar{\hat{\phi}}||_{B^\rho_{1\infty}} = \frac{C_{\rho m}(\phi, Q)}{|y|^\rho}, \tag{B.21}$$

where $C_{\rho m}(\phi, Q)$ is a constant depending on ρ, m and functions ϕ and Q. Note that from (B.20) and (B.21) we get

$$\langle (\delta v_{m_0}^{(1)}(x))^2 \rangle \leq \frac{C'_{ml}}{b_0^{2\rho - 1}}.$$

The same arguments are used to estimate the terms in the right hand side of (B.2) related with $\delta v_m^{(2)}$.

The proof of Proposition 2 is complete.

Chapter 3

Stochastic Lagrangian models of turbulent flows: Relative dispersion of a pair of fluid particles

Stochastic Lagrangian models of relative motion of two fluid particles in one and three dimensions for locally isotropic incompressible turbulent flow are presented. A principle of consistency of statistics between the Eulerian and Lagrangian velocity fields for general random forcing models of a diffusion type is proposed. This enables us to analyze and improve some well known models. An analog of the well-mixed condition for the relative dispersion is proposed. An explicit form of an ∞-order consistency model is given. While there is a series of applications (both in theory and applied turbulence) of the relative motion of two particles, a relation to the problem of concentration fluctuations reflect our motivation by the problem of pollutant scattering in the turbulent atmosphere.

3.1 Introduction

There are two main approaches in stochastic modeling of turbulent motion of fluid particles. The first one uses models of Eulerian velocity field $U_E(x,t)$, for instance, randomized Monte Carlo models [191, 197], or models based on numerical methods for solving the Navier–Stokes equation [24, 160]. With such models, one simulates an ensemble of Lagrangian stochastic trajectories and calculates the desired statistical characteristics. A second approach uses different approximations of the Lagrangian velocities of a set of particles. The first model of this type was introduced by G. Taylor [235] in 1921, and was then developed by many authors (see e. g., [91, 108, 133, 157, 214, 236]).

The first approach is rigorous, but it requires a lot of computer time. The second is much more effective, but its correct utilization needs special justifications. Both approaches are useful for modeling the pollutant scattering in turbulent flows, and especially for calculating the concentration fluctuations (see e. g., [144, 146, 214]). Most developed are the 1-particle Lagrangian models which allow the calculation of only the mean concentration. To find the concentration fluctuations, or, more generally, the covariance of the concentration, it is necessary to have 2-particle Lagrangian models. Unfortunately, such models are not well developed.

The motion of two particles can often be regarded as a motion of one particle plus the relative motion of the second particle with respect to the first one. Compared to the absolute diffusion of two particles, it is believed that the relative diffusion can be treated, due to the Kolmogorov similarity hypotheses, using a simpler and universal description. Therefore, models of relative diffusion attract much attention (see e. g., [44, 45, 108, 155, 236]). We mention also 2-particle Gaussian models [123, 214] and the model [20].

We give here two examples to motivate why the relative diffusion is of particular interest. Assume that we describe a pollutant transport in a homogeneous velocity field, and $c(x,t)$, the instantaneous concentration of the pollutant is also statistically homogeneous. Then $B_c(r,t)$, the covariance of the concentration

$$B_c(r,t) = \langle c(x,t)\, c(x+r,t)\rangle$$

is represented as (see e. g., [135])

$$B_c(r,t) = \int p(r,t \mid r_0) B_c(r_0,0)\, dr_0. \tag{3.1}$$

Here $p(r,t \mid r_0)$ is the probability density function (p. d. f.)

$$p(r,t \mid r_0) = \langle \delta(r - r_L(t,r_0))\rangle,$$

where $r_L(t,r_0)$ is the instantaneous separation (at the time instant t) between the particles initially separated by r_0. In the derivation of (3.1) it is supposed that the initial concentration field $c(x,0)$ is statistically independent of the velocity field.

Thus, to evaluate the covariance of concentration, it is necessary (and sufficient, in this case) to know the statistics of the relative motion of the two particles.

The second example comes from the Smolouchovsky equation governing the coagulation of particles in homogeneous turbulent flows [191], where the coagulation rates depend on the relative velocity between two colliding particles.

In this chapter we deal with the problem of the relative motion of two particles. We suggest different stochastic models (3D and quasi-1-dimensional) which are consistent in a sense with the Eulerian velocity field. Let us explain this point in more detail.

In statistical fluid mechanics [146] the Eulerian turbulent velocity field is considered as a 3D random field $U_E(x,t)$ depending on a spatial coordinate x and time t. By fluid particles we mean mathematical points moving with the local velocity of the flow. The Lagrangian trajectory $X(t,x_0)$, $t \geq 0$ of a fluid particle originating at position x_0 at time $t = 0$ is found from the equation of motion:

$$\frac{\partial X}{\partial t} = U_E(X,t) \tag{3.2}$$

subject to the initial condition

$$X(t=0,x_0) = x_0. \tag{3.3}$$

The Lagrangian velocity $V(t,x_0)$ is related to the Eulerian velocity $U_E(x,t)$ by

$$V(t,x_0) = U_E(X(t,x_0),t). \tag{3.4}$$

Equation (3.2) can be integrated numerically in time; (3.4) means that the instantaneous particle velocity is the same as the fluid velocity at the instantaneous particle position.

Each Eulerian quantity $f(x,t)$ can be related to the Lagrangian quantity by

$$F(x_0,t) = f(X(x_0,t),t), \quad x_0 \in R^3, \quad t \geq 0. \tag{3.5}$$

A Lagrangian description allows us to analyze directly the motion of material fluid elements. Unfortunately, the Lagrangian statistics are extremely difficult to measure accurately. Numerical solution of (3.2) based on different representations of the random field U_E involves nontrivial difficulties which must be overcome before we can have confidence in the accuracy of the result.

The second approach we mentioned above is based on direct approximations of the stochastic Lagrangian velocity $V(t,x_0)$, say $\hat{V}(t,x_0)$. The relevant approximate Lagrangian trajectory is then obtained from the ordinary stochastic differential equation

$$\frac{d\hat{X}}{dt} = \hat{V}(t,x_0) \tag{3.6}$$

with $\hat{X}(x_0,0) = x_0$. Note that this is generally a nonlinear equation since the random process $\hat{V}(t,x_0)$ itself depends on the random position $\hat{X}(t)$. Most frequently, models of the type (3.6) are written in the form of stochastic differential equations (SDE) of the Ito type. These equations are much easier to solve numerically than the general equations of fluid particle motion (3.2) in random fields. Numerical methods for solving SDE are well developed (see e. g., [233].

The models (3.2) and (3.6) define one and the same flow if $\hat{V}(t,x_0) = V(t,x_0)$ for all Lagrangian trajectories. It would be an ideal approximation if we could find $V(t,x_0)$. Of course, it is impossible to do even for the simplest flows. However, when constructing models of the type (3.6), it is reasonable to use some relations between the statistics of the Eulerian velocity field $U_E(X,t)$ and of the 6-dimensional Lagrangian field $Y(t,x_0) = (X(t,x_0,V(t,x_0))$. For instance, in the case of incompressible flows there is an integral relation between the probability density functions of $U_E(x,t)$ and $Y(t,x_0)$ (see [153]; the more general situation is treated in [154]).

In addition to this kind of consistency, we will use also some well-known properties of the stationary local isotropic incompressible turbulence such as isotropy and the Kolmogorov similarity hypotheses, which assume that there exists a parameter $\bar{\varepsilon}$ (called the mean rate of dissipation of kinetic energy) such that the statistical characteristics of the increments of the Eulerian velocity after normalizing by a quantity v_η become universal functions (for all kinds of fully developed turbulence) of a nondimensional time t/τ_η and nondimensional spatial distance r/η. Here η is Kolmogorov's spatial microscale which is related to the velocity microscale v_η and the time microscale τ_η by

$$\eta = (\nu^3/\bar{\varepsilon})^{1/4}, \quad v_\eta = (\nu\,\bar{\varepsilon})^{1/4}, \quad \tau_\eta = \eta/v_\eta,$$

where ν is the kinematic viscosity of the flow.

3.2 Criticism of 2-particle models

We start with a critical analysis of some known 2-particle stochastic models. Assumably, the first 2-particle model appeared in the early 60s, in [129, 152]. In [129], it was assumed that the random process in the 6-dimensional phase space (r, v) (r is the separation vector between the two particles, and v is the relative velocity) is Markovian. Using then the similarity and isotropy assumptions, it is possible to derive an equation governing the p. d. f. [17]. Novikov [152] assumed that the motion of two particles is governed by two independent Langevin equations:

$$\frac{dX^{(i)}(t)}{dt} = u^{(i)}, \tag{3.7}$$

$$du^{(i)} = -\frac{1}{T_L}u^{(i)}dt + \sigma_u\sqrt{2/T_L}\,dW^{(i)}, \quad i = 1, 2, 3. \tag{3.8}$$

Here T_L is the Lagrangian time scale, σ_u^2 is the variance of $u_1^{(i)}$, and $W^{(i)}$, $i = 1, 2, 3$ are independent standard Wiener processes. Subtracting the two equations yields

$$\frac{dr}{dt} = v, \quad dv = -\frac{1}{T_L}vdt + 2\sigma_u\sqrt{1/T_L}\,dW, \tag{3.9}$$

where $r = X^{(2)} - X^{(1)}$, $v = u^{(2)} - u^{(1)}$. Note that the p. d. f. of this equation coincides with the p. d. f. of the Markovian model [129].

By construction, this model correctly describes the motion of particles only for large separations, where the correlation between the velocities vanishes. A detailed analysis of this model is given in [17]. Models which take into account the statistical structure of the velocity in the inertial subrange were developed in [44, 45, 108, 155, 236]. Note that the 1-dimensional model [44] and its modification [236] have no adequate physical interpretation (see the introductory part in Section 3.3). However, quasi-1-dimensional models which describe the evolution of the distance between two fluid particles in 3D space have clear physical meaning. The motion of two particles in the framework of a quasi-1-dimensional model [45] is governed by the equation

$$\frac{d\Delta}{dt} = R_E^{1/2}(\Delta)\,U(t) \tag{3.10}$$

with the initial condition: $\Delta(0) = \Delta_0$. Here $\Delta = \Delta(t, \Delta_0)$ is the distance between the particles at the time instant t, and

$$R_E(\Delta) = \left(\frac{\Delta^2}{\Delta^2 + L^2}\right)^{1/3}, \tag{3.11}$$

L is the integral spatial scale of the Eulerian velocity field, and $U(t)$ is the Uhlenbeck–Ornstein process (UO):

$$dU = -\frac{1}{T_L} U dt + \sigma_U \sqrt{2/T_L}\, dW. \tag{3.12}$$

Obviously,

$$\frac{d\Delta}{dt}(t = 0,\ |\Delta_0|) = u_E(|\Delta_0|) \tag{3.13}$$

for all initial distances Δ_0, where $u_E(|\Delta_0|)$ is the Eulerian longitudinal relative velocity.

The coefficient $R_E^{1/2}(\Delta)$ in (3.10) was introduced to ensure that

$$\left\langle \left\{ \frac{d\Delta}{dt}(t = 0,\ |\Delta_0|) \right\}^2 \right\rangle = \langle u_E^2(|\Delta_0|) \rangle = \sigma_U^2\, R_E(|\Delta_0|). \tag{3.14}$$

Note that at large separations the model (3.9) behaves like the model (3.7) ($R_E(\Delta) \to 1$ as $\Delta \to \infty$), while in the inertial subrange the model should have reflected the Kolmogorov law, $R_E(\Delta) \sim \Delta^{2/3}$. However it appears [108] that the particles, starting at a distance from the inertial subrange, will collide with probability $1/2$ at a point after time of the order of $(\Delta_0^2/\bar{\varepsilon})^{1/3}$, which is much less than T_L. This makes the Durbin model questionable in practical usage. A criticism of Durbin's model was given in [46]. Thomson [236] has modified Durbin's model by adding an additional term $\frac{dR^{1/2}(\Delta)}{d\Delta}$. The same could be done with the quasi-1-dimensional model:

$$\frac{d\Delta}{dt} = \tilde{R}_E^{1/2}(\Delta)\, U(t), \tag{3.15}$$

$$dU = -\frac{U}{T_L}\, dt + \left[\frac{2}{T_L}\right]^{1/2} dW(t) + \frac{d\tilde{R}_E^{1/2}}{d\Delta}\, dt, \tag{3.16}$$

where

$$\tilde{R}_E(\Delta) = 2\sigma^2 \left(\frac{\Delta^2}{\Delta^2 + L^2}\right)^{1/3},$$

and σ^2 is the variance of a component of the velocity. The shortcoming of the Durbin model mentioned above seems to disappear in this model since the new term increases the divergence between the two particles as Δ tends to zero. However, we will see later on that this modified model contradicts the integral relation (3.31) which is a consequence of incompressibility. We will further modify this model to provide some improvement ensuring a relevant consistency.

An interesting approach was proposed by Novikov in [155] where the Lagrangian relative diffusion model consistent with some features of the inertial subrange was constructed. In the next section we will use this approach; therefore, let us give some details.

The general 3D model has the form

$$\begin{aligned} dv_i &= g_i(t, r, v)\, dt + d\mu_i, \\ dr_i &= v_i\, dt, \quad i = 1, 2, 3, \end{aligned} \tag{3.17}$$

where $d\mu_i$ stands for the random forcing, and $r(0) = r_0$, $v(0) = \xi$, ξ being a random vector with p. d. f

$$\langle \delta(v - \xi) \rangle = p_E(v|\, r_0).$$

Here $p_E(v|\, r_0)$ is the p. d. f. of the Eulerian relative velocity between two points separated by r_0. Then the Lagrangian probability density

$$p_L(t, v, r|r_0) = \langle \delta(v - v(t, r_0))\, \delta(r - r(t, r_0)) \rangle$$

for the relative velocity and distance is governed by the master equation [24]

$$\begin{aligned} &\frac{\partial p_L}{\partial t} + v_i \frac{\partial p_L}{\partial r_i} + \frac{\partial}{\partial v_i}(g_i\, p_L) \\ &\qquad = \int p_L(t, v - u|\, r_0)\, q(u|\, v - u, r)\, du - p_L(t, v|r_0)\, m_0(v, r), \end{aligned} \tag{3.18}$$

where

$$m_0(v, r) = \int q(u|\, v, r)\, du,$$

and $q(u|\, v, r)$ is the transition probability per unit time of the jump (from v to $v + u$). Here and all through the chapter we will use (if not specially noted) the summation convention taking a sum over the repeated subscript.

The Kramers–Moyal expansion of (3.18) gives (see e. g., [248])

$$\frac{\partial p_L}{\partial t} + v_i \frac{\partial p_L}{\partial r_i} + \frac{\partial}{\partial v_i}(a_i\, p_L) = \sum_{n=2}^{\infty} \frac{(-1)^n}{n!} \frac{\partial^n}{\partial v_{i_1} \dots \partial v_{i_n}} (\, m^{(n)}_{i_1,\dots,i_n}\, p_L), \tag{3.19}$$

where $a_i = g_i + m_i^{(1)}$, and

$$m^{(n)}_{i_1,\dots,i_n}(v, r) = \int u_{i_1} \dots u_{i_n}\, q(u|\, v, r)\, du.$$

In the diffusion approximation, when (3.17) is a stochastic differential equation of the Ito type

$$dv_i = g_i(t,r,v)\,dt + b_{ij}\,dW, \quad dr_i = v_i\,dt, \quad i = 1,2,3, \tag{3.20}$$

equation (3.19) reduces to

$$\frac{\partial p_L}{\partial t} + v_i \frac{\partial p_L}{\partial r_i} + \frac{\partial}{\partial v_i}(a_i\; p_L) = \frac{1}{2}\frac{\partial^2}{\partial v_i \partial v_j}(m^{(2)}_{i,j}\, p_L), \tag{3.21}$$

where the matrix $(b_{ij})^3_{i,j=1}$ is the square root of the matrix $(m^{(2)}_{ij})^3_{i,j=1}$.

Under the incompressibility hypothesis, the following relation between p_L and p_E, the Eulerian probability density function, can be derived [155]:

$$p_E(t,v|\,r) = \int p_L(t,v,r|\,r_0)\,dr_0. \tag{3.22}$$

Integrating (3.21) over r_0 and using (2.14) one gets the same equation for p_E:

$$\frac{\partial p_E}{\partial t} + v_i \frac{\partial p_E}{\partial r_i} + \frac{\partial}{\partial v_i}(a_i\; p_E) = \frac{1}{2}\frac{\partial^2}{\partial v_i \partial v_j}(m^{(2)}_{i,j}\, p_E). \tag{3.23}$$

From this it is not difficult to derive equations for the statistical moments.

Thus we conclude that this attempt to construct a model of type (3.17), specifying the three first moments of p_E, leads to some constraints on a_i and $m^{(2)}_{ij}$. This promising approach needs detailing of the strategy of finding the functions g_i and the statistical structure of the random forcing $d\mu_i$. We do not know any detailed model satisfying the constraints mentioned above. In the last section of the chapter we suggest (in the framework of diffusion approximation) a model which satisfies (3.23).

Note that if one wishes to include into a model important features of the turbulence, for instance, its structure in the inertial subrange, then it is necessary to give a formulation of the Lagrangian analog of these details. For example, in accordance with the well-known second Kolmogorov similarity hypothesis [146] there exists a universal stationary stochastic process $\xi_E(\tau)$ defined by

$$\frac{u_E(r,t)}{(\bar{\varepsilon}|r|)^{1/3}} = \xi_E(\tau), \tag{3.24}$$

$$\frac{t}{(r^2/\bar{\varepsilon})^{1/3}} = \tau, \tag{3.25}$$

where $u_E(r,t)$ is the relative longitudinal Eulerian velocity between two points whose separation is r, at the time instant t; τ is a nondimensional time.

In [108], an analog of (3.24, (3.25) in Lagrangian formulation was supposed:

$$\frac{1}{(\bar{\varepsilon}\Delta)^{1/3}}\frac{d\Delta}{dt} = \xi_L(\tau), \tag{3.26}$$

$$\frac{1}{(\Delta^2/\bar{\varepsilon})^{1/3}}\,dt = d\tau, \tag{3.27}$$

where Δ is the distance between the Lagrangian fluid particles, and $\xi_L(\tau)$ is assumed to be a universal stationary process.

Note that the model (3.26, 3.27)

$$\frac{d\Delta}{dt} = (\bar{\varepsilon}\Delta)^{1/3}\xi_L(\tau) \qquad (3.28)$$

looks like the Durbin model [45] in the inertial subrange. The difference lies in the temporal variable: in (3.9), the UO process depends on the physical time t while in (3.28) the stationary random process $\xi_L(\tau)$ is a function of a nondimensional time which is related to t through the local relation (3.27). This is a crucial feature of our model which overcomes the drawback mentioned above.

Note that in the model (3.26, 3.27) we are free to choose the stationary stochastic process $\xi_L(\tau)$. In Section 3.3 we give a special choice of this process which makes the model consistent with the Eulerian velocity statistics.

In conclusion we mention the Gaussian 2-particle models [123, 213, 214] and the model [20] adjusted to the viscous subrange. In the inertial subrange the velocity distribution is not Gaussian; therefore, the Gaussian models [123, 213] could be applied to describe the dispersion of particles by energy containing subranges. Finally, there are also approaches intermediate between the Lagrangian and Eulerian descriptions (see e. g., [91]) which are beyond the scope of this section.

3.3 The quasi-1-dimensional Lagrangian model of relative dispersion

Note that there are 1-dimensional models which deal with the description of one of the components of the separation vector $r(t)$ (e. g., [44, 214]); these models do not adequately describe important features of the real turbulence. As an example, note that there are events where the distance between particles is large (and hence the relative velocity should also be large) while a projection of the separation vector might nevertheless be small which implies that the 1-dimensional model predicts a small relative velocity.

We call a model of relative dispersion quasi-1-dimensional if it describes the evolution of the distance $\Delta(t)$ between two fluid particles in the space R^3.

Since the quasi-1-dimensional model describes the motion of two particles in 3D, it is reasonable to require that this model be consistent with the integral relation (3.22) in a sense given in the next subsection.

In this section we give an approach to make a choice of $\xi_L(\tau)$ in (3.26, 3.27) which ensures that an important integral relation between the Eulerian and Lagrangian statistics mentioned above is valid. In the next subsection we derive this relation which is a consequence (not a straightforward one) of the Novikov integral relation (3.22).

3.3.1 Quasi-1-dimensional analog of formula (2.14a)

Assumptions 3.1. *We assume that $U_E(x,t)$ is a stationary locally isotropic* 3D *incompressible velocity field* [146].

We use the following notations: $X(t,x_0)$ is the Lagrangian coordinate, and $V(t,x_0)$ is the Lagrangian velocity of a fluid particle at the time instant t, which started at the point x_0, i. e.,

$$V = \frac{\partial X}{\partial t} = U_E(X,t), \quad X(0) = x_0. \tag{3.29}$$

The relative Eulerian velocity is

$$u_E(r,t) = U_E(x+r,t) - U_E(x,t)$$

between two fixed points x and $x+r$. The distribution density function of u_E is

$$p_E(u|\, r) = \langle \delta(u - u_E(r,t)) \rangle. \tag{3.30}$$

The Lagrangian relative velocity is

$$v_L(t,r_0) = V(t,x_0+r_0) - V(t,x_0)$$

between two fluid particles which were separated initially by the vector r_0, and the p. d. f. of v_L:

$$p_L(t,v,r|\, r_0) = \langle \delta(v - v_L(t,r_0))\, \delta(r - r_L(t,r_0)) \rangle,$$

where $r_L(t,r_0) = X(t,x_0+r_0) - X(t,x_0)$ is the separation vector.

Remark 3.1. Note that the field $u_E(r,t)$ depends on x, while v_L depends on x_0. However, since U_E is locally isotropic, the densities p_E and p_L do not depend on x and x_0, respectively.

We will also use a longitudinal Eulerian velocity which is defined as the scalar product

$$u_E^{\|}(\rho,t) = (u_E(\rho\omega,t),\omega),$$

where $\rho = |r|$ is the distance between the two points, and $\omega = \frac{r}{|r|}$ is a unit directional vector. The p. d. f. of $u_E^{\|}(\rho,t)$ is denoted as

$$p_E^{\|}(u|\, \rho) = \langle \delta(u - u_E^{\|}(\rho,t)) \rangle.$$

The longitudinal Lagrangian velocity is defined by the following scalar product

$$v_L^{\|}(t,\Delta_0) = \frac{(v_L(t,\Delta_0\,\omega_0), r_L(t,\Delta_0\,\omega_0))}{|r_L(t_0,\Delta_0\,\omega_0)|},$$

and the distance between two fluid particles is

$$\Delta_L(t, \Delta_0) = |r_L(t, \Delta_0\, \omega_0)|,$$

where $\Delta_0 = |r_0|$, and $\omega_0 = r_0/|r_0|$.

The density of $v_L^{\|}$ is

$$p_L^{\|}(t, v, \Delta |\, \Delta_0) = \langle \delta(v - v_L^{\|}(t, \Delta_0))\, \delta(\Delta - \Delta_L(t, \Delta_0))\rangle.$$

By definition, the instantaneous velocities $u_E^{\|}$ and $v_L^{\|}$ depend on ω and ω_0, respectively, while the p. d. f.'s p_E and p_L are independent of these arguments.

We now present the following statement which is an analog of (3.22) in the quasi-1-dimensional case.

Proposition 3.1. *Under the* assumptions *given above, the following integral relation holds:*

$$\int_0^\infty \Delta_0^2\, p_L^{\|}(t, v, \Delta |\, \Delta_0)\, d\Delta_0 = \Delta^2 p_E^{\|}(v |\, \Delta). \tag{3.31}$$

Proof. Let us consider two Lagrangian trajectories $\bar{X}(t, x_0)$, $\bar{X}(t, x_0 + \Delta_0\, \omega_0)$ starting at x_0 and $x_0 + \Delta_0\, \omega_0$, respectively, where ω_0 is a unit isotropic random vector independent of the velocity field, and let us introduce a p. d. f. by

$$\tilde{p}_L(t, v, r |\, \Delta_0) = \langle \delta(v - \tilde{v}_L(t, \Delta_0))\, \delta(r - \tilde{r}_L(t, \Delta_0))\rangle, \tag{3.32}$$

where $\tilde{v}_L(t, \Delta_0)$ is the relative velocity, and $\tilde{r}_L(t, \Delta_0)$ is the separation between the two fluid particles. By the formula of conditional expectations, the function $\tilde{p}_L(t, v, r |\, \Delta_0)$ is related to $p_L(t, v, r |\, r_0)$ by

$$\tilde{p}_L(t, v, r |\, \Delta_0) = \int_\Omega p_L(t, v, r | \Delta_0\, \omega_0)\, d\Omega(\omega_0), \tag{3.33}$$

where $d\Omega(\omega_0) = (\sin\theta\; d\theta)(d\varphi/4\pi)$ is the normalized surface measure on the unit sphere Ω.

We introduce the density

$$f_L'(t, v, \Delta, \omega |\, \Delta_0) = \left\langle \delta(v - \tilde{v}_L(t, \Delta_0))\, \delta(\Delta - |\tilde{r}_L(t, \Delta_0)|)\, \delta\left(\omega - \frac{\tilde{r}_L(t, \Delta_0)}{|\tilde{r}_L(t, \Delta_0)|}\right)\right\rangle \tag{3.34}$$

and the conditional density

$$f_L(t, v, \Delta |\, \omega, \Delta_0) = \frac{f_L'(t, v, \Delta, \omega |\, \Delta_0)}{\int f_L'(t, v, \Delta, \omega |\, \Delta_0)\, dv\, d\Delta}. \tag{3.35}$$

The integral in (3.35) is equal to $1/4\pi$, since the integrand is a p. d. f. of the vector $\tilde{r}_L(t,\Delta_0)/|\tilde{r}_L(t,\Delta_0)|$, which is isotropic.

Simple geometric arguments show that

$$f_L'(t,v,\Delta,\omega|\,\Delta_0) = \Delta^2\,\tilde{p}_L(t,v,\Delta\omega|\,\Delta_0). \tag{3.36}$$

From (3.35) and (3.36) we get

$$f_L(t,v,\Delta|\,\omega,\Delta_0) = 4\pi\Delta^2\tilde{p}_L(t,v,\Delta\omega|\,\Delta_0). \tag{3.37}$$

By the definition,

$$p_L^{\|}(t,v_3,\Delta|\,\Delta_0) = \iint f_L(t,v,\Delta|\,e_3,\Delta_0)\,dv_1\,dv_2, \tag{3.38}$$

where $v = (v_1,v_2,v_3)$ and $e_3 = (0,0,1)$.

Multiplying the last equality by Δ_0^2 and integrating over Δ_0 we get, using (3.37) and (3.33),

$$\begin{aligned}
&\int_0^\infty \Delta_0^2\,p_L^{\|}(t,v_3,\Delta|\,\Delta_0)\,d\Delta_0\\
&= \int_0^\infty \Delta_0^2\,d\Delta_0 \int_{-\infty}^\infty dv_1 \int_{-\infty}^\infty dv_2\,4\pi\Delta^2\,\tilde{p}_L(t,v,\Delta e_3|\,\Delta_0)\\
&= \int_0^\infty \Delta_0^2\,d\Delta_0 \int_{-\infty}^\infty dv_1 \int_{-\infty}^\infty dv_2\,4\pi\Delta^2 \int_\Omega p_L(t,v,\Delta e_3|\,\Delta_0\omega_0)\,d\Omega_0(\omega_0)\\
&= \Delta^2 \int_{-\infty}^\infty dv_1 \int_{-\infty}^\infty dv_2 \left(4\pi \int_0^\infty \Delta_0^2\,d\Delta_0 \int_\Omega p_L(t,v,\Delta e_3|\,\Delta_0\omega_0)\,d\Omega_0\right)\\
&= \Delta^2 \int_{-\infty}^\infty dv_1 \int_{-\infty}^\infty dv_2 \left(\int_{R^3} p_L(t,v,\Delta e_3|\,r_0)\,dr_0\right)\\
&= \Delta^2 \int_{-\infty}^\infty dv_1 \int_{-\infty}^\infty dv_2\,p_E(v|\,\Delta e_3) = \Delta^2\,p_E^{\|}(v_3|\,\Delta).
\end{aligned} \tag{3.39}$$

This completes the proof. □

This result will be used in Section 3.3.3 when constructing the stochastic process $\xi_L(\tau)$ in (3.26).

3.3.2 Models with a finite-order consistency

Let us consider the following system of stochastic differential equations modeling the evolution of the distance $\Delta_L(t,\Delta_0)$ between two fluid particles in $3D$ space:

$$\frac{d\hat{\Delta}_L}{dt} = \hat{v}_L^{\|}, \tag{3.40}$$

$$d\hat{v}_L^{\|} = a(\hat{\Delta}_L,\hat{v}_L^{\|})dt + b(\hat{\Delta}_L,\hat{v}_L^{\|})\,dW(t). \tag{3.41}$$

We will use the notation $\hat{\Delta}_L(t, \Delta_0)$ and $\hat{v}_L^{\|}(t, \Delta_0)$ for the solution to (3.40, 3.41) satisfying the initial conditions

$$\hat{\Delta}_L(t = 0, \Delta_0) = \Delta_0, \quad \hat{v}_L^{\|}(t = 0, \Delta_0) = \xi_0, \tag{3.42}$$

where ξ_0 is a random variable with p. d. f. $\langle \delta(v - \xi_0) \rangle$ equal to $p_E^{\|}(v | \Delta_0)$.

Our general objective is to determine the coefficients in (3.41) in such a way that the density

$$\hat{p}_L^{\|}(t, v, \Delta | \Delta_0) = \langle \delta(v - \hat{v}_L^{\|}(t, \Delta_0)) \, \delta(\Delta - \hat{\Delta}_L^{\|}(t, \Delta_0)) \rangle$$

when substituting into (3.31) (replacing $p_L^{\|}$) will satisfy this integral relation. If so, then, as in the case of derivation of (3.23), we arrive at the following PDE:

$$v \frac{\partial [\Delta^2 p_E^{\|}(v | \Delta)]}{\partial \Delta} + \frac{\partial}{\partial v} [\Delta^2 a(\Delta, v) p_E^{\|}(v | \Delta)] = \frac{1}{2} \frac{\partial^2}{\partial v^2} \left(\Delta^2 b^2(\Delta, v) p_E^{\|}(v | \Delta) \right). \tag{3.43}$$

Let

$$\mu_n(\Delta) = \int v^n p_E^{\|}(v | \Delta) \, dv, \quad n = 0, 1, \dots.$$

It is obvious that

$$\mu_0 = 1, \; \mu_1 = 0, \; \mu_2(\Delta) = D_{LL}(\Delta), \tag{3.44}$$

and in the inertial subrange

$$\mu_2(\Delta) = C(\bar{\varepsilon}\Delta)^{2/3}, \quad \mu_3(\Delta) = -\frac{4}{5}\bar{\varepsilon}\Delta. \tag{3.45}$$

The function D_{LL} and the constant C are defined below in Section 3.5.3.

Multiplying (3.43) by v^n and integrating over v yields

$$\begin{aligned} \frac{d}{d\Delta}[\Delta^2 \mu_{n+1}(\Delta)] - n\Delta^2 \int_{-\infty}^{\infty} v^{n-1} a(\Delta, v) p_E^{\|}(v | \Delta) \, dv \\ = \frac{n(n-1)}{2} \Delta^2 \int_{-\infty}^{\infty} v^{n-2} b^2(\Delta, v) p_E^{\|}(v | \Delta) \, dv \end{aligned} \tag{3.46}$$

provided the functions

$$v^n a(\Delta, v) p_E^{\|}(v | \Delta), \; v^n \frac{\partial}{\partial v}[b^2(\Delta, v) p_E^{\|}(v | \Delta)], \; v^{n-1} b^2(\Delta, v) p_E^{\|}(v | \Delta)$$

tend to zero as $|v| \to \infty$.

We say that the model (3.40), (3.41) satisfies the *m-order consistency* condition if (3.46) is true for all $n \le m$.

Note that the ∞-order consistency implies that (3.43) is fulfilled.

Let us analyze the consistency property of the models (3.9) and (3.14). For simplicity, we use the notation Δ and v instead of $\hat{\Delta}$ and $\hat{v}_L$, respectively. Let us express the Durbin quasi-1-dimensional model (3.9) and the Thomson modified model (3.14) in the general form

$$\frac{d\Delta}{dt} = R^{1/2}(\Delta)U(t),$$
$$dU = \left[-\frac{U}{T_L} + \gamma\frac{dR^{1/2}}{d\Delta}\right]dt + \left(\frac{2}{T_L}\right)^{1/2} dW(t). \tag{3.47}$$

For the sake of simplicity we use here the notation $R(\Delta) = \tilde{R}_E(\Delta)$; γ is a dimensionless parameter.

From the Ito formula we arrive at the standard form of the type (3.40), (3.41) with

$$a(\Delta, v) = \frac{v^2}{2R(\Delta)}\frac{dR(\Delta)}{d\Delta} - \frac{v}{T_L} + \frac{\gamma}{2}\frac{dR(\Delta)}{d\Delta}, \quad b(\Delta, v) = \left[\frac{2R(\Delta)}{T_L}\right]^{1/2}. \tag{3.48}$$

In the inertial subrange $R(\Delta) \sim \Delta^{2/3}$, and we get from (3.45), (3.48), and (3.46) with $n = 1$ that the first-order consistency condition is satisfied only if $\gamma = 7$. Hence the Durbin model ($\gamma = 0$) and the Thomson modification ($\gamma = 1$) do not satisfy the first-order consistency condition. Note that there is no value of γ providing the second-order consistency.

We now suggest a model for the inertial subrange with a second-order consistency:

$$\frac{d\Delta}{dt} = R^{1/2}(\Delta)\xi(t),$$
$$d\xi = \left[-\frac{\xi}{\tau(\Delta)} + \gamma\frac{dR^{1/2}(\Delta)}{d\Delta}\right]dt + \alpha\left(\frac{2}{\tau(\Delta)}\right)^{1/2} dW(t), \tag{3.49}$$

where $\tau(\Delta) = \beta(\Delta^2/\bar{\varepsilon})^{1/3}$, and $R(\Delta) = C(\bar{\varepsilon}\Delta)^{2/3}$; α, β and γ are nondimensional constants.

From (3.49) we get, using the Ito formula, the standard form (3.41) with

$$a(\Delta, v) = \frac{v^2}{3\Delta} - \frac{\bar{\varepsilon}^{1/3}}{\beta\Delta^{2/3}}\, v + \frac{C\gamma\bar{\varepsilon}^{2/3}}{3\Delta^{1/3}}, \quad b(\Delta, v) = \alpha[2C\bar{\varepsilon}/\beta]^{1/2}. \tag{3.50}$$

If we suppose that $p_E(v|\ \Delta)$ satisfies the conditions at the infinity mentioned above then the zero-order consistency condition is fulfilled automatically. To provide the first-order and second-order consistency we found the constraints

$$\gamma = 7, \ \alpha^2 = 1 - \frac{14\beta}{15C}. \tag{3.51}$$

Here we have one free parameter β satisfying the inequality $0 < \beta < 15C/14$.

Note that the last model reflects properly the well known Kolmogorov second similarity hypothesis (compare the functions b in (3.50) and (3.48)).

When $\Delta \gg L$, then the motion of two particles can be considered as independent; hence, it is reasonable to assume that, due to isotropy,

$$\left\langle \left(\frac{d\Delta}{dt} \right)^2 \right\rangle = 2\sigma^2, \tag{3.52}$$

where σ^2 is the variance of a component of the Eulerian velocity. Note that the Durbin model and the Thomson modification both possess this property. It is not difficult to generalize the model (3.49) in such a way that (3.52) will be satisfied. Indeed, let us take the constants α, β and γ as mentioned above (see (3.23)), and let us choose $\tau(\Delta)$ as

$$\tau(\Delta) = \beta \left[\frac{(L\Delta)^2}{(\Delta^2 + L^2)\bar{\varepsilon}} \right]^{1/3}. \tag{3.53}$$

Define a function

$$\phi(x) = 1 + \frac{1-\alpha}{\alpha}(1 - \exp(-x^2)), \quad x \geq 0.$$

The modified model

$$\begin{aligned} \frac{d\Delta}{dt} &= R^{1/2}(\Delta)\xi(t), \\ d\xi &= \left[-\frac{\xi}{\tau(\Delta)} + \gamma \frac{dR^{1/2}}{d\Delta} \right] dt + \alpha\phi\left(\frac{\Delta}{L}\right) \left(\frac{2}{\tau(\Delta)} \right)^{1/2} dW(t), \end{aligned} \tag{3.54}$$

satisfies the desired condition (3.52) if in (3.53) the relation

$$\bar{\varepsilon} = \left(\frac{2\sigma^2}{C} \right)^{3/2} \frac{1}{L}$$

is taken into account.

In the next subsection we suggest a model with an ∞-order consistency.

3.3.3 Explicit form of the model (3.26, 3.27)

We will work here with the model

$$\frac{d\hat{\Delta}_L}{dt} = (\bar{\varepsilon}\hat{\Delta}_L)^{1/3}\xi_L(\tau), \tag{3.55}$$

$$\frac{d\tau}{dt} = \left(\frac{\bar{\varepsilon}}{\hat{\Delta}_L^2} \right)^{1/3}, \tag{3.56}$$

$$d\xi_L(\tau) = a(\xi_L)\, d\tau + \sigma(\xi_L)\, dW(\tau). \tag{3.57}$$

Here $W(\tau)$ is the standard Wiener process, and a, σ are some functions to be defined. Note that we use here the notation $\hat{\Delta}_L$ to distinguish between the model distance $\hat{\Delta}_L$ and the true distance Δ_L.

To complete the construction of the model it remains to give the explicit form of a and σ. It is natural to postulate that the p. d. f. $\hat{p}_L(t, v, \Delta|\, \Delta_0)$ of the vector

$$\left(\frac{d\hat{\Delta}_L(t, \Delta_0)}{dt},\ \hat{\Delta}_L(t, \Delta_0)\right)$$

satisfies, as the true density $p_L(t, v, \Delta|\, \Delta_0)$ does, the equality

$$\int_0^\infty \Delta_0^2\, \hat{p}_L^{\|}(t, v, \Delta|\, \Delta_0)\, d\Delta_0 = \Delta^2 p_E^{\|}(v|\, \Delta). \tag{3.58}$$

Thus we can use the information about the Eulerian velocity field, namely, we suppose that $p_E(v|\, \Delta)$ is given. Starting from (3.55)–(3.58), we derive the desired coefficients a and b.

First, we rewrite the system, using a random change of variables. Let $\tilde{W}(t)$ be a standard Wiener process, and let $\tilde{\xi}_L(t) = \xi_L(\tau(t))$. Then, using the relation

$$dW(\tau) = \sqrt{\frac{d\tau}{dt}}\, d\tilde{W}(t),$$

from (3.55)–(3.58) we get

$$\frac{d\hat{\Delta}_L}{dt} = (\bar{\varepsilon}\hat{\Delta}_L)^{1/3}\tilde{\xi}_L(t), \tag{3.59}$$

$$d\tilde{\xi}_L = \left(\frac{\bar{\varepsilon}}{\hat{\Delta}_L^2}\right)^{1/3} a(\tilde{\xi}_L)\, dt + \left(\frac{\bar{\varepsilon}}{\hat{\Delta}_L^2}\right)^{1/6} \sigma(\tilde{\xi}_L)\, d\tilde{W}(t). \tag{3.60}$$

We introduce a p. d. f.:

$$\hat{f}_L(t, \xi, \Delta|\, \Delta_0) = \langle \delta(\xi - \tilde{\xi}_L(t, \Delta_0))\, \delta(\Delta - \hat{\Delta}_L(t, \Delta_0))\rangle, \tag{3.61}$$

where the pair $\tilde{\xi}_L(t, \Delta_0)$, $\tilde{\Delta}_L(t, \Delta_0)$ is the solution to (3.59), (3.60) with the initial conditions

$$\hat{\Delta}_L(t = 0, \Delta_0) = \Delta_0, \quad \tilde{\xi}_L(t = 0, \Delta_0) = \xi_E,$$

and ξ_E is a random variable with the p. d. f.:

$$f_E(\xi) = (\bar{\varepsilon}\Delta_0)^{1/3}\, p_E^{\|}\left((\bar{\varepsilon}\Delta_0)^{1/3}\xi|\, \Delta_0\right), \quad -\infty < \xi < \infty. \tag{3.62}$$

Note that f_E does not depend on Δ_0 in the inertial subrange, in accordance with the second Kolmogorov hypothesis.

The function (3.61) satisfies the Kolmogorov–Fokker–Planck equation

$$\frac{\partial \hat{f}_L}{\partial t} + \frac{\partial}{\partial \Delta}\left[(\bar{\varepsilon}\Delta)^{1/3}\xi \hat{f}_L\right] + \frac{\partial}{\partial \xi}\left[\left(\frac{\bar{\varepsilon}}{\Delta^2}\right)^{1/3} a(\xi)\hat{f}_L\right] = \frac{1}{2}\left(\frac{\bar{\varepsilon}}{\Delta^2}\right)^{1/3}\frac{\partial^2}{\partial \xi^2}\left[\sigma^2(\xi)\hat{f}_L\right], \tag{3.63}$$

with $\hat{f}_L(t = 0, \xi, \Delta|\ \Delta_0) = \delta(\Delta - \Delta_0) f_E(\xi)$. We also assume that the following condition is satisfied:

$$\left[\frac{1}{2}\frac{\partial}{\partial \xi}(\sigma^2 \hat{f}_L) - a\hat{f}_L\right]\Bigg|_{\xi=\pm\infty} = 0 \tag{3.64}$$

and that the problem (3.63), (3.64) has a unique solution. The equality (3.58) can be rewritten in $\hat{f}_L$-notation, using (3.62):

$$\int_0^\infty \Delta_0^2\, \hat{f}_L(t, \xi, \Delta|\ \Delta_0)\, d\Delta_0 = \Delta^2 f_E(\xi). \tag{3.65}$$

Multiplying (3.63) by Δ_0^2 and integrating over Δ_0 we get

$$\frac{\partial}{\partial \Delta}\left[\Delta^2(\bar{\varepsilon}\Delta)^{1/3}\xi f_E(\xi)\right] + \frac{\partial}{\partial \xi}\left[\Delta^2\left(\frac{\bar{\varepsilon}}{\Delta^2}\right)^{1/3} a(\xi) f_E(\xi)\right] = \frac{\Delta^2}{2}\left(\frac{\bar{\varepsilon}}{\Delta^2}\right)^{1/3}\frac{\partial^2}{\partial \xi^2}[\sigma^2(\xi) f_E(\xi)],$$

which can be simplified to

$$\frac{1}{2}\frac{d^2}{d\xi^2}[\sigma^2(\xi) f_E(\xi)] = \frac{d}{d\xi}[a(\xi) f_E(\xi)] + \frac{7}{3}\xi f_E(\xi). \tag{3.66}$$

Let

$$F(\xi) = \int_{-\infty}^{\xi} \eta f_E(\eta)\, d\eta, \tag{3.67}$$

and let

$$\Phi(\xi) = \frac{1}{2}\frac{d}{d\xi}(\sigma^2 f_E(\xi)) - a(\xi) f_E(\xi). \tag{3.68}$$

Integrating (3.66) we get

$$\Phi(\xi) = \Phi(-\infty) + \frac{7}{3}F(\xi). \tag{3.69}$$

The conditions (3.64) imply that $\Phi(-\infty) = 0$; hence we come to the system

$$\frac{1}{2}\frac{d}{d\xi}[\sigma^2(\xi) f_E(\xi)] - a(\xi) f_E(\xi)] = \frac{7}{3}F(\xi). \tag{3.70}$$

We now choose a function ϕ and make the following assumptions:

(a) The SDE

$$d\eta(\theta) = \phi(\eta)\, d\theta + dW(\theta) \tag{3.71}$$

has a unique solution defined for all the time instances, and

(b)

$$I_f \equiv \int_0^\infty |F(t)| \exp\left\{-2\int_0^t \phi(\tau)\, d\tau\right\} dt < \infty. \tag{3.72}$$

We will also assume that

$$a(\xi) = \phi(\xi)\sigma^2(\xi). \tag{3.73}$$

Let

$$\chi(\xi) = \frac{1}{2}\sigma^2(\xi) f_E(\xi). \tag{3.74}$$

Then from (3.70), (3.73), and (3.74) we get

$$\frac{d\chi}{d\xi} = 2\phi(\xi)\chi(\xi) + \frac{7}{3}F(\xi). \tag{3.75}$$

We seek a positive solution to (3.75), since, from (3.74), we see that $\chi \geq 0$. We assume that

$$\chi(0) = \chi_0 > \frac{7}{3}I_f. \tag{3.76}$$

The explicit solution to (3.75) is

$$\chi(\xi) = \exp\left\{2\int_0^\xi \phi(\eta)\, d\eta\right\}\left[\chi_0 + \frac{7}{3}\int_0^\xi d\eta\, F(\eta) \exp\left\{-2\int_0^\eta \phi(\tau)\, d\tau\right\}\right], \tag{3.77}$$

which is positive, in view of (3.76).

Proposition 3.2. *Under the assumptions* (a), (b), (3.76) *and* (3.73), *the solution to* (3.57) *with*

$$a(\xi) = 2\phi(\xi)\chi(\xi)/f_E(\xi) \tag{3.78}$$

and

$$\sigma(\xi) = \{2\chi(\xi)/f_E(\xi)\}^{1/2}, \tag{3.79}$$

where $\chi(\xi)$ *is given by* (3.77), *is unique for each initial condition* $\xi_L(0) = \xi_0$ *and is defined for all the time instances* τ. *The solution to* (3.57) *with these coefficients is ergodic.*

Proof. We use the following random transformation of time:

$$\frac{d\theta}{d\tau} = \frac{f_E(\xi(\tau))}{2\chi(\xi(\tau))}, \tag{3.80}$$

and introduce a function $\tilde{\xi}(\theta) = \xi(\tau(\theta))$. From (3.57) we get by the Ito formula that

$$d\tilde{\xi}(\theta) = \frac{2a(\tilde{\xi})\chi(\tilde{\xi})}{f_E(\tilde{\xi})}\,d\theta + \sigma(\tilde{\xi})\left[\frac{2\chi(\tilde{\xi})}{f_E(\tilde{\xi})}\right]^{1/2} d\tilde{W}(\theta), \tag{3.81}$$

where $\tilde{W}(\theta)$ is a standard Wiener process. Substituting the coefficients (3.78), (3.79) into (3.81) we come to (3.71) with $\eta = \tilde{\xi}$. Hence the random transformation (3.80) leads to the transformation of (3.57) to (3.71); therefore, the solution to (3.57) exists for all the time instances and is unique.

To prove that the solution to (3.57) is ergodic, it is sufficient to verify that the following condition holds [71]:

$$I = \int_{-\infty}^{\infty} \frac{1}{\sigma^2(\xi)} \exp\left\{\int_0^{\xi} \frac{2a(\tau)}{\sigma^2(\tau)}\,d\tau\right\} d\xi < \infty. \tag{3.82}$$

Substituting the coefficients a and σ from (3.78, 3.79) into (3.82) we get

$$\begin{aligned} I &= \int_{-\infty}^{\infty} \frac{f_E(\xi)}{2\chi(\xi)} \exp\left\{2\int_0^{\xi} \phi(\tau)\,d\tau\right\} d\xi \\ &\leq \int_{-\infty}^{\infty} \frac{f_E(\xi)\,d\xi}{\chi_0 - \frac{7}{3}\int_0^{\xi} d\eta |F(\eta)| \exp\left\{-2\int_0^{\eta} \phi(\tau)\,d\tau\right\}} \\ &\leq \frac{1}{c_0}\int_{-\infty}^{\infty} f_E(\xi)\,d\xi = \frac{1}{c_0}, \end{aligned} \tag{3.83}$$

where $c_0 = \chi_0 - 7I_f/3$. □

Remark 3.2. By construction of the model (3.55)–(3.57) it is clear that it was actually supposed that the inertial subrange expands from 0 to ∞. In realistic models, however, this range is finite. It is not difficult to improve the model (3.55)–(3.57) by writing it in a more general form:

$$\frac{d\hat{\Delta}_L}{dt} = V(\hat{\Delta}_L, t)\xi_L, \quad \frac{d\tau}{dt} = \frac{1}{T(\hat{\Delta}_L, t)},$$
$$d\xi_L(\tau) = a(\xi_L)\,d\tau + \sigma(\xi_L)\,dW(\tau),$$

where $V(\Delta, t)$ and $T(\Delta, t)$ are the characteristic velocity and the life time of the turbulent eddies of size Δ, respectively. Note that the time t appears here since the flow, in general, is not stationary.

3.3.4 Example

We present an example of the model (3.57) with

$$f_E(\xi) = \frac{1}{\sqrt{2\pi\sigma_1^2}} \exp\{-\xi^2/2\sigma_1^2\}.$$

In this case, from (3.67) we get

$$F(\xi) = \int_{-\infty}^{\xi} \eta f_E(\eta)\, d\eta = -\frac{\sigma_1}{\sqrt{2\pi}} \exp\{-\xi^2/2\sigma_1^2\}.$$

For ϕ, we choose the function

$$\phi(\xi) = -\lambda\xi, \quad -\infty < \xi < \infty,$$

where λ is a positive parameter, satisfying the condition

$$\lambda < \frac{1}{2\sigma_1^2}, \tag{3.84}$$

and

$$\chi_0 > \frac{7}{6}\sigma_0\sigma_1, \tag{3.85}$$

where σ_0 is determined from

$$\frac{1}{2\sigma_0^2} = \frac{1}{2\sigma_1^2} - \lambda. \tag{3.86}$$

Assumptions (a) and (b) are satisfied. Indeed, assumption (a) is straightforward, since (3.71) is in our case a Langevin equation which has a unique solution for all the time instances. To the assumption (b):

$$I_f = \int_0^{\infty} \frac{1}{\sqrt{2\pi}} \exp\left\{-\frac{t^2}{2\sigma_1^2}\right\} \exp(\lambda t^2)\, dt = \frac{\sigma_1\sigma_0}{2} < \infty$$

in view of (3.84) and (3.86). The condition (3.76) is equivalent to (3.85).

It remains to give the explicit form of the coefficients a and σ. The function χ is

$$\begin{aligned}\chi(\xi) &= \exp(-\lambda\xi^2)\left[\chi_0 - \frac{7}{3}\sigma_0\sigma_1 \int_0^{\xi} \frac{\exp(-t^2/2\sigma_0^2)}{\sqrt{2\pi}\sigma_0}\, dt\right] \\ &= \exp(-\lambda\xi^2)\left[\chi_0 - \frac{7}{3}\sigma_0\sigma_1 erf\left(\frac{\xi}{\sigma_0}\right)\right],\end{aligned}$$

where

$$\mathrm{erf}(x) = \int_0^{x} \frac{\exp(-t^2/2)}{\sqrt{2\pi}}\, dt.$$

Thus,

$$\sigma^2(\xi) = \frac{2\chi(\xi)}{f_E(\xi)} = 2\sqrt{2\pi}\sigma_1 \exp[\xi^2/2\sigma_0^2]\Big[\chi_0 - \frac{7}{3}\sigma_0\sigma_1 \mathrm{erf}\Big(\frac{\xi}{\xi_0}\Big)\Big],$$

and $a(\xi) = \phi(\xi)\sigma^2(\xi)$.

Note that we have here three parameters of the model: λ, σ_1, and χ_0. It is natural to take $\sigma_1^2 = C$, the constant in the Kolmogorov law of two thirds: $D_{LL}(r) = C(\bar{\varepsilon})^{2/3}r^{2/3}$, $(C \approx 1.9)$; λ is an arbitrary constant in the interval $(0, 1/2C)$, and χ_0 is an arbitrary constant satisfying (3.85).

Another practically interesting example presents a superposition of two Gaussian probability densities:

$$f_E(\xi) = p\frac{1}{\sqrt{2\pi}\sigma_1}\exp\left\{-\frac{(\xi - a_1)^2}{2\sigma_1^2}\right\} + q\frac{1}{\sqrt{2\pi}\sigma_2}\exp\left\{-\frac{(\xi - a_2)^2}{2\sigma_2^2}\right\}, \quad (3.87)$$

with $p \geq 0,\ q \geq 0,\ p + q = 1$.

We come to a series of constraints which follow from the known relations in the inertial subrange [146]:

$$\int u\ p_E(u|\ \Delta)\,du = 0, \qquad \int u^2\ p_E(u|\ \Delta)\,du = C(\bar{\varepsilon}\Delta)^{2/3},$$

$$\int u^3\ p_E(u|\ \Delta)\,du = -\frac{4}{5}\bar{\varepsilon}\Delta, \qquad \int u^4\ p_E(u|\ \Delta)\,du = (E + 3)\{C(\bar{\varepsilon}\Delta)^{2/3}\}^2,$$

where E is the excess of p_E.

In the inertial subrange the following relation is true (see (3.62)):

$$p_E^{\|}(u|\ \Delta) = \frac{1}{(\bar{\varepsilon}\Delta)^{1/3}} f_E\left(\frac{u}{(\bar{\varepsilon}\Delta)^{1/3}}\right).$$

Hence, the constraints are

$$\int \xi\ f_E(\xi)\,d\xi = 0, \qquad \int \xi^2\ f_E(\xi)\,d\xi = C,$$

$$\int \xi^3\ f_E(\xi)\,d\xi = -\frac{4}{5}, \qquad \int \xi^4\ f_E(\xi)\,d\xi = (E + 3)C^2,$$

which implies

$$p\,a_1 + q\,a_2 = 0, \qquad p\sigma_1^2 + q\sigma_2^2 = C,$$

$$p\,[3a_1\sigma_1^2 + a_1^3\,] + q\,[3a_2\sigma_2^2 + a_2^3\,] = -\frac{4}{5},$$

$$p\,[3\sigma_1^4 + 4a_1^2\sigma_1^2 + a_1^4\,] + q\,[3\sigma_2^4 + 4\sigma_2^2 a_2^2 + a_2^4\,] = (E + 3)C^2.$$

Thus, from 5 parameters, $p,\ a_1,\ , a_2\ , \sigma_1,\ \sigma_2$, we have one free parameter, e. g., p.

Evaluation of $F(\xi)$ gives

$$F(\xi) = \int_{-\infty}^{\xi} t f_E(t)\,dt$$
$$= p\,\mathrm{erf}\left\{\frac{\xi - a_1}{\sigma_1}\right\} + q\,\mathrm{erf}\left\{\frac{\xi - a_2}{\sigma_2}\right\} - \frac{p\sigma_1}{\sqrt{2\pi}}\exp\left\{-\frac{(\xi - a_1)^2}{2\sigma_1^2}\right\}$$
$$- \frac{q\sigma_2}{\sqrt{2\pi}}\exp\left\{-\frac{(\xi - a_2)^2}{2\sigma_2^2}\right\}. \tag{3.88}$$

Let

$$\phi(\xi) = -\lambda\xi, \quad 0 < \lambda < \lambda_0 = \min\left\{\frac{1}{2\sigma_1^2}, \frac{1}{2\sigma_2^2}\right\}.$$

As in the previous example, assumption (a) is obviously satisfied. To assumption (b): using the known inequality

$$\int_x^{\infty} \exp(-t^2/2)\,dt \le \frac{1}{x}\exp(-x^2/2).$$

we get

$$|F(\xi)| \le \left(1 + \frac{1}{|\xi - a_1|}\right)\frac{p\sigma_1}{\sqrt{2\pi}}\exp\left\{-\frac{(\xi - a_1)^2}{2\sigma_1^2}\right\}$$
$$+ \left(1 + \frac{1}{|\xi - a_2|}\right)\frac{q\sigma_2}{\sqrt{2\pi}}\exp\left\{-\frac{(\xi - a_2)^2}{2\sigma_2^2}\right\}.$$

This yields

$$I_f = \int_0^{\infty} |F(\xi)|\exp(\lambda\xi^2)\,d\xi < \infty.$$

Thus, all the assumptions are satisfied, and we can find, numerically, from (3.88) and (3.77) the function χ; next from (3.78), (3.79) we find the functions $a(\xi)$, $\sigma(\xi)$.

3.4 A 3-dimensional model of relative dispersion

In realistic problems of turbulent diffusion, one needs 3D models of relative dispersion. An attempt to construct such models was made in [21]. It seems, however, that there are no 3D models providing a realistic description of a relative dispersion.

We will now present a 3D analog of the Durbin model. We recall that a stationary incompressible locally isotropic turbulence is considered.

Let r be a 3D separation vector (for simplicity, we omit here the "Lagrangian subscript" L). We construct a model in the following general form:

$$\frac{dr}{dt} = A(r)\,U(t), \quad r(0) = r_0, \tag{3.89}$$

where $U(t) = (U_1(t), U_2(t), U_3(t))$, U_i, $i = 1,2,3$ are independent UO processes defined in (3.11), and $A(r)$ is an isotropic tensor:

$$(A(r))_{ij} = f(|r|)\frac{r_i r_j}{|r|^2} + g(|r|)\delta_{ij}, \quad i,j = 1,2,3. \tag{3.90}$$

Here $f(|r|)$ and $g(|r|)$ are nondimensional functions.

Since

$$\frac{dr}{dt}(t = 0; r_0) = u_E(r_0, t = 0),$$

it is in the spirit of Durbin's work [10] to require that

$$\begin{aligned} D_{ij}(r_0) &\equiv \left\langle \left(u_E(r_0, t = 0)\, u_E^T(r_0, t = 0)\right)_{ij} \right\rangle \\ &= \left\langle \left\{ \frac{dr}{dt}(0, r_0) \left(\frac{dr}{dt}(0, r_0) \right)^T \right\}_{ij} \right\rangle \\ &= \left\langle \left\{ A(r_0)U(0)[A(r_0)U(0)]^T \right\}_{ij} \right\rangle \\ &= \sigma^2 \left[(f^2 + 2fg)\frac{r_i r_j}{|r_0|^2} + g^2\delta_{ij} \right]. \end{aligned}$$

Using the well-known relation [146]

$$D_{ij}(r_0) = [D_{LL}(|r_0|) - D_{NN}(|r_0|)]\frac{r_{0i} r_{0j}}{|r_0|^2} + D_{NN}(|r_0|)\delta_{ij}$$

we find

$$g(\Delta) = \frac{\sqrt{D_{NN}(\Delta)}}{\sigma}, \quad f(\Delta) = \frac{\sqrt{D_{LL}(\Delta)} - \sqrt{D_{NN}(\Delta)}}{\sigma}.$$

Thus we arrive at the model

$$\frac{dr_i}{dt} = \left\{ (\sqrt{D_{LL}(|r|)} - \sqrt{D_{NN}(|r|)})\frac{r_i r_j}{|r|^2} + \sqrt{D_{NN}(|r|)}\delta_{ij} \right\} \frac{1}{\sigma} U_j(t), \tag{3.91}$$

$$dU_j(t) = -\frac{1}{T_L}\, U_j(t)\, dt + \sigma\sqrt{\frac{2}{T_L}}\, dW_j(t) \qquad j = 1,2,3. \tag{3.92}$$

We recall that in (3.91), a summation over the subscript j is assumed. Through some algebra we can see that this model does not satisfy the integral relation (3.22). However, the model (3.91), (3.32) can be generalized and improved, as described in Section 3.3.3. To do this, we introduce a random time transformation, and rewrite (3.91), (3.32) in the form:

$$\begin{aligned} \frac{dr_i}{dt} &= \left\{ \left(\sqrt{D_{LL}(|r|)} - \sqrt{D_{NN}(|r|)} \right) \frac{r_i r_j}{|r|^2} + \sqrt{D_{NN}(|r|)}\delta_{ij} \right\} \xi_j(\tau), \\ \frac{d\tau}{dt} &= \left(\frac{\bar{\varepsilon}}{|r|^2} \right)^{1/3}, \end{aligned} \tag{3.93}$$

where $\xi(\tau) = (\xi_1(\tau), \xi_2(\tau), \xi_3(\tau))$ is a stationary isotropic stochastic process. It is convenient to define it as a solution to the system of SDE:

$$d\xi_i(\tau) = \alpha(|\xi|)\xi_i \, d\tau + \left[\beta(|\xi|)\frac{\xi_i \xi_j}{|\xi|^2} + \gamma(|\xi|)\delta_{ij} \right] dW_j(\tau), \tag{3.94}$$

where α, β, γ are nondimensional functions of $|\xi|$. We assume that these coefficients are chosen so that there exists a stationary solution to (3.94). It is convenient to rewrite the system (3.93), (3.94) in the form

$$d\tilde{\xi}_i = \left(\frac{\bar{\varepsilon}}{r^2}\right)^{1/3} \alpha(|\tilde{\xi}|)\tilde{\xi}_i \, dt + \left(\frac{\bar{\varepsilon}}{r^2}\right)^{1/6} \left[\beta(|\tilde{\xi}|)\frac{\tilde{\xi}_i \tilde{\xi}_j}{|\tilde{\xi}|^2} + \gamma(|\tilde{\xi}|)\delta_{ij} \right] d\tilde{W}_j(t), \tag{3.95}$$

$$dr_i = \left\{ \left(\sqrt{D_{LL}(|r|)} - \sqrt{D_{NN}(|r|)} \right) \frac{r_i r_j}{|r|^2} + \sqrt{D_{NN}(|r|)}\delta_{ij} \right\} \tilde{\xi}_j(t), \tag{3.96}$$

where $\tilde{\xi}_i(t) = \xi(\tau(t))$, $\tilde{W}_i(t)$ are standard Wiener processes, $i = 1, 2, 3$. The authors plan to construct such a model consistent with the integral relation (3.22).

3.5 Lagrangian models consistent with the Eulerian statistics

In this section we will examine the general model (3.17), namely, with the diffusion approximation (3.20). Our goal is to present a way of deriving the coefficients in (3.20) which ensures that the integral relation (3.22) holds.

We recall the assumptions: the Eulerian turbulent velocity field is assumed to be stationary, locally isotropic, and incompressible.

3.5.1 Diffusion approximation

The relative motion of two fluid particles in the framework of a diffusion approximation is governed by

$$d\hat{v}_i = a_i(t, \hat{r}, \hat{v}) \, dt + b_{ij}(t, \hat{r}, \hat{v}) \, dW_j(t), \tag{3.97}$$

where W_j, $j = 1, 2, 3$ are standard independent Wiener processes. The assumptions formulated above imply that the coefficients a_i and b_{ij} should be chosen as isotropic tensors which do not depend on time. It means that

$$a_i(t, r, v) = \alpha_0 \frac{r_i}{|r|} + \alpha_1 \frac{v_i}{|v|}, \tag{3.98}$$

and

$$b_{ij}(t,r,v) = \beta_1 \frac{r_i r_j}{|r|^2} + \beta_2 \frac{v_i v_j}{|v|^2} + \beta_3 \frac{r_i v_j + v_i r_j}{|r||v|} + \beta_0 \delta_{ij}, \qquad (3.99)$$

where $i, j = 1, 2, 3$ and α_i, β_i are functions of $|r|$, $|v|$, and $\mu = (r, v)/(|r||v|)$.

Let $\hat{r}(t, r_0)$, $\hat{v}(t, r_0)$ be a solution to (3.97) with the initial condition

$$\hat{r}(t = 0, r_0) = r_0, \quad \hat{v}(t = 0, r_0) = \xi_0, \qquad (3.100)$$

where ξ_0 is a random vector in 3D, with p. d. f. $\langle \delta(v - \xi_0) \rangle$ taken equal to $p_E(v|\, r_0)$, where $p_E(v|\, r)$ is the p. d. f. of the original Eulerian relative velocity. Then the p. d. f.

$$\hat{p}_L(t, v, r|\, r_0) = \langle \delta(v - \hat{v}(t, r_0))\, \delta(r - \hat{r}(t, r_0)) \rangle \qquad (3.101)$$

satisfies the Kolmogorov–Fokker–Planck equation

$$\frac{\partial \hat{p}_L}{\partial t} + v_i \frac{\partial \hat{p}_L}{\partial r_i} + \frac{\partial}{\partial v_i}(a_i \hat{p}_L) = \frac{1}{2} \frac{\partial^2}{\partial v_i \partial v_j}(\sigma_{ij} \hat{p}_L) \qquad (3.102)$$

with the initial conditions

$$\hat{p}_L(t = 0, v, r|\, r_0) = \delta(r - r_0) p_E(v|\, r_0). \qquad (3.103)$$

Here $\sigma_{ij} = b_{ik} b_{jk}$.

Let us define a function $\hat{p}_E(t, v|\, r)$ by

$$\hat{p}_E(t, v|\, r) = \int \hat{p}_L(t, v, r|\, r_0)\, dr_0. \qquad (3.104)$$

Integrating the relations (3.102), (3.103) over r_0 yields

$$\frac{\partial \hat{p}_E}{\partial t} = L \hat{p}_E, \qquad (3.105)$$

$$\hat{p}_E(t = 0, v|\, r) = p_E(v|\, r), \qquad (3.106)$$

where

$$Lp = \frac{1}{2} \frac{\partial^2}{\partial v_i \partial v_j}(\sigma_{ij} p) - v_i \frac{\partial p}{\partial r_i} - \frac{\partial}{\partial v_i}(a_i p_L).$$

Assume that it is possible to choose the coefficients (3.98), (3.99) in such a way that

$$L p_E = 0. \qquad (3.107)$$

Then obviously p_E solves the problem (3.105), (3.106), since $\frac{\partial p_E}{\partial t} = 0$ (the Eulerian velocity is stationary), and the condition (3.106) is automatically satisfied. We assume also that the chosen coefficients ensure that the problem (3.105), (3.106) has a unique solution in a class of functions vanishing at infinity. Then it is clear that

$$p_E(v|\, r) = \hat{p}_E(t, v|\, r). \qquad (3.108)$$

From (3.104) and (3.108) we conclude that p. d. f. $\hat{p}_L(t, v, r|r_0)$ of our model satisfies the desired condition (3.22).

Thus the integral relation (3.22) is in this case equivalent to the relation (3.107). Using the latter relation, it is possible to find the desired coefficients a_i and b_{ij} provided p_E is given.

3.5.2 Relation to the well-mixed condition

Note that our relation (3.107) looks like the well-mixed condition formulated by Thomson [15] for the 1-particle model. Let us show the relation between the Novikov formula and the well-mixed condition.

Let us consider the model equation for one particle motion (3.6) where the velocity $\hat{V}$ is defined by SDE:

$$d\hat{V}_i = A_i(t, \hat{X}, \hat{V})\, dt + B_{ij}(t, \hat{X}, \hat{V})\, dW_j(t), \quad i = 1, 2, 3, \tag{3.109}$$

with the initial conditions $\hat{X}(0) = x_0$, $\hat{V}(0) = \xi_0$ where ξ_0 is a 3D random vector with the p. d. f.

$$\langle \delta(\xi_0 - V) \rangle = P_E(t = 0, V|\, x_0).$$

Here

$$P_E(t, V|\, X) = \langle \delta(V - U_E(X, t)) \rangle$$

is the p. d. f. of the Eulerian velocity.

The well-mixed condition [237] is formulated for an ensemble of particles from an instantaneous source $S(x_0)$ which move independently according to (3.6), (3.109). Then the p. d. f. of all tracer particles in the phase space (V, X) can be represented by the conditional expectation formula as

$$G(t, V, X) = \int \hat{P}_L(t, V, X|\, x_0) S(x_0)\, dx_0, \tag{3.110}$$

where

$$\hat{P}_L(t, V, X|\, x_0) = \langle \delta(V - \hat{V}(t, x_0))\, \delta(X - \hat{X}(t, x_0)) \rangle.$$

If the particles are uniformly distributed in space ($S(x_0) = const = S_0$) then, by definition,

$$G(t = 0, V, X) = P_E(t = 0, V|\, X) S_0.$$

The well-mixed condition [15] requires that this relation will be satisfied for all time instances:

$$G(t, V, X) = P_E(t, V|\, X) S_0. \tag{3.111}$$

From (3.109) and (3.6) it follows that the p. d. f. $\hat{P}_L(t, V, X|\, x_0)$ satisfies the Kolmogorov–Fokker–Planck equation

$$\frac{\partial \hat{P}_L}{\partial t} + M \hat{P}_L = 0, \tag{3.112}$$

where the operator M is defined by

$$Mf(t,V,X) = V_i\frac{\partial f}{\partial X_i} + \frac{\partial}{\partial V_i}(A_i f) - \frac{1}{2}\frac{\partial^2}{\partial V_i \partial V_j}(D_{ij} f),$$

with $D_{ij} = b_{ik}b_{jk}$.

Integrating this equation, by (3.110) we get

$$\frac{\partial G}{\partial t} + MG = 0. \tag{3.113}$$

From the well-mixed condition (3.111) we conclude by (3.113) that

$$\frac{\partial P_E}{\partial t} + MP_E = 0. \tag{3.114}$$

Note that this formula (and the well-mixed condition as well) follows from the Novikov formula [153]:

$$P_E(t,V|\,X) = \int P_L(t,V,X|\,x_0)\,dx_0, \tag{3.115}$$

where P_L is the true Lagrangian 1-particle p. d. f. Indeed, if we require that the model p. d. f. $\hat{P}_L$ satisfies (3.115) then, integrating (3.112) over x_0, we obtain equation (3.114) governing P_E. This is the same strategy we used above in the case of the 2-particle model.

3.5.3 A choice of the coefficients a_i and b_{ij}

As mentioned above, in the 1-dimensional case (Section 3.3.2) we can find different options of coefficients a_i, b_{ij} ensuring (3.107). Let us demonstrate a method of a consistent choice of such coefficients.

Let us introduce a function $\phi(v,r)$ through the relation

$$\frac{r_i}{|r|}\frac{\partial \phi}{\partial v_i} = -v_i\frac{\partial p_E}{\partial r_i}. \tag{3.116}$$

Then if we choose the functions a_i, b_{ij} from the condition

$$a_i p_E = \frac{\partial}{\partial v_j}\left(\frac{1}{2}\sigma_{ij} p_E\right) + \frac{r_i}{|r|}\phi, \tag{3.117}$$

then (3.107) holds.

Note that this strategy is similar to that used by Thomson [237]. Isotropy explains why the term $r_i/|r|$ appears. Integrating (3.116) yields

$$\phi(v,r) = \phi_0(r) - \int_0^{s_*}\left[v_i + \frac{r_i}{|r|}(s-s_*)\right]\frac{\partial p_E}{\partial r_i}\left(v + \frac{r}{|r|}(s-s_*)|\,r\right)ds, \tag{3.118}$$

where $s_* = \frac{(v,r)}{|r|}$, and ϕ_0 is an arbitrary function (e. g., $\phi_0 = 0$). Then (3.118) can be rewritten as

$$\phi(v,r) = -\frac{\partial}{\partial r_i} \int_0^{s_*} \left[v_i - \frac{r_i}{|r|} s \right] p_E\Big(v - \frac{r}{|r|} s | \, r\Big) \, ds$$
$$- \int_0^{s_*} \frac{2\,s}{|r|} p_E\Big(v - \frac{r}{|r|} s | \, r\Big) \, ds. \tag{3.119}$$

As a special example, let us concentrate on the Gaussian form of p_E,

$$p_E(v | \, r) = A(r) \exp\left\{ -\alpha(r) v^2 - \beta(r) \frac{(v,r)^2}{|r|^2} \right\}, \tag{3.120}$$

where

$$A(r) = \frac{1}{(2\pi)^{3/2} D_{NN}(|r|) \sqrt{D_{LL}(|r|)}},$$

$$\alpha(r) = \frac{1}{2D_{NN}(|r|)}, \qquad \beta(r) = \frac{1}{2D_{LL}(|r|)} - \frac{1}{2D_{NN}(|r|)},$$

where $D_{LL}(|r|)$, $D_{NN}(|r|)$ are the longitudinal and transversal structure functions of the Eulerian velocity, respectively [146]. In particular, in the inertial subrange [146] we have

$$D_{LL} = C(\bar{\varepsilon}|r|)^{2/3}, \quad D_{NN} = C'(\bar{\varepsilon}|r|)^{2/3}, \quad C' = \frac{4}{3} C,$$

and C is the Kolmogorov universal constant ($C \approx 1.9$).

We list here the following expressions:

$$p_E\left(v - s \frac{r}{|r|} | \, r\right) = A \exp\big\{ -\alpha(v^2 - s_*^2)\big\} \exp\big\{ -(\alpha+\beta)(s - s_*)^2 \big\}, \tag{3.121}$$

$$I_0 = \int_0^{s_*} p_E(v - s \frac{r}{|r|} | \, r) \, ds$$
$$= A \sqrt{\frac{\pi}{\alpha+\beta}} \exp\{-\alpha(v^2 - s_*^2)\} erf\{\sqrt{2(\alpha+\beta)}\, s_*\}, \tag{3.122}$$

$$I_1 = \int_0^{s_*} s \, p_E(v - s \frac{r}{|r|} | \, r) \, ds$$
$$= s_* I_0 + \frac{A}{2(\alpha+\beta)} \exp\{-\alpha(v^2 - s_*^2)\}\{1 - \exp[-s_*^2(\alpha+\beta)]\}. \tag{3.123}$$

From (3.119)–(3.123) we get

$$\phi(v,r) = -\frac{\partial}{\partial r_i} \left(v_i I_0 - \frac{r_i}{|r|} I_1 \right) - \frac{2}{|r|} I_1.$$

If we wish to describe an asymmetric non-Gaussian statistics we can take

$$p_E(v|\,r) = \frac{\sqrt{C}\exp\left\{-\frac{v^2-s_*^2}{2D_{NN}(|r|)}\right\}}{(2\pi)^{3/2}D_{NN}(|r|)\sqrt{D_{LL}(|r|)}}\left[\frac{p}{\sigma_1}\exp\left\{-\frac{1}{2\sigma_1^2}\left(\frac{s_*}{(\bar{\varepsilon}|r|)^{1/3}}-a_1\right)^2\right\}\right.$$
$$\left.+\frac{q}{\sigma_2}\exp\left\{-\frac{1}{2\sigma_2^2}\left(\frac{s_*}{(\bar{\varepsilon}|r|)^{1/3}}-a_2\right)^2\right\}\right].$$

In this case, it is also possible to find explicitly the function $\phi(v,r)$ through the functions D_{NN} and D_{LL}. In the inertial subrange, the similarity arguments and dimensional analysis show that $\sigma_{ij} = \sqrt{c_0\bar{\varepsilon}}\,\delta_{ij}$, where $c_0 = O(1)$ is a nondimensional constant. The functions a_i are obtained from (3.117).

3.6 Conclusions

We have considered 2-particle stochastic models of relative dispersion in locally isotropic incompressible turbulence. A critical analysis of the relevant models was given. Starting with Novikov's formula (3.22) in the diffusion approximation, we gave a rigorous formulation of the criteria for consistency between the Lagrangian and Eulerian statistics of the relative velocity in explicit form (3.107). For the quasi-1-dimensional case (the distance between the fluid particles is modeled) we get an analog (3.31) of the Novikov formula (3.22) which allows us to get, in the diffusion approximation, a variant of the model [108] satisfying the consistency criteria. A 3-dimensional variant of the Durbin model was developed. We presented in Section 3.5 a way of evaluating the coefficients of the 3D random forcing model (3.97) which is consistent with the criterion (3.107).

In conclusion, we should mention that there arises a series of problems related to the models discussed in this chapter. We list here some of them:

1. the existence of a global solution to the stochastic differential equations (3.40)–(3.42), (3.95)–(3.96) and (3.97) and the uniqueness of these solutions;
2. the uniqueness of the solution of Kolmogorov–Fokker–Planck equations relevant to SDE mentioned in the previous point;
3. since the SDEs constructed are stiff (which is related to the physical nature of the turbulent dispersion), a problem of constructing effective numerical methods for solving such SDE is very important;
4. construction of stochastic Lagrangian models based on relations between the Eulerian and Lagrangian statistical characteristics which are more general than that used in this chapter
5. to find SDE models based on recent similarity hypotheses which take into account the intermittency phenomena.

Chapter 4

A new Lagrangian model of 2-particle relative turbulent dispersion

A qualitative analysis of Durbin's stochastic 2-particle dispersion model [215] is given. We conclude that this model essentially contradicts the Lagrangian description. A new model of 2-particle relative dispersion free of these shortcomings is proposed.

4.1 Introduction

There are several Lagrangian 2-particle models of dispersion in turbulent flows. A survey of literature on the given subject can be found in [146, 215]. Of these, the most studied is Durbin's nonlinear model [44]. There exist some qualitative justifications of consistency of this model's prediction with the experimental data (see [215]).

According to Durbin's nonlinear model an instantaneous distance $l(t)$ between two fixed fluid particles at time t, is governed by

$$\frac{dl}{dt} = (R(l))^{1/2} U(t), \quad l(0) = l_0, \tag{4.1}$$

where l_0 is the initial distance between the particles, $R(l) = (l^2/(l^2 + L^2))^{1/3}$ is an approximation to the Eulerian structure function, L is the Eulerian integral length scale, and $U(t)$ is the Uhlenbeck–Ornstein (UO) process, i. e., a stationary Gaussian–Markov process with zero mean and such that

$$\langle U(t)U(t+\tau)\rangle = \sigma_u^2 \exp(-\tau/T_L).$$

Here σ_u is the root mean square (r.m.s.) fluctuating velocity, T_L is the Lagrangian integral time scale, and $\langle\cdot\rangle$ indicates the ensemble average.

In this chapter we first undertake a qualitative analysis of Durbin's model (4.1). We conclude from this analysis that this model essentially contradicts the Lagrangian description. Therefore, we suggest an improvement of the model (4.1) to overcome these shortcomings.

4.2 An examination of Durbin's nonlinear model

Let us consider a local isotropic high-Reynolds number turbulence. By utilizing the well-known semiempirical relation $\sigma_u^3 = L\bar{\varepsilon}/A$, where A is a dimensionless constant of order $O(1)$, $\bar{\varepsilon}$ is the mean rate of dissipation of kinetic energy, the UO process, $U(t)$, can be represented in the form

$$U(t) = (L\bar{\varepsilon})^{1/3} U_1(t).$$

Here $U_1(t)$ is a dimensionless UO process with

$$\langle U_1(t) U_1(t+\tau) \rangle = A^{-1/3} \exp(-\tau/T_L)\,. \tag{4.2}$$

Therefore, (4.1) becomes

$$\frac{dl}{dt} = (\bar{\varepsilon} l)^{1/3} U_1(t), \quad l(0) = l_0, \tag{4.3}$$

provided that $l(t) \ll L$. Thus (4.3) describes the process of relative dispersion of 2-fluid particles in the inertial range (IR) of the local isotropic high-Reynolds number turbulence.

To show a contradiction of (4.1) with the Lagrangian description, we need the property

$$P\{\lim_{t\to\infty} S_t = -\infty\} = 1 \tag{4.4}$$

of the function $S_t = \int_0^t U(s)ds$. The proof of (4.4) will be given in Section 4.4 (see Corollary 4.1).

Further let

$$F(x) = \int_0^x \frac{dl}{R^{1/2}(l)}$$

and let $G(y) = F^{-1}(y)$ be the inverse function specified by the function $y = F(x)$ (i. e., $G(F(x)) = x,\ F(G(y)) = y$). Then (4.1) yields

$$l(t) = G(F(l_0) + S_t). \tag{4.5}$$

Due to almost certain continuity of S_t and (4.4)–(4.5) the existence of a time $\tau = \tau(l_0)$ such that

$$l(\tau(l_0)) = 0, \quad P\{\tau(l_0) < \infty\} = 1. \tag{4.6}$$

immediately follows. Therefore by (4.6), at time $\tau(l_0)$ the two particles will appear at the same point, which essentially contradicts the basis of the Lagrangian description.

To carry out a more detailed analysis of this circumstance we need the following property of a Gaussian stochastic function (see, e. g., [29]).

Proposition 4.1. [6] *Let $\xi(t)$, $t \in [0,1]$ be a separable Gaussian stochastic process such that for some α, $0 < \alpha \le 1$,*

$$\sup_{0\le t,s\le 1} \frac{\langle |\xi(t) - \xi(s)|^2 \rangle}{|t-s|^{2\alpha}} < \infty. \tag{4.7}$$

Then for every β such that $0 < \beta < \alpha$ the process $\xi(t)$ satisfies the property:

$$P\left\{\sup_{0\le t,s\le 1}\frac{|\xi(t)-\xi(s)|}{|t-s|^{\beta}} < \infty\right\} = 1. \tag{4.8}$$

It is evident that the UO process, $U(t)$, satisfies the condition (4.7) for $\alpha = 1$. Consequently, for $\beta = 1/2$ in view of (4.8) there exists a random variable η, such that $P\{0 < \eta < \infty\} = 1$ and

$$|\xi(t)-\xi(s)| \le \eta\,|t-s|^{1/2}, \quad t,s\in[0,1].$$

Therefore we get

$$S_t = \int_0^t U(s)ds = \int_0^t [U(0) + (U(s)-U(0))]ds \le U(0)t + \frac{2}{3}\eta t^{3/2}, \quad t\in[0,1]. \tag{4.9}$$

Likewise,

$$S_t \ge U(0)t - \frac{2}{3}\eta t^{3/2}. \tag{4.10}$$

For given $l_0 > 0$, (provided $U(0) < 0$), we denote by $\tau_+ = \tau_+(l_0)$, $\tau_- = \tau_-(l_0)$ the roots of equations

$$F(l_0) + U(0)\tau_+ + \frac{2}{3}\eta\tau_+^{3/2} = 0, \quad F(l_0) + U(0)\tau_- - \frac{2}{3}\tau_-^{3/2} = 0, \tag{4.11}$$

respectively. By continuity of S_t and (5), it follows from (4.9)–(4.10) that the random time $\tau = \tau(l_0)$ satisfies

$$l(\tau(l_0)) = 0, \quad P\{\tau_-(l_0) < \tau(l_0) < \tau_+(l_0)\} = 1, \tag{4.12}$$

provided $U(0) < 0$. From (4.11), (4.12) it follows that

$$\tau(l_0) \simeq \frac{\int_0^{l_0}\frac{dl}{R^{1/2}(l)}}{|\,U(0)\,|} \simeq \frac{\int_0^{l_0}(\frac{L}{l})^{1/3}dl}{(L\bar{\varepsilon})^{1/3}\,|\,U_1(0)\,|} = \frac{3}{2\,|\,U_1(0)\,|}\left(\frac{l_0^2}{\bar{\varepsilon}}\right)^{1/3},$$

provided that $l_0 \ll L$, $U_1(0) < 0$. Thus the random time $\tau(l_0)$ behaves like $\tau(l_0) \sim (\frac{l_0^2}{\bar{\varepsilon}})^{1/3}$ as $l_0 \to 0$ (provided that $U_1(0) < 0$).

4.3 Mathematical formulation of a new model

Again, let us consider a local isotropic high-Reynolds number turbulence. Inertial range (IR) is an important ingredient of such flows. Let $l(t)$ be the instantaneous distance between two fixed fluid particles at time t. It is well known (see, e. g., [146]), that essentially the eddies of size comparable to the separation of the particles are responsible for their relative dispersion. Let v_l, τ_l be a typical value of the velocity and the life-time of eddies of size l, respectively. Then in accordance with the above-stated argument we formulate the following hypothesis:

Let τ be a dimensionless time defined by the local relation

$$d\tau = \frac{dt}{\tau_{l(t)}}$$

under the initial condition $\tau(0) = 0$. Then the dimensionless relative velocity of particles (more exactly, the longitudinal component of the relative velocity)

$$\xi_\tau = \frac{\frac{dl}{dt}}{v_{l(t)}}$$

is a universal stationary stochastic process (i. e., the statistical characteristics of this process do not depend on the turbulence parameters $\bar{\varepsilon}$, L, T_L, etc.).

In IR, this hypothesis takes more precise form. Indeed, in accordance with the second similarity Kolmogorov's hypothesis (see [146]) and using dimensional considerations we have $v_l \sim (\bar{\varepsilon}l)^{1/3}$, $\tau_l \sim (\frac{l^2}{\bar{\varepsilon}})^{1/3}$. Therefore, by the above-stated hypothesis in IR of the local isotropic high-Reynolds number turbulence the relative distance between two fluid particles, $l(t)$, satisfies the relation

$$\frac{dl}{dt} = (\bar{\varepsilon}l)^{1/3}\xi(\tau), \quad \frac{d\tau}{dt} = \left(\frac{\bar{\varepsilon}}{l^2}\right)^{1/3}, \quad t > 0, \tag{4.13}$$

with the initial conditions $l(0) = l_0$, $\tau(0) = 0$; here l_0 is the initial distance between the particles.

In spite of some similarity between (4.3) and the first equation of (4.13) there is an essential difference between them. In (4.3), the instantaneous velocity $\frac{dl}{dt}$ depends only on $l(t)$ while in accordance with (4.13)

$$\frac{dl}{dt} = (\bar{\varepsilon}l(t))^{1/3}\xi\left(\int_0^t \left(\frac{\bar{\varepsilon}}{l^2(s)}\right)^{1/3} ds\right)$$

i. e., it depends on the whole trajectory $\{l(s), 0 \le s \le t\}$.

In what follows we assume that $\xi(\tau)$ is a measurable, separable stochastic process almost certainly having continuous sample functions (for definitions see, e. g., [29]). Without loss of generality we also assume that $\bar{\varepsilon} = 1$, $l_0 = 1$ and consider the next dimensionless problem

$$\frac{dX}{dt} = X^{1/3}\xi(\tau), \quad \frac{d\tau}{dt} = X^{-2/3}, \quad t > 0 \tag{4.14}$$

$$X(0) = 1, \quad \tau(0) = 0. \tag{4.15}$$

4.4 A qualitative analysis of the problem (4.14) for symmetric $\xi(\tau)$

4.4.1 Analysis of the problem (4.14) in the deterministic case

First, let us consider the problem

$$\frac{dX}{dt} = X^{1/3} f(Y), \quad \frac{dY}{dt} = X^{-2/3}, \tag{4.16}$$

$$0 < X, Y < \infty, \quad t > 0, \tag{4.17}$$

$$X(0) = 1, \quad Y(0) = 0, \tag{4.18}$$

where $f : (0, \infty) \to (-\infty, \infty)$ is a continuous function. In accordance with the theory of ordinary differential equations, there exists a solution $X = X(t)$, $Y = Y(t)$ to (4.16) which is defined in some time-interval $[0, t_*)$ and such that the following alternative cases happen:

1. $t_* = \infty$;
2. $t_* < \infty$ and the solution $X = X(t), Y = Y(t)$ satisfies at least one of the conditions

$$\text{2a.} \quad \lim_{t \to t_*} X(t) = 0, \qquad \text{2b.} \quad \lim_{t \to t_*} X(t) = \infty, \tag{4.19}$$

$$\text{2c.} \quad \lim_{t \to t_*} Y(t) = 0, \qquad \text{2d.} \quad \lim_{t \to t_*} Y(t) = \infty. \tag{4.20}$$

This solution is called the maximal solution to (4.16), and the interval $[0, t_*)$ is called the maximal interval.

Lemma 4.1. *There exists a unique maximal solution $X = X(t)$, $Y = Y(t)$, $t \in [0, t_*)$ to (4.16) with*

$$t_* = \int_0^\infty \exp(\frac{2}{3} S_\tau) d\tau, \quad S_\tau = \int_0^\tau f(s) ds, \tag{4.21}$$

and

$$\lim_{t \to t_*} Y(t) = \infty, \quad \overline{\lim}_{t \to t_*} X(t) = \overline{\lim}_{\tau \to \infty} \exp(S_\tau), \tag{4.22}$$

$$\underline{\lim}_{t \to t_*} X(t) = \underline{\lim}_{\tau \to \infty} \exp(S_\tau). \tag{4.23}$$

Proof. Let us suppose on the contrary, that there exist two maximal solutions $X_i(t)$, $Y_i(t)$, $t \in [0, t_i^*)$ $(i = 1, 2)$. It is obvious that the function $Y_i : [0, t_i^*) \to [0, Y_i^*)$ is monotonically increasing. Here,

$$Y_i^* = \lim_{t \to t_i^*} Y_i(t), \quad i = 1, 2.$$

Therefore, there exists a converse function $Y_i^{-1} : [0, Y_i^*) \to [0, t_i^*)$ $(i = 1, 2)$.

Let

$$\bar{X}_i(Y) = X_i(Y_i^{-1}(Y)), \quad \bar{Y}_i(Y) = Y_i^{-1}(Y), \quad Y \in [0, Y_i^*) \quad (i = 1, 2).$$

Then, it is not difficult to show that $\bar{X}_i(Y), \bar{Y}_i(Y)$ $(i = 1, 2)$ obey

$$\frac{d\bar{X}}{dY} = \bar{X} f(Y), \quad \frac{d\bar{Y}}{dY} = \bar{X}^{2/3}, \quad \bar{X}(0) = 1, \quad \bar{Y}(0) = 0. \tag{4.24}$$

Moreover, by definition,

$$\lim_{Y \to Y_i^*} \bar{X}_i(Y) = \lim_{t \to t_i^*} X_i(t), \quad \lim_{t \to t_i^*} Y_i(t) = \lim_{t \to t_i^*} Y_i^*, \quad \lim_{Y \to Y_i^*} \bar{Y}_i(Y) = t_i^*. \tag{4.25}$$

Now, let us show that $Y_i^* = \infty$ $(i = 1, 2)$. Indeed, otherwise, if $Y_i^* < \infty$ then by (4.24), (4.25)

$$\lim_{t \to t_i^*} X_i(t) = \exp\left\{\int_0^{Y_i^*} f(Y) dY\right\}, \quad t_i^* = \int_0^{Y_i^*} \exp\left\{\frac{2}{3}\int_0^Y f(y')dy'\right\} dY. \tag{4.26}$$

Since $f(\cdot)$ is continuous, (4.26) contradicts (4.19). Thus we have $Y_i^* = \infty$, $i = 1, 2$. Further, in view of the obvious uniqueness of the solution to (4.24),

$$\bar{X}_1(Y) = \bar{X}_2(Y), \quad \bar{Y}_1(Y) = \bar{Y}_2(Y), \quad Y \in [0, \infty). \tag{4.27}$$

Since $t_i^* = \lim_{Y \to \infty} \bar{Y}_i(Y)$, $i = 1, 2$, then (22) yields $t_1^* = t_2^*$ and $Y_1(t) = Y_2(t)$, $t \in [0, t_*)$. Here $t_* = t_1^* = t_2^*$.

Thus uniqueness of the maximal solution $X = X(t)$, $Y = Y(t)$, $t \in [0, t_*)$ to (4.16) is shown; (4.21) is an evident consequence of (4.26). Finally, by $Y_i^* = \lim_{t \to t_*} Y_i(t) = \infty$ and the first formula in (4.26) we get (4.22). The proof is completed. □

4.4.2 Analysis of the problem (4.14) for stochastic $\xi(\tau)$

Now let us consider the problem (4.14) involving the stochastic function $\xi(\tau)$.

The stochastic process $\{\xi(\tau), \tau \geq 0\}$ is called symmetric if it has the same statistic characteristics as the process $\{-\xi(\tau), \tau \geq 0\}$. For given real numbers α, β let M_α^β be a σ–algebra generated by a family of random variables $\{\xi(\tau), \alpha \leq \tau \leq \beta\}$. A stochastic process $\{\xi(\tau), \tau \geq 0\}$ is called regular if the set $M_\infty^\infty = \bigcap_{\tau > 0} M_\tau^\infty$ is trivial (i. e., M_∞^∞ consists of events having probability one or zero).

It is well known that $\{\xi(\tau), \tau \geq 0\}$ is regular if and only if

$$\sup_{A \in M_\tau^\infty} \mid P(AB) - P(A)P(B) \mid \to 0, \quad \tau \to \infty$$

for every $B \in M_0^\infty$.

Theorem 4.1. *Let $\{\xi(\tau), \tau \geq 0\}$ be a symmetric regular stochastic process. Then with probability 1, the solution to* (4.14) *is defined on the half-line $R_+ = [0, \infty)$ and*

$$P\{\lim_{t\to\infty} X(t) = 0\} = P\{\lim_{t\to\infty} X(t) = \infty\} = 0. \tag{4.28}$$

Proof. Let us define an event A, generated by the random variable

$$t_* = \int_0^\infty \exp\left\{\frac{2}{3}S_\tau\right\}d\tau,$$

$$S_\tau = \int_0^\tau \xi(s)ds$$

through $A = \{\omega : t_*(\omega) < \infty\}$. It is obvious that $A \in M_\infty^\infty$ and $P(A)$ is equal to 0 or 1 due to regularity of $\{\xi(\tau), \tau \geq 0\}$. Let us show that $P(A) = 0$. Indeed, let $A' = \{\omega : t'_* < \infty\}$, where

$$t'_* = \int_0^\infty \exp\left\{-\frac{2}{3}S_\tau\right\}d\tau.$$

Then we have $P(A) = P(A')$ due to symmetry of $\{\xi(\tau), \tau \geq 0\}$. Further, by virtue of the inequality $x + \frac{1}{x} \geq 2$, for positive x we have

$$\int_0^\infty \left[\exp\left(\frac{2}{3}S_\tau\right) + \exp\left(-\frac{2}{3}S_\tau\right)\right]d\tau = \infty$$

i. e., $P(AA') = 0$. Therefore $P(A) = P(A') = 0$, i. e., $P\{t_* = \infty\} = 1$. By Lemma 4.1 the last equality means that every solution of (4.14) is almost certainly defined on R_+.

To establish (4.28) let us consider two events:

$$B = \{\lim_{t\to\infty} X(t) = 0\}, \quad \text{and} \quad C = \{\lim_{t\to\infty} X(t) = \infty\}.$$

By Lemma 4.1 we have

$$B = \{\lim_{\tau\to\infty} S_\tau = -\infty\}, \quad C = \{\lim_{\tau\to\infty} S_\tau = \infty\}.$$

It is obvious that $B, C \in M_\infty^\infty$. By symmetry and regularity properties of $\{\xi(\tau), \tau \geq 0\}$ we have $P(B) = P(C) = 0$ or 1. Since the events B and C are inconsistent, it is obvious that $P(B) = P(C) = 0$. This completes the proof. □

Let us give two examples which show that without the regularity or symmetry conditions, the assertion of Theorem 4.1 may be wrong.

Example 4.1. Let ξ be a symmetric real-valued random variable. Let $\xi(\tau) \equiv \xi, \tau \geq 0$. Then it is obvious that

$$t_* = \begin{cases} \infty, & \xi \geq 0 \\ -\frac{2}{3}\xi, & \xi < 0. \end{cases}$$

Therefore,

$$P\{t_* < \infty\} = P\{\xi < 0\} = \frac{1}{2},$$
$$P\{\lim_{t \to t_*} X(t) = 0\} = P\{\xi < 0\} = \frac{1}{2}.$$

In this example the regularity condition is broken.

Example 4.2. Let

$$\psi(\tau) = \begin{cases} \tau(1-\tau), & \tau \in [0,1], \\ 0, & \tau > 1. \end{cases}$$

Let $\{\eta_k\}_{k=1}^{\infty}$ – be a sequence of mutually independent random variables having the following distributions:

$$P\{\eta_k = k\} = \frac{k^2}{(k^2+1)}, \quad P\{\eta_k = -k^3\} = \frac{1}{(k^2+1)}, \quad k = 1, 2, \ldots.$$

Let us define a stochastic process $\{\xi(\tau), \tau \geq 0\}$ as

$$\xi(\tau) = (\eta_{[\tau+1]} - \eta_{[\tau]})\psi(\tau - [\tau]), \quad \tau \geq 0,$$

where $[\tau]$ is the integral part of τ, $\eta_0 \equiv 0$. It is evident that the process $\{\xi(\tau), \tau \geq 0\}$ is regular but not symmetric. From $\sum_k P\{\eta_k = -k^3\} < \infty$ we get by the Borel–Cantelly lemma that there exists an integer-valued random variable ν such that $P\{\nu < \infty\} = 1$, and

$$\eta_k = k, \quad k \geq \nu. \tag{4.29}$$

Since

$$S_{k+\theta} = -\frac{\eta_k}{6} + (\eta_{k+1} - \eta_k)\frac{\theta^2(3-2\theta)}{6}, \quad k = 1, 2, \ldots, \quad 0 \leq \theta < 1, \tag{4.30}$$

then from (4.29) we get

$$S_{k+\theta} = -\frac{k}{6} + \frac{\theta^2(3-2\theta)}{6}, \quad k \geq \nu, \quad 0 \leq \theta < 1. \tag{4.31}$$

From (4.31) it follows that

$$P\left\{\int_0^\infty \exp\left\{\frac{2}{3}S_\tau\right\}d\tau < \infty\right\} = 1,$$

and

$$P\{\lim_{\tau\to\infty} S_\tau = -\infty\} = 1.$$

Consequently, by Lemma 4.1 $P\{t_* < \infty\} = 1$, $P\{\lim_{t\to t_*} X(t) = 0\} = 1$, i. e., the assertions of Theorem 4.1 are broken.

The following theorem makes the asymptotic behavior of $X = X(t)$ as $t \to \infty$ more precise.

Theorem 4.2. *Let $\{\xi(\tau),\ \tau \geq 0\}$ be a symmetric regular stochastic process such that there exists a sequence $\{\varepsilon_n\}_{n=1}^{\infty}$ of real-valued positive numbers and an increasing sequence $\{\tau_n\}_{n=1}^{\infty}$ of real-valued positive numbers satisfying the following conditions:*

$$\sum_{n=1}^{\infty} P\{\mid \hat{S}_n \mid < \varepsilon_n\} < \infty, \text{ and } \varepsilon_n\sigma_n \to \infty, \text{ as } n \to \infty, \tag{4.32}$$

where $\sigma_n^2 = MS_{\tau_n}^2$, $\hat{S}_n = \frac{1}{\sigma_n} S_{\tau_n}$, $S_\tau = \int_0^\tau \xi(s)ds$. Then

$$P\{\lim_{t\to\infty} X(t) = 0\} = P\{\overline{\lim_{t\to\infty}}\, X(t) = \infty\} = 1. \tag{4.33}$$

Proof. Let

$$A = \{\lim_{t\to\infty} X(t) = 0\}, \qquad B = \{\overline{\lim_{t\to\infty}}\, X(t) = \infty\},$$
$$A' = \{\lim_{\tau\to\infty} S_\tau = -\infty\}, \quad B' = \{\overline{\lim_{\tau\to\infty}}\, S_\tau = \infty\}.$$

Then by Lemma 4.1 we have $P(A\Delta A') = 0$, $P(B\Delta B') = 0$, where $A\Delta A' = (A\backslash A')\bigcup(A'\backslash A)$. Therefore it is sufficient to establish that $P(A') = P(B') = 1$. Indeed, by (4.32) and the Borel–Cantelly lemma there exists an integer-valued random variable ν, $P\{\nu < \infty\} = 1$ such that

$$\mid \hat{S}_n \mid \geq \varepsilon_n, \quad n \geq \nu,$$

i. e.,

$$\mid S_{\tau_n} \mid \geq \varepsilon_n\sigma_n, \quad n \geq \nu. \tag{4.34}$$

Therefore, since $\varepsilon_n\sigma_n \to \infty$ as $n \to \infty$, we have $P(A' \cup B') = 1$. Because of the symmetry of $\{\xi(\tau),\ \ \tau \geq 0\}$, it is obvious that $P(A') = P(B')$. Finally, from $A', B' \in M_\infty^\infty$ and using the regularity of $\{\xi(\tau),\ \tau \geq 0\}$ we get $P(A') = P(B') = 1$. This completes the proof. □

The following example shows that in the general case the conditions (4.32) can not be ignored.

Example 4.3. Let $\{\xi(\tau),\ \tau \geq 0\}$ be a stochastic process defined as in Example 4.2, but with

$$P\{\eta_k = k^2\} = P\{\eta_k = -k^2\} = \frac{1}{2k^2},$$

$$P\{\eta_k = 1\} = P\{\eta_k = -1\} = \frac{1}{2}\left(1 - \frac{1}{k^2}\right).$$

Then by the Borel–Cantelly lemma there exists an integer-valued random variable ν, $P\{\nu < \infty\} = 1$ such that

$$|\ \eta_k \ | = 1, \quad k \geq \nu.$$

Therefore from (4.30) and by symmetry of $\{\xi(\tau), \tau \geq 0\}$ we have

$$P\left\{\overline{\lim_{\tau\to\infty}}\ S_\tau = \frac{1}{6}\right\} = P\left\{\underline{\lim_{\tau\to\infty}}\ S_\tau = -\frac{1}{6}\right\} = 1,$$

and hence by Lemma 4.1 we get

$$P\{\overline{\lim_{t\to\infty}}\ X(t) = e^{1/6}\} = P\{\underline{\lim_{t\to\infty}}\ X(t) = e^{-1/6}\} = 1.$$

Thus, the assertion (4.33) of Theorem 4.2 is not true. This is a consequence of violation of the condition (4.32).

Now, to prove (4.4) we first obtain the following.

Corollary 4.1. *Let $\{\xi(t), t \geq 0\}$ be a UO process. Then* (4.33) *is valid.*

Proof. Since $\langle \xi(t)\xi(t+\tau)\rangle = \sigma^2 \exp(-\alpha\tau)$, the spectral density of $\xi(t)$, $p(\lambda)$, is equal to (see, e. g., [146]):

$$p(\lambda) = \frac{2\sigma^2\alpha}{\pi(\alpha^2 + \lambda^2)}.$$

It is known (see, e. g., [85]) that a stationary Gaussian process with the spectral density $p(\lambda)$ is regular if and only if

$$\int_0^\infty \frac{|\ \ln p(\lambda)\ |}{(1+\lambda^2)} d\lambda < \infty.$$

It is obvious that for UO process $\xi(t)$ this condition is satisfied. Now, $\langle \xi(t)\rangle = 0$ implies that $\xi(t)$ is symmetric. To show (4.32), we choose $\tau_n = n,\ \varepsilon_n = 1,\ n = 1, 2, \ldots$. Then

$$\sigma_n^2 = \left\langle \left(\int_0^n \xi(t)dt \right)^2 \right\rangle = \frac{2\sigma^2}{\alpha^2}(e^{-\alpha n} - 1 + \alpha n),$$

i. e., σ_n behaves like $\sigma_n \sim n^{1/2}$ as $n \to \infty$. Therefore, the condition (4.32) is obviously true. Thus all the sufficient conditions of the Theorem 4.2 are true and therefore (4.33) implies the assertion of the corollary. The proof is completed. □

In the next section we generalize the results obtained for nonsymmetric processes $\xi(\tau)$.

4.5 Qualitative analysis of the problem (4.14) in the general case

A stochastic process $\{\xi(\tau),\ \tau \geq 0\}$ is said to be a process satisfying the strong mixed condition if

$$\alpha(\tau) = \sup_{\theta \geq 0} \sup_{A \in M_0^\theta, B \in M_{\theta+\tau}^\infty} |P(AB) - P(A)P(B)| \to_{\tau\to\infty} 0.$$

When it satisfies the following stronger condition

$$\varphi(\tau) = \sup_{\theta \geq 0} \sup_{A \in M_0^\theta, P(A) \neq 0, B \in M_{\theta+\tau}^\infty} |P(B/A) - P(B)| \to_{\tau\to\infty} 0,$$

then it is said that $\{\xi(\tau), \tau \geq 0\}$ satisfies the uniformly strong mixed condition.

Lemma 4.2. *Let $\{\xi(\tau),\ \tau \geq 0\}$ be a stochastic process satisfying the strong mixed condition; let $\{\tau_k\}_{k=0}^\infty$ be a monotone increasing sequence of real-valued positive numbers; let $\{A_k\}_{k=0}^\infty$ be a sequence of events such that $A_k \in M_{\tau_{2k}}^{\tau_{2k+1}}$, $k = 0, 1, 2, \dots$ and*

$$\sum_{k=0}^{\infty} P(A_k) = \infty, \quad \sum_{k=1}^{\infty} \alpha(\tau_{2k} - \tau_{2k-1}) < \infty. \tag{4.35}$$

Then

$$P\Big(\bigcap_{n=0}^{\infty} \bigcup_{k=n}^{\infty} A_k\Big) = 1,$$

i. e., almost surely an infinite number of events of sequence A_k, $k = 0, 1, 2, \dots$ happens.

Proof. Let $A_k^c = \Omega \backslash A_k$. Then

$$P\Big\{\Omega\backslash\Big(\bigcap_{n=0}^{\infty}\bigcup_{k=n}^{\infty} A_k\Big)\Big\} = P\Big\{\bigcup_{n=0}^{\infty}\bigcap_{k=n}^{\infty} A_k^c\Big\} = \lim_{n\to\infty} P\Big(\bigcap_{k=n}^{\infty} A_k^c\Big)$$
$$= \lim_{n\to\infty}\lim_{m\to\infty} P\Big(\bigcap_{k=n}^{n+m} A_k^c\Big). \tag{4.36}$$

From the strong mixed condition we get

$$
\begin{aligned}
P\Big(\bigcap_{k=n}^{n+m} A_k^c\Big) &\le \prod_{k=n}^{n+m} P(A_k^c) + \sum_{k=n+1}^{n+m} \alpha(\tau_{2k} - \tau_{2k-1}) \\
&= \prod_{k=n}^{n+m} (1 - P(A_k)) + \sum_{k=n+1}^{n+m} \alpha(\tau_{2k} - \tau_{2k-1}).
\end{aligned}
$$

Consequently,

$$
\lim_{m\to\infty} P\Big(\bigcap_{k=n}^{n+m} A_k^c\Big) \le \prod_{k=n}^{\infty} (1 - P(A_k)) + \sum_{k=n+1}^{\infty} \alpha(\tau_{2k} - \tau_{2k-1}).
$$

The properties (4.35) show that the *rhs* of the last inequality tends to 0 as $n \to \infty$. Therefore by (4.36) we get $P\{\Omega \backslash (\bigcap_{n=0}^{\infty} \bigcup_{k=n}^{\infty} A_k)\} = 0$. This completes the proof. □

Lemma 4.3. *Let* $\{\xi(\tau),\ \tau \ge 0\}$ *– be a stochastic process satisfying the following conditions:*

1. $M\xi(\tau) = 0,\ \tau \ge 0,\ \sup_{\tau\ge 0} M\xi^2(\tau) < \infty,\ \sigma_\tau^2 = MS_\tau^2 \to \infty$, *as* $\tau \to \infty$, *where* $S_\tau = \int_0^\tau \xi(s)ds$;
2. *there exists a real-valued random variable* η, *such that* $P\{\eta < 0\} > 0,\ P\{\eta > 0\} > 0$;
$$
\hat{S}_\tau = \frac{S_\tau}{\sigma_\tau} \xrightarrow{P} \eta, \text{ as } \tau \to \infty; \qquad (4.37)
$$
3. *there exist positive real numbers* C , δ *such that*
$$
P\{\sup_{m\le\tau\le m+1} \mid \xi(\tau) \mid < C\} > \delta, \quad m = 0, 1, 2, \ldots; \qquad (4.38)
$$
4. *the strong mixed condition is valid.*

Then for every real number $a > 0$

$$
P\Big\{\int_0^\infty \exp\{aS_\tau\} d\tau = \infty\Big\} = 1,
$$

and

$$
P\{\underline{\lim}_{\tau\to\infty} S_\tau = -\infty\} = P\{\overline{\lim}_{\tau\to\infty} S_\tau = +\infty\} = 1. \qquad (4.39)
$$

Proof. Let $S(\theta,\theta+\tau) = S_{\theta+\tau} - S_\theta$, $\sigma^2(\theta,\theta+\tau) = MS^2(\theta,\theta+\tau)$. Then the conditions 1–2 of the theorem yield

$$\sigma^2(\theta,\theta+\tau) \to \infty, \quad \hat{S}(\theta,\tau+\theta) = \frac{S(\theta,\tau+\theta)}{\sigma(\theta,\tau+\theta)} \xrightarrow{P} \eta, \text{ as } \tau \to \infty \tag{4.40}$$

for every $\theta \geq 0$. Therefore by the conditions 3)–4) of the theorem there exists a monotone increasing sequence

$$\cdots < n_k < n_{k+\frac{1}{4}} < n_{k+\frac{1}{2}} < n_{k+\frac{3}{4}} < n_{k+1} < \cdots \tag{4.41}$$

of positive integers such that for every $k = 1, 2, \ldots$

$$\begin{aligned}
&\sigma^2(n_k, n_{k+\frac{1}{4}}) \geq (2\sigma^2_{n_k})^4, \quad \sigma^2_{n_{k+\frac{1}{2}}} \geq k^2,\\
&\alpha(n_{k+\frac{1}{2}} - n_{k+\frac{1}{4}}) \leq \frac{1}{k^2}, \quad \sigma^2(n_{k+\frac{1}{2}}, n_{k+\frac{3}{4}}) \geq (2\sigma^2_{n_{k+\frac{1}{2}}})^4,\\
&\sigma^2_{n_{k+1}} \geq (k+1)^2, \quad \alpha(n_{k+1} - n_{k+\frac{3}{4}}) \leq \frac{1}{k^2},\\
&P\{\hat{S}(n_k, n_{k+\frac{1}{4}}) \geq \frac{1}{\sqrt{\sigma(n_k, n_{k+\frac{1}{4}})}}\} \geq \epsilon > 0,\\
&P\{\hat{S}(n_{k+\frac{1}{2}}, n_{k+\frac{3}{4}}) \leq \frac{-1}{\sqrt{\sigma(n_{k+\frac{1}{2}}, n_{k+\frac{3}{4}})}}\} \geq \epsilon > 0.
\end{aligned} \tag{4.42}$$

Here ϵ is a positive real number (depending on the distribution of the random variable η).

For the sake of simplicity we suppose that $a \geq 0$. Let us define the following sequences of events:

$$\begin{gathered}
A_k = \{S(n_k, n_{k+\frac{1}{4}}) \geq \sigma^{\frac{1}{2}}(n_k, n_{k+\frac{1}{4}}),\\
S(n_{k+\frac{1}{2}}, n_{k+\frac{3}{4}}) \leq -\sigma^{\frac{1}{2}}(n_{k+\frac{1}{2}}, n_{k+\frac{3}{4}})\}, \quad B_k = \{| S_{n_k} | \geq \sigma^2_{n_k}\},\\
C_k = \{| S_{n_{k+\frac{1}{2}}} | \geq \sigma^2_{n_{k+\frac{1}{2}}}\}, \quad D_k = \{\sup_{n_{k+\frac{1}{4}} \leq \tau \leq n_{k+\frac{1}{4}}+1} | \xi(\tau) | < \sigma^2_{n_k}\},\\
k = 1, 2, \ldots.
\end{gathered}$$

Then by the Chebyshev inequality we have

$$P(B_k) \leq \frac{1}{k^2}, \quad P(C_k) \leq \frac{1}{k^2}, \quad k = 1, 2, \ldots,$$

and consequently

$$\sum_k P(B_k) < \infty, \quad \sum_k P(C_k) < \infty. \tag{4.43}$$

Due to (4.38) there exists a positive integer k_0 such that

$$P(D_k) \geq \delta, \quad k \geq k_0. \tag{4.44}$$

By the strong mixed condition and by the last two inequalities of the set (4.42) we get

$$\begin{aligned} P(A_k) \geq P\{S(n_k, n_{k+\frac{1}{4}}) \geq \sigma^{\frac{1}{2}}(n_k, n_{k+\frac{1}{4}})\} P\{S(n_{k+\frac{1}{2}}, n_{k+\frac{3}{4}}) \\ \leq -\sigma^{\frac{1}{2}}(n_{k+\frac{1}{2}}, n_{k+\frac{3}{4}})\} - \alpha(n_{k+\frac{1}{2}} - n_{k+\frac{1}{4}}) \geq \epsilon^2 - \frac{1}{k^2}, \\ k = 1, 2, \ldots . \end{aligned} \tag{4.45}$$

By (4.43) and the Borel–Cantelly Lemma there exists an integer-valued random variable ν, $P\{\nu < \infty\} = 1$, such that for every $k \geq \nu$ the following events $B_k^c = \Omega \backslash B_k$ and $C_k^c = \Omega \backslash C_k$ happen, i. e.,

$$| S_{n_k} | \leq \sigma_{n_k}^2, \quad | S_{n_{k+\frac{1}{2}}} | < \sigma_{n_{k+\frac{1}{2}}}^2, \quad k \geq \nu. \tag{4.46}$$

Next from the strong mixed condition, the inequality $\alpha(n_{k+1} - n_{k+\frac{3}{4}}) \leq \frac{1}{k^2}$, the set of inequalities (4.42) and by Lemma 4.2 it follows that almost surely there is an infinite number of events from all sequences $\{A_k\}_{k=1}^{\infty}$, $\{D_k\}_{k=1}^{\infty}$. For the sake of simplicity of notation, we suppose that almost surely there occur all events in the sequences $\{A_k\}_{k=1}^{\infty}, \{D_k\}_{k=1}^{\infty}$. Then, denoting $n'_k = n_{k+\frac{1}{4}}$, $k \geq \nu$ we get

$$\begin{aligned} S_{n'_k} &= S_{n_{k+\frac{1}{4}}} = S(n_k, n_{k+\frac{1}{4}}) + S_{n_k} \geq \sigma^{\frac{1}{2}}(n_k, n_{k+\frac{1}{4}}) - \sigma_{n_k}^2 \geq \sigma_{n_k}^2, \\ S_{n_{k+\frac{3}{4}}} &= S(n_{k+\frac{1}{2}}, n_{k+\frac{3}{4}}) + S_{n_{k+\frac{1}{2}}} \leq -\sigma^{\frac{1}{2}}(n_{k+\frac{1}{2}}, n_{k+\frac{3}{4}}) + \sigma_{n_{k+\frac{1}{2}}}^2 \leq -\sigma_{n_{k+\frac{1}{2}}}^2, \\ S_{n'_k+\theta} &= S_{n'_k} + \int_{n'_k}^{n'_k+\theta} \xi(s) ds \geq \sigma_{n_k}^2 - \sup_{n'_k \leq \tau \leq n'_k+1} | \xi(s) | \geq 0, \\ & \qquad k \geq \nu,\ 0 \leq \theta \leq 1. \end{aligned}$$

From the condition $\sigma_\tau^2 \to \infty$ as $\tau \to \infty$ and the last conditions we have the assertions of the lemma. The case $a < 0$ can be considered likewise. Hence, the lemma is proved. □

Here we make the following remark:

Remark 4.1. For a stationary (in the narrow sense) process $\{\xi(\tau), \tau \geq 0\}$ the condition (4.38) is always valid. This is a consequence of almost certain continuity of a sample function of $\xi(\tau)$.

The direct consequence of the proved theorem is the following assertion.

Theorem 4.3. *Let* $\{\xi(\tau),\ \tau \geq 0\}$ *be a stochastic process satisfying all the conditions of Lemma 4.3. Then almost certainly there exists a unique solution of the problem* (4.14) *defined on* R_+ *and satisfying the condition* (4.33).

In what follows we assume that $M\xi(\tau) = 0,\ \tau \geq 0$.

Corollary 4.2. *Let* $\{\xi(\tau),\ \tau \geq 0\}$, *be a stochastic process such that* (i) *it satisfies the uniformly strong mixed condition,* (ii) *the condition* (4.38) *is valid, and* (iii) *for some positive number* $\delta > 0$

$$\sup_{\tau \geq 0} M \mid \xi(\tau) \mid^{2+\delta} < \infty, \quad \sigma_n^{2+\delta} n^{-1} \to \infty, \text{ as } n \to \infty.$$

Then almost certainly the solution to (4.14) *is unique, it is defined on* R_+ *and* (4.33) *is valid.*

This assertion is a consequence of the Theorem 4.3 and the sufficient conditions of the central limit theorem (CLT) [8]. Using the sufficient conditions of the CLT [9] and by Theorem 4.3 we get the following assertion:

Corollary 4.3. *Let* $\{\xi(\tau),\ \tau \geq 0\}$ *be a stationary (in the narrow sense) stochastic process such that* (i) *it obeys the uniformly strong mixed condition,* (ii) *there exists some* $\delta > 0$ *such that* $M \mid \xi(0) \mid^{2+\delta} < \infty$, *and* (iii) $\sigma_n^2 \to \infty,\ n \to \infty$. *Then all the assertions of Corollary 4.2 are valid.*

Thus we have established several sufficient conditions for $\xi(\tau)$ under which the stochastic model (4.14) has a unique solution defined for all positive time instances.

Chapter 5

The combined Eulerian–Lagrangian model

5.1 Introduction

The commonly used approach to the well-developed turbulence is the statistical description where the Eulerian velocity $\mathbf{U}_E(x,t)$ is considered to be a 3D-random field. Having the samples of $\mathbf{U}_E(x,t)$, one determines the Lagrangian trajectory $\mathbf{X}(t,\mathbf{x}_0)$, $t \geq 0$ of a fluid particle originating at position $\mathbf{x}_0$ at time $t = 0$ from the equation of motion:

$$\frac{\partial \mathbf{X}}{\partial t} = \mathbf{U}_E(\mathbf{X},t) \tag{5.1}$$

subject to the initial condition

$$\mathbf{X}(t = 0, \mathbf{x}_0) = \mathbf{x}_0. \tag{5.2}$$

The Lagrangian velocity $\mathbf{V}(t,\mathbf{x}_0)$ is related to the Eulerian velocity $\mathbf{U}_E(\mathbf{x},t)$ by

$$\mathbf{V}(t,\mathbf{x}_0) = \mathbf{U}_E(\mathbf{X}(t,\mathbf{x}_0),t). \tag{5.3}$$

Equation (5.1) can be integrated numerically in time; (5.3) means that the instantaneous particle velocity is the same as the fluid velocity at the instantaneous particle position.

Each Eulerian quantity $f(\mathbf{x},t)$ can be related to the Lagrangian quantity by

$$F(\mathbf{x}_0,t) = f(\mathbf{X}(t,\mathbf{x}_0),t), \quad \mathbf{x}_0 \in R^3, \quad t \geq 0.$$

A Lagrangian description allows us to analyze directly the motion of material fluid elements. Importance of the Lagrangian trajectories is that the quantities of practical interest are expressed through the n-particle statistical characteristics [152, 157]. In particular, the mean concentration of a passive scalar and its covariance are defined through the 1-particle and 2-particle statistical characteristics, respectively [146,214]:

$$\langle c(\mathbf{x},t)\rangle = \int p_{1L}(\mathbf{x},t;\mathbf{x}_0)S(\mathbf{x}_0)d\mathbf{x}_0,$$

$$\langle c(\mathbf{x},t)c(\mathbf{x}',t)\rangle = \int p_{2L}(\mathbf{x},\mathbf{x}',t;\mathbf{x}_0,\mathbf{x}'_0)S(\mathbf{x}_0)S(\mathbf{x}'_0)d\mathbf{x}_0 d\mathbf{x}'_0,$$

where $c(\mathbf{x},t)$ is the concentration of the passive scalar released from an instantaneous deterministic source $S(\mathbf{x}_0)$. The functions p_{1L} and p_{2L} are the 1-point and

2-point p. d. f.'s:

$$p_{1L}(\mathbf{x}, t; \mathbf{x}_0) = \langle \delta(\mathbf{x} - \mathbf{X}(t, \mathbf{x}_0)) \rangle,$$

$$p_{2L}(\mathbf{x}, \mathbf{x}', t; \mathbf{x}_0, \mathbf{x}'_0) = \langle \delta(\mathbf{x} - \mathbf{X}(t, \mathbf{x}_0)) \delta(\mathbf{x}' - \mathbf{X}(t, \mathbf{x}'_0)) \rangle.$$

Here and throughout all the chapter the symbol $\langle \cdot \rangle$ means the average over the realizations of the Eulerian velocity field.

We are dealing here with *homogeneous stationary incompressible high-Reynolds number* turbulence. According to the Kolmogorov's similarity hypotheses [146], in this case there exists a parameter $\bar{\varepsilon}$ (called mean rate of dissipation of kinetic energy) such that the statistical characteristics of the increments of the Eulerian velocity after normalyzing by a quantity v_η become universal functions (for all kind of fully developed turbulence) of a dimensionalless time t/τ_η and dimensionalless spatial distance r/η provided $r \ll L, t \ll T$. Here L and T are the external spatial and temporal scales, respectively; η is the Kolmogorov's spatial microscale which is related to the velocity microscale v_η and the time microscale τ_η by

$$\eta = (\nu^3/\bar{\varepsilon})^{1/4}, \quad v_\eta = (\nu\, \bar{\varepsilon})^{1/4}, \quad \tau_\eta = \eta / v_\eta,$$

where ν is the kinematic viscosity of the flow.

Another important property of the velocity field is that the 1-point distributions are approximately Gaussian while the structure of the 2-point distributions is more complicated. In particular, the p. d. f. of the velocity increments at two points separated by r is essentially non-Gaussian if $r \approx \eta$. The same is true in the lower part of the inertial subrange [160].

Unfortunately, there are no adequate models which satisfactorily govern the turbulent velocity field described. Therefore, different approximations are constructed. For instance, in [102, 113, 198], spectral models of Gaussian random velocity fields were used to numerically solve the problem (5.1). In [160, 170, 262], the Eulerian velocity field was obtained by the numerical solution of the Navier–Stokes equation.

All these Eulerian approximations have drawbacks: the models [102, 113, 198] are pure Gaussian, while the models [160, 170, 262] are restricted to moderate values of the Reynolds number. In addition, the latter models need a great deal of computer time.

Thus after choosing an appropriate approximation to $\mathbf{U}_E(\mathbf{x}, t)$, and numerically constructing the set of Lagrangian trajectories, one calculates the desired Lagrangian statistical characteristics by averaging. This method we call for short the E-scheme.

There is another method for simulation of Lagrangian trajectories which we call the Lagrangian scheme (for short, L-scheme). This approach is based on approximations of the stochastic Lagrangian velocity $\mathbf{V}(t, \mathbf{x}_0)$, say $\hat{\mathbf{V}}(t, \mathbf{x}_0)$. The relevant approximate Lagrangian trajectory is then obtained from the ordinary stochastic differential equation

$$\frac{d\hat{\mathbf{X}}}{dt} = \hat{\mathbf{V}}(t, \mathbf{x}_0), \tag{5.4}$$

with $\hat{\mathbf{X}}(\mathbf{x}_0, 0) = \mathbf{x}_0$. Note that this is generally a nonlinear equation, since the random process $\hat{\mathbf{V}}(t, \mathbf{x}_0)$ itself depends on the random position $\hat{\mathbf{X}}(t)$. Most frequently models of the type (5.4) are written in the form of stochastic differential equations (SDE) of the Ito type. These equations are much easier to solve numerically than the general equations of fluid particle motion (5.1) in random fields.

The models (5.1) and (5.4) define one and the same flow if $\hat{\mathbf{V}}(t, \mathbf{x}_0) = \mathbf{V}(t, \mathbf{x}_0)$ for all Lagrangian trajectories. It would be an ideal approximation if we could find $\mathbf{V}(t, \mathbf{x}_0)$. Of course, this is impossible to do even for the simplest flows. However, when constructing models of the type (5.4), it is reasonable to use some relations between the statistics of the Eulerian velocity field $\mathbf{U}_E(\mathbf{X}, t)$ and of the 6-dimensional Lagrangian field $\mathbf{Y}(t, \mathbf{x}_0) = (\mathbf{X}(t, \mathbf{x}_0, \mathbf{V}(t, \mathbf{x}_0))$. For instance, in the case of incompressible flows there is an integral relation between the probability density functions of $\mathbf{U}_E(\mathbf{x}, t)$ and $\mathbf{Y}(t, \mathbf{x}_0)$ (see [153]; a more general situation of compressible flows is treated in [154]). The E-scheme is rigorous; in this scheme the particles are moving in a self-consistent way. In particular, the following relation holds:

$$P_{1L}(\mathbf{x}, \vec{v}, t; \mathbf{x}_0) = \int P_{2L}(\mathbf{x}, \vec{v}, \mathbf{x}', \vec{v}', t; \mathbf{x}_0, \mathbf{x}_0') d\mathbf{x}' d\vec{v}', \tag{5.5}$$

where P_{1L}, P_{2L} are the 1- and 2-particle Lagrangian p. d. f.'s of the E-scheme:

$$P_{1L}(\mathbf{x}, \vec{v}, t; \mathbf{x}_0) = \langle \delta(\mathbf{x} - \mathbf{X}(t, \mathbf{x}_0)) \delta(\vec{v} - \mathbf{V}(t, \mathbf{x}_0)) \rangle,$$

and

$$\begin{aligned} &P_{2L}(\mathbf{x}, \vec{v}, \mathbf{x}', \vec{v}', t; \mathbf{x}_0, \mathbf{x}_0') \\ &\quad = \langle \delta(\mathbf{x} - \mathbf{X}(t, \mathbf{x}_0)) \delta(\vec{v} - \mathbf{V}(t, \mathbf{x}_0)) \delta(\mathbf{x}' - \mathbf{X}(t, \mathbf{x}_0')) \delta(\vec{v}' - \mathbf{V}(t, \mathbf{x}_0')) \rangle. \end{aligned}$$

However in the L-scheme, this relation is generally not satisfied. Therefore, in [238] it was noted that when constructing 2-particle Lagrangian models by the L-scheme, it is desired to satisfy the relation (5.5). This principle is called in [238] a "two-to-one" reduction. As far as we know, there is no L-scheme which satisfies this principle.

The incompressibility of $\mathbf{U}_E$ leads to certain restrictions on the L-scheme. In particular, the following integral relation was derived by Novikov [153]:

$$\begin{aligned} &P_{nE}(\vec{v}_1, \dots, \vec{v}_n; \mathbf{x}_1, \dots, \mathbf{x}_n) \\ &\quad = \int P_{nL}(\mathbf{x}_1, \vec{v}_1, \dots, \mathbf{x}_n, \vec{v}_n, t; \mathbf{x}_{01}, \dots, \mathbf{x}_{0n}) d\mathbf{x}_{01} \dots d\mathbf{x}_{0n}, \end{aligned}$$

where P_{nE}, P_{nL} are the n-point Eulerian and the n-particle Lagrangian p. d. f.'s, respectively:

$$P_{nE}(\vec{v}_1, \dots, \vec{v}_n; \mathbf{x}_1, \dots, \mathbf{x}_n) = \Big\langle \prod_{i=1}^{n} \delta(\vec{v}_i - \mathbf{U}_E(\mathbf{x}_i, t)) \Big\rangle,$$

$$P_{nL}(\mathbf{x}_1, \vec{v}_1, \dots, \mathbf{x}_n, \vec{v}_n, t; \mathbf{x}_{01}, \dots, \mathbf{x}_{0n}) = \Big\langle \prod_{i=1}^{n} \delta(\mathbf{x}_i - \mathbf{X}(t, \mathbf{x}_{0i})) \prod_{j=1}^{n} \delta(\vec{v}_j - \mathbf{V}(t, \mathbf{x}_{0j})) \Big\rangle .$$

In [169] this restriction is introduced as a consistency condition. In the cases $n = 1, 2$, this restriction was used by Thomson [237,238] (called by him a well-mixed condition) when constructing 1- and 2-particle L-schemes. Note also that Novikov [155] used his integral relation for $n = 2$ to construct stochastic Lagrangian models of the relative dispersion of two particles.

Remark 5.1. It should be noted that the practical interest of the L-scheme comes from the fact that it provides the desired behavior of the statistics of relative dispersion in the inertial subrange. In the E-scheme, this a one of the main problems.

Let us give a brief overview of the work on the 2-particle models. There are many 1-dimensional 2-particle models (e. g., see [44, 45, 123, 129, 213–218, 237], etc.). As mentioned in [114,238], these models do not adequately describe the 3D turbulent dispersion. As far as 2-particle 3D models in high-Reynolds-number isotropic stationary turbulence is concerned, we mention the Eulerian model presented in [62, 113, 198], and two Lagrangian models: Thomson's model [238], the Borgas-Sawford model [15] and the combined Eulerian–Lagrangian model proposed in [91].

Let us consider some features of these models. In the Eulerian model [62, 113] n-particle Lagrangian characteristics can be calculated in a self-consistent way in the sense that a "m to k" reduction principle holds which means that the k-particle p. d. f.'s (for $k < m$) are the marginal p. d. f.'s of the m-particle p. d. f.'s. However, its Gaussian nature leads to the failure of the relative dispersion statistics in the inertial subrange. In the models [15,238] this is not the case, but another difficult problem arises: there is no unique choice of the coefficients specifying the model, and the "two-to-one" reduction principle does not hold.

Models which take into account the statistical structure of the velocity in the inertial subrange were developed in [44,45,108,155,236]. Note that the 1-dimensional models have no adequate physical interpretation. However quasi-1-dimensional models which describe the evolution of the distance between two fluid particles in 3D space have clear physical meaning.

We call a model of relative dispersion quasi-1-dimensional if it describes the evolution of the distance $\Delta(t)$ between two fluid particles in the space R^3 [114].

Since the quasi-1-dimensional model describes the motion of two particles in 3D, it is reasonable to require that this model is consistent with the integral relation (5.6) in a sense given in the next subsection.

As to the Eulerian model, the Kaplan–Dinnar's model [91] also deals with n particles and satisfies the "m to k" reduction principle. However, the well-mixed condition for the 2-particle model is not satisfied (see the Appendix to this chapter).

In this chapter, we construct a new combined Eulerian-Lagrangian model which is believed to be free of the drawbacks of all the models mentioned above. It is based on a modification of our model (see [109, 113]) and a Lagrangian 3D model of relative dispersion [110].

5.2 2-particle models

5.2.1 Eulerian stochastic models of high-Reynolds-number pseudoturbulence

The numerical models of high-Reynolds-number pseudoturbulence were used in [102]. In [144], a randomized model was constructed for a simplified case, and in [153, 191], a randomized model of 3-dimensional incompressible Gaussian stationary velocity with the Kolmogorov spectral density frozen in time was created. In this section we modify this model by introducing correlation in time. The correlation is chosen in accordance with the Kolmogorov similarity hypotheses.

Let us assume that the Eulerian velocity field $\mathbf{U}_E(\mathbf{x}, t)$ has the following partial energy spectrum:

$$E(k,t) = E(k)\varphi\left(\frac{t}{\tau(k)}\right), \tag{5.6}$$

where $E(k) = E(k,0)$ is the 3-dimensional energy spectrum; φ, τ are given functions such that $\varphi(0) = 1$, and $\varphi(t)$ is an even positive definite function of t, i. e.,

$$\tilde{\varphi}(\omega) = \int_{-\infty}^{\infty} \exp(i\omega t)\varphi(t)dt \geq 0.$$

In addition it is assumed that $\tau(k) > 0$.

These assumptions come from the condition that

$$\Psi_{jl}(\mathbf{k}, t) = \frac{E(k,t)}{4\pi k^2}\left(\delta_{jl} - \frac{k_j k_l}{k^2}\right), \quad j, l = 1, 2, 3, \tag{5.7}$$

is a partial spatial-temporal spectral tensor of an isotropic random field. The energy spectrum is defined by

$$E(k) = \begin{cases} C_1 \bar{\varepsilon}^{2/3} k^{-5/3}, & k_{\min} \leq k \leq k_{\max}, \\ 0, & \text{otherwise} \end{cases} \tag{5.8}$$

with the normalization

$$\int_0^{\infty} E(k)dk = \frac{3}{2}u_0^2. \tag{5.9}$$

For the frequency part, we use four models. Model 1 is

$$\tau(k) = \frac{1}{k_0 u_0}, \quad \varphi\left(\frac{t}{\tau(k)}\right) = \exp\left\{-\frac{1}{2}k_0^2 u_0^2 t^2\right\}. \tag{5.10}$$

In the next three models we choose

$$\tau(k) = (\bar{\varepsilon}k^2)^{-1/3}, \tag{5.11}$$

so that in Model 2

$$\varphi\left(\frac{t}{\tau(k)}\right) = \exp\left\{-\alpha\frac{t}{\tau(k)}\right\} = e^{-\alpha\bar{\varepsilon}^{1/3}k^{2/3}t}, \tag{5.12}$$

and in Model 3

$$\varphi\left(\frac{t}{\tau(k)}\right) = \exp\left\{-\alpha\left(\frac{t}{\tau(k)}\right)^2\right\} = e^{-\alpha\bar{\varepsilon}^{2/3}k^{4/3}t^2}. \tag{5.13}$$

Finally, Model 4 is:

$$\varphi\left(\frac{t}{\tau(k)}\right) = \exp\left\{-\alpha\frac{t}{\tau(k)}\right\}\cos\left(\mu\frac{t}{\tau(k)}\right) = e^{-\alpha\bar{\varepsilon}^{1/3}k^{2/3}t}\cos(\mu\bar{\varepsilon}^{1/3}k^{2/3}t). \tag{5.14}$$

In formulas (5.12)–(5.14), $\alpha \geq 0$, $\mu \geq 0$ are some parameters of the model. The quantity $\bar{\varepsilon}$ is the rate of energy dissipation, $k_{\min}$ and $k_{\max}$ are the minimal and maximal wave numbers, respectively. Thus $k_{\min} = 2\pi/L$ where L is the external length scale, C_1 is a universal constant ($C_1 = 1.4$).

It is easy to verify that for all four models the relevant correlation tensor is positive definite.

The input parameters of the models are: $\bar{\varepsilon}$, L, $k_{\max}$, u_0, and k_0 (Model 1), plus the quantity α in Models 2, 3 and plus μ in Model 4. These parameters are related by the condition (5.9), so we consider $\bar{\varepsilon}$ to be a dependent parameter.

To construct the simulation formula for the velocity field, it is necessary to have the spatial-temporal spectra $\Phi_{lm}(\mathbf{k}, \omega)$ (see Section 1.2.4 in [191]). From

$$\Phi_{lm}(\mathbf{k}, \omega) = \frac{1}{2\pi}\int_{\infty}^{\infty} \Psi_{lm}(\mathbf{k}, t)e^{-i\omega t}\,dt. \tag{5.15}$$

From (5.6), (5.7) we find for the model $p = 1, 2, 3, 4$:

$$\Phi_{lm}(\mathbf{k}, \omega) = F_{lm}(\mathbf{k})q_p(k, \omega), \quad k = |\mathbf{k}| = \sqrt{k_1^2 + k_2^2 + k_3^2}, \tag{5.16}$$

where

$$F_{lm}(\mathbf{k}) = \frac{E(k)}{4\pi k^2}\left(\delta_{lm} - \frac{k_l k_m}{k^2}\right), \quad l, m = 1, 2, 3,$$

and

$$q_1(k,\omega) = \frac{1}{\sqrt{2\pi}k_0u_0}\exp\left\{-\frac{\omega^2}{2k_0^2u_0^2}\right\},$$

$$q_2(k,\omega) = \frac{\alpha\bar{\varepsilon}^{1/3}k^{2/3}}{\pi(\alpha^2\varepsilon^{2/3}k^{4/3}+\omega^2)},$$

$$q_3(k,\omega) = \frac{1}{\varepsilon^{1/3}k^{2/3}\sqrt{\alpha\pi}}\exp\left\{-\frac{\omega^2}{4\alpha\varepsilon^{2/3}k^{4/3}}\right\},$$

$$q_4(k,\omega) = \frac{\alpha\bar{\varepsilon}^{1/3}k^{2/3}(\alpha^2\bar{\varepsilon}^{2/3}k^{4/3}+\mu^2\bar{\varepsilon}^{2/3}k^{4/3}+\omega^2)}{\pi[\alpha^2\bar{\varepsilon}^{2/3}k^{4/3}+(\mu\bar{\varepsilon}^{1/3}k^{2/3}-\omega)^2]} \times \frac{1}{[\alpha^2\bar{\varepsilon}^{2/3}k^{4/3}+(\mu\bar{\varepsilon}^{1/3}k^{2/3}+\omega)^2]}.$$

Now we can write the randomized model of the homogeneous stationary isotropic Gaussian random field $\mathbf{U}_E(X,t)$ specified by the spatial-temporal spectral tensor $\Phi^p_{lm}(\mathbf{k},\omega)$ given by (5.16) (for the derivation of the simulation formula, see [191]):

$$\mathbf{U}_E(\mathbf{x},t) = \sum_{j=1}^{n}\sqrt{E_n}\Big\{(\boldsymbol{\xi}_j\times\boldsymbol{\Omega}_j)\cos(\theta_j^p)+(\boldsymbol{\eta}_j\times\boldsymbol{\Omega}_j)\sin(\theta_j^p)\Big\}, \tag{5.17}$$

where $\theta_j^p = k_j(\boldsymbol{\Omega}_j,\mathbf{x})+\omega_j^p t$, and $\boldsymbol{\Omega}_j = (\Omega_j^{(1)},\Omega_j^{(2)},\Omega_j^{(3)})$, $j=1,\dots,n$ are independent 3-dimensional isotropic unit vectors for models $p=1,2,3,4$; $\boldsymbol{\xi}_j = (\xi_j^{(1)},\xi_j^{(2)},\xi_j^{(3)})$ and $\boldsymbol{\eta}_j = (\eta_j^{(1)},\eta_j^{(2)},\eta_j^{(3)})$ are mutually independent standard Gaussian vectors; k_j, $j=1,\dots n$ are random variables with the densities

$$p_j(k) = \begin{cases}\dfrac{1}{E_n}E(k), & k\in\Delta_j,\\ 0, & \text{otherwise,}\end{cases}$$

with

$$E_n = \int_{\Delta_j}E(k)dk, \quad j=1,\dots,n,$$

and Δ_j, $j=1,\dots,n$ are nonoverlapping segments which compose a partition of $\Delta=(0,\infty)$, the support of the spectrum. For each fixed $k=k_j$, $j=1,\dots,n$, the random quantities ω_j^p, $p=1,\dots,4$, $j=1,\dots,n$ are distributed with the densities $q_p(k_j,\omega)$, $j=1,\dots,n$.

The partition of the spectrum support

$$\Delta = \bigcup_{j=1}^{n}\Delta_j, \quad \Delta_j\bigcap\Delta_l=\emptyset, \quad j\neq l,$$

where $\Delta_j = [\tilde{k}_j, \tilde{k}_{j+1})$, $j = 1, \dots, n$, $k_1 = k_{\min}$, $\tilde{k}_{n+1} = k_{\max}$, is chosen so that (e. g., see [191])

$$\int_{\Delta_j} E(k)dk = \frac{1}{n}\int_{\Delta} E(k)dk.$$

From this, we find

$$\tilde{k}_{j+1} = \left[k_{\min}^{-2/3}\left(1-\frac{j}{n}\right) + \frac{j}{n}k_{\max}^{-2/3}\right]^{-3/2}, \quad j = 1, \dots, n.$$

The random numbers k_j are simulated by

$$k_j = \left[\tilde{k}_j^{-2/3} - \frac{u_0^2}{nC_1\bar{\varepsilon}^{2/3}}\gamma_{j,1}\right]^{-3/2}, \quad j = 1, \dots, n, \tag{5.18}$$

and the random frequencies ω_j^p are simulated as

$$\begin{aligned}
\omega_j^1 &= k_0 u_0 \sqrt{-2\ln\gamma_{j,2}}\sin(2\pi\gamma_{j,3}),\\
\omega_j^2 &= \alpha\bar{\varepsilon}^{1/3}k_j^{2/3}\tan\left(\gamma_{j,4}\pi - \frac{\pi}{2}\right),\\
\omega_j^3 &= \sqrt{2\alpha}\bar{\varepsilon}^{1/3}k_j^{2/3}\sqrt{-2\ln(\gamma_{j,5}}\sin(2\pi\gamma_{j,6}),
\end{aligned}$$

and

$$\omega_j^4 = \begin{cases} \bar{\varepsilon}^{1/3}k_j^{2/3}\left(\mu - \alpha\tan\left(\frac{\pi}{2} - \gamma_{j,8}\pi\right)\right) & \text{if } \gamma_{j,7} < 0.5\\ -\bar{\varepsilon}^{1/3}k_j^{2/3}\left(\mu + \alpha\tan\left(\frac{\pi}{2} + \gamma_{j,8}\pi\right)\right) & \text{if } \gamma_{j,7} \geq 0.5. \end{cases}$$

In the above formulas, $\gamma_{j,l}$, $j = 1, \dots, n$, $l = 1, \dots, 8$ are mutually independent random numbers uniformly distributed in [0, 1].

5.3 A new 2-particle Eulerian–Lagrangian stochastic model

5.3.1 Formulation of 2-particle Eulerian–Lagrangian model

The motion of two particles can often be regarded as a motion of one particle plus the relative motion of a second particle with respect to the first one. Thus the motion is expanded into absolute and relative dispersion. This expansion can be carried out in different ways. For instance, it is possible to model the 1-particle coordinate $X^{(1)}$, and then to simulate the motion of the relative vector $\mathbf{X}^{(2)} - \mathbf{X}^{(1)}$.

Another description uses the center of the 2-particle coordinates $\mathbf{R}(t) = [\mathbf{X}^{(1)}(t) + \mathbf{X}^{(2)}(t)]/2$ and the difference between the coordinates:

$$\mathbf{r}(t) = \mathbf{X}^{(2)}(t) - \mathbf{X}^{(1)}(t).$$

The motivation of such a description is that at small separation, when $r(t) = |\mathbf{X}^{(2)} - \mathbf{X}^{(1)}| \ll L$, the random processes $\mathbf{R}(t)$ and $\mathbf{r}(t)$ can be simulated independently (e. g., see [146]). However, when studying a long-time behavior, the correlation between $\mathbf{R}(t)$ and $\mathbf{r}(t)$ should be properly taken into account. This is a difficult problem in the framework of Lagrangian models (e. g., see [15, 238]).

In this section, we present a reasonable expansion using both the Eulerian and the Lagrangian stochastic models, which we call the combined Eulerian–Lagrangian model.

Ideally, if we had at hand the Eulerian velocity field $\mathbf{U}_E$, we could write

$$\frac{d\mathbf{R}}{dt} = \frac{1}{2}\Big\{\mathbf{U}_E\Big(\mathbf{R} + \frac{\mathbf{r}}{2}, t\Big) + \mathbf{U}_E\Big(\mathbf{R} - \frac{\mathbf{r}}{2}, t\Big)\Big\}, \tag{5.19}$$

$$\frac{d\mathbf{r}}{dt} = \mathbf{U}_E\Big(\mathbf{R} + \frac{\mathbf{r}}{2}, t\Big) - \mathbf{U}_E\Big(\mathbf{R} - \frac{\mathbf{r}}{2}, t\Big). \tag{5.20}$$

In practice, one deals with approximate models of $\mathbf{U}_E$ (for instance, the DNS models [160], or the randomized spectral models described in the previous section). These models are reasonable for large scales, but they are unacceptable for small values of r. This means that in (5.19), the sum in the right-hand side of the first equation can be well simulated by these approximate models, while for the difference of the second equation, this is not the case if $r \ll L$. Note that for $r \sim L$, both the difference and the sum of velocities in (5.19) are well approximated by the models mentioned above.

Thus we have to construct a different model for $\frac{d\mathbf{r}}{dt}$ at $r \ll L$. To this end, we will use Lagrangian models of relative motion. Note that the models of relative diffusion are of high interest in many applied fields (e. g., see [15,20,45,108,114,155,216,238]). As mentioned in the introduction, these models are not uniquely defined. We use a unique closed model proposed by one of the authors [110].

Thus let us stress that the motivation to turn to more sophisticated models for the velocity difference comes from the fact that the above models ignore a fundamental small-scale characteristic of turbulence (e. g., see [62]).

We recall that we deal with flows with extremely large Reynolds numbers. Therefore, at scales which are much less than the external scale L, the $\mathbf{U}_E$ can be considered as a random locally isotropic field. The relative velocity is governed by a generalized Langevin equation [110]

$$\frac{d\mathbf{r}}{dt} = \vec{v}(t), \tag{5.21}$$

$$d\vec{v} = \mathbf{a}(\mathbf{r}, \vec{v})\, dt + \sqrt{2C_0\bar{\varepsilon}}\, d\mathbf{W}(t), \tag{5.22}$$

where $\mathbf{W}$ is a standard 3D Wiener process and

$$\mathbf{a}(\mathbf{r}, \vec{v}) = \Big(\frac{\bar{\varepsilon}^2}{r}\Big)^{1/3}\Big\{a_0(\xi_\parallel, \xi_\perp)\frac{\mathbf{r}}{r} + a_1(\xi_\parallel, \xi_\perp)\frac{\vec{v}}{v}\Big\}. \tag{5.23}$$

Here,

$$\xi_\| = \frac{v_\|}{(\bar{\varepsilon} r)^{1/3}}, \qquad \xi_\perp = \frac{v_\perp}{(\bar{\varepsilon} r)^{1/3}}, \tag{5.24}$$
$$v_\| = \left(\vec{v}, \frac{\mathbf{r}}{r}\right), \qquad v_\perp = \sqrt{\vec{v}^2 - v_\|^2},$$

are the longitudinal component of relative velocity vector and the modulus of the transverse component, $v = |\vec{v}|$, and $a_0(\xi_\|, \xi_\perp), a_0(\xi_\|, \xi_\perp)$ are dimensionless functions of dimensionless arguments, $-\infty < \xi_\| < \infty$, $\xi_\| \geq 0$. These functions are defined through the Eulerian p. d. f.

$$p_E^{(\mathrm{rel})}(\vec{v}; \mathbf{r}) = \Big\langle \delta\Big(\vec{v} - \big(\mathbf{U}_E(\mathbf{x} + \mathbf{r}, t) - \mathbf{U}_E(\mathbf{x}, t)\big)\Big)\Big\rangle$$

which is usually assumed to be known.

Due to the isotropy, the function $p_E^{(\mathrm{rel})}$ depends only on $(v_\|, v_\perp; r)$.

Moreover, by the second Kolmogorov hypothesis, it has the form [146]

$$p_E^{(\mathrm{rel})}(v_\|, v_\perp; r) = \frac{1}{\bar{\varepsilon} r} f_E\left(\frac{v_\|}{(\bar{\varepsilon} r)^{1/3}}, \frac{v_\perp}{(\bar{\varepsilon} r)^{1/3}}\right), \tag{5.25}$$

where f_E is a dimensionless function of dimensionless arguments normalized so that

$$\int_{-\infty}^{\infty} d\xi_\| \int_0^\infty 2\pi \xi_\perp f_E(\xi_\|, \xi_\perp) d\xi_\perp = 1.$$

In the consideration which follows, the function f_E is restricted to be factorized:

$$f_E(\xi_\|, \xi_\perp) = f_E^\|(\xi_\|) f_E^\perp(\xi_\perp). \tag{5.26}$$

This means that the longitudinal and the modulus of the transverse components of the relative velocity are independent. Note that the function $f_E^\perp(\xi_\perp)$ is a density of the transverse component of the dimensionless relative velocity:

$$\int_{-\infty}^{\infty} \int_{-\infty}^{\infty} f_E^\perp\left(\sqrt{\xi_1^2 + \xi_2^2}\right) d\xi_1 d\xi_2 = 1.$$

In this case, the functions a_0, a_1 are defined as (see formulas (5.10)–(5.11) in [110])

$$a_0(\xi_\|, \xi_\perp) = C_0 \frac{d}{d\xi_\|} \ln[f_E^\|(\xi_\|)] - C_0 \frac{\xi_\|}{\xi_\perp} \frac{d}{d\xi_\perp} \ln[f_E^\perp(\xi_\perp)] \tag{5.27}$$
$$- (\xi_\|^2 + \xi_\perp^2) - \frac{7}{3 f_E^\|(\xi_\|)} \int_{-\infty}^{\xi_\|} y f_E^\|(y) dy \tag{5.28}$$

and

$$a_1(\xi_\parallel, \xi_\perp) = \frac{C_0\sqrt{\xi_\parallel^2 + \xi_\perp^2}}{\xi_\perp} \frac{d}{d\xi_\perp} \ln[f_E^\perp(\xi_\perp)] + \frac{4}{3}\xi_\parallel\sqrt{\xi_\parallel^2 + \xi_\perp^2}. \tag{5.29}$$

Now we are ready to write the combined Eulerian–Lagrangian 2-particle model. Let us define a function $\theta_{\alpha,\beta}(y)$ for $y > 0$:

$$\theta(y) = \theta_{\alpha,\beta}(y) = \begin{cases} 0, & 0 < y < \alpha, \\ \gamma(y), & y \in (\alpha, \beta), \\ 1, & y > \beta, \end{cases}$$

where $0 < \alpha \ll \beta < 1$, and γ is an arbitrary monotonically increasing function so that θ is smooth on $(0, \infty)$. The combined Eulerian–Lagrangian 2-particle model in the variables $\mathbf{R}$, $\mathbf{r}$ has the form

$$\begin{aligned} \frac{d\mathbf{R}}{dt} &= \frac{1}{2}\left\{\mathbf{U}_E\left(\mathbf{R} + \frac{\mathbf{r}}{2}, t\right) + \mathbf{U}_E\left(\mathbf{R} - \frac{\mathbf{r}}{2}, t\right)\right\}, \\ \frac{d\mathbf{r}}{dt} &= \theta\left(\frac{r}{L}\right)\left\{\mathbf{U}_E\left(\mathbf{R} + \frac{\mathbf{r}}{2}, t\right) - \mathbf{U}_E\left(\mathbf{R} - \frac{\mathbf{r}}{2}, t\right)\right\} + \left\{1 - \theta\left(\frac{r}{L}\right)\right\}\vec{v}(t), \\ d\vec{v}(t) &= \mathbf{a}(\mathbf{r}, \vec{v})dt + \sqrt{2C_0\bar{\varepsilon}}\, d\mathbf{W}(t), \end{aligned}$$

where $\mathbf{U}_E$ is the pseudoturbulent velocity taken in the form as described in Section 5.2, and $\mathbf{a}(\mathbf{r}, \vec{v})$ is defined in (5.23). The function θ is defined so that for small r ($r < \alpha L$), the function $\frac{d\mathbf{r}}{dt}$ is governed by (5.21), while for large r ($r > \beta L$), we come to the system (5.19). Thus, α and β are free parameters of the model.

5.3.2 Models for the p. d. f. of the Eulerian relative velocity

Thus to specify our model, we need the p. d. f. $p_E^{(\mathrm{rel})}$. Unfortunately, the information about the Eulerian p. d. f. $p_E^{(\mathrm{rel})}$ is usually approximate, taken from measurements or calculations. Therefore, we have to use some analytical formula approximating the Eulerian p. d. f. For simplicity, we construct the p. d. f. f_E which is defined in (5.25) so that (5.26) is satisfied. The function $f_E^\parallel$ is chosen in a form of a sum of three Gaussian densities

$$\begin{aligned} f_E^\parallel(\xi) = \frac{(1-q_0)}{\sqrt{2\pi C}} \exp\left(-\frac{\xi^2}{2C}\right) + q_0\left\{\frac{(1+x)}{2}\frac{1}{\sqrt{2\pi}\sigma}\exp\left(-\frac{(\xi - a_1)^2}{2\sigma^2}\right)\right. \\ \left. + \frac{(1-x)}{2}\frac{1}{\sqrt{2\pi}\sigma}\exp\left(-\frac{(\xi - a_2)^2}{2\sigma^2}\right)\right\}, \end{aligned} \tag{5.30}$$

where C is the universal constant in Kolmogorov's law of two thirds, and

$$a_1 = \left\{\frac{2(1-x)^2}{5q_0x(1+x)}\right\}^{1/3}, \quad a_2 = -\frac{(1+x)}{(1-x)}a_1,$$

$$\sigma = \left\{C - \frac{(1+x)}{2}a_1^2 - \frac{(1-x)}{2}a_2^2\right\}^{1/2}. \tag{5.31}$$

The parameters are chosen so that $0 \le q_0 < 1$, and $x \in (x_0, 1)$, where $x_0 = x_0(q_0) = (1 + \frac{25}{4}q_0^2C^3)^{-1/2}$. This choice ensures that the moments satisfy the known relations [114, 146]

$$\int_{-\infty}^{\infty} f_E^{\|}(\xi)\, d\xi = 1, \qquad \int_{-\infty}^{\infty} \xi f_E^{\|}(\xi)\, d\xi = 0,$$

$$\int_{-\infty}^{\infty} \xi^2 f_E^{\|}(\xi)\, d\xi = C, \quad \int_{-\infty}^{\infty} \xi^3 f_E^{\|}(\xi)\, d\xi = -\frac{4}{5}.$$

Note that the free parameters q_0 and x can be chosen to satisfy some additional constrains for the fourth and fifth cumulants. Indeed, let us consider the excess E and a quantity R defined by

$$\int_{-\infty}^{\infty} \xi^4 f_E^{\|}(\xi)d\xi = C^2(E+3), \quad \int_{-\infty}^{\infty} \xi^5 f_E^{\|}(\xi)d\xi = C^{5/2}(R+10S),$$

where $S = 4/(5C^{3/2})$ is the asymmetry. From (5.30)–(5.31) we find

$$E = \left(\frac{2}{5}\right)^{4/3} \frac{1}{C^2 q_0^{1/3}} \frac{2(3x^2-1)}{(1-x^2)^{1/3}x^{4/3}},$$

$$R = -\left(\frac{2}{5}\right)^{5/3} \frac{1}{C^{5/2} q_0^{2/3}} \frac{4(x^2+1)}{(x(1-x^2))^{2/3}}.$$

Thus if the quantities E and R are given, the values of x and q_0 can be found from the latter relations.

We assume that the transverse component of the relative velocity is Gaussian. From the Kolmogorov two thirds law, $\langle v_\perp^2\rangle = 2C'(\bar{\varepsilon}r)^{2/3}$ where $C' = 4C/3$. The function $f_E^{\perp}$ takes in this case the form

$$f_E(\xi) = \frac{1}{2\pi C'}\exp\left\{-\frac{\xi^2}{2C'}\right\}.$$

Remark 5.2. In the case of Gaussian $p_E^{(rel)}$, the coefficient $\mathbf{a}(\mathbf{r}, \vec{v})$ can be found explicitly. Indeed, in this case we have

$$f_E^{\|}(\xi) = \frac{1}{\sqrt{2\pi C}}\exp\left\{-\frac{\xi^2}{2C}\right\}, \quad f_E^{\perp}(\xi) = \frac{1}{2\pi C'}\exp\left\{-\frac{\xi^2}{2C'}\right\}.$$

From (5.27), (5.29), and (5.23) we find

$$\mathbf{a}(\mathbf{r},\vec{v}) = \left(\frac{\bar{\varepsilon}^2}{r}\right)^{1/3}\left\{\left[\frac{7}{3}C - \frac{C_0}{4C}\frac{v_\parallel}{(\bar{\varepsilon}r)^{1/3}} - \frac{v^2}{(\bar{\varepsilon}r)^{2/3}}\right]\frac{\mathbf{r}}{r} + \left[\frac{4}{3}\frac{v_\parallel}{(\bar{\varepsilon}r)^{1/3}} - \frac{C_0}{C'}\right]\frac{\vec{v}}{(\bar{\varepsilon}r)^{1/3}}\right\}.$$

5.4 Appendix

The 2-particle model [91] is given in a finite-difference form:

$$\vec{v}^{(k)}(t+\Delta t) = R_L(\Delta t)\vec{v}^{(k)}(t) + \sqrt{1 - R_L^2(\Delta t)}\,\boldsymbol{\theta}(\mathbf{x}^{(k)}(t)),$$
$$\mathbf{x}^{(k)}(t+\Delta t) = \mathbf{x}^{(k)}(t) + \vec{v}^{(k)}(t)\Delta t, \quad k = 1,2,$$

with the initial conditions $\vec{v}^{(k)}(0) = \boldsymbol{\theta}(\mathbf{x}^{(k)}(0))$, $k = 1,2$.

Its continuous analog can be written as

$$\frac{d\mathbf{x}^{(k)}}{dt} = \vec{v}^{(k)}(t), \tag{A.1}$$

$$\frac{d\vec{v}^{(k)}}{dt} = -\frac{\vec{v}^{(k)}}{T_L} + \sqrt{\frac{2}{T_L}}\,\boldsymbol{\theta}(\mathbf{x}^{(k)}(t),t), \quad k = 1,2. \tag{A.2}$$

Here $\boldsymbol{\theta}$ is a Gaussian random field with

$$\langle\theta_i(\mathbf{x},t)\theta_j(\mathbf{x}',t')\rangle = C_{ij}(\mathbf{r})\delta(t-t'), \quad i,j = 1,2,3,$$

$$C_{ij}(\mathbf{r}) = \sigma^2\left\{\mu(r)\delta_{ij} + \nu(r)\frac{r_i r_j}{r^2}\right\}, \quad \mathbf{r} = x - x',$$

where μ and ν are functions of the distance $r = |\mathbf{x}' - \mathbf{x}|$:

$$\mu = g(r), \quad \nu = f(r) - g(r), \quad g(r) = f(r) + \frac{1}{2}rf'(r),$$

$$f(r) = 1 - \left(\frac{r^2}{r^2 + L^2}\right)^{1/3}.$$

Let us introduce the column vectors:

$$\mathbf{X} = (\mathbf{x}^{(1)},\mathbf{x}^{(2)})^T, \quad \mathbf{V} = (\vec{v}^{(1)},\vec{v}^{(2)})^T.$$

Then the pair of systems (16.47) is written as

$$\frac{d\mathbf{X}}{dt} = \mathbf{V},$$

$$\frac{d\mathbf{V}}{dt} = -\frac{\mathbf{V}}{T_L} + \boldsymbol{\Theta}(\mathbf{X},t),$$

where

$$\boldsymbol{\Theta}(\mathbf{X},t) = \sqrt{\frac{2}{T_L}}\begin{pmatrix}\boldsymbol{\theta}(\mathbf{x}^{(1)},t)\\ \boldsymbol{\theta}(\mathbf{x}^{(2)},t)\end{pmatrix}.$$

Hence,

$$\langle \boldsymbol{\Theta}(\mathbf{X},t)\boldsymbol{\Theta}^T(\mathbf{X}',t')\rangle = \frac{2}{T_L}\begin{bmatrix} C(\mathbf{x}^{(1)}-\mathbf{x}'^{(1)}) & C(\mathbf{x}^{(1)}-\mathbf{x}'^{(2)})\\ C(\mathbf{x}^{(2)}-\mathbf{x}'^{(1)}) & C(\mathbf{x}^{(2)}-\mathbf{x}'^{(2)})\end{bmatrix},$$

where C is the matrix with the entries $C_{ij},\ i,j = 1,2,3$.

We noe need to derive the Fokker–Planck equation. We present here the general case given in [96]. Let us consider the system of random equations

$$\frac{dx_i}{dt} = f_i(\mathbf{x},t) + \xi_i(\mathbf{x},t), \quad (i = 1,\dots,n), \tag{A.3}$$

where f_i are deterministic functions and ξ_i are Gaussian random fields such that

$$\langle \xi_i(\mathbf{x},t)\xi_j(\mathbf{x}',t')\rangle = K_{ij}(\mathbf{x},\mathbf{x}',t)\delta(t-t') \quad (i,j = 1,\dots,n).$$

Let $p(\mathbf{x},t;\mathbf{x}_0) = \langle \delta(\mathbf{x}-\mathbf{x}(t;\mathbf{x}_0))\rangle$ be the transition density, where $\mathbf{x}(t;\mathbf{x}_0)$ is the solution of the system of random equations (A.3) with the initial conditions $\mathbf{x}(0;\mathbf{x}_0) = \mathbf{x}_0$. Then the Fokker–Planck equation for the density p has the form

$$\frac{\partial p}{\partial t} + \frac{\partial\,[A_i(\mathbf{x},t)\,p]}{\partial x_i} = \frac{1}{2}\frac{\partial^2}{\partial x_i \partial x_j}\{B_{ij}(\mathbf{x},t)p\},$$

where

$$A_i(\mathbf{x},t) = f_i(\mathbf{x},t) + \frac{1}{2}\frac{K_{ij}(\mathbf{x},\mathbf{x}',t)}{\partial x_j}\bigg|_{\mathbf{x}'=\mathbf{x}},$$

and $B_{ij}(\mathbf{x},t) = K_{ij}(\mathbf{x},\mathbf{x},t)$.

Thus in our case the Fokker–Planck equation reads

$$\frac{\partial p_L}{\partial t} + V_\alpha\frac{\partial p_L}{\partial X_\alpha} - \frac{1}{T_L}\frac{\partial}{\partial V_\alpha}(V_\alpha p_L) = \frac{\sigma^2}{T_L}\frac{\partial^2}{\partial V_\alpha \partial V_\beta}(\Sigma_{\alpha\beta}\,p_L), \tag{A.4}$$

where

$$\|\Sigma_{\alpha\beta}\|_{\alpha,\beta=1}^6 = \begin{pmatrix} I & R\\ R & I\end{pmatrix}, \quad R = \frac{1}{\sigma^2}C$$

and I is a 3×3-identity matrix.

As commonly used, we consider the case

$$p_E(V) = (2\pi\sigma)^{-3}\sqrt{\lambda}\exp\left\{-\frac{1}{2\sigma^2}\lambda_{\alpha\beta}V_\alpha V_\beta\right\},$$

where $\lambda_{\alpha\beta}$ are the entries of the matrix

$$\Lambda = \|\lambda_{\alpha\beta}\|_{\alpha,\beta=1}^{6} = \left(\|\Sigma_{\alpha\beta}\|_{\alpha,\beta=1}^{6}\right)^{-1} = \begin{pmatrix} I & R \\ R & I \end{pmatrix}^{-1},$$

and

$$\lambda = \det \Lambda.$$

Let us introduce the notation

$$\xi = 1 - \mu^2, \quad \eta = 1 - (\mu + \nu)^2.$$

Then through some algebra we find

$$\lambda = \frac{1}{\eta\xi^2},$$

$$\Lambda = \begin{pmatrix} A & B \\ B & A \end{pmatrix}, \quad A = (I - R^2)^{-1}, \quad B = -RA,$$

where

$$A = \|\alpha_{ij}\|_{i,j=1}^{3}; \quad B = \|\beta_{ij}\|_{i,j=1}^{3},$$

with

$$\alpha_{ij} = \frac{1}{\xi}\delta_{ij} + \frac{2\mu\nu + \nu^2}{\xi\eta}\frac{r_i r_j}{r^2},$$

$$\beta_{ij} = -\frac{\mu}{\xi}\delta_{ij} - \frac{\mu\nu(\mu + \nu) + \nu}{\xi\eta}\frac{r_i r_j}{r^2}.$$

To derive the well-mixed condition, we need the following representations which can be found by differentiating the function p_E:

$$\frac{\partial p_E}{\partial X_\alpha} = \frac{1}{2} p_E \left\{ \frac{\partial \ln \lambda}{\partial X_\alpha} - \frac{1}{\sigma^2}\frac{\partial \lambda_{\beta\gamma}}{\partial X_\alpha} V_\beta V_\gamma \right\},$$

$$\frac{\partial p_E}{\partial V_\alpha} = -\frac{1}{\sigma^2} p_E \lambda_{\alpha\beta} V_\beta,$$

$$\frac{\partial^2 p_E}{\partial V_\alpha \partial V_\beta} = -\frac{1}{\sigma^2} p_E \left\{ -\lambda_{\alpha\beta} + \frac{1}{\sigma^2}\lambda_{\alpha\gamma} V_\gamma \lambda_{\beta\delta} V_\delta \right\}.$$

The well-mixed condition requires that p_E satisfies (16.48). Since p_E does not depend on t, this implies that

$$\frac{1}{2} V_\alpha p_E \left[\frac{\partial \ln \lambda}{\partial X_\alpha} - \frac{1}{\sigma^2}\frac{\partial \lambda_{\beta\gamma}}{\partial X_\alpha} V_\beta V_\gamma \right] - \frac{6}{T_L} p_E - \frac{V_\alpha}{T_L}\frac{1}{\sigma^2} p_E \lambda_{\alpha\beta} V_\beta = \frac{p_E\, \Sigma_{\alpha\beta}}{T_L} \left\{ -\lambda_{\alpha\beta} + \frac{1}{\sigma^2}\lambda_{\alpha\gamma} V_\gamma \lambda_{\beta\delta} V_\delta \right\}. \quad \text{(A.5)}$$

When considering the terms with equal powers of the velocity we note that the "zero-power terms" are equal: $6p_E/T_L = p_E \Sigma_{\alpha\beta}\lambda_{\alpha\beta}/T_L$. However the linear and cubic terms can not vanish, hence, the well-mixed condition is not satisfied for this model.

Chapter 6

Stochastic Lagrangian models for 2-particle relative dispersion in high-Reynolds-number turbulence

A new stochastic model (a diffusion approximation) for the relative dispersion of pair of particles in high-Reynolds-number incompressible turbulence is proposed. An attempt is made to uniquely define the coefficients of SDE governing the relative dispersion process under a closure assumption about the quasi-1-dimensional model. An approach for constructing a diffusion approximation of the relative dispersion taking into account the intermittency is proposed.

6.1 Introduction

To model processes such as mixing and combustion we need statistical information about the relative displacement of particle pairs in a turbulent flow (e. g., [44, 214]). In this chapter we will consider the process of relative dispersion of two particles (briefly, the relative dispersion process) in small spatial and temporal regions compared to the external scale of turbulence. For the high-Reynolds number flows, in these scales the Kolmogorov similarity hypotheses hold [146]. Therefore, universal laws are true for the relative dispersion process.

Classical results due to Batchelor in the problem under discussion are well known [10]. However a series of unknown functions and constants in this theory is yet not explicitly known. Therefore, there is a much interest in constructing models which enable to find the desired statistical characteristics of the relative dispersion process.

There are several stochastic Lagrangian 2-particle models of turbulent dispersion. survey of literature on the given subject can be found in [114, 146]. In [237], a new approach to construct stochastic Lagrangian models which are consistent with the incompressibility was proposed. This approach deals with a generalized Langevin equation governing the dynamics of the particle's motion whose coefficients satisfy the so-called well mixed condition. It should be noted that this condition does not define uniquely the stochastic model even in the case of isotropic stationary turbulence [15]. Therefore, it is important to find out physically plausible constraints on these coefficients.

In this chapter we propose an assumption which, together with the well-mixed condition due to Thomson, lead to a uniquely defined stochastic model of relative dispersion in the isotropic or locally isotropic turbulence. This approach treats the case of high-Reynolds-number turbulence with intermittency in a general manner.

6.2 Preliminaries

Throughout this chapter we follow the statistical approach to turbulent flows, i. e., we regard the flow as a member of an ensemble of flows with identical external conditions (e. g., [146]) and consider only ensemble average quantities. Such averages will be denoted by angled brackets. Thus the velocity field $\mathbf{U}_E(\mathbf{x}, t)$ is considered as a sample from 3D random field. Let $\mathbf{U}_E(\mathbf{x}, t)$ be a velocity field in a high-Reynolds-number turbulent flow. As commonly used, we assume that the relative dispersion process $(\mathbf{r}(t), \vec{v}(t))$ can be approximated by a 6D Markov diffusion process $(\hat{\mathbf{r}}(t), \hat{\vec{v}}(t))$ governed by the SDE

$$d\hat{\mathbf{r}} = \hat{\vec{v}}\, dt, \quad d\hat{\vec{v}} = \mathbf{a}(t, \hat{\mathbf{r}}, \hat{\vec{v}})\, dt + \sqrt{2C_0\bar{\varepsilon}}\, d\mathbf{W}(t), \tag{6.1}$$

where $\mathbf{a} = (a^1, a^2, a^3)$ is a $3D$ vector function to be defined; C_0 is a universal constant in the linear law for the structure function of the Lagrangian velocity [146], $\bar{\varepsilon}$ is the mean rate of dissipation of turbulence kinetic energy, $\mathbf{W}(t)$ is the standard $3D$ Wiener process. The choice of the diffusion coefficient $(C_0\bar{\varepsilon})^{1/2}$ is commonly used (e. g., see 15, 237, 238).

We briefly present the Thomson's approach for the determinig the drift coefficient $\mathbf{a}$ in (6.1). In [153], the following relation was derived:

$$p_E(\vec{v}; \mathbf{r}, t) = \int p_L(\mathbf{r}, \vec{v}; t, \mathbf{r}_0)\, d\mathbf{r}_0, \tag{6.2}$$

where

$$p_E(\vec{v}; \mathbf{r}, t) = \langle \delta(\vec{v} - \mathbf{U}_E(\mathbf{x} + \mathbf{r}, t) + \mathbf{U}_E(\mathbf{x}, t)) \rangle$$

is the probability density function (p. d. f.) of the relative Eulerian velocity $\mathbf{U}_E(\mathbf{x} + \mathbf{r}, t) - \mathbf{U}_E(\mathbf{x}, t)$ and

$$p_L(\mathbf{r}, \vec{v}; t, \mathbf{r}_0) = \langle \delta(\mathbf{r} - \mathbf{r}(t; \mathbf{r}_0)) \delta(\vec{v} - \vec{v}(t; \mathbf{r}_0)) \rangle$$

is the joint p. d. f. for the separation $\mathbf{r}(t; \mathbf{r}_0)$ and relative velocity $\vec{v}(t; \mathbf{r}_0)$. Here $\delta(\cdot)$ is the Dirac delta function, $\mathbf{r}(t) = \mathbf{r}(t; \mathbf{r}_0)$ is the separation vector between two particles which initially were separated by $\mathbf{r}_0$, and $\vec{v}(t) = \vec{v}(t; \mathbf{r}_0)$ is the relative velocity.

We denote by $(\hat{\mathbf{r}}(t; \mathbf{r}_0), \hat{\vec{v}}(t; \mathbf{r}_0))$ the solutions to (6.1) under the initial conditions

$$\hat{\mathbf{r}}(t = t_0) = \mathbf{r}_0, \quad \hat{\vec{v}}(t = t_0) = \vec{v}_0, \tag{6.3}$$

where $\vec{v}_0$ is the 3D random variable with the density $\langle \delta(\vec{v} - \vec{v}_0) \rangle = p_E(\vec{v}; \mathbf{r}_0, t_0)$. Then the p. d. f.

$$\hat{p}_L(\mathbf{r}, \vec{v}; t, \mathbf{r}_0) = \langle \delta(\mathbf{r} - \hat{\mathbf{r}}(t; \mathbf{r}_0)) \delta(\vec{v} - \hat{\vec{v}}(t; \mathbf{r}_0)) \rangle$$

solves the Kolmogorov–Fokker–Planck equation

$$\frac{\partial \hat{p}_L}{\partial t} + v^i \frac{\partial \hat{p}_L}{\partial r^i} + \frac{\partial}{\partial v^i}(a^i \hat{p}_L) = C_0\bar{\varepsilon} \frac{\partial^2 \hat{p}_L}{\partial v^i \partial v^i}.$$

Here and in what follows we use the summation convention. Since the true p. d. f. p_L satisfies Novikov's relation (6.2), it is natural to require that the p. d. f. of our model, $\hat{p}_L$, also satisfies this relation. From this, integrating the last equality with respect to $\mathbf{r}_0$, and changing $\int \hat{p}_L \, d\mathbf{r}_0$ with p_E we come to the well-mixed condition by Thomson [153]:

$$\frac{\partial p_E}{\partial t} + v^i \frac{\partial p_E}{\partial r^i} + \frac{\partial}{\partial v^i}(a^i p_E) = C_0 \bar{\varepsilon} \frac{\partial^2 p_E}{\partial v^i \partial v^i}. \tag{6.4}$$

This equation can be considered as a restriction on $\mathbf{a}$, p_E fixed. It is clear that (6.4) does not define $\mathbf{a}$ uniquely. As mentioned in [15], the problem of the unique choice of $\mathbf{a}$ is nontrivial. For, even in the case of stationary isotropic turbulence it is possible to find at least two models of the type (6.1) with different coefficients $\mathbf{a}$ satisfying the well-mixed condition (6.4) which lead to essentially different behavior of the mean squared separation between two particles in the inertial subrange. Therefore it is of interest to formulate new constraints on $\mathbf{a}$ originating from physically plausible arguments. Such an attempt will be made in the next section.

6.3 A closure of the quasi-1-dimensional model of relative dispersion

In [114], a quasi-1-dimensional process of relative dispersion (for short: Q1D-process) is defined as a stochastic process simulating the distance $r(t)$ between two particles. A diffusion model for Q1D-process in [45] was proposed for analysis of the temperature fluctuations in decaying isotropic turbulence. Further study of this process was undertaken in [108, 109, 114]. It turns out that for Q1D models a Q1D analog of the 3D well-mixed condition is true [114]. This enables us to construct a diffusion approximation of the Q1D process in the form

$$d\hat{r} = \hat{v}_\| dt, \; d\hat{v}_\| = X(t, \hat{r}, \hat{v}_\|)\, dt + Y(t, \hat{r}, \hat{v}_\|) dW(t), \tag{6.5}$$

where $X(\cdot), Y(\cdot)$ are some coefficients, $W(t)$ is a 1-dimensional standard Wiener process (see [109, 114]).

Let us consider the Q1D-process $\hat{r}(t) = (\hat{\mathbf{r}}(t), \hat{\mathbf{r}}(t))^{1/2}$ determined by the 3D model (6.1). Then the 2D process $(\hat{r}(t), \hat{v}_\|(t))$ satisfies the SDE

$$d\hat{r} = \hat{v}_\| dt, \quad d\hat{v}_\| = (\hat{a}_\| + \frac{\hat{v}_\perp^2}{r}) dt + \sqrt{2C_0\bar{\varepsilon}}\, dW_\|(t), \tag{6.6}$$

where

$$\hat{a}_\| = a_\|(t, \hat{\mathbf{r}}, \hat{\vec{v}}) = (\mathbf{a}(t, \hat{\mathbf{r}}, \hat{\vec{v}}), \frac{\hat{\mathbf{r}}(t)}{\hat{r}(t)}), \quad \hat{v}_\perp^2 = (\hat{\vec{v}}, \hat{\vec{v}}) - \hat{v}_\|^2,$$

and $W_\|(t)$ is the $1D$ Wiener process which is related to $\mathbf{W}(t)$ through $dW_\|(t) = (d\mathbf{W}(t), \frac{\hat{\mathbf{r}}(t)}{\hat{r}(t)})$.

Note that in general case the system (6.6) is not closed with respect to the variables $\hat{r}$, $\hat{v}_\parallel$. For instance, in [15, 238] the models are not closed in this sense.

In this chapter we make the following assumption about the model (6.1):

Assumption 6.1. *The Q1D model* (6.6) *is closed relative to the variables* $\hat{r}$, $\hat{v}_\parallel$, *in the sense that*

$$a_\parallel(t, \mathbf{r}, \vec{v}) + \frac{v_\perp^2}{r} = X(t, r, v_\parallel) \tag{6.7}$$

for some function X depending only on the shown arguments.

Remark 6.1. Note that there is a full analogy between this assumption and the hypothesis of a closed description of the 1-particle vertical dispersion in vertically inhomogeneous surface layer of the atmosphere (e. g., see [237]).

In the next section we show that in the case of isotropic turbulence Assumption 6.1 together with the well-mixed condition (6.4), lead to a unique choice of the coefficient **a** in (6.1).

6.4 Choice of the model (6.1) for isotropic turbulence

Thus we assume that the Eulerian velocity field $\mathbf{U}_E(\mathbf{x}, t)$ is an incompressible isotropic random field. In this case, $p_E(\vec{v}; \mathbf{r}, t) = p_E(v_\parallel, v_\perp; r, t)$, and the coefficient $\mathbf{a}(t, \mathbf{r}, \vec{v})$ has the general form

$$\mathbf{a}(t, \mathbf{r}, \vec{v}) = \varphi(v_\parallel, v_\perp; r, t)\frac{\mathbf{r}}{r} + \psi(v_\parallel, v_\perp; r, t)\frac{\vec{v}}{v}, \tag{6.8}$$

where

$$r = (\mathbf{r}, \mathbf{r})^{1/2}, \quad v = (\vec{v}, \vec{v})^{1/2}, \quad v_\parallel = (\vec{v}, \frac{\mathbf{r}}{r}), \quad v_\perp = \sqrt{v^2 - v_\parallel^2}.$$

Under Assumption 6.1 we have

$$\varphi + \psi\frac{v_\parallel}{v} + \frac{v_\perp^2}{r} = X(t, r, v_\parallel). \tag{6.9}$$

Then the well-mixed condition (6.4) takes the form

$$\frac{\partial p_E}{\partial t} + \frac{\partial}{\partial v_\parallel}(X p_E) + v_\parallel \frac{\partial p_E}{\partial r} + \frac{2v_\parallel}{r} p_E + \frac{1}{v_\perp}\frac{\partial}{\partial v_\perp}\left(v_\perp^2 \frac{\psi}{v} - \frac{v_\parallel v_\perp^2}{r}\right) p_E$$
$$= C_0\bar{\varepsilon}\left\{\frac{\partial^2 p_E}{\partial v_\parallel^2} + \frac{1}{v_\perp}\frac{\partial}{\partial v_\perp}\left(v_\perp \frac{\partial p_E}{\partial v_\perp}\right)\right\}. \tag{6.10}$$

Integrating this equality over $v_\perp$ with a weight $2\pi v_\perp$ and taking into account that X does not depend on $v_\perp$, we find

$$\frac{\partial p_E^{\|}}{\partial t} + \frac{\partial}{\partial v_\|}(X p_E^{\|}) + v_\| \frac{\partial p_E^{\|}}{\partial r} + \frac{2v_\|}{r} p_E^{\|} = C_0 \bar{\varepsilon} \frac{\partial^2 p_E^{\|}}{\partial v_\|^2}, \tag{6.11}$$

where

$$p_E^{\|}(v_\|; r, t) = \int_0^\infty 2\pi v_\perp p_E(v_\|, v_\perp; r, t)\, dv_\perp$$

is the p. d. f. of the longitudinal Eulerian relative velocity $u_E^{\|}(r,t) = (\mathbf{U}_E(\mathbf{x}+\mathbf{r},t) - \mathbf{U}_E(\mathbf{x},t), \frac{\mathbf{r}}{r})$:

$$p_E^{\|}(v_\|; r, t) = \langle \delta(v_\| - u_E^{\|}(r,t)) \rangle.$$

Here we assumed that

$$\left(v_\perp^2 \frac{\psi}{v} - \frac{v_\| v_\perp^2}{r} \right) p_E - C_0 \bar{\varepsilon} v_\perp \frac{\partial p_E}{\partial v_\perp} \bigg|_{|v_\perp| \to \infty} = 0. \tag{6.12}$$

Under the additional assumption

$$\left(\varphi + \psi \frac{v_\|}{v} + \frac{v_\perp^2}{r} \right) p_E^{\|} \bigg|_{v_\| \to -\infty} = 0 \tag{6.13}$$

we obtain

$$X(t, r, v_\|) = C_0 \bar{\varepsilon} \frac{\partial}{\partial v_\|} \ln p_E^{\|}(v_\|; r, t) - \frac{F(v_\|; r, t)}{p_E^{\|}(v_\|; r, t)} \tag{6.14}$$

by integrating (6.11) over $v_\|$. Here,

$$F(v_\|; r, t) = \int_{-\infty}^{v_\|} \left\{ \frac{\partial p_E^{\|}}{\partial t} + v_\|' \frac{\partial p_E^{\|}}{\partial r} + \frac{2v_\|'}{r} p_E^{\|}(v_\|'; r, t) \right\} dv_\|'. \tag{6.15}$$

Thus the coefficient $X(\cdot)$ in (6.6) is determined.

We find the coefficients φ and ψ from (6.9) and (6.10). Integrating (6.10) over $v_\perp$ with the weight $v_\perp$ we get

$$\begin{aligned} \psi(v_\|, v_\perp; r, t) = {} & C_0 \bar{\varepsilon} \frac{v}{v_\perp} \frac{\partial}{\partial v_\perp} \ln p_E(v_\|, v_\perp; r, t) + \frac{v_\| v}{r} \\ & - \frac{v\, G(v_\|, v_\perp; r, t)}{v_\perp^2 p_E(v_\|, v_\perp; r, t)}, \end{aligned} \tag{6.16}$$

where

$$\begin{aligned} & G(v_\|, v_\perp; r, t) \\ & = \int_0^{v_\perp} \left\{ \frac{\partial p_E}{\partial t} + \frac{\partial}{\partial v_\|}(X p_E) + v_\| \frac{\partial p_E}{\partial r} + \frac{2v_\|}{r} p_E(v_\|, v_\perp'; r, t) \right\} v_\perp' \, dv_\perp'. \end{aligned} \tag{6.17}$$

From (6.9), (6.14), and (6.16) we get

$$\varphi(v_{\|}, v_{\perp}; r, t) = C_0\bar{\varepsilon}\frac{\partial}{\partial v_{\|}} \ln p_E^{\|}(v_{\|}; r, t) - C_0\bar{\varepsilon}\frac{v_{\|}}{v_{\perp}}\frac{\partial}{\partial v_{\perp}} \ln p_E(v_{\|}, v_{\perp}; r, t)$$
$$- \frac{F(v_{\|}; r, t)}{p_E^{\|}(v_{\|}; r, t)} + \frac{v_{\|} G(v_{\|}, v_{\perp}; r, t)}{v_{\perp}^2 p_E(v_{\|}, v_{\perp}; r, t)} - \frac{v^2}{r}. \tag{6.18}$$

Thus Assumption 6.1 leads to the unique choice of the coefficient **a** through the Eulerian p. d. f. p_E under quite general conditions (6.12)–(6.13).

Let us consider the case when

$$p_E(v_{\|}, v_{\perp}; r, t) = p_E^{\|}(v_{\|}; r, t) p_E^{\perp}(v_{\perp}; r, t). \tag{6.19}$$

Under this assumption the formulas (6.16)–(6.18) can be simplified. Indeed, from (6.11) we find

$$G(v_{\|}, v_{\perp}; r, t) = p_E^{\|}(v_{\|}; r, t) F_1(v_{\|}, v_{\perp}; r, t), \tag{6.20}$$

where

$$F_1(v_{\|}, v_{\perp}; r, t) = \int_0^{v_{\perp}} \left\{ \frac{\partial p_E^{\perp}}{\partial t}(v'_{\perp}; r, t) + v_{\|}\frac{\partial p_E^{\perp}}{\partial r}(v'_{\perp}; r, t) \right\} v'_{\perp}\, dv'_{\perp}. \tag{6.21}$$

Substituting the value G from (6.20) in (6.18) we get by (6.19)

$$\varphi(v_{\|}, v_{\perp}; r, t) = C_0\bar{\varepsilon}\frac{\partial}{\partial v_{\|}} \ln p_E^{\|}(v_{\|}; r, t) - \frac{F(v_{\|}; r, t)}{p_E^{\|}(v_{\|}; r, t)} \tag{6.22}$$
$$- C_0\bar{\varepsilon}\frac{v_{\|}}{v_{\perp}}\frac{\partial}{\partial v_{\perp}} \ln p_E^{\perp}(v_{\perp}; r, t) + \frac{v_{\|} F_1(v_{\|}, v_{\perp}; r, t)}{v_{\perp}^2 p_E^{\perp}(v_{\perp}; r, t)} - \frac{v^2}{r}.$$

Analogously, from (6.20), (6.16), and (6.19) it follows

$$\psi(v_{\|}, v_{\perp}; r, t) = C_0\bar{\varepsilon}\frac{v}{v_{\perp}}\frac{\partial}{\partial v_{\perp}} \ln p_E^{\perp}(v_{\perp}; r, t) + \frac{v_{\|} v}{r}$$
$$- \frac{v F_1(v_{\|}, v_{\perp}; r, t)}{v_{\perp}^2 p_E^{\perp}(v_{\perp}; r, t)}. \tag{6.23}$$

Thus we have finished the construction of the model (6.1) for the isotropic turbulence.

6.5 The model of relative dispersion of two particles in a locally isotropic turbulence

6.5.1 Specification of the model

In this section we deal with incompressible flow with a large Reynolds number. Let us consider the relative dispersion process in the scales from the inertial subrange which are much less than the external turbulence scale. In this case one uses an approximation called local isotropic which means that the distributions of the velocity increments are isotropic and stationary [146]. Since in the inertial subrange the unique external parameter is $\bar{\varepsilon}$, we have the following representation for the Eulerian p. d. f. p_E:

$$p_E(\vec{v};\mathbf{r},t) = p_E(\vec{v};\mathbf{r}) = \frac{1}{\bar{\varepsilon}r} f_E\left(\frac{v_\parallel}{(\bar{\varepsilon}r)^{1/3}}, \frac{v_\perp}{(\bar{\varepsilon}r)^{1/3}}\right), \tag{6.24}$$

where $f_E(\xi_\parallel, \xi_\perp)$ is a universal dimensionless function of dimensionless arguments $\xi_\parallel \in (-\infty, \infty)$, and $\xi_\perp \geq 0$. Due to the local isotropy, we seek the coefficient $\mathbf{a}$ in (6.1) in the form

$$\mathbf{a} = \mathbf{a}(\mathbf{r},\vec{v}) = \left(\frac{\bar{\varepsilon}^2}{r}\right)^{1/3} \left\{\tilde{\varphi}\left(\frac{v_\parallel}{(\bar{\varepsilon}r)^{1/3}}, \frac{v_\perp}{(\bar{\varepsilon}r)^{1/3}}\right)\frac{\mathbf{r}}{r} + \tilde{\psi}\left(\frac{v_\parallel}{(\bar{\varepsilon}r)^{1/3}}, \frac{v_\perp}{(\bar{\varepsilon}r)^{1/3}}\right)\frac{\vec{v}}{v}\right\}, \tag{6.25}$$

where $\tilde{\varphi}(\xi_\parallel, \xi_\perp)$, $\tilde{\psi}(\xi_\parallel, \xi_\perp)$ are universal dimensionless functions. We find these functions assuming the function f_E to be given. We restrict our considerations to the case (6.19). The general case can be also treated analogously. The condition (6.19) can be written, due to (6.24), in the form

$$f_E(\xi_\parallel, \xi_\perp) = f_E^\parallel(\xi_\parallel) f_E^\perp(\xi_\perp), \tag{6.26}$$

where $f_E^\parallel(\xi_\parallel)$, $f_E^\perp(\xi_\perp)$ are dimensionless universal functions normalized so that

$$\int_{-\infty}^{\infty} f_E^\parallel(\xi)\, d\xi = 1, \quad \int_0^\infty 2\pi\xi f_E^\perp(\xi)\, d\xi = 1.$$

The dimensional functions $p_E^\parallel$, $p_E^\perp$ in (6.19) satisfy the relations

$$p_E^\parallel(v_\parallel;r) = \frac{1}{(\bar{\varepsilon}r)^{1/3}} f_E^\parallel\left(\frac{v_\parallel}{(\bar{\varepsilon}r)^{1/3}}\right), \quad p_E^\perp(v_\perp;r) = \frac{1}{(\bar{\varepsilon}r)^{2/3}} f_E^\perp\left(\frac{v_\perp}{(\bar{\varepsilon}r)^{1/3}}\right). \tag{6.27}$$

To find the coefficient $\mathbf{a}$ we first derive the coefficient $X = a_\parallel + v_\perp^2/r$ in (6.6). By the definition of F (see (6.15)) and from the relations (6.27) it is easy to get

$$F(v_\parallel;r,t) = \frac{(\bar{\varepsilon}r)^{1/3}}{r}\left\{-\frac{\xi_\parallel^2}{3} f_E^\parallel(\xi_\parallel) + \frac{7}{3}\int_{-\infty}^{\xi_\parallel} \xi f_E^\parallel(\xi)\, d\xi\right\}, \quad \xi_\parallel = \frac{v_\parallel}{(\bar{\varepsilon}r)^{1/3}}. \tag{6.28}$$

Consequently, by (6.14),

$$X(v_\parallel; r,t) = \frac{\bar{\varepsilon}^{2/3}}{r^{1/3}} \tilde{X}\left(\frac{v_\parallel}{(\bar{\varepsilon} r)^{1/3}}\right),$$
$$\tilde{X}(\xi_\parallel) = C_0 \frac{d \ln f_E^\parallel}{d\xi_\parallel} + \frac{\xi_\parallel^2}{3} - \frac{\frac{7}{3}\int_{-\infty}^{\xi_\parallel} \xi f_E^\parallel(\xi)\, d\xi}{f_E^\parallel(\xi_\parallel)} . \qquad (6.29)$$

When dealing with the quasi-1-dimensional model (6.6) it is convenient to use the dimensionless velocity $\hat{\xi}_\parallel = \hat{v}_\parallel/(\bar{\varepsilon}\hat{r})^{1/3}$. Then from (6.29) and (6.6) we obtain

$$d\hat{r} = (\bar{\varepsilon}\hat{r})^{1/3}\hat{\xi}_\parallel dt, \;\; d\hat{\xi}_\parallel = (\frac{\bar{\varepsilon}}{\hat{r}^2})^{1/3} a_0(\hat{\xi}_\parallel) dt + (\frac{\bar{\varepsilon}}{\hat{r}^2})^{1/6}\sqrt{2C_0}\, dW(t), \qquad (6.30)$$

where

$$a_0(\xi_\parallel) = \tilde{X}(\xi_\parallel) - \frac{\xi_\parallel^2}{3} = C_0 \frac{d \ln f_E^\parallel}{d\xi_\parallel} - \frac{\frac{7}{3}\int_{-\infty}^{\xi_\parallel} \xi f_E^\parallel(\xi)\, d\xi}{f_E^\parallel(\xi_\parallel)} . \qquad (6.31)$$

To determine the coefficients φ and ψ, we first find F_1. By the definition (6.21) and the relation (6.27) it follows that

$$F_1(v_\parallel, v_\perp; r,t) = -\frac{v_\parallel v_\perp^2}{3r} p_E^\perp(v_\perp; r).$$

Substituting this expression for F_1 and F given by (6.28) into (6.22)–(6.23) we find

$$\varphi(v_\parallel, v_\perp; r) = \left(\frac{\bar{\varepsilon}^2}{r}\right)^{1/3} \tilde{\varphi}\left(\frac{v_\parallel}{(\bar{\varepsilon} r)^{1/3}}, \frac{v_\perp}{(\bar{\varepsilon} r)^{1/3}}\right),$$
$$\psi(v_\parallel, v_\perp; r) = \left(\frac{\bar{\varepsilon}^2}{r}\right)^{1/3} \tilde{\psi}\left(\frac{v_\parallel}{(\bar{\varepsilon} r)^{1/3}}, \frac{v_\perp}{(\bar{\varepsilon} r)^{1/3}}\right), \qquad (6.32)$$

where

$$\tilde{\varphi}(\xi_\parallel, \xi_\perp) = C_0 \frac{d \ln f_E^\parallel}{d\xi_\parallel} - \frac{\frac{7}{3}\int_{-\infty}^{\xi_\parallel} \xi f_E^\parallel(\xi)\, d\xi}{f_E^\parallel(\xi_\parallel)} - C_0 \frac{\xi_\parallel}{\xi_\perp}\frac{d \ln f_E^\perp}{d\xi_\perp} - (\xi_\parallel^2 + \xi_\perp^2), \qquad (6.33)$$

and

$$\tilde{\psi}(\xi_\parallel, \xi_\perp) = C_0 \frac{\sqrt{\xi_\parallel^2 + \xi_\perp^2}}{\xi_\perp}\frac{d \ln f_E^\perp}{d\xi_\perp} + \frac{4\xi_\parallel}{3}\sqrt{\xi_\parallel^2 + \xi_\perp^2} . \qquad (6.34)$$

In the case of Gaussian p. d. f. p_E these formulas have a very simple form. The condition (6.19) is automatically satisfied in this case. Note that the integral in (6.33) can be explicitly found. After some algebra one can see that $\mathbf{a}(\mathbf{r}, \vec{v})$ is quadratically dependent on $\vec{v}$.

6.5.2 Numerical analysis of the Q1D-model (6.30)

In this section we compare the model (6.30) against the models [15, 238] and analyse how the non-Gaussianity of p_E affects the constant g in the cubic Richardson law. As $f_E^{\parallel}$, we choose a sum of three Gaussian densities, specified by two parameters, q_0 and x:

$$f_E^{\parallel}(\xi) = \frac{(1-q_0)}{\sqrt{2\pi C}} \exp\left(-\frac{\xi^2}{2C}\right) + q_0 \left\{ \frac{(1+x)}{2} \frac{1}{\sqrt{2\pi}\sigma} \exp\left(-\frac{(\xi - a_1)^2}{2\sigma^2}\right) + \frac{(1-x)}{2} \frac{1}{\sqrt{2\pi}\sigma} \exp\left(-\frac{(\xi - a_2)^2}{2\sigma^2}\right)\right\}, \tag{6.35}$$

where C is a universal constant in the Kolmogorov law of two-thirds and

$$a_1 = \left\{\frac{2(1-x)^2}{5q_0 x(1+x)}\right\}^{1/3}, \quad a_2 = -\frac{(1+x)}{(1-x)} a_1,$$
$$\sigma = \left\{C - \frac{(1+x)}{2} a_1^2 - \frac{(1-x)}{2} a_2^2\right\}^{1/2}. \tag{6.36}$$

The parameters q_0 and x are chosen so that $0 \le q_0 \le 1$, and $x \in (x_0, 1)$, where $x_0 = x_0(q_0) = (1 + \frac{25}{4} q_0^2 C^3)^{-1/2}$. The choice of quantities (6.36) ensures that

$$\int_{-\infty}^{\infty} f_E^{\parallel}(\xi)\, d\xi = 1, \qquad \int_{-\infty}^{\infty} \xi f_E^{\parallel}(\xi)\, d\xi = 0,$$
$$\int_{-\infty}^{\infty} \xi^2 f_E^{\parallel}(\xi)\, d\xi = C, \quad \int_{-\infty}^{\infty} \xi^3 f_E^{\parallel}(\xi)\, d\xi = -\frac{4}{5}, \tag{6.37}$$

provided $q_0 > 0$. The reason of this choice is that the true $f_E^{\parallel}$ satisfies (6.37) (see [114, 146]). Note that choosing appropriately the two free parameters q_0 and x, it is possible to control the excess and the fifth moment of $f_E^{\parallel}$. In our numerical calculations, we studied the behavior of the constant g in the Richardson law

$$\langle r^2(t_0 + \tau)\rangle = g\bar{\varepsilon}\tau^3\left(1 + o\left(\frac{\tau}{\tau_0}\right)\right), \quad \tau \gg \tau_0 = \frac{r_0^{2/3}}{\bar{\varepsilon}^{1/3}}, \tag{6.38}$$

for different models of $f_E^{\parallel}$ and the values of C_0. This dependence for the Gaussian p_E is well studied in [15].

In Figure 6.1 we show the dependence $g = g(C_0)$ for the Gaussian p_E ($q_0 = 0$), and compare it against the relevant results obtained in [15] for the Thomson and Borgas–Sawford models. In the range $C_0 \sim 2-7$, the difference between our Gaussian model and the latter models appear to be more than 10 times. Thus Assumption 6.1 results in models which qualitatively differ from the models [15, 238].

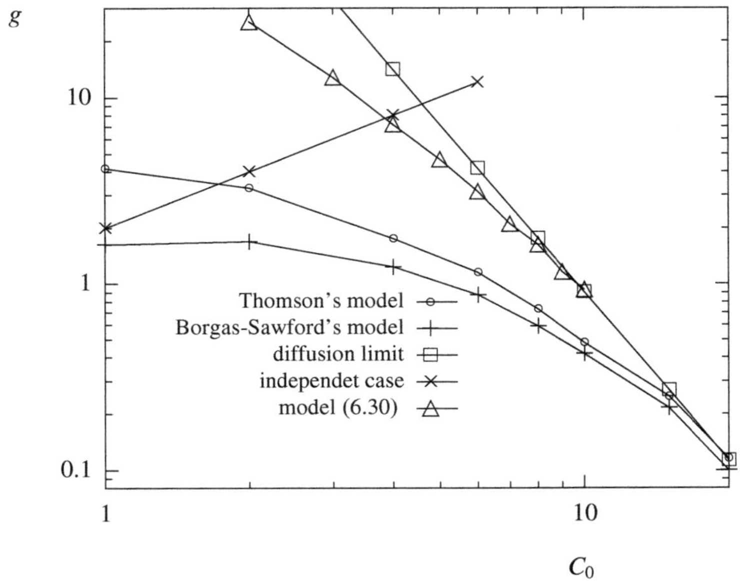

Figure 6.1. Inertial subrange dispersion constant g as a function of C_0 for the model (6.30),bn(6.35) with $q_0 = 0$, for Thomson's model (see [15], the model (4.3)), for Borgas–Sawford's model (see [15], the model (7.6) with $\varphi = -0.4$), for the diffusion limit (see [15]), and for the independent-particle case.

In Figure 6.2, the dependence of g on C_0 is shown for the Gaussian case (i. e., $q_0 = 0.$), and for five non-Gaussian models (6.35)–(6.36), for five values of the parameter pairs $(q_0, x(q_0))$: $(1.0, 0.929), (0.8, 0.917), (0.6, 0.9), (0.4, 0.874), (0.2, 0.827)$. This choice ensures that all the five models have the same excess $E = 0.5$.

The data in Figure 6.2 show that for $C_0 \leq 3$ the dependence $g = g(C_0)$ violates the kinematic constrain t$g(C_0) \leq 4C_0$. Therefore, it seems that our model is applicable for $C_0 \geq 3$.

Note that the Gaussian and non-Gaussian cases do not qualitatively differ. In the more accepted range of $C_0 \sim$ 2–7, the function $g = g(C_0, q_0)$ depends on C_0 (q_0 fixed) much more strongly than on q_0, while C_0 is fixed. Therefore, since we have no precise information about the values of C_0, there is no evidence to prefer non-Gaussian p_E.

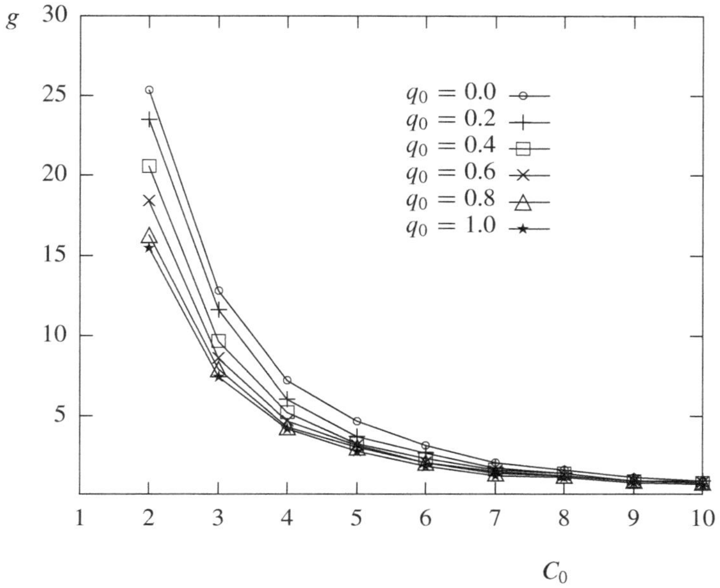

Figure 6.2. Inertial subrange dispersion constant g as a function of C_0 for the model (6.30), (6.35), for different values of q_0.

6.6 Model of the relative dispersion in intermittent locally isotropic turbulence

In this section we attempt to take account of intermittency.

In the intermittent turbulence, the relation (6.24) for the Eulerian p. d. f. p_E takes the form (e. g., see [3, 146])

$$p_E(\vec{v};\mathbf{r}) = \frac{1}{\bar{\varepsilon}r} f_E\left(\frac{v_\parallel}{(\bar{\varepsilon}r)^{1/3}}, \frac{v_\perp}{(\bar{\varepsilon}r)^{1/3}}, \frac{r}{L}\right), \tag{6.39}$$

where $f_E(\xi_\parallel, \xi_\perp, \zeta)$ is a dimensionless function of dimensionless arguments $\xi_\parallel \in (-\infty, \infty)$, $\xi_\perp \geq 0$, $\zeta > 0$. As in the case considered above, we assume that the process of relative dispersion $(\mathbf{r}(t), \vec{v}(t))$ can be approximated by a Markov diffusion process (6.1). By Assumption 6.1 and the well-mixed condition (6.10) we obtain

$$\mathbf{a}(\mathbf{r}, \vec{v}) = \left(\frac{\bar{\varepsilon}^2}{r}\right)^{1/3} \left\{\tilde{\varphi}\left(\frac{v_\parallel}{(\bar{\varepsilon}r)^{1/3}}, \frac{v_\perp}{(\bar{\varepsilon}r)^{1/3}}, \frac{r}{L}\right)\frac{\mathbf{r}}{r} + \tilde{\psi}\left(\frac{v_\parallel}{(\bar{\varepsilon}r)^{1/3}}, \frac{v_\perp}{(\bar{\varepsilon}r)^{1/3}}, \frac{r}{L}\right)\frac{\vec{v}}{v}\right\}, \tag{6.40}$$

where $\tilde{\varphi}$ and $\tilde{\psi}$ are some dimensionless functions uniquely defined by f_E. For simplicity, we consider the case

$$f_E(\xi_\parallel, \xi_\perp, \zeta) = f_E^\parallel(\xi_\parallel, \zeta) f_E^\perp(\xi_\perp, \zeta). \tag{6.41}$$

Then by (6.22),(6.23) we find

$$\begin{aligned}\tilde{\varphi}(\xi_\parallel, \xi_\perp, \zeta) = {} & C_0 \frac{\partial \ln f_E^\parallel}{\partial \xi_\parallel} - \frac{\frac{7}{3} I_\parallel(\xi_\parallel, \zeta)}{f_E^\parallel(\xi_\parallel)} - C_0 \frac{\xi_\parallel}{\xi_\perp} \frac{\partial \ln f_E^\perp}{\partial \xi_\perp} - (\xi_\parallel^2 + \xi_\perp^2) \\ & + \zeta \left(\frac{\xi_\parallel^2}{\xi_\perp^2} \frac{1}{f_E^\perp} \frac{\partial I_\perp}{\partial \zeta} - \frac{1}{f_E^\parallel} \frac{\partial I_\parallel}{\partial \zeta} \right),\end{aligned} \tag{6.42}$$

$$\begin{aligned}\tilde{\psi}(\xi_\parallel, \xi_\perp, \zeta) = {} & C_0 \frac{\sqrt{\xi_\parallel^2 + \xi_\perp^2}}{\xi_\perp} \frac{\partial \ln f_E^\perp}{\partial \xi_\perp} + \frac{4\xi_\parallel}{3} \sqrt{\xi_\parallel^2 + \xi_\perp^2} \\ & - \zeta \frac{\xi_\parallel \sqrt{\xi_\parallel^2 + \xi_\perp^2}}{\xi_\perp^2 f_E^\perp} \frac{\partial I_\perp}{\partial \zeta},\end{aligned} \tag{6.43}$$

where

$$I_\parallel(\xi_\parallel, \zeta) = \int_{-\infty}^{\xi_\parallel} \xi f_E^\parallel(\xi, \zeta)\, d\xi, \quad I_\perp(\xi_\perp, \zeta) = \int_0^{\xi_\perp} \xi f_E^\perp(\xi, \zeta)\, d\xi.$$

For the coefficient $X(v_\parallel, r) = X(v_\parallel, r, \bar{\varepsilon}, L)$ in the model (6.6) we have

$$X(v_\parallel; r) = \frac{\bar{\varepsilon}^{2/3}}{r^{1/3}} \tilde{X}\left(\frac{v_\parallel}{(\bar{\varepsilon} r)^{1/3}}, \frac{r}{L}\right), \tag{6.44}$$

where

$$\tilde{X}(\xi_\parallel, \zeta) = C_0 \frac{\partial \ln f_E^\parallel}{\partial \xi_\parallel} + \frac{\xi_\parallel^2}{3} - \frac{\frac{7}{3} I_\parallel}{f_E^\parallel} - \zeta \frac{1}{f_E^\parallel} \frac{\partial I_\parallel}{\partial \zeta} \quad .$$

Remark 6.2. Note that for the intermittent turbulence, in addition to to $\bar{\varepsilon}$, the parameter L enters the model. Hence, this model seems to give results which should differ from the classical Batchelor's theory of relative dispersion (see [10, 146]). In particular, when considering the normalized longitudinal relative velocity

$$v_\tau^\parallel = \frac{1}{(\bar{\varepsilon}\tau)^{1/2}} (v_\parallel(t_0 + \tau, \mathbf{r}_0) - v_\parallel(t_0, \mathbf{r}_0))$$

then, for $\tau \gg \frac{r_0^2}{\bar{\varepsilon}^{1/3}}$ the p. d. f. $\langle \delta(\xi_\parallel - v_\tau^\parallel) \rangle$ of this velocity, due to the Batchelor theory, depends on $\xi_\parallel$ and does not depend on τ. In our model (6.1), (6.40) the p. d. f. depends both on $\xi_\parallel$ and τ/T, where $T = L^{2/3}/\bar{\varepsilon}^{1/3}$. Note that the model described in Section 6.5 fits Batchelor' theory.

6.7 Conclusions

A new stochastic model (a diffusion approximation) for the relative dispersion of a pair of particles in high-Reynolds-number incompressible turbulence was proposed. An attempt was made to uniquely define the coefficients of SDE governing the relative dispersion process under a closure assumption about the quasi-1-dimensional model. The coefficients of the diffusion approximation (6.1) are uniquely expressed through the Eulerian p. d. f. p_E. The relative dispersion in the inertial subrange of the locally isotropic turbulence was treated in detail. Using the stationarity and the second Kolmogorov similarity hypothesis, simple expressions for the coefficients of the diffusion model were obtained. In this case, a numerical study of the influence of the non-Gaussinity of p_E on the statistical characteristics of the model, namely, on the constant $g = g(C_0)$ in the Richardson cubic law (6.38) was carried out. Qualitatively, the Gaussian and non-Gaussian models do not lead to a remarkable difference; the variation of the constant g when varying different models of p_E, C_0 fixed, is much less than its change when varying C_0 in an acceptable range $2 \leq C_0 \leq 7$. We conclude that there is no evidence for preferring non-Gaussian models since there is no exact information about the true value of C_0. In addition, an approach to constructing a diffusion approximation of the relative dispersion taking into account the intermittency ws proposed. This approximation requires further investigation and comparison with the new results in the problem under study (e. g., see [155, 156]).

Chapter 7

Stochastic Lagrangian models for 2-particle motion in turbulent flows. Numerical results

It is shown that the relative diffusion in a stationary incompressible Gaussian isotropic random field does not exhibit the Richardson cubic law. A 2-particle combined Eulerian–Lagrangian stochastic model which correctly reflects the behavior in the inertial subrange is developed. In addition, in this model, the "two-to-one" reduction due to Thomson is satisfied with high accuracy except for a small initial time interval where the error is slightly higher.

7.1 Introduction

In this chapter we use the notion of *pseudoturbulence* which is understood as a *stationary incompressible Gaussian isotropic* random field with a given spatial-temporal spectral tensor [146]. Throughout the chapter we assume that the mean velocity is zero. For brevity, we will call the pseudoturbulence with the Kolmogorov–Obukhov energy spectrum in the inertial subrange a *classical pseudoturbulence*. It is well known that the random velocity fields are often used to model different types of turbulent flows. For instance, S. Orszag has used a numerical scheme based on cubic formulas of the relevant spectral representation in the form of integrals with respect to the Gaussian measure. In [102], a randomized model of incompressible Gaussian isotropic random velocity field with a given energy spectrum was constructed for simulating Lagrangian trajectories and calculating some Eulerian and Lagrangian statistical characteristics. This approach was then used and developed by many authors (e. g., see [62, 63, 113, 128, 139, 191, 198]). In all these papers, spectral representations of the Gaussian random field were used. We mention also a wavelet expansion of the pseudoturbulence which was tried in [49].

It should be noted that it was generally believed that when choosing the Kolmogorov–Obukhov energy spectrum in the inertial subrange, this model gives a correct behavior of the 1- and many-particle Lagrangian statistical characteristics. For instance, it was assumed that the pseudoturbulence model will properly simulate the relative diffusion of two particles in the inertial subrange, and in particular, the Richardson cubic law

$$\langle r^2(t)\rangle \simeq g\bar{\varepsilon}t^3, \tag{7.1}$$

is true, provided that

$$\eta \ll \langle r^2(t)\rangle^{1/2} \ll L, \quad \text{and} \quad r_0 \ll \langle r^2(t)\rangle^{1/2}.$$

Here $r(t)$ is the distance between two particles, r_0 is the initial distance, $\bar{\varepsilon}$ is the mean rate of dissipation of kinetic energy, η is the internal and L is the external characteristic spatial scales of turbulence; g is a universal constant. Attempts to confirm numerically the Richardson law were made in [63, 198]. In these papers, the numerical results have not shown clear evidence of the reproduction of the universal Richardson law. In [51] it was reported that in 2D pseudoturbulence, the cubic law expands over even many decades of scaling behavior.

Our calculations, however, have shown that for the classical pseudoturbulence the Richardson law does not hold. It turns out that the mean square separation can be approximated by the power law

$$\langle r^2(t)\rangle \simeq g\bar{\varepsilon}t^3[k_0(\varepsilon t^3)^{1/2}]^{\gamma}, \tag{7.2}$$

where $k_0 = 2\pi/L$, γ and g are dimensionless constants which generally depend on the time correlation structure. This confirms our assumption that the Gaussian nature of the pseudoturbulence leads to the failure of the relative dispersion statistics in the inertial subrange. In turn, this leads to incorrect microscale behavior of the concentration fluctuations. Note that the study of the strong influence of the small scale relative dispersion on the intensity fluctuations is of current interest [44,214]. To construct a 2-particle stochastic model which correctly reflects the behavior in the inertial subrange we made an attempt [199], suggesting a combined Eulerian–Lagrangian model. In this chapter we present the calculation details, which confirm that the classical pseudoturbulence model does not mimick the Richardson law (at least for the model Reynolds numbers about 10^9), while the combined model reproduces the Richardson law with high accuracy. In addition, the new combined model manifests another important property due to Thomson [238], the so-called "two-to-one" reduction principle. We applied both stochastic models to the calculation of the intensity of concentration fluctuations for an instantaneous 3D Gaussian source, which gives considerably different results for times larger than the characteristic external time scale.

7.2 Classical pseudoturbulence model

7.2.1 Randomized model of classical pseudoturbulence

Let us assume that the Eulerian pseudoturbulent velocity field $\mathbf{U}_E(\mathbf{x}, t)$ has the following partial spatial-temporal spectral tensor (e. g., see [146]):

$$\Psi_{jl}(\mathbf{k}, t) = \frac{E(k)}{4\pi k^2}\Big(\delta_{jl} - \frac{k_j k_l}{k^2}\Big)e^{-a\bar{\varepsilon}^{1/3}k^{2/3}t}, \quad j,l = 1,2,3, \tag{7.3}$$

where $E(k)$ is the energy spectrum, a is a dimensionless constant which controls the time decorrelation, and $k = |\mathbf{k}|$.

The energy spectrum is defined by

$$E(k) = \begin{cases} C_1 \bar{\varepsilon}^{2/3} k^{-5/3}, & k_0 \le k \le k_{\max}, \\ 0, & \text{otherwise}, \end{cases} \tag{7.4}$$

with the normalization

$$\int_0^\infty E(k)dk = \frac{3}{2} u_0^2, \tag{7.5}$$

where $C_1 \simeq 1.4$ is the universal constant in the Kolmogorov–Obukhov five-thirds law and $3u_0^2 = \langle |\mathbf{U}_E|^2 \rangle$ is the energy of turbulence.

In the model, the following input parameters are involved: $\bar{\varepsilon}$ is the mean rate of dissipation of kinetic energy, k_0, $k_{\max}$ are the minimal and maximal wave numbers, respectively, and a is a dimensionless parameter characterizing the velocity decorrelation in time. The inner and external spatial scales of our model are $\eta = 2\pi/k_{\max}$, and $L = 2\pi/k_0$, respectively. Therefore, since the Reynolds number is expressed by (e. g., see [61, 146]) $Re \sim \left(\frac{L}{\eta}\right)^{4/3}$, it is naturally in our case to define the model Reynolds number as $\hat{Re} = \left(\frac{k_{\max}}{k_0}\right)^{4/3}$.

In [199], we presented four variants of the spatial-temporal spectral tensor with different decorrelation functions. The first variant is that used by Kraichnan [102], and in the three other cases the decorrelation time scale was dependent on the wave number. All these three cases have not shown a qualitative difference; therefore in the present chapter we have chosen the partial spatial-temporal spectral tensor in the form (7.3). The general simulation formula of the pseudoturbulent velocity field with the tensor (7.3) is [191, 199]:

$$\mathbf{U}_E(\mathbf{x}, t) = \sum_{j=1}^{n} \sqrt{E_j} \Big\{ (\boldsymbol{\xi}_j \times \boldsymbol{\Omega}_j) \cos(\theta_j) + (\boldsymbol{\eta}_j \times \boldsymbol{\Omega}_j) \sin(\theta_j) \Big\}, \tag{7.6}$$

where $\theta_j = k_j(\boldsymbol{\Omega}_j, \mathbf{x}) + \omega_j t$, and $\boldsymbol{\Omega}_j = (\Omega_j^{(1)}, \Omega_j^{(2)}, \Omega_j^{(3)})$, $j = 1, \dots, n$ are independent 3-dimensional random isotropic unit vectors ; $\boldsymbol{\xi}_j = (\xi_j^{(1)}, \xi_j^{(2)}, \xi_j^{(3)})$ and $\boldsymbol{\eta}_j = (\eta_j^{(1)}, \eta_j^{(2)}, \eta_j^{(3)})$ are mutually independent standard Gaussian random vectors; $k_j, j = 1, \dots n$ are random variables with the densities

$$p_j(k) = \begin{cases} \frac{1}{E_j} E(k), & k \in \Delta_j, \\ 0, & \text{otherwise}, \end{cases}$$

with

$$E_j = \int_{\Delta_j} E(k)dk, \quad j = 1, \dots, n$$

and Δ_j, $j = 1, \dots, n$ are nonoverlapping intervals which compose a partition of $\Delta = (k_0, k_{\max})$, the support of the spectrum. For each fixed $k = k_j$, $j = 1, \dots, n$, the random quantities ω_j, $j = 1, \dots, n$ are distributed with the densities $q(k_j, \omega)$, $j = 1, \dots, n$ where

$$q(k,\omega) = \frac{a\bar{\varepsilon}^{1/3}k^{2/3}}{\pi(a^2\varepsilon^{2/3}k^{4/3} + \omega^2)}, \qquad \omega \in (-\infty, \infty).$$

In [199], we have chosen the partition of the spectrum support

$$\Delta = \bigcup_{j=1}^{n} \Delta_j, \quad \Delta_j \bigcap \Delta_l = \emptyset, \ j \neq l,$$

where $\Delta_j = [\tilde{k}_j, \tilde{k}_{j+1})$, $j = 1, \dots, n$, $\tilde{k}_1 = k_0$, $\tilde{k}_{n+1} = k_{\max}$, so that

$$\int_{\Delta_j} E(k)dk = \frac{1}{n}\int_{\Delta} E(k)dk. \tag{7.7}$$

From this,

$$\tilde{k}_{j+1} = \left[k_0^{-2/3}\left(1 - \frac{j}{n}\right) + \frac{j}{n}k_{\max}^{-2/3}\right]^{-3/2}, \quad j = 1, \dots, n.$$

The random numbers k_j are simulated by

$$k_j = \left[\tilde{k}_j^{-2/3} - \frac{u_0^2}{nC_1\bar{\varepsilon}^{2/3}}\gamma_{j,1}\right]^{-3/2}, \quad j = 1, \dots, n, \tag{7.8}$$

and the random frequencies ω_j are simulated as

$$\omega_j = a\bar{\varepsilon}^{1/3}k_j^{2/3}\tan\left\{\pi\left(\gamma_{j,2} - 0.5\right)\right\}. \tag{7.9}$$

In the above formulas, $\gamma_{j,l}$, $(j = 1, \dots, n, l = 1, 2)$ are mutually independent random numbers uniformly distributed in [0, 1].

Modified algorithm

In this chapter we use a different version of the simulation formula. First, we divide the interval $[k_0, k_{\max})$ into n_0 parts $[\hat{k}_i, \hat{k}_{i+1})$, $(i = 0, \dots, n_0 - 1)$ uniformly in logarithmic scale: $\hat{k}_i = k_0 Q^i$, $i = 0, \dots, n_0$, where Q is chosen so that $\hat{k}_{n_0} = k_{\max}$. Then in each subinterval $[\hat{k}_i, \hat{k}_{i+1})$ we apply the same subdivision (energy uniformly) as in the formula (7.7). This algorithm provides better statistics in all parts of the energy spectrum. The number of simulated harmonics in this model equals $n_0 n$ and is proportional to $\ln(\hat{R}e)$. In the next section we present the calculations of the mean square separation $\langle r^2(t)\rangle$.

7.2.2 Mean square separation of two particles in classical pseudoturbulence

Having the samples of $\mathbf{U}_E(x,t)$ constructed by the modified algorithm, we determine a pair of Lagrangian trajectories $\mathbf{X}_i(t) = \mathbf{X}(t;\mathbf{x}_{0i},t_0)$, $t \in (-\infty,\infty)$ of fluid particles originating at positions $\mathbf{x}_{0i}$ at time $t = t_0$, $i = 1, 2$, from the equation of motion:

$$\frac{\partial \mathbf{X}_i}{\partial t} = \mathbf{U}_E(\mathbf{X}_i, t) \tag{7.10}$$

subject to the initial condition

$$\mathbf{X}_i(t_0) = \mathbf{x}_{0i}. \tag{7.11}$$

Let $r(t) = |\mathbf{X}_2(t) - \mathbf{X}_1(t)|$ be the distance between two particles initially separated by $r_0 = r(t_0)$. Generally, $\langle r^2(t)\rangle$ depends on $\bar{\varepsilon}$, a, k_0, $k_{\max}$, the turbulence characteristics, and on t, r_0. Keeping in mind the Richardson cubic law, it is natural to expect that if the condition

$$\eta = \frac{2\pi}{k_{\max}} \ll \langle r^2(t)\rangle^{1/2} \ll L = \frac{2\pi}{k_0}, \quad \text{and} \quad r_0 \ll \langle r^2(t)\rangle^{1/2}, \tag{7.12}$$

is satisfied, the mean square separation $\langle r^2(t)\rangle$ will depend only on t, $\bar{\varepsilon}$, and the dimensionless parameter a. In this case, by dimensional analysis we come to the Richardson cubic law (7.1) with the constant $g = g(a)$. Hence, by fixing a, we can try to check whether or not the Richardson law holds for classical pseudoturbulence.

In what follows we use the CGS units, and fix $\bar{\varepsilon} = 1\ \text{cm}^2/\text{s}^3$. Then, for simplicity, we omit the dimension units (e. g., we simply write $\bar{\varepsilon} = 1$). In Figure 7.1 we present $\rho(t) = (\langle r^2(t)\rangle)^{1/2}$ in the log-log scales as a function of time, for three different values of r_0: 10^{-4}, 10^{-3}, and 10^{-2} (curves 1, 2, and 3) with the value of $k_{\max}$ equal to 10^5, 10^4, and 10^3, respectively. The following parameters are here fixed: $a = 1$, $k_0 = 0.1$. The value of n_0 was chosen equal to the number of decades in the interval $[k_0, k_{\max})$: $n_0 = 6$, $n_0 = 5$, and $n_0 = 4$ for the curves 1, 2, and 3, respectively. The number of harmonics in each subinterval is $n = 25$. The number of simulated pairs of trajectories is: $N = 13{,}000$, $N = 38{,}000$ and $N = 10{,}000$ for the curves 1, 2, and 3, respectively.

To make easy the comparison with the cubic behavior, we also show the straight line 4 which represents the cubic law.

As is clearly seen from Figure 7.1, all three curves tend to the power law $\rho^2 = at^{\beta}$ with $a = 0.0127$ and $\beta = 3.75$ (see the straight line 5, which represents the power law $\rho^2 \propto t^{3.75}$).

Note that the smaller the initial distance is, the broader the time interval where the power law (the corresponding part of the curve is parallel to the straight line 5) approximates the relevant curve (e. g., for curve 1 this is the interval $(0.1, 10)$). If we tried to find an interval for each curve where $\rho^2(t)$ behaves cubically, it would imply that the cubic law depends on the initial distance, which is not true.

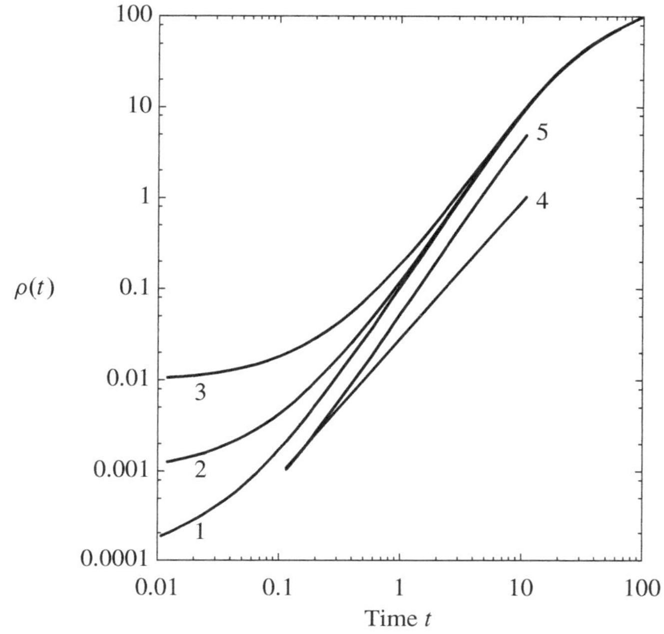

Figure 7.1. The function $\rho(t) = \langle r^2(t)\rangle^{1/2}$ for different initial separations: r_0: 10^{-4}, 10^{-3}, and 10^{-2} (curves 1, 2, and 3) with the value of $k_{\max}$ equal to 10^5, 10^4, and 10^3, respectively, and $k_0 = 0.1$, $a = 1$. The cubic law $\rho^2 \propto t^3$ (curve 4) and the power law $\rho^2 \propto t^{3.75}$ (curve 5).

Note that in [198] we reported calculations showing that it is likely that the cubic law exists; but the initial distance was relatively large ($r_0 \simeq 0.01$) (compare with curves 3 and 4 in Figure 7.1). In [198], we had no chance to carry out calculations for smaller values of r_0, since the direct randomization method (without the modification described above) would require some thousands of harmonics to provide stable calculations.

To illustrate the accuracy achieved by our simulation technique, we could also plot the relative statistical errors (measured via the standard Monte Carlo 3σ law) whose values are too small to be clearly seen in the picture: they vary along time from less than 5 % in curve 2 to a maximum of 10 % in the curves 1 and 3.

Note that generally, the dependence

$$\langle r^2(t)\rangle \simeq g\bar{\varepsilon}t^3 f(k_0(\bar{\varepsilon}t^3)^{1/2})$$

is true, provided that

$$\langle r^2(t)\rangle \gg r_0^2, \quad \langle r^2(t)\rangle \gg \eta^2. \tag{7.13}$$

From Figure 7.1, we can estimate that $f(\xi) \propto \xi^\gamma$ with $\gamma \simeq 0.5$ (for $0.03 \le \xi \le 3$), which yields (7.2).

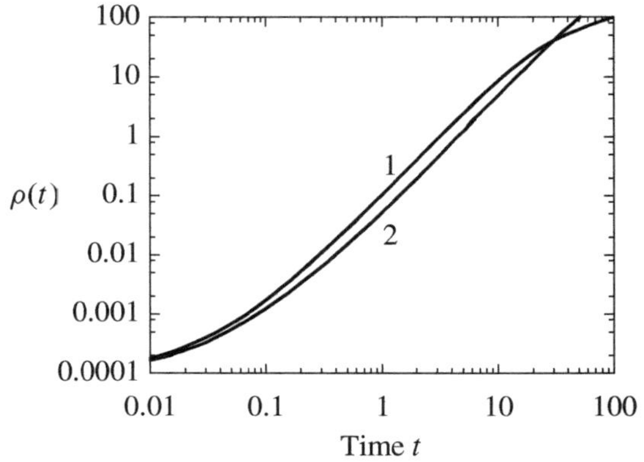

Figure 7.2. The function $\rho(t) = \langle r^2(t)\rangle^{1/2}$ for $k_0 = 0.1$ (curve 1) and $k_0 = 0.01$ (curve 2); $r_0 = 10^{-4}$, $k_{\max} = 10^5$, $a = 1$.

Concerning the dependence of $\rho(t)$ on k_0 plotted in Figure 7.2 for $k_0 = 0.1$ (curve 1) and $k_0 = 0.01$ (curve 2), the results show that $\rho^2 \simeq 0.013t^{3.75}$ and $\rho^2 \simeq 0.003t^{3.75}$, respectively.

For both curves, $r_0 = 10^{-4}$ and $a = 1$, the number of trajectories $N = 15{,}000$.

From this we conclude that the dependence (7.2) holds under the conditions (7.12).

It is interesting to study whether or not the power dependence (7.2) holds when varying the value of a. Remarkably, the change of a in $(0, 1)$ does not affect the behavior of $\langle r^2(t)\rangle$. For larger values of a the situation is different. In Figure 7.3 we present $\rho(t)$ for $a = 10$. Here, by 1 we indicate the curve for $r_0 = 10^{-4}$, $k_0 = 0.1$, $k_{\max} = 10^5$, and by 2 we denote the curve with $r_0 = 0.01$, $k_0 = 10^{-4}$, $k_{\max} = 10^3$. The number of trajectories $N = 15{,}000$.

From these curves we can conclude that the power low (7.2) is approximately true with $\gamma = 0.16$. Note that from these results the dependence on k_0 in (7.2) is also confirmed (with accuracy to within 10 %). The number of harmonics in Figures 7.2 and 7.3 was taken as the number of decades in the interval $(k_0, k_{\max})$ multiplied by 25.

Thus we conclude that the classical pseudoturbulence model gives a stable power law in the inertial subrange which is different from the classical Richardson cubic law.

It should be stressed that we are dealing here with high model Reynolds numbers in the range of $10^6 \div 10^9$.

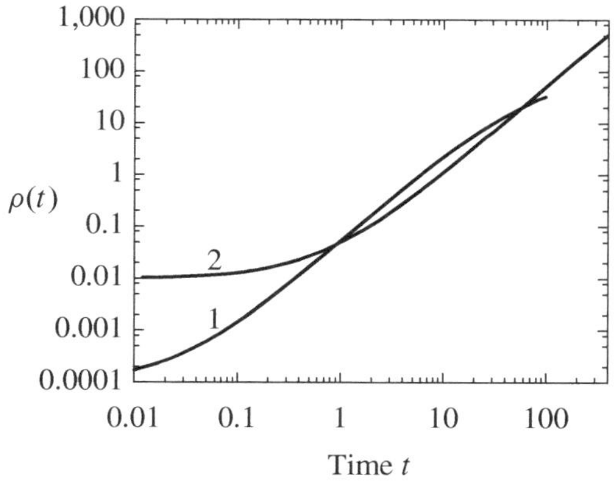

Figure 7.3. The function $\rho(t) = \langle r^2(t)\rangle^{1/2}$ for $a = 10$ with $k_0 = 0.1$, $k_{\max} = 10^5$, $r_0 = 10^{-4}$ (curve 1); and $k_0 = 10^{-4}$, $k_{\max} = 10^3$ and $r_0 = 0.01$ (curve 2).

7.3 Calculations by the combined Eulerian–Lagrangian stochastic model

7.3.1 Mean square separation of two particles

In this section we present numerical results for the mean square separation of two particles using the combined Eulerian–Lagrangian stochastic model. We show that the Richardson law is well satisfied in this model.

We use the following notation: $\mathbf{R}(t) = [\mathbf{X}_1(t) + \mathbf{X}_2(t)]/2$ is the center of the two particle coordinates, and $\mathbf{r}(t) = \mathbf{X}_2(t) - \mathbf{X}_1(t)$ is the separation vector.

The combined Eulerian–Lagrangian 2-particle model [199] has the form

$$\begin{aligned} \frac{d\mathbf{R}}{dt} &= \frac{1}{2}\Big\{\mathbf{U}_E\Big(\mathbf{R} + \frac{\mathbf{r}}{2}, t\Big) + \mathbf{U}_E\Big(\mathbf{R} - \frac{\mathbf{r}}{2}, t\Big)\Big\}, \\ \frac{d\mathbf{r}}{dt} &= \theta\Big(\frac{r}{L}\Big)\Big\{\mathbf{U}_E\Big(\mathbf{R} + \frac{\mathbf{r}}{2}, t\Big) - \mathbf{U}_E\Big(\mathbf{R} - \frac{\mathbf{r}}{2}, t\Big)\Big\} + \Big\{1 - \theta\Big(\frac{r}{L}\Big)\Big\}\vec{v}(t), \\ d\vec{v}(t) &= \mathbf{a}(\mathbf{r}, \vec{v})dt + \sqrt{2C_0\bar{\varepsilon}}\, d\mathbf{W}(t), \end{aligned} \tag{7.14}$$

where $\mathbf{U}_E$ is the pseudoturbulent velocity taken in the form (7.6) with the energy uniform subdivision (7.7), and $\mathbf{a}(\mathbf{r}, \vec{v})$ is defined by [199]:

$$\mathbf{a}(\mathbf{r}, \vec{v}) = \frac{\bar{\varepsilon}^{2/3}}{r^{1/3}}\left\{\left[\frac{7}{3}C - \frac{C_0}{4C}\frac{v_\parallel}{(\bar{\varepsilon}r)^{1/3}} - \frac{v^2}{(\bar{\varepsilon}r)^{2/3}}\right]\frac{\mathbf{r}}{r} + \left[\frac{4}{3}\frac{v_\parallel}{(\bar{\varepsilon}r)^{1/3}} - \frac{C_0}{C'}\right]\frac{\vec{v}}{(\bar{\varepsilon}r)^{1/3}}\right\}.$$

Here $v_\parallel = \big(\vec{v}, \frac{\mathbf{r}}{r}\big)$ is the longitudinal component of the relative velocity vector, $v = |\vec{v}|$, $C \simeq 2$ is the universal constant in the Kolmogorov two-thirds law, $C' = 4C/3$, and

C_0 is the universal constant in the linear law for the structure function of the Lagrangian velocity [146]. The function θ is defined through

$$\theta(y) = \theta_{\alpha,\beta}(y) = \begin{cases} 0, & 0 < y < \alpha, \\ \phi(y), & y \in (\alpha, \beta), \\ 1, & y > \beta, \end{cases}$$

where $0 < \alpha < \beta < 1$, and ϕ is an arbitrary monotonically increasing function so that θ is continuous on $(0, \infty)$. We recall that in (7.14), $\mathbf{W}(t)$ is a standard $3D$ Wiener process.

Thus in this model, two additional parameters are involved: the constant C_0 and the function θ. We choose θ as a piecewise linear function defined by the constants α and β. These adjusting parameters were chosen as $\alpha = 0.05$ and $\beta = 0.1$. Some remarks concerning this choice will be given in the next section.

The exact value of the constant C_0 is not known, it belongs to the interval (2, 8) (e. g., see [170, 262]). Note also that results [261] obtained by DNS suggest that C_0 lies between 4 and 8.

In Figure 7.4 we show the function $\rho(t) = \langle r^2(t)\rangle^{1/2}$ in log-log coordinates obtained for $C_0 = 4$ (left panel) and $C_0 = 6$ (right panel). The parameters were chosen as follows: $k_0 = 0.01$, $k_{\max} = 1{,}000$, the number of harmonics in the spectral representation (7.6) was taken to be equal to 10. The number of samples was $N = 4{,}000$. In Figure 7.4 the upper curve 1 corresponds to the initial distance $r_0 = 0.05$, and the lower curve 2 to $r_0 = 0.005$. For convenience, we also show the straight line 3, which represents the cubic law $\rho^2(t) \propto t^3$.

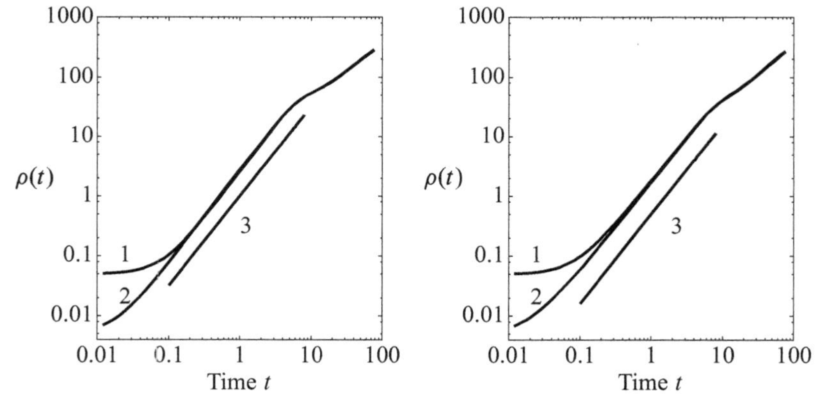

Figure 7.4. Left panel: The function $\rho(t) = \langle r^2(t)\rangle^{1/2}$ for two different initial distances: $r_0 = 0.05$ (curve 1) and $r_0 = 0.005$ (curve 2); $C_0 = 4$, $k_0 = 0.01$. Curve 3 represents the cubic law $\rho^2 \propto t^3$. Right panel: the same curves, but for $C_0 = 6$.

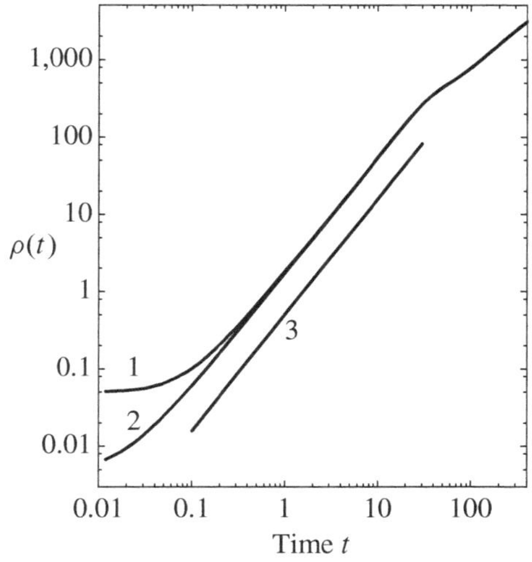

Figure 7.5. The same as in Figure 7.4, for $C_0 = 6$ and $k_0 = 0.001$.

From these curves it is clearly seen that the Richardson cubic law $\langle r^2(t)\rangle = g\bar{\varepsilon}t^3$ with $g = 6.65$ for $C_0 = 4$ and $g = 2.92$ for $C_0 = 6$ is satisfied with high accuracy in the range of about two decades (see the relevant intervals where the curves are well approximated by linear dependencies in Figure 7.4).

In contrast to the power laws discussed in Section 7.2 (see Figures 7.1–7.3) the value of the constant g in (7.1) does not change (to within 3 % of accuracy) when the value of k_0 varies. Indeed, when $k_0 = 0.001$, we find from Figure 7.5 that $g = 2.96$. The value of C_0, the number of harmonics, the number of trajectories, and the initial distances were the same as in Figure 7.4. In addition, the inertial subrange is extended, as expected, when increasing the integral scale of turbulence. Finally we note that the stabilization of numerical results happens when the number of harmonics in (7.6) reaches the value 10: further increase does not noticeably change the value of $\rho(t)$.

Remark 7.1. Thus our combined Eulerian–Lagrangian stochastic model confirms the Richardson cubic law (7.1). The constant g in this law essentially depends on the universal constant C_0. The calculations presented above suggest that $g \simeq 2.9$ for $C_0 = 6$ and $g \simeq 6.6$ for $C_0 = 4$. These values agree with the results given by Kraichnan [101] ($g = 2.42$), Lundgren [135] ($g = 3$), and Larcheveque and Lesieur [125] ($g = 3.5$). In [149] the effective Hamiltonian method was used to derive the Richardson cubic law. From this approach it is possible to estimate that g varies from 3.5 to 9 when the decorrelation parameter a changes from 2 to 0.5. Thomson's Lagrangian model [238] suggests that g varies between 1 and 2 when C_0 varies between 4 and 6. Values close to these were obtained by Borgas and Sawford [15].

7.3.2 Thomson's "two-to-one" reduction principle

We recall the following obvious relation:

$$P_{1L}(\mathbf{x}, \vec{v}, t; \mathbf{x}_0) = \int P_{2L}(\mathbf{x}, \vec{v}, \mathbf{x}', \vec{v}', t; \mathbf{x}_0, \mathbf{x}_0') d\mathbf{x}' d\vec{v}', \tag{7.15}$$

where P_{1L}, P_{2L} are the 1- and 2-particle Lagrangian p. d. f.'s:

$$P_{1L}(\mathbf{x}, \vec{v}, t; \mathbf{x}_0) = \langle \delta(\mathbf{x} - \mathbf{X}(t, \mathbf{x}_0)) \delta(\vec{v} - \mathbf{V}(t, \mathbf{x}_0)) \rangle,$$

and

$$\begin{aligned} P_{2L}(\mathbf{x}, \vec{v}, \mathbf{x}', \vec{v}', t; \mathbf{x}_0, \mathbf{x}_0') = \langle &\delta(\mathbf{x} - \mathbf{X}(t, \mathbf{x}_0)) \delta(\vec{v} - \mathbf{V}(t, \mathbf{x}_0)) \\ &\times \delta(\mathbf{x}' - \mathbf{X}(t, \mathbf{x}_0')) \delta(\vec{v}' - \mathbf{V}(t, \mathbf{x}_0')) \rangle. \end{aligned}$$

Here we denote for simplicity

$$\mathbf{X}(t, \mathbf{x}_0) = \mathbf{X}(t; \mathbf{x}_0, t_0), \quad \mathbf{V}(t, \mathbf{x}_0) = \mathbf{V}(t; \mathbf{x}_0, t_0) = \mathbf{U}_E(\mathbf{X}(t, \mathbf{x}_0), t).$$

However, in the Lagrangian models this relation is generally not satisfied. Therefore, in [238] it was noted that when constructing 2-particle Lagrangian models, it is desired to satisfy the relation (7.15). This principle is called in [238] a "two-to-one" reduction. In [15], some new criteria were derived from Thomson's "two-to-one" reduction. As far as we know, there is no Lagrangian model exactly satisfying this principle.

It is difficult to verify the relation (7.15) for a particular Lagrangian model. Therefore, more simple consequences of this relation can be studied. For instance, the 1-particle mean square velocity component $\sigma^2(t) = \frac{1}{3}\langle |\mathbf{V}(t)|^2 \rangle$ was calculated in [15]. Note that in our case of stationary incompressible isotropic turbulence, σ^2 is a constant equal to the Eulerian 1-point mean square velocity component $\frac{1}{3}\langle |\mathbf{U}_E|^2 \rangle$. If a 2-particle stochastic model satisfies this property, we say that the "two-to-one" reduction principle in a broad sense is satisfied. Note that there is no Lagrangian model exactly satisfying the "two-to-one" reduction principle even in this broad sense.

In Figure 7.6 we plot the mean square Lagrangian velocities for both particles $\sigma_i^2(t) = \frac{1}{3}\langle \mathbf{V}_i^2(t) \rangle$, $i = 1, 2$, at $C_0 = 6$, $k_0 = 1$, $k_{\max} = 10^4$, and for the initial separation $r_0 = 10^{-3}$ which is much less than the characteristic spatial scale $L = 2\pi$. From these curves we see that the "two-to-one" reduction principle in a broad sense is approximately satisfied. Note that $\sigma^2(t)$ deviates from the true value $\sigma = 1.4$ to within 7 %, for statistics about $N = 4{,}000$. When increasing the accuracy (i. e., $N = 50{,}000$), it turns out that in the initial time interval $(0, 0.5)$ the deviation from the constant remains the same, while it decreases to $\approx 1\%$ for $t \geq 0.5$ (see Figure 7.6, right panel). Note that the characteristic external time scale of turbulence estimated as $L^{2/3}/\bar{\varepsilon}^{1/3} = (2\pi)^{2/3}$. Hence, in our calculations, the time interval where we see the 7 % deviation is much less than this external time scale.

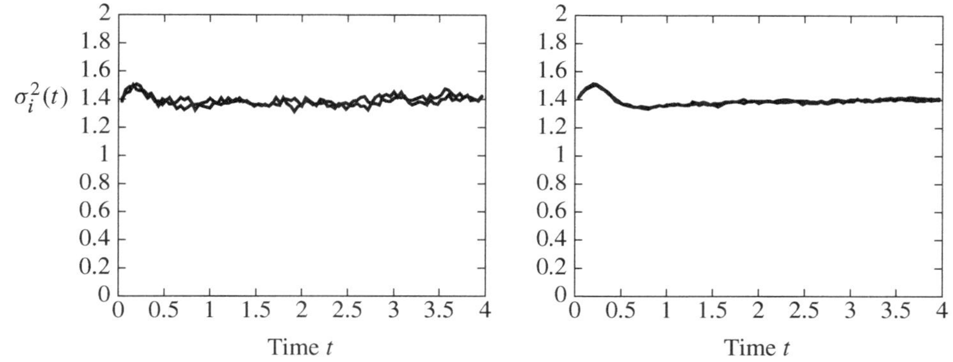

Figure 7.6. The functions $\sigma_i^2(t) = \frac{1}{3}\langle \mathbf{V}_i^2(t)\rangle$, $i = 1, 2$, for $N = 4{,}000$ (left panel) and $N = 50{,}000$ (right panel) samples.

Note that the behavior of this deviation is not surprising. Indeed, for small times, when the instantaneous distance $r(t)$ is less than αL, as seen from the definition of the function θ, the motion equation is

$$\frac{d\mathbf{R}}{dt} = \Sigma\mathbf{V}, \quad \frac{d\mathbf{r}}{dt} = \vec{v}, \tag{7.16}$$

where

$$\Sigma\mathbf{V} = \frac{1}{2}\left\{\mathbf{U}_E\left(\mathbf{R}(t) + \frac{\mathbf{r}(t)}{2}, t\right) + \mathbf{U}_E\left(\mathbf{R}(t) - \frac{\mathbf{r}(t)}{2}, t\right)\right\}.$$

Therefore, the Lagrangian velocities of the first and second particles can be written as

$$\mathbf{V}_1 = \Sigma\mathbf{V} - \frac{\vec{v}}{2}, \quad \mathbf{V}_2 = \Sigma\mathbf{V} + \frac{\vec{v}}{2},$$

respectively. By definition, for small times, $\Sigma\mathbf{V}$ and $\vec{v}$ are independent, and $\langle|\Sigma\mathbf{V}(t)|^2\rangle \approx \langle|\mathbf{U}_E|^2\rangle$. We conclude from this that both particles have a small contribution of the order $\langle|\vec{v}(t)|^2\rangle$ to their mean square Lagrangian velocities. Since $\langle|\vec{v}(t)|^2\rangle$ increases linearly in time, these arguments explain why the curves in Figure 7.6 slightly increase at the beginning. For times when $r(t) > \beta L$, our equation of motion of two particles coincides with the equation of motion in the pseudoturbulence $\mathbf{U}_E$ which results in Figure 7.6 in the small deviation of the curves from the exact constant. When choosing the parameters α and β we have to balance between two situations: on the one hand, taking smaller values of α and β, we ensure that the "two-to-one" reduction principle will be better satisfied. On the other hand, for a high-Reynolds-number turbulence, the physically natural model (7.16) is reasonable for $r(t)/L \leq \alpha_0$, with α_0 being a constant whose value depends on the velocity features. Therefore, if we take the value of α too small, we ignore this physical picture.

In all the above considerations, we assumed that the initial distance between two particles is small compared to the characteristic spatial scale of turbulence. This is the worst case, and for larger initial distances, the "two-to-one" reduction principle is satisfied with even higher accuracy. Moreover, these arguments make it clear that for times $\rho(t) > \beta L$ the "two-to-one" reduction principle will be satisfied in our model even in the strong formulation (7.15).

7.3.3 Concentration fluctuations

As we have seen from the calculations of the function $\rho(t)$, the pseudoturbulent model and the combined Eulerian–Lagrangian stochastic model lead to qualitatively different results in the inertial subrange. Another important quantity which is commonly considered as a fine test (e. g., see [44, 214, 238]) is the intensity of the concentration fluctuations of particles in a fixed point x defined as

$$I_c(x,t) = \frac{\sqrt{\langle c^2(x,t)\rangle - \langle c(x,t)\rangle^2}}{\langle c(x,t)\rangle},$$

where $c(x,t)$ is the instantaneous concentration at time t in the point x. In this section we use this test to compare the pseudoturbulent model and the combined Eulerian–Lagrangian stochastic model.

We take an instantaneous deterministic source of particles in the Gaussian form

$$S(\mathbf{x}) = \frac{1}{(2\pi)^{3/2}\sigma_0^3} \exp\left\{-\frac{|\mathbf{x}|^2}{2\sigma_0^2}\right\}, \tag{7.17}$$

where σ_0 is the characteristic size of the puff. The mean and the second moment of concentration can be represented as [146]

$$\langle c(\mathbf{x},t)\rangle = \int p_{1L}(\mathbf{x},t;\mathbf{x}_0,t_0) S(\mathbf{x}_0) d\mathbf{x}_0,$$

$$\langle c(\mathbf{x},t)c(\mathbf{x}',t)\rangle = \int p_{2L}(\mathbf{x},\mathbf{x}',t;\mathbf{x}_0,\mathbf{x}_0',t_0) S(\mathbf{x}_0) S(\mathbf{x}_0') d\mathbf{x}_0 d\mathbf{x}_0'.$$

The functions p_{1L} and p_{2L} are the 1-point and 2-point p. d. f.'s:

$$p_{1L}(\mathbf{x},t;\mathbf{x}_0,t_0) = \langle \delta(\mathbf{x} - \mathbf{X}(t;\mathbf{x}_0,t_0))\rangle,$$

$$p_{2L}(\mathbf{x},\mathbf{x}',t;\mathbf{x}_0,\mathbf{x}_0',t_0) = \langle \delta(\mathbf{x} - \mathbf{X}(t;\mathbf{x}_0,t_0))\delta(\mathbf{x}' - \mathbf{X}(t;\mathbf{x}_0',t_0))\rangle,$$

where $\mathbf{X}(t;\mathbf{x}_0,t_0)$ is the Lagrangian trajectory started at time $t_0 < t$ in the point $\mathbf{x}_0$. We call this path a direct Lagrangian trajectory. Backward Lagrangian trajectory starting at t at the point x is defined as $\mathbf{X}(t_0; \mathbf{x}, t)$, the solution of the equation

$$\frac{\partial \mathbf{X}}{\partial t_0} = \mathbf{U}_E(\mathbf{X}, t_0), \quad t_0 < t$$

with the initial condition $\mathbf{X}(t) = x$.

Note that in the incompressible velocity field the direct and backward p. d. f.'s are equal (e. g., see [135]):

$$\begin{aligned} p_{1L}(\mathbf{x}, t; \mathbf{x}_0, t_0) &= p_{1L}(\mathbf{x}_0, t_0; \mathbf{x}, t), \\ p_{2L}(\mathbf{x}, \mathbf{x}', t; \mathbf{x}_0, \mathbf{x}_0', t_0) &= p_{2L}(\mathbf{x}_0, \mathbf{x}_0', t_0; \mathbf{x}, \mathbf{x}', t). \end{aligned}$$

These relations are the basis for the backward trajectory technique (e. g., see [214]) which uses the equalities

$$\begin{aligned} \langle c(\mathbf{x}, t)\rangle &= \langle S(\mathbf{X}(t_0; \mathbf{x}, t))\rangle, \\ \langle c^2(\mathbf{x}, t)\rangle &= \langle S(\mathbf{X}(t_0; \mathbf{x}_1, t)) S(\mathbf{X}(t_0; \mathbf{x}_2, t))\rangle. \end{aligned}$$

The latter relation involves two backward trajectories, $\mathbf{X}(t_0; \mathbf{x}_1, t)$ and $\mathbf{X}(t_0; \mathbf{x}_2, t)$ and the equality is understood in the sense that in the right-hand side the limit is taken as $\mathbf{x}_2, \mathbf{x}_1 \to \mathbf{x}$ with the rate satisfying $|\mathbf{x}_2 - \mathbf{x}_1| \propto \eta$ (e. g., see [44]).

This approach is convenient when it is necessary to calculate the concentrations at a fixed point provided the source has relatively large spatial extension.

Thus to calculate the mean of $c(\mathbf{x}, t)$, we start at the point $\mathbf{x}$ at the time t, and simulate the Lagrangian trajectory backward in time till the time instant t_0, then calculate the random estimator $S(\mathbf{X}(t_0; \mathbf{x}, t))$. After averaging over large number of trajectories we get the mean concentration.

The same is true for $\langle c^2(\mathbf{x}, t)\rangle$: we start two backward trajectories at points $\mathbf{x}_1$ and $\mathbf{x}_2$ at the time t separated by very small distance (which is of the order of the Kolmogorov inner scale η), and calculate the random estimator $S(\mathbf{X}(t_0; \mathbf{x}_1, t))\, S(\mathbf{X}(t_0; \mathbf{x}_2, t))$. After averaging over a large number of pairs of trajectories we get the second moment of the concentration.

We used the backward trajectory technique to calculate the intensity of concentration fluctuations for the source (7.17) with $\sigma_0 = 0.1 u_0^3/\bar{\varepsilon}$, where $3u_0^2$ is the energy of turbulence (see (7.5)).

In Figure 7.7, we show $I_c(0, t)$ as a function of the dimensionless time $\tau = \frac{(t-t_0)\bar{\varepsilon}}{u_0^2}$ in log-log coordinates. The upper curve (2) is obtained by the pseudoturbulent model where $N = 16{,}000$ pairs of trajectories were used, and the lower curve (1) is obtained by the combined Eulerian–Lagrangian stochastic model with $N = 30{,}000$ pairs. In the latter case, $C_0 = 6$, $\alpha = 0.05$, $\beta = 0.1$. In both cases, the wave interval from $k_0 = 2\pi$ to $k_{\max} = 10^3$ was used. In the pseudoturbulent model 50 harmonics were needed, while in the combined model 30 harmonics were sufficient to give stable numerical results.

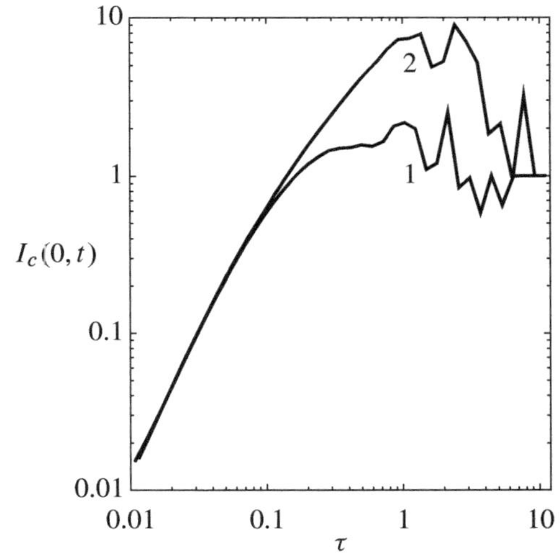

Figure 7.7. Intensity of concentration fluctuations as a function of dimensionless time $\tau = (t-t_0)\bar{\varepsilon}/u_0^2$. Upper curve 2: the pseudoturbulent model ($N = 16{,}000$ samples); lower curve 1: the combined Eulerian–Lagrangian stochastic model ($N = 30{,}000$ samples).

Calculations of $I_c(0,t)$ by both models are of relatively high accuracy in the interval $(0.01, 1)$. Then the error increases because of the large variance of the second moment $\langle c^2 \rangle$. Note that the mean concentration is calculated with small error in all the interval $\tau \in (0.01, 10)$. Both models give close results for $I_c(0,t)$ in the interval $\tau \in (0.01, 0.1)$. Further the difference between the results of two models is increasing to about of 500 %. This shows that the combined Eulerian–Lagrangian stochastic model qualitatively differs from the pseudoturbulent stochastic model. Note that the combined Eulerian–Lagrangian stochastic model agrees fairly well with the results reported by Thomson [238].

7.4 Technical remarks

In this section we give some details of the numerical schemes applied both to the pseudoturbulent stochastic model described in Section 7.2, and to the combined Eulerian–Lagrangian stochastic model of Section 7.3.

Theoretically, we are dealing with the equation of motion in a Gaussian random velocity field with a given continuous spectral tensor. However, numerically we are dealing with some approximate random field where only a finite number of harmonics is involved. Relevant convergence results can be found in [143]. In practice, we increase the number of harmonics until the numerical results become stable. It should be noted that for the pseudoturbulent stochastic model of Section 7.1, when calculating the mean square separation, it was sufficient to take 25 harmonics per decade of the interval $(k_0, k_{\max})$ where the energy spectrum is located. Then the total number of harmonics in this model approximately equals $25\log_{10}(\hat{Re})$. Note that the model of [62, 63] requires $\propto (\hat{Re})^{3/4}$ harmonics. In our calculations the cost of the combined Eulerian–Lagrangian stochastic model is much lower: the total number of harmonics providing stable numerical results for the mean square separation was about 10. In the case of the intensity of concentration fluctuations, the needed number of harmonics was about 30.

Another important issue of the numerical scheme of integration of the motion equation is the choice of the time step of discretization. We used the standard Euler scheme

$$\mathbf{X}(t+\Delta t) = \mathbf{X}(t) + \Delta t \mathbf{U}_E(\mathbf{X}(t), t) \tag{7.18}$$

for the pseudoturbulent stochastic model, and

$$\begin{aligned} \mathbf{R}(t+\Delta t) &= \mathbf{R}(t) + \Delta t \Sigma \mathbf{V}, \\ \mathbf{r}(t+\Delta t) &= \mathbf{r}(t) + \Delta t \left\{ \theta\left(\frac{r(t)}{L}\right) \Delta \mathbf{V} + \left[1 - \theta\left(\frac{r(t)}{L}\right)\right] \vec{v}(t) \right\}, \\ \vec{v}(t+\Delta t) &= \vec{v}(t) + \mathbf{a}(\mathbf{r}(t), \vec{v}(t))\Delta t + \sqrt{2C_0\bar{\varepsilon}\Delta t}\, \boldsymbol{\eta}_t, \end{aligned} \tag{7.19}$$

where

$$\Sigma \mathbf{V} = \frac{1}{2}\left\{ \mathbf{U}_E\left(\mathbf{R}(t) + \frac{\mathbf{r}(t)}{2}, t\right) + \mathbf{U}_E\left(\mathbf{R}(t) - \frac{\mathbf{r}(t)}{2}, t\right) \right\},$$

$$\Delta \mathbf{V} = \mathbf{U}_E\left(\mathbf{R}(t) + \frac{\mathbf{r}(t)}{2}, t\right) - \mathbf{U}_E\left(\mathbf{R}(t) - \frac{\mathbf{r}(t)}{2}, t\right),$$

and $\boldsymbol{\eta}_t$ are standard 3D Gaussian vectors mutually independent at different time instances.

We choose a variable time step

$$\Delta t = \delta \left(\frac{r^2(t)}{\bar{\varepsilon}}\right)^{1/3},$$

where δ is a small adjusting parameter. This choice was also used in [238].

The choice of the parameter δ in the scheme (7.18) should be consistent with the initial distance r_0. In our case, for $r_0 = 0.01$ it was sufficient to take $\delta = 0.05$. When $r_0 = 10^{-4}$, it requires a much smaller time step. However, we took $\delta = 0.05$ which cannot be considered sufficient and could be further diminished at a small initial time interval. We have actually carried out such high-accuracy calculations with $\Delta = 0.01$ and $r_0 = 10^{-4}$ for small time intervals ($t \le 0.1$). From these calculations we conclude that the parameter γ in (7.2) cannot be less than that given in Section 7.1.

In (7.19) we put $\delta = 0.01$. These values of δ ensure stable numerical results to within the statistical error of the Monte Carlo calculations.

7.5 Conclusion

A 2-particle combined Eulerian–Lagrangian stochastic model which correctly reflects the behavior in the inertial subrange was developed. In this chapter, which develops the results described in the previous chapter, we presented the calculation details, which confirm that the classical pseudoturbulence model does not mimick the Richardson law (at least for the model Reynolds numbers about 10^9), while the combined model reproduces the Richardson law with high accuracy already for $\hat{Re} \sim 10^6$. However the classical pseudoturbulence model leads to a stable power law (7.2), which depends on the parameter $\bar{\varepsilon}$ and the characteristic external spatial scale of turbulence.

The new combined model manifests the "two-to-one" reduction principle.

The cost of the Monte Carlo technique proposed for simulation of the pseudo-turbulent dispersion depends logarithmically on the model Reynolds number. The combined Eulerian–Lagrangian stochastic model is even more efficient, since the variable time step of integration increases in this model faster, compared to the case of the pseudoturbulent stochastic model.

In the combined Eulerian–Lagrangian stochastic model two parameters, α and β, are introduced which have clear physical meaning: when $r(t) < \alpha L$, the processes of relative and absolute motions are approximately statistically independent, while for $r(t) > \beta L$, these processes are strongly correlated. We adjusted these parameters from the "two-to-one" reduction principles. Generally, more physics should be taken into account to make a proper choice of α and β.

In conclusion we note that the combined Eulerian–Lagrangian stochastic model presented for classical pseudoturbulence can be easily extended to more general turbulent flows. Indeed, having models with a given large-scale statistical structure $\mathbf{U}_E$, say, obtained by DNS or large eddy simulation methods, we can put it into the model (7.10).

Chapter 8

The 1-particle stochastic Lagrangian model for turbulent dispersion in horizontally homogeneous turbulence

A 1-particle stochastic Lagrangian model in 2 and 3 dimensions is constructed for transport of particles in horizontaly homogeneous turbulent flows with arbitrary 1-point probability density function. It is shown that in the case of anisotropic turbulence with Gaussian p. d. f., this model essentially differs from the known Thomson model. The results of calculations according to our model in the case of neutrally-stratified atmospheric surface layer agree satisfactorily with the measurements known from the literature.

8.1 Introduction

This chapter deals with 1-particle stochastic Lagrangian models for 2D and 3D turbulent transport. Here we treat the fully developed turbulence (i. e., a flow with a very high Reynolds number) as a random velocity field (u, v, w) which is assumed to be incompressible. Therefore, the trajectories of particles in such flows are stochastic processes. To simulate these stochastic processes, two different approaches are known in the literature. The first one is based on the numerical solution of the system of random equations

$$\begin{aligned}\frac{\partial X}{\partial t} &= u(X, Y, Z, t),\\ \frac{\partial Y}{\partial t} &= v(X, Y, Z, t),\\ \frac{\partial Z}{\partial t} &= w(X, Y, Z, t).\end{aligned} \tag{8.1}$$

Here $X(t), Y(t), Z(t)$ are the coordinates of the Lagrangian trajectory at time t. The random fields u, v, w are simulated by Monte Carlo methods (e. g., see [42, 63, 102, 191, 198, 244]), and the random trajectories are then obtained by the numerical solution of (8.1) with the relevant initial data.

In the second approach the true trajectory $X(t), Y(t), Z(t)$ is assumed to be approximated by a model trajectory $\hat{X}(t), \hat{Y}(t), \hat{Z}(t)$, a solution to a stochastic differential equation of the Ito type (e. g., see [215, 237]:

$$
\begin{aligned}
d\hat{X} &= \hat{U}\,dt, \quad d\hat{Y} = \hat{V}\,dt, \quad d\hat{Z} = \hat{W}\,dt,\\
d\hat{U} &= a_u(t,\hat{X},\hat{Y},\hat{Z},\hat{U},\hat{V},\hat{W})dt + b_u(t,\hat{X},\hat{Y},\hat{Z},\hat{U},\hat{V},\hat{W})\,dB_u(t),\\
d\hat{V} &= a_v(t,\hat{X},\hat{Y},\hat{Z},\hat{U},\hat{V},\hat{W})dt + b_v(t,\hat{X},\hat{Y},\hat{Z},\hat{U},\hat{V},\hat{W})\,dB_v(t),\\
d\hat{W} &= a_w(t,\hat{X},\hat{Y},\hat{Z},\hat{U},\hat{V},\hat{W})dt + b_w(t,\hat{X},\hat{Y},\hat{Z},\hat{U},\hat{V},\hat{W})\,dB_w(t). \quad (8.2)
\end{aligned}
$$

Here $\hat{U}, \hat{V}, \hat{W}$ are the components of the model Lagrangian velocity, $B_u(t)$, $B_v(t)$, $B_w(t)$ are three standard independent Wiener processes.

Ideally, one would have an approximation such that the true and model Lagrangian velocities coincide:

$$
\begin{aligned}
\hat{U}(t) &= u(\hat{X}(t), \hat{Y}(t), \hat{Z}(t), t),\\
\hat{V}(t) &= v(\hat{X}(t), \hat{Y}(t), \hat{Z}(t), t),\\
\hat{W}(t) &= w(\hat{X}(t), \hat{Y}(t), \hat{Z}(t), t), \quad (8.3)
\end{aligned}
$$

which would assure that the true and model trajectories are the same. However it is unrealistic to satisfy (8.3); therefore one uses different consistency principles. Namely, the general consistency principle says that the statistics of the model process

$$\hat{X}(t), \hat{Y}(t), \hat{Z}(t), \hat{U}(t), \hat{V}(t), \hat{W}(t)$$

satisfies the same relations which are satisfied by the true process

$$X(t), Y(t), Z(t),\ U(t),\ V(t),\ W(t),$$

where $U(t) = u(X(t), Y(t), Z(t), t)$, $V(t) = v(X(t), Y(t), Z(t), t)$, and $W(t) = w(X(t), Y(t), Z(t), t)$ are the components of the true Lagrangian velocity.

Two consistency criteria used in the literature are:

(a) consistency with the Kolmogorov similarity theory;
(b) consistency with Novikov's integral relation.

Here (a) reads

$$\langle (dU)^2 \rangle = \langle (dV)^2 \rangle = \langle (dW)^2 \rangle = C_0 \varepsilon dt,$$

and

$$\langle dU\, dV \rangle = \langle dU\, dW \rangle = \langle dW\, dV \rangle = 0,$$

where dU, dV, dW are the components of the increments of the Lagrangian velocity, ε is the mean rate of the dissipation of turbulence energy, C_0 is the universal constant (e. g., see [146, 215, 237]); here, and in the following, the angle brackets stand for the ensemble average over the samples of the random velocity field.

Note that (A) implies (e. g., see [237]) that in (8.2), all the terms b_u, b_v, b_w are equal to $\sqrt{C_0\varepsilon}$:

$$b_u = b_v = b_w = \sqrt{C_0\varepsilon}. \tag{8.4}$$

Novikov's integral relation has the form [153]

$$p_E(u, v, w; x, y, z, t) = \int_{R^3} p_L(x, y, z, u, v, w; x_0, y_0, z_0, t)dx_0dy_0dz_0. \tag{8.5}$$

Here p_E is the probability density function (pdf) of the Eulerian velocity u, v, w in the fixed point x, y, z at the time t, and p_L is the joint p. d. f. of the true Lagrangian phase point (X, Y, Z, U, V, W) defined by the trajecory started at x_0, y_0, z_0.

Thus the consistency with the Novikov relation (8.5) means that the p. d. f. of the model phase point governed by (8.2), say $\hat{p}_L$, satisfies

$$p_E(u, v, w; x, y, z, t) = \int_{R^3} \hat{p}_L(x, y, z, u, v, w; x_0, y_0, z_0, t)dx_0dy_0dz_0. \tag{8.6}$$

Note that (8.6), the Focker–Planck–Kolmogorov equation for $\hat{p}_L$ and (8.4) lead to the well-mixed condition due to D. Thomson [237]:

$$\begin{aligned}\frac{\partial p_E}{\partial t} + u\frac{\partial p_E}{\partial x} + v\frac{\partial p_E}{\partial y} + w\frac{\partial p_E}{\partial z} + \frac{\partial}{\partial u}(a_u p_E) + \frac{\partial}{\partial v}(a_v p_E) + \frac{\partial}{\partial w}(a_w p_E) \\ = \frac{C_0\varepsilon}{2}\left\{\frac{\partial^2 p_E}{\partial u^2} + \frac{\partial^2 p_E}{\partial v^2} + \frac{\partial^2 p_E}{\partial w^2}\right\}.\end{aligned} \tag{8.7}$$

In this chapter we study a horizontally-homogeneous turbulent flow which implies that p_E does not depend on x, y. Therefore, in the left-hand side of (8.7) the second and third terms vanish. Here the main problem is that (8.7) does not uniquely define the coefficients a_u, a_v, and a_w of the model (8.2). Indeed, even for homogeneous turbulence, in [217] two different choices of a_u, a_v, a_w are presented, both satisfying the well-mixed condition (8.7), but whose statistical characteristics are different. For the Gaussian form of p_E, one of the appropriate techniques for getting the coefficients a_u, a_v, a_w is given in [237]. In the non-Gaussian 3D case, to the authors knowledge there is no appropriate choice of these coefficients. In 2D a non-Gaussian case was treated by Flesch and Wilson in [56]. These authors mentioned that the two different models do not lead to essentially different results in the case of Gaussian p_E. As reported in [56], the same is true for two models considered in [217].

In this chapter we suggest a proper choice of the coefficients a_u, a_v, a_w in a general case of p. d. f. p_E. Our derivation is based on some assumptions which ensure a unique choice of the model. It should be stressed that even in the Gaussian case our model essentially differs, as shown below (Section 8.4), from the model given by Thomson [237]. This confirms our opinion that it is necessary along theoretical studies to extract additional information from experiments.

8.2 Choice of the coefficients in the Ito equation

Let us formulate the main assumptions about the Lagrangian model of the type (8.2). We consider a horizontally homogeneous incompressible high-Reynolds-number turbulent flow in the space R^3. Thus the mean velocity has no vertical component. Without loss of generality we assume in addition that the mean velocity is directed along the x-axis. Thus the mean velocity vector is $(\bar{u}(x,y,z,t),0,0)$, while p_E and $\bar{u}$ do not depend on x, y. We will write the p. d. f. p_E in the form

$$p_E(u,v,w;z,t) = p'_E(u',v',w';z,t)$$

where $u' = u - \bar{u}(z,t)$, $v' = v$ and $w' = w$.

By (8.4), equation (8.2) in these variables has the form

$$\begin{aligned}
d\hat{X} &= (\hat{U}' + \bar{u}(\hat{Z},t))dt,\ d\hat{Y} = \hat{V}'dt,\ d\hat{Z} = \hat{W}'dt,\\
d\hat{U}' &= a'_u(t,\hat{Z},\hat{U}',\hat{V}',\hat{W}')dt + \sqrt{C_0\varepsilon}\,dB_u(t),\\
d\hat{V}' &= a'_v(t,\hat{Z},\hat{U}',\hat{V}',\hat{W}')dt + \sqrt{C_0\varepsilon}\,dB_v(t),\\
d\hat{W}' &= a'_w(t,\hat{Z},\hat{U}',\hat{V}',\hat{W}')dt + \sqrt{C_0\varepsilon}\,dB_w(t).
\end{aligned} \tag{8.8}$$

The well-mixed condition in new variables is

$$\begin{aligned}
\frac{\partial p'_E}{\partial t} + w'\frac{\partial p'_E}{\partial z} + \frac{\partial}{\partial u'}(a'_u p'_E) + \frac{\partial}{\partial v'}(a'_v p'_E) + \frac{\partial}{\partial w'}(a'_w p'_E)&\\
= \frac{C_0\varepsilon}{2}\left\{\frac{\partial^2 p'_E}{\partial(u')^2} + \frac{\partial^2 p'_E}{\partial(v')^2} + \frac{\partial^2 p'_E}{\partial(w')^2}\right\}&.
\end{aligned} \tag{8.9}$$

Assumption 8.1. *We assume in addition that a'_u does not depend on v' while a'_w does not depend on u', v', i. e., $a'_u = a'_u(t,z,u',w')$, $a'_w = a'_w(t,z,w')$.*

Then the model (8.8) reads

$$\begin{aligned}
d\hat{X} &= (\hat{U}' + \bar{u}(\hat{Z},t))dt,\ d\hat{Y} = \hat{V}'dt,\ d\hat{Z} = \hat{W}'dt,\\
d\hat{U}' &= a'_u(t,\hat{Z},\hat{U}',\hat{W}')dt + \sqrt{C_0\varepsilon}\,dB_u(t),\\
d\hat{V}' &= a'_v(t,\hat{Z},\hat{U}',\hat{V}',\hat{W}')dt + \sqrt{C_0\varepsilon}\,dB_v(t),\\
d\hat{W}' &= a'_w(t,\hat{Z},\hat{W}')dt + \sqrt{C_0\varepsilon}\,dB_w(t).
\end{aligned} \tag{8.10}$$

Integrating (8.9) over u' and v' yields

$$\frac{\partial p'_{1E}}{\partial t} + w'\frac{\partial p'_{1E}}{\partial z} + \frac{\partial}{\partial w'}(a'_w(t,z,w')p'_{1E}) = \frac{C_0\varepsilon}{2}\frac{\partial^2 p'_{1E}}{\partial(w')^2}, \tag{8.11}$$

where

$$p'_{1E} = p'_{1E}(w';z,t) = \int_{-\infty}^{\infty}\int_{-\infty}^{\infty} p'_E(u',v',w';z,t)\,du'dv'. \tag{8.12}$$

Here we have assumed that

$$a'_u p'_E,\ a'_v p'_E,\ \frac{\partial p'_E}{\partial u'},\ \frac{\partial p'_E}{\partial v'} \quad \text{all tend to zero as } (u')^2 + (v')^2 \to \infty.$$

Similarly, the integration of (8.9) over v' leads to

$$\frac{\partial p'_{2E}}{\partial t} + w'\frac{\partial p'_{2E}}{\partial z} + \frac{\partial}{\partial u'}(a'_u(t,z,u',w')p'_{2E}) + \frac{\partial}{\partial w'}(a'_w(t,z,w')p'_{2E}) = \frac{C_0\varepsilon}{2}\left(\frac{\partial^2 p'_{2E}}{\partial(u')^2} + \frac{\partial^2 p'_{2E}}{\partial(w')^2}\right), \tag{8.13}$$

where

$$p'_{2E} = p'_{2E}(u',w';z,t) = \int_{-\infty}^{\infty} p'_E(u',v',w';z,t)\,dv'. \tag{8.14}$$

Now, under the assumption made about the behavior at infinity, it is possible to uniquely define the coefficients a'_u, a'_v and a'_w. Indeed, from (8.11) one gets a'_w, then from (8.13) one finds a'_u, and from (8.9) one obtains a'_v. This yields

$$a'_w(t,z,w) = \frac{1}{p'_{1E}(w;z,t)}\left\{\frac{C_0\varepsilon}{2}\frac{\partial p'_{1E}}{\partial w} - \left(\frac{\partial f_{1E}}{\partial t} + \frac{\partial F_{1E}}{\partial z}\right)\right\}, \tag{8.15}$$

where

$$f_{1E}(w;z,t) = \int_{-\infty}^{w} p'_{1E}(w';z,t)\,dw',$$

$$F_{1E}(w;z,t) = \int_{-\infty}^{w} w'p'_{1E}(w';z,t)\,dw',$$

and

$$a'_u(t,z,u,w) = \frac{1}{p'_{2E}}\left\{\frac{C_0\varepsilon}{2}\left(\frac{\partial p'_{2E}}{\partial u} + \frac{\partial^2 f_{2E}}{\partial w^2}\right) - \left(\frac{\partial f_{2E}}{\partial t} + w\frac{\partial f_{2E}}{\partial z}\right) - \frac{\partial}{\partial w}\left(a'_w f_{2E}\right)\right\}, \tag{8.16}$$

where

$$f_{2E}(u,w;z,t) = \int_{-\infty}^{u} p'_{2E}(u',w;z,t)\,du'.$$

Finally,

$$
\begin{aligned}
a'_v(t,z,u,w) = \frac{1}{p'_E}\Bigg\{ & \frac{C_0\varepsilon}{2}\left(\frac{\partial^2 f_E}{\partial u^2} + \frac{\partial p'_E}{\partial v} + \frac{\partial^2 f_E}{\partial w^2}\right) - \left(\frac{\partial f_E}{\partial t} + w\,\frac{\partial f_E}{\partial z}\right) \\
& - \frac{\partial}{\partial u}(a'_u f_E) - \frac{\partial}{\partial w}\left(a'_w\, f_E\right)\Bigg\},
\end{aligned} \tag{8.17}
$$

where

$$
f_E(u,v,w;z,t) = \int_{-\infty}^{v} p'_E(u,v',w;z,t)\,dv',
$$

Thus the coefficients (8.15)–(8.17) define a unique stochastic model (8.10) through the p. d. f. p'_E.

Remark 8.1. This model is a natural extension of the 1-dimensional (in z direction) Thomson model [237] in the sense that the vertical coordinates (z, w) are governed in our model by a SDE which coincides with the Thomson model. Note that in Thomson's 3D model (with Gaussian p'_E) this is not the case: the statistics of (z, w) in his 3D model essentially differ from those of (z, w) in his 1-dimensional model (see Section 8.4).

8.3 2D stochastic model with Gaussian p. d. f.

In this section we present concrete expressions for the coefficients in the case of Gaussian p. d. f.. We extract a 2D model from a 3D model as

$$
\begin{aligned}
d\hat{X} &= (\hat{U}' + \bar{u}(\hat{Z},t))dt,\; d\hat{Z} = \hat{W}'dt, \\
d\hat{U}' &= a'_u(t,\hat{Z},\hat{U}',\hat{W}')dt + \sqrt{C_0\varepsilon}\,dB_u(t), \\
d\hat{W}' &= a'_w(t,\hat{Z},\hat{W}')dt + \sqrt{C_0\varepsilon}\,dB_w(t).
\end{aligned} \tag{8.18}
$$

In the Gaussian case

$$
p'_{2E}(u,w;z,t) = \frac{1}{2\pi\sigma_{u/w}\sigma_w}\exp\left\{-\frac{1}{2\sigma_{u/w}^2}(u-\mu)^2 - \frac{w^2}{2\sigma_w^2}\right\}, \tag{8.19}
$$

where

$$
\sigma_{u/w} = \frac{\Delta^{1/2}}{\sigma_w}, \quad \mu = \frac{\overline{uw}}{\sigma_w^2}\,w, \quad \Delta = \sigma_u^2\sigma_w^2 - (\overline{uw})^2,
$$

and σ_u^2, σ_w^2 are the variances of the x- and z-velocity components, respectively. From (8.19)

$$
p'_{1E}(w;z,t) = \frac{1}{\sqrt{2\pi}\sigma_w}\exp\left\{-\frac{w^2}{2\sigma_w^2}\right\}, \tag{8.20}
$$

then,

$$f_{1E}(w;z,t) = \int_{-\infty}^{\frac{w}{\sigma_w}} \frac{1}{\sqrt{2\pi}} \exp(-t^2/2)\, dt = \Phi\left(\frac{w}{\sigma_w}\right),$$

$$F_{1E}(w;z,t) = -\sigma_w^2\, p'_{1E}(w;z,t).$$

Note that

$$\frac{1}{p'_{1E}} \frac{\partial p'_{1E}}{\partial w} = -\frac{w}{\sigma_w^2}, \quad -\frac{1}{p'_{1E}} \frac{\partial F_{1E}}{\partial z} = \frac{1}{2}(w^2+1)\frac{\partial \sigma_w^2}{\partial z}$$

and

$$\frac{\partial f_{1E}}{\partial t} = -\frac{w}{\sigma_w^2} \frac{\partial \sigma_w}{\partial t} \dot{\Phi}(w/\sigma_w),$$

where $\dot{\Phi}(\tau) = \frac{d\Phi}{d\tau}$, and

$$\Phi(\tau) = \int_{-\infty}^{\tau} \frac{1}{\sqrt{2\pi}} \exp(-t^2/2)\, dt.$$

From (8.15) we find

$$a'_w(t,z,w) = -\left(\frac{C_0 \varepsilon}{2\sigma_w^2} - \frac{1}{\sigma_w} \frac{\partial \sigma_w}{\partial t}\right) w + \frac{1}{2} \frac{\partial \sigma_w^2}{\partial z} \left(\frac{w^2}{\sigma_w^2} + 1\right). \tag{8.21}$$

Note that this coincides with Thomson's relevant expression in his 1D model [237].

By the definition,

$$f_{2E}(u,w,z,t) = p'_{1E}(w;z,t)\, \Phi\left(\frac{u-\mu}{\sigma_{u/w}}\right).$$

To find a'_u from (8.16), we need the expressions for

$$\frac{\partial f_{2E}}{\partial t}, \quad \frac{\partial f_{2E}}{\partial z}, \quad \frac{\partial f_{2E}}{\partial w}, \quad \frac{\partial p'_{2E}}{\partial u}, \quad \frac{\partial^2 f_{2E}}{\partial w^2}.$$

By definition we get

$$\begin{aligned}
\frac{\partial p'_{2E}}{\partial u} &= -\frac{(u-\mu)}{\sigma_{u/w}^2} p'_{2E} \quad \frac{\partial f_{2E}}{\partial t} = f_{2E}\left\{\frac{1}{2\sigma_w^2} \frac{\partial \sigma_w^2}{\partial t}\left(\frac{w^2}{\sigma_w^2} - 1\right) + \Psi(\xi)\frac{\partial \xi}{\partial t}\right\},\\
\frac{\partial f_{2E}}{\partial z} &= f_{2E}\left\{\frac{1}{2\sigma_w^2} \frac{\partial \sigma_w^2}{\partial z}\left(\frac{w^2}{\sigma_w^2} - 1\right) + \Psi(\xi)\frac{\partial \xi}{\partial z}\right\},\\
\frac{\partial f_{2E}}{\partial w} &= f_{2E}\left\{-\frac{w}{\sigma_w^2} - \Psi(\xi)\eta\right\},\\
\frac{\partial^2 f_{2E}}{\partial w^2} &= f_{2E}\left\{\left[-\frac{w}{\sigma_w^2} - \Psi(\xi)\eta\right]^2 - \frac{1}{\sigma_w^2} + \eta^2 \dot{\Psi}(\xi)\right\},
\end{aligned} \tag{8.22}$$

where

$$\Psi(\tau) = \frac{d}{d\tau}\ln\Phi(\tau), \quad \dot{\Psi}(\tau) = \frac{d\Psi(\tau)}{d\tau}, \quad \xi = \frac{u-\mu}{\sigma_{u/w}}, \quad \eta = \frac{\overline{uw}}{\sigma_{u/w}\sigma_w^2}.$$

Substituting (8.22) in (8.16) yields

$$\begin{aligned}
a'_u &= -\frac{C_0\varepsilon}{2\sigma_{u/w}^2}(u-\mu) \\
&\quad + \frac{1}{p'_{2E}}\left\{-\frac{\partial f_{2E}}{\partial t} - w\frac{\partial f_{2E}}{\partial z} - f_{2E}\frac{\partial a'_w}{\partial w} - a'_w\frac{\partial f_{2E}}{\partial w} + \frac{C_0\varepsilon}{2}\frac{\partial^2 f_{2E}}{\partial w^2}\right\} \\
&= -\frac{C_0\varepsilon}{2\sigma_{u/w}^2}(u-\mu) + \frac{f_{2E}}{p'_{2E}}\left\{-\frac{1}{2\sigma_w^2}\frac{\partial\sigma_w^2}{\partial t}\left(\frac{w^2}{\sigma_w^2}-1\right) - \Psi(\xi)\frac{\partial\xi}{\partial t}\right. \qquad (8.23)\\
&\quad - w\left[\frac{1}{2\sigma_w^2}\frac{\partial\sigma_w^2}{\partial z}\left(\frac{w^2}{\sigma_w^2}-1\right) + \Psi(\xi)\frac{\partial\xi}{\partial z}\right] - \frac{\partial a'_w}{\partial w} - a'_w\left[-\frac{w}{\sigma_w^2} - \Psi(\xi)\eta\right] \\
&\quad \left. + \frac{C_0\varepsilon}{2}\left[\left(-\frac{w}{\sigma_w^2} - \Psi(\xi)\eta\right)^2 - \frac{1}{\sigma_w^2} + \eta^2\dot{\Psi}(\xi)\right]\right\}.
\end{aligned}$$

Since

$$\frac{f_{2E}\Psi}{p'_{2E}} = \sigma_{u/w}, \quad \dot{\Psi}(\xi) = -\Psi(\xi)(\xi + \Psi(\xi)),$$

we find from (8.23)

$$\begin{aligned}
a'_u(t,z,u,w) = &-\frac{C_0\varepsilon(1+\rho^2)}{2\sigma_{u/w}^2}(u-\rho w) + \frac{\rho}{2\sigma_w^2}\left(C_0\varepsilon + \frac{\partial\sigma_w^2}{\partial t}\right)w \\
&+ \frac{\rho}{2}\frac{\partial\sigma_w^2}{\partial z}\left(\frac{w^2}{\sigma_w^2}+1\right) - \sigma_{u/w}\left(\frac{\partial\xi}{\partial t} + w\frac{\partial\xi}{\partial z}\right). \qquad (8.24)
\end{aligned}$$

Here

$$\rho = \frac{\overline{uw}}{\sigma_w^2}, \quad \xi = \frac{u-\rho w}{\sigma_{u/w}}.$$

Note that in the stationary case these expressions can be simplified to

$$\begin{aligned}
a'_w(t,z,w) &= -\frac{C_0\varepsilon}{2\sigma_w^2}w + \frac{1}{2}\frac{\partial\sigma_w^2}{\partial z}\left(\frac{w^2}{\sigma_w^2}+1\right), \\
a'_u(t,z,u,w) &= -\frac{C_0\varepsilon(1+\rho^2)}{2\sigma_{u/w}^2}(u-\rho w) + \frac{\rho\, C_0\varepsilon}{2\sigma_w^2}w \qquad (8.25)\\
&\quad + \frac{\rho}{2}\frac{\partial\sigma_w^2}{\partial z}\left(\frac{w^2}{\sigma_w^2}+1\right) - \sigma_{u/w}\frac{\partial\xi}{\partial z}w.
\end{aligned}$$

In the next section we present some numerical experiments which show that the model presented essentially differs from Thomson's model [237].

8.4 Numerical experiments

In this section we compare our model against Thomson's model in the case of a 2D stationary turbulence. First we consider the case of a 2D homogeneous turbulence specified by

$$\sigma_u = b_u u_*, \quad \sigma_w = b_w u_*, \tag{8.26}$$

where u_* is defined by $u_*^2 = -\overline{uw}$; b_u and b_w are some dimensionless constants. The mean velocity field is zero.

Thomson's model in the stationary homogeneous case reads [56, 237]

$$d\hat{X} = \hat{U}'dt,\ d\hat{Z} = \hat{W}'dt, \tag{8.27}$$

$$\begin{aligned} d\hat{U}' &= a'_u(\hat{U}', \hat{W}')dt + \sqrt{C_0\varepsilon}\, dB_u(t), \\ d\hat{W}' &= a'_w(\hat{U}', \hat{W}')dt + \sqrt{C_0\varepsilon}\, dB_w(t) \end{aligned} \tag{8.28}$$

with

$$a'_u(u, w) = -\frac{C_0\varepsilon}{2\Delta}(\sigma_w^2 u + u_*^2 w), \quad a'_w(u, w) = -\frac{C_0\varepsilon}{2\Delta}(\sigma_u^2 w + u_*^2 u). \tag{8.29}$$

where $\Delta = \sigma_u^2\sigma_w^2 - u_*^4$.

We calculated the dimensionless vertical eddy diffusivity

$$k(\tau) = \frac{\varepsilon\langle\hat{Z}\hat{W}\rangle}{u_*^4}, \tag{8.30}$$

where $\tau = t/T_L$. Here $T_L = \frac{2\sigma_w^2}{C_0\varepsilon}$ is the Lagrangian time scale in z-direction. Since this characteristic depends only on z, w, it is sufficient to take in our model (8.18) only the equation governing z, w:

$$\begin{aligned} d\hat{Z} &= \hat{W}'dt, \\ d\hat{W}' &= a'_w(\hat{U}', \hat{W}')dt + \sqrt{C_0\varepsilon}\, dB_w(t), \end{aligned} \tag{8.31}$$

where

$$a'_w(U, W) = -\frac{C_0\varepsilon}{2\sigma_w^2} W.$$

We note again (see Remark 2.1) that (8.31) is exactly the Thomson 1D model [237]. In our numerical experiments, we have fixed $b_u = 2.3$, and have made calculations for $b_w = 0.8, 1, 1.2$, and 1.4. These values characterize the anisotropy in the neutrally-stratified surface layer of the atmosphere [19]. The values of C_0 and ε were taken as 4 and 1 m^2/s^3, respectively, while $u_* = 0.4$ m/s.

In Figure 8.1 we show the function $k(\tau)$ (see (8.30)) obtained by Thomson's 2D model (8.27) (solid curves) and Thomson's 1D model (8.31) (dashed curve). The

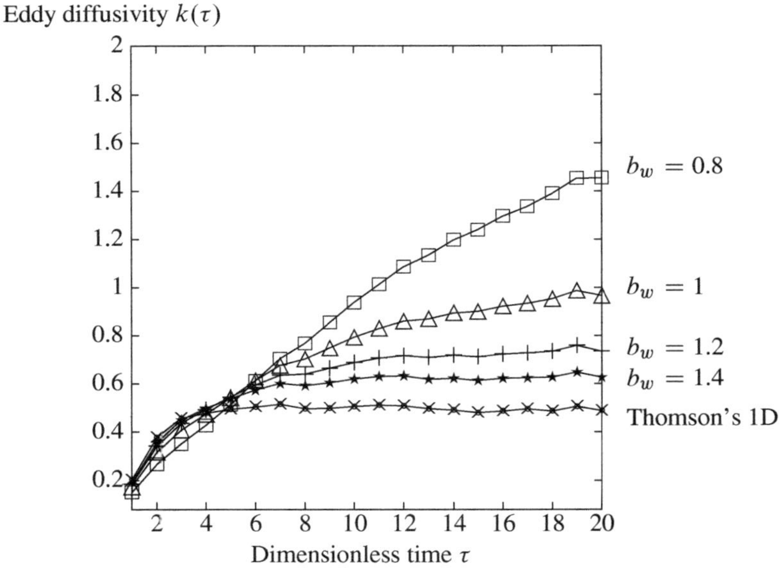

Figure 8.1. Dimensionless vertical eddy diffusivity coefficient as a function of τ, for different values of b_w.

lower solid line corresponds to $b_w = 1.4$, the next one to $b_w = 1.2$, the next to $b_w = 1.0$, and the very upper curve to $b_w = 0.8$.

It is clearly seen that the diffusivity coefficients of this two models essentially differ, e. g., for $b_w = 0.8$, this difference is about a factor of 3 at the steady-state values of $k(\tau)$. With decreasing anisotropy (i. e., the ratio b_u/b_w decreases), this difference becomes smaller. Thus we conclude that even in the homogeneous turbulence (which is, however, anisotropic) the two studied models may give essentially different results. To choose a proper case, one would need relevant measurements. We have no such experimental results in the homogeneous case, while in the neutrally-stratified surface layer (NSSL) the measurements are at hand (e. g., see [19]); therefore, it is interesting to compare the 2D Thomson model with our model defined by (8.18), (8.25).

Thomson's 2D model of 1-particle dispersion in horizontally homogeneous stationary turbulent flow reads [56, 237]

$$\begin{aligned} d\hat{X} &= (\hat{U}' + \bar{u}(\hat{Z},t))dt, \; d\hat{Z} = \hat{W}'dt, \\ d\hat{U}' &= a'_u(\hat{Z},\hat{U}',\hat{W}')dt + \sqrt{C_0\varepsilon}\, dB_u(t), \\ d\hat{W}' &= a'_w(\hat{Z},\hat{U}',\hat{W}')dt + \sqrt{C_0\varepsilon}\, dB_w(t), \end{aligned} \tag{8.32}$$

where

$$
\begin{aligned}
a'_u(z,u,w) = & -\frac{C_0\varepsilon}{2\Delta}\left[\sigma_w^2 u - \overline{uw}w\right] + \frac{1}{2}\frac{d\overline{uw}}{dz} \\
& + \frac{w}{2\Delta}\left\{\frac{d\sigma_u^2}{dz}(\sigma_w^2 u - \overline{uw}w) + \frac{d\overline{uw}}{dz}(-\overline{uw}u + \sigma_u^2 w)\right\}, \\
a'_w(z,u,w) = & -\frac{C_0\varepsilon}{2\Delta}\left[\sigma_u^2 w - \overline{uw}u\right] + \frac{1}{2}\frac{d\sigma_w^2}{dz} \\
& + \frac{w}{2\Delta}\left\{\frac{d\overline{uw}}{dz}(\sigma_w^2 u - \overline{uw}w) + \frac{d\sigma_w^2}{dz}(-\overline{uw}u + \sigma_u^2 w)\right\}.
\end{aligned}
$$

Here $\Delta = \sigma_u^2\sigma_w^2 - \overline{uw}^2$.

For the NSSL, the coefficients in this model can be taken as follows [19, 146]:

$$
\varepsilon(z) = \frac{u_*^3}{\kappa z}, \qquad \bar{u}(z) = \frac{u_*}{\kappa}\ln(z/z_0),
$$

and σ_u and σ_w are given by (8.26) with $u_*^2 = -\overline{uw}$ =const; $\kappa = 0.4$, z_0 is the roughness height.

Hence, the Thomson model in this case is specified by

$$
a'_u(z,u,w) = -\frac{C_0\varepsilon(z)}{2\Delta}(\sigma_w^2 u + u_*^2 w), \quad a'_w(z,u,w) = -\frac{C_0\varepsilon(z)}{2\Delta}(\sigma_u^2 w + u_*^2 u).
$$

Our model (8.18), (8.25) in the case of NSSL is determined by

$$
\begin{aligned}
a'_u(z,u,w) &= -\frac{C_0\varepsilon(z)(1+\rho^2)}{2\sigma_{u/w}^2}(u - \rho w) + \frac{\rho C_0\varepsilon(z)}{2\sigma_w^2}w, \\
a'_w(z,w) &= -\frac{C_0\varepsilon(z)}{2\sigma_w^2}w.
\end{aligned}
\tag{8.33}
$$

In the comparative calculations, we have calculated the following dimensionless Lagrangian characteristics:

$$
\begin{aligned}
A(t) &= \frac{\sqrt{\langle Z^2(t)\rangle}}{u_* t}, \quad B(t) = \frac{\langle Z(t)\rangle}{u_* t}, \\
C(t) &= \frac{z_0}{u_* t}\exp\left\{\frac{\kappa\langle X(t)\rangle}{u_* t} + 1\right\}.
\end{aligned}
\tag{8.34}
$$

It is known (e. g., see [19, p. 77]) that these functions tend, as $t \to \infty$, to constant values a, b and c, respectively, provided h_s and z_0 are much less than $u_* t$. Here h_s is the height at which the Lagrangian trajectory starts. The experimental measurements of the constants a, b, and c are scattered to a certain amount. However as can be

Table 8.1. Steady-state values of the functions (8.34), for $b_w = 1.2$.

Model	a	b	c
Thomson's	0.65 ± 0.01	0.48 ± 0.02	0.21 ± 0.03
ours	0.52 ± 0.02	0.38 ± 0.01	0.14 ± 0.01

Table 8.2. Steady-state values of the functions (8.34), for $b_w = 1.3$.

Model	a	b	c
Thomson's	0.78 ± 0.01	0.59 ± 0.02	0.25 ± 0.04
ours	0.66 ± 0.02	0.49 ± 0.02	0.18 ± 0.01

extracted from [19, Tables 3.6 and 3.8], we conclude that the values of a, b, and c lie in the intervals $(0.32, 0.58)$, $(0.28, 0.49)$, and $(0.14, 0.30)$, respectively.

In our calculations, we have chosen $u_* = 0.4$ m/s, $z_0 = 0.1$ m, $b_u = 2.3$, and two variants of b_w: $b_w = 1.2$ and $b_w = 1.3$; the results are given in Tables 8.1 and 8.2.

Here we present the results which correspond to $b_w = 1.2$ in Table 8.1, and $b_w = 1.3$ in Table 8.2. The tables show that in both cases the results for a and b obtained by our model agree slightly better with the experiments compared to those obtained by Thomson's model. As to the quantity c, both models agree satisfactorily with the measurements.

In conclusion we stress again that even in the case of homogeneous but anisotropic turbulence, the well-mixed condition (8.7) does not define the 1-particle model uniquely. Here we compared our model against Thomson's model, which gives two significantly different values of the eddy diffusivity coefficient (8.30). To choose between these models, more accurate measurements are needed.

Chapter 9

Direct and adjoint Monte Carlo for the footprint problem

Lagrangian stochastic models and algorithms are constructed and justified for solving the footprint problem, namely, the problem of calculation of the mean concentration and the flux of particles at a fixed point released from a source arbitrarily situated in space. The direct and adjoint Monte Carlo algorithms are suggested, and rigorous justifications are given. Two different backward trajectory algorithms are considered: Thomson's method, and a method based on probabilistic representations of the relevant initial value problem. The cost of the latter algorithm may increase with time, but it allows treating the general situation when a set of reacting species is scattered by the flow.

9.1 Introduction

The footprint problem as formulated in the literature (e. g., see [55–57, 221]) essentially deals with the calculation of the contribution to the mean concentration and its flux at a fixed point from an arbitrary given source of particles. There are mainly two different approaches: (i) conventional deterministic methods based on the semiempirical turbulent diffusion equation and closure assumptions (e. g., see [221] and (ii), the stochastic approach, which utilizes trajectory simulations (e. g., see [108, 191], [198–256]).

The deterministic approach directly deals with the equation governing the mean concentration, but it is restricted by the use of the Bousinesque hypothesis, whose applicability should be additionally studied (e. g., see [19]). For instance, this hypothesis can not be true if the concentration is calculated close to the sources [19, 146]. More generally, the high-order closure methods are developed, a but different closure hypothesis also should be made [146] .

Stochastic models do not require any closure hypotheses, and the main difficulty is in constructing adequate Lagrangian trajectories with the desired statistical characteristics.

There are two main approaches in constructing stochastic methods. The first one is based on the Monte Carlo simulation of Eulerian random velocity fields (e. g., see [102, 191, 244]). The second approach treats the stochastic Lagrangian trajectories as solutions to the stochastic generalized Langevin equation (e. g., see [198, 215, 237]).

The first approach is more rigorous, but generally requires a lot of computer time. In addition it needs detailed information about the statistical characteristics of the whole velocity field. In contrast, the second approach needs only a 1-pont probability density function (p. d. f.) of the Eulerian velocity field, and is much more efficient in numerical calculations. It should be noted however, that this approach is rigorously justified only in the case of stationary isotropic turbulent flow. Even in the case of homogeneous but nonisotropic turbulence, the justification problem remains unsolved; in particular there are several different stochastic models which satisfy the well-mixed condition [200, 237].

In this chapter, we suggest direct and backward Monte Carlo algorithms for solving the footprint problem by simulation of Lagrangian trajectories as solutions to generalized Langevin equations. We derive random estimators for the mean concentration and its flux, both for direct and backward schemes. Two different methods are compared: the adjoint scheme, which is based on probabilistic representations of the relevant PDE [59], and the backward Thomson scheme [237]. The latter approach is extended to general stochastic differential equations.

9.2 Formulation of the problem

Let us consider a passive scalar dispersed by the turbulent velocity field in the surface layer of the atmosphere. The passive scalar is assumed to follow the streamlines of the flow. We assume that the source of particles is quite arbitrary; for instance, it might be situated on the surface or in the space, or even at given points. Let us denote by $q(\mathbf{x}, t)$ the spatial-temporal density distribution function of the source, i. e, the number of emitted particles per unit volume in a unit time interval at the phase point $(\mathbf{x}, t)$. Initially, the spatial density of particles is given by $q_0(\mathbf{x})$. The particles are transported by the 3D turbulent velocity field $\vec{u}(\mathbf{x}, t)$ in the surface layer $D = \{\mathbf{x} = (x_1, x_2, x_3) : x_3 \geq 0\}$. Let us denote by $\mathbf{X}(t; \mathbf{x}_0, t_0)$ and $\mathbf{V}(t; \mathbf{x}_0, t_0)$ the Lagrangian spatial coordinates and the velocity, respectively.

The mean concentration at $(\mathbf{x}, t)$ is defined by [146]

$$\begin{aligned} \langle c(\mathbf{x}, t)\rangle &= \bar{c}(\mathbf{x}, t) \qquad (9.1) \\ &= \int_0^t dt_0 \int_D d\mathbf{x}_0\, q(\mathbf{x}_0, t_0) p_L(\mathbf{x}, t; \mathbf{x}_0, t_0) + \int_D d\mathbf{x}_0\, q_0(\mathbf{x}_0) p_L(\mathbf{x}, t; \mathbf{x}_0, 0), \end{aligned}$$

where

$$p_L(\mathbf{x}, t; \mathbf{x}_0, t_0) = \langle \delta(\mathbf{x} - \mathbf{X}(t; \mathbf{x}_0, t_0))\rangle$$

is the probability density function (p. d. f.) of the particle's coordinate at the time t which was started in the point $\mathbf{x}_0$ at the time t_0, and $\delta(\cdot)$ is the Dyrac delta-function. Here and throughout the chapter we use the notation $\langle\cdot\rangle$ for the averaging over the samples of the turbulent velocity field. We define also the concentration fluxes by

$$F_i(\mathbf{x}, t) = \langle u_i(\mathbf{x}, t) c(\mathbf{x}, t)\rangle, \quad i = 1, 2, 3,$$

where $c(\mathbf{x}, t)$ is the instant concentration.

As in the case (9.1), the fluxes can be represented in the integral form (see Appendix A):

$$F_i(\mathbf{x},t) = \int_{\mathrm{IR}^3} d\vec{u} \int_0^t dt_0 \int_D d\mathbf{x}_0\, u_i q(\mathbf{x}_0,t_0) p_L(\mathbf{x},\vec{u},t;\mathbf{x}_0,t_0)$$
$$+ \int_{\mathrm{IR}^3} d\vec{u} \int_D d\mathbf{x}_0\, u_i q_0(\mathbf{x}_0) p_L(\mathbf{x},\vec{u},t;\mathbf{x}_0,0). \tag{9.2}$$

Here

$$p_L(\mathbf{x},\vec{u},t;\mathbf{x}_0,t_0) = \langle \delta(\mathbf{x}-\mathbf{X}(t;\mathbf{x}_0,t_0))\delta(\vec{u}-\mathbf{V}(t;\mathbf{x}_0,t_0))\rangle \tag{9.3}$$

is the p. d. f. of the spatial-velocity phase point.

In the analysis, it is convenient to deal with a general quantity, the spatial-velocity distribution of an ensemble of particles:

$$p(\mathbf{x},\vec{u},t) = \int_0^t dt_0 \int_D d\mathbf{x}_0\, q(\mathbf{x}_0,t_0) p_L(\mathbf{x},\vec{u},t;\mathbf{x}_0,t_0)$$
$$+ \int_D d\mathbf{x}_0\, q_0(\mathbf{x}_0) p_L(\mathbf{x},\vec{u},t;\mathbf{x}_0,0). \tag{9.4}$$

From (9.1), (9.2), and (9.4) we find

$$\bar{c}(\mathbf{x},t) = \int_{\mathrm{IR}^3} p(\mathbf{x},\vec{u},t)d\vec{u}\ , \quad F_i(\mathbf{x},t) = \int_{\mathrm{IR}^3} u_i\, p(\mathbf{x},\vec{u},t)d\vec{u}, \quad i = 1,2,3.$$

It is of practical interest to calculate the mean concentration and relevant fluxes for arbitrarily situated surface sources. In the literature, this problem is called the footprint problem (e. g., see [55, 57, 221]. Note that in this problem the mean concentration and fluxes are evaluated at a fixed point. We consider a more general function:

$$(p,h) = \int_{\mathrm{IR}^3} d\vec{u} \int_0^T dt \int_D d\mathbf{x}\ p(\mathbf{x},\vec{u},t)h(\mathbf{x},\vec{u},t),$$

where $h(\mathbf{x},\vec{u},t)$ is an arbitrary function which can be chosen relevant to the quantity of interest. For instance, in the case $h(\mathbf{y},\vec{u},s) = \delta(\mathbf{y}-\mathbf{x})\delta(s-t)$ we have $(p,h) = \bar{c}(\mathbf{x},t)$. If $h(\mathbf{y},\vec{u},s) = u_i\delta(\mathbf{y}-\mathbf{x})\delta(s-t)$, then $(p,h) = F_i(\mathbf{x},t)$. Thus we concentrate on the problem of calculation of the function (p,h).

9.3 Stochastic Lagrangian algorithm

To construct algorithms based on the representations given above, we need samples of the Lagrangian trajectories $\mathbf{X}(t) = \mathbf{X}(t;\mathbf{x}_0,t_0)$, $t \geq t_0$. Ideally, if we had samples of the velocity $\vec{u}(\mathbf{x},t)$, the trajectories could be simulated by solving the problem

$$\frac{d\mathbf{X}(t)}{dt} = \vec{u}(\mathbf{X}(t),t), \quad t > t_0 \quad \mathbf{X}(t_0) = \mathbf{x}_0\ .$$

In practice one uses approximate models of the velocity field. For instance, randomized models of the Gaussian velocity fields are used (e.g., see [191]). This approach is well developed and justified only in the case of homogeneous turbulence, while the inhomogeneous case needs further development. In the general nonhomogeneous case one uses another approach based on a stochastic differential equation of the Langevin type directly governing the Lagrangian trajectory. This equation has the form [200, 215, 237]

$$\begin{aligned} d\mathbf{X}(t) &= \mathbf{V}(t)dt, \\ d\mathbf{V}(t) &= \mathbf{a}(t, \mathbf{X}(t), \mathbf{V}(t))dt + \sqrt{C_0 \bar{\varepsilon}(\mathbf{X}(t), t)}\, d\mathbf{W}(t), \end{aligned} \tag{9.5}$$

where the function $\mathbf{a}$ is to be specified in each specific situation, C_0 is the universal Kolmogorov constant ($C_0 \approx 4 \sim 6$), $\bar{\varepsilon}(\mathbf{x}, t)$ is the mean dissipation rate of the kinetic energy of turbulence, and $\mathbf{W}(t)$ is the standard 3D Wiener process. In this section, we deal with the general scheme.

Remark 9.1. Note that to complete the description of the Lagrangian stochastic model, we need to define the behavior of $(\mathbf{X}(t), \mathbf{V}(t))$ in the neighborhood of the boundary $\Gamma = \{\mathbf{x} = (x_1, x_2, x_3) : x_3 = 0\}$. We assume that the boundary is impenetrable, i.e., that $u_3(\mathbf{x})|_{\mathbf{x} \in \Gamma} = 0$. This implies that the Lagrangian trajectories satisfying (9.5) never do reach Γ. Therefore it is reasonable to require that the same property holds for $\mathbf{X}(t)$, the solutions to (9.5). This can be realized by special choice of the function $\bar{\varepsilon}(\mathbf{x}, t)$ (see for details Sec. 4).

9.3.1 Direct Monte Carlo algorithm

Let $\mathbf{X}(t; \mathbf{x}_0, t_0)$, $\mathbf{V}(t; \mathbf{x}_0, t_0)$, $t \geq t_0$ be solutions to (9.5) satisfying the initial condition $\mathbf{X}(t_0) = \mathbf{x}_0$ with $\mathbf{V}(t_0) = \vec{u}_0$, where $\vec{u}_0$ is the random velocity whose p. d. f. coincides with $p_E(\vec{u}_0; \mathbf{x}_0, t_0)$, the p. d. f. of the Eulerian velocity $\vec{u}(\mathbf{x}_0, t_0)$.

It is convenient to use the representation $(p, h) = I + I_0$, where

$$\begin{aligned} I &= \int_{\mathrm{IR}^3} d\vec{u} \int_0^T dt \int_D d\mathbf{x} h(\mathbf{x}, \vec{u}, t) \int_0^t dt_0 \int_D d\mathbf{x}_0\, q(\mathbf{x}_0, t_0) p_L(\mathbf{x}, \vec{u}, t; \mathbf{x}_0, t_0) \\ &= \int_0^T dt_0 \int_D d\mathbf{x}_0 q(\mathbf{x}_0, t_0) \int_{\mathrm{IR}^3} d\vec{u} \int_{t_0}^T dt \int_D d\mathbf{x} p_L(\mathbf{x}, \vec{u}, t; \mathbf{x}_0, t_0), \end{aligned}$$

and

$$I_0 = \int_D d\mathbf{x}_0 q_0(\mathbf{x}_0) \int_{\mathrm{IR}^3} d\vec{u} \int_0^T dt \int_D d\mathbf{x} h(\mathbf{x}, \vec{u}, t) p_L(\mathbf{x}, \vec{u}, t; \mathbf{x}_0, t_0).$$

From this representations we can write down the Monte Carlo estimators for I and I_0. Let

$$Q = \int_D d\mathbf{x} \int_0^T dt\, q(\mathbf{x}, t), \quad Q_0 = \int_D q_0(\mathbf{x}) d\mathbf{x},$$

and let $(\tilde{\mathbf{x}}, \tilde{t})$ be a random point distributed in $D \times [0, T]$ with the p. d. f. $q(\mathbf{x}, t)/Q$, and $\tilde{\mathbf{x}}_0$ be a random point distributed in D with $q_0(\mathbf{x})/Q_0$. Standard arguments of the Monte Carlo theory [191] yield

$$I = Q\mathbf{E}_{(\tilde{\mathbf{x}},\tilde{t})}\mathbf{E}_{W(\cdot)} \int_{\tilde{t}}^{T} h(\mathbf{X}(t; \tilde{\mathbf{x}}, \tilde{t}), \mathbf{V}(t; \tilde{\mathbf{x}}, \tilde{t}), t)dt, \tag{9.6}$$

$$eI_0 = Q_0\mathbf{E}_{\tilde{\mathbf{x}}_0}\mathbf{E}_{W(\cdot)}h(\mathbf{X}(t; \tilde{\mathbf{x}}_0, 0), \mathbf{V}(t; \tilde{\mathbf{x}}_0, 0), t). \tag{9.7}$$

Here $\mathbf{E}_{(\tilde{\mathbf{x}},\tilde{t})}\mathbf{E}_{W(\cdot)}$ means averaging, first, over all starting points $(\tilde{\mathbf{x}}, \tilde{t})$ and, second, over all solutions of (9.5); these two averagings are taken independently. Similar notation is used in (9.7).

From the probabilistic representations (9.6), (9.7) we can construct the direct Monte Carlo algorithm. For this we need a numerical scheme for solving the stochastic differential equation (9.5). For simplicity, we choose the Euler scheme.

The algorithm can be described as follows. The mathematical expectation in (9.6) is usually calculated as

$$I \simeq \frac{1}{N} \sum_{i=1}^{N} \xi_i,$$

where N is the number of samples and ξ_i, $i = 1, \dots, N$, are independent samples of the random estimator

$$\xi = Q \int_{\tilde{t}}^{T} h(\mathbf{X}(t; \tilde{\mathbf{x}}, \tilde{t}), \mathbf{V}(t; \tilde{\mathbf{x}}, \tilde{t}), t)dt.$$

First we choose the time step of integration as $\Delta t = T/N_t$, N_t is the total number of steps. Then the calculation of ξ_i is as follows.

Step 0.
$S := 0$, and sample the random point $(\tilde{\mathbf{x}}, \tilde{t})$ in $D \times [0, T]$ from the density $q(\mathbf{x}, t)/Q$;
sample $\tilde{\vec{u}}_0$ in IR^3 from the density $p_E(\tilde{\vec{u}}_0; \tilde{\mathbf{x}}, \tilde{t})$;
put $t := \tilde{t}.$, $\mathbf{X} := \tilde{\mathbf{x}}$, $\mathbf{V} := \tilde{\vec{u}}$;

Step 1.
$S := S + Q\Delta t h(\mathbf{X}, \mathbf{V}, t)$;
Sample a 3D standard Gaussian random vector η;
put $\mathbf{X} := \mathbf{X} + \mathbf{V}\Delta t$, $\mathbf{V} = \mathbf{V} + \mathbf{a}(t, \mathbf{X}, \mathbf{V})\Delta t + \sqrt{C_0\bar{\varepsilon}(\mathbf{X}, t)\Delta t}\eta$; $t := t + \Delta t$;
if $t > T$, then go to step 2. Otherwise go back to start the step 1.

Step 2.
put $\xi_i := S$;

The calculation of I_0 is treated analogously:

$$I_0 \simeq \frac{1}{N}\sum_{i=1}^{N}\chi_i$$

where N is the number of samples, $\chi_i, i = 1,\ldots,N$ are independent samples of the random estimator

$$\chi = Q_0 h(\mathbf{X}(t;\tilde{\mathbf{x}}_0,0),\mathbf{V}(t;\tilde{\mathbf{x}}_0,0),t).$$

The calculation of χ_i:

Step 0.
Sample a random point $\tilde{\mathbf{x}}_0$ in D from the density $q_0(\mathbf{x})/Q_0$;
sample $\tilde{\vec{u}}_0$ in IR^3 from the density $p_E(\tilde{\vec{u}}_0;\tilde{\mathbf{x}}_0,0)$;
put $\mathbf{X} := \tilde{\mathbf{x}}_0$, $\mathbf{V} := \tilde{\vec{u}}_0$; $t := 0$.

Step 1.
Sample a 3D standard Gaussian random vector η;
put $\mathbf{X} := \mathbf{X} + \mathbf{V}\Delta t$, $\mathbf{V} = \mathbf{V} + \mathbf{a}(t,\mathbf{X},\mathbf{V})\Delta t + \sqrt{C_0\bar{\varepsilon}(\mathbf{X}(t),t)\Delta t}\;\eta$; $t := t + \Delta t$;
if $t \geq T$, then go to step 2. Otherwise go to start the step 1.

Step 2.
put $\chi_i := Q_0 h(t,\mathbf{X},\mathbf{V})$;

9.3.2 Adjoint algorithm

Clearly, the direct algorithm is not practically applicable in the general situation if the function $h(\mathbf{x},\vec{u},t)$ is concentrated on small domains. In this case, an adjoint scheme is preferable. For this purpose we need the probabilistic representation for $p(\mathbf{x},\vec{u},t)$. In what follows we use the summation convention. The function $p_L(\mathbf{x},\vec{u},t;\mathbf{x}_0,t_0)$ defined in (9.3) satisfies the Kolmogorov–Fokker–Planck equation for the stochastic differential equation (9.5):

$$\frac{\partial p_L}{\partial t} + \frac{\partial(u_i p_L)}{\partial x_i} + \frac{\partial(a_i p_L)}{\partial u_i} = \frac{1}{2}C_0\bar{\varepsilon}(\mathbf{x},t)\frac{\partial^2}{\partial u_i\partial u_i}p_L(\mathbf{x},\vec{u},t;\mathbf{x}_0,t_0)$$

with the initial conditions

$$p_L(\mathbf{x},\vec{u},t;\mathbf{x}_0,t_0)\big|_{t=t_0} = \delta(\mathbf{x}-\mathbf{x}_0)p_E(\vec{u};\mathbf{x}_0,t_0).$$

From this, using (9.4) we find

$$\frac{\partial p}{\partial t} + \frac{\partial(u_i p)}{\partial x_i} + \frac{\partial(a_i p)}{\partial u_i} = \frac{1}{2}C_0\bar{\varepsilon}(\mathbf{x},t)\frac{\partial^2 p}{\partial u_i\partial u_i}p(\mathbf{x},\vec{u},t) + q(\mathbf{x},t)p_E(\vec{u};\mathbf{x},t)$$

and

$$p(\mathbf{x},\vec{u},t)\big|_{t=0} = q_0(\mathbf{x})p_E(\vec{u};\mathbf{x},0).$$

Using the probabilistic representation presented in Appendix B we get

$$p(\mathbf{x},\vec{u},t) = \mathbf{E}_{(\mathbf{x},\vec{u},t)}\Bigg\{ \int_0^t q(\hat{\mathbf{X}}_{t_0}^{\mathbf{x},\vec{u},t}, \hat{\mathbf{V}}_{t_0}^{\mathbf{x},\vec{u},t}, t_0)\, p_E(\hat{\mathbf{V}}_{t_0}^{\mathbf{x},\vec{u},t}; \hat{\mathbf{X}}_{t_0}^{\mathbf{x},\vec{u},t}, t_0) \\ \times \exp\left(-\int_{t_0}^t \rho(\hat{\mathbf{X}}_{t_0'}^{\mathbf{x},\vec{u},t}, \hat{\mathbf{V}}_{t_0'}^{\mathbf{x},\vec{u},t}, t_0')dt_0'\right)dt_0 + q_0(\hat{\mathbf{X}}_0^{\mathbf{x},\vec{u},t}) \\ \times p_E(\hat{\mathbf{V}}_0^{\mathbf{x},\vec{u},t}; \hat{\mathbf{X}}_0^{\mathbf{x},\vec{u},t}, 0) \exp\left(-\int_0^t \rho(\hat{\mathbf{X}}_{t_0'}^{\mathbf{x},\vec{u},t}, \hat{\mathbf{V}}_{t_0'}^{\mathbf{x},\vec{u},t}, t_0')dt_0'\right)\Bigg\}, \tag{9.8}$$

where $\rho(\mathbf{x},\vec{u},t) = \frac{\partial a_i(\mathbf{x},\vec{u},t)}{\partial u_i}$, and $\hat{\mathbf{X}}(t_0) \equiv \hat{\mathbf{X}}_{t_0}^{\mathbf{x},\vec{u},t}$, $\hat{\mathbf{V}}(t_0) \equiv \hat{\mathbf{V}}_{t_0}^{\mathbf{x},\vec{u},t}$, $t_0 \le t$ is the adjoint trajectory determined as the solution to the equation

$$d\hat{\mathbf{X}}(t_0) = \hat{\mathbf{V}}(t_0)dt_0,$$
$$d\hat{\mathbf{V}}(t_0) = \mathbf{a}(t_0, \hat{\mathbf{X}}(t_0), \hat{\mathbf{V}}(t_0))dt_0 + \sqrt{C_0\bar{\varepsilon}(\hat{\mathbf{X}}(t_0), t_0)}\, d\hat{\mathbf{W}}(t_0),$$

with the initial conditions

$$\hat{\mathbf{X}}(t_0)\Big|_{t_0=t} = \mathbf{x}, \quad \hat{\mathbf{V}}(t_0)\Big|_{t_0=t} = \vec{u}.$$

Note that here $\mathbf{a}(\cdot)$ is the same function as in (9.5), and $\hat{\mathbf{W}}(t_0)$ is a standard 3D Wiener process.

From the probabilistic representation (9.8) we get

$$(p,h) = \int_{\mathrm{IR}^3} d\vec{u} \int_0^T dt \int_D d\mathbf{x}\, h(\mathbf{x},\vec{u},t)\mathbf{E}_{(\mathbf{x},\vec{u},t)} \\ \times \Bigg\{ \int_0^t q(\hat{\mathbf{X}}_{t_0}^{\mathbf{x},\vec{u},t}, \hat{\mathbf{V}}_{t_0}^{\mathbf{x},\vec{u},t}, t_0).p_E(\hat{\mathbf{V}}_{t_0}^{\mathbf{x},\vec{u},t}; \hat{\mathbf{X}}_{t_0}^{\mathbf{x},\vec{u},t}, t_0) \\ \times \exp\left(-\int_{t_0}^t \rho(\hat{\mathbf{X}}_{t_0'}^{\mathbf{x},\vec{u},t}, \hat{\mathbf{V}}_{t_0'}^{\mathbf{x},\vec{u},t}, t_0')dt_0'\right)dt_0 \\ + q_0(\hat{\mathbf{X}}_0^{\mathbf{x},\vec{u},t})\, p_E(\hat{\mathbf{V}}_0^{\mathbf{x},\vec{u},t}; \hat{\mathbf{X}}_0^{\mathbf{x},\vec{u},t}, 0) \exp\left(-\int_0^t \rho(\hat{\mathbf{X}}_{t_0'}^{\mathbf{x},\vec{u},t}, \hat{\mathbf{V}}_{t_0'}^{\mathbf{x},\vec{u},t}, t_0')dt_0'\right)\Bigg\}.$$

Now we are in position to write the Monte Carlo estimator for $\hat{I}$ and $\hat{I}_0$.

Let $\pi(\mathbf{x},\vec{u},t)$ be an arbitrary probability density function in $D \times \mathrm{IR}^3 \times [0,T]$ which satisfies the condition

$$\pi(\mathbf{x},\vec{u},t) \neq 0 \quad \text{if } h(\mathbf{x},\vec{u},t) \neq 0.$$

Let $(\hat{\mathbf{x}}, \hat{\vec{u}}, \hat{t})$ be a random point distributed in $D \times \mathrm{IR}^3 \times [0,T]$ with density π. Then

$$(p,h) = \mathbf{E}_{(\hat{\mathbf{x}},\hat{\vec{u}},\hat{t})}\mathbf{E}_{\hat{\mathbf{W}}(\cdot)}\hat{\xi},$$

where

$$\hat{\xi} = \frac{h(\hat{\mathbf{x}}, \hat{\vec{u}}, \hat{t})}{\pi(\hat{\mathbf{x}}, \hat{\vec{u}}, \hat{t})} \left[\int_0^{\hat{t}} \left\{ q(\hat{\mathbf{X}}_{t_0}, t_0) p_E(\hat{\mathbf{V}}_{t_0}; \hat{\mathbf{X}}_{t_0}, t_0) \exp\left(- \int_{t_0}^{\hat{t}} \rho(\hat{\mathbf{X}}_{t_0'}, \hat{\mathbf{V}}_{t_0'}, t_0') dt_0' \right) \right\} dt_0 \right.$$
$$\left. + q_0(\hat{\mathbf{X}}_0) p_E(\hat{\mathbf{V}}_0; \hat{\mathbf{X}}_0, 0) \exp\left(- \int_0^{\hat{t}} \rho(\hat{\mathbf{X}}_{t_0'}, \hat{\mathbf{V}}_{t_0'}, t_0') dt_0' \right) \right].$$

Here we used a brief notation

$$\hat{\mathbf{X}}_{t_0} \equiv \hat{\mathbf{X}}_{t_0}^{\hat{\mathbf{x}}, \hat{\vec{u}}, \hat{t}}, \quad \hat{\mathbf{V}}_{t_0} \equiv \hat{\mathbf{V}}_{t_0}^{\hat{\mathbf{x}}, \hat{\vec{u}}, \hat{t}}.$$

The calculations are carried out according to

$$(p, h) \simeq \frac{1}{N} \sum_{i=1}^{N} \hat{\xi}_i,$$

where N is the number of samples, $\hat{\xi}_i$, $i = 1, \dots, N$, are independent samples of the random estimator $\hat{\xi}$.

The algorithm of calculation of the sample $\hat{\xi}_i$:

Step 0.
$S := 0$; sample the random point $(\hat{\mathbf{x}}, \hat{\vec{u}}, \hat{t})$ in $D \times \mathrm{IR}^3 \times [0, T]$ from the density $\pi(\mathbf{x}, \vec{u}, t)$;
$Q := \frac{h(\hat{\mathbf{x}}, \hat{\vec{u}}, \hat{t})}{\pi(\hat{\mathbf{x}}, \hat{\vec{u}}, \hat{t})}$; put $t := \hat{t}$, $\mathbf{X} := \hat{\mathbf{x}}$, $\mathbf{V} := \hat{\vec{u}}$; $\mu := 1$.

Step 1.
$S := S + \Delta t q(\mathbf{X}, t) p_E(\mathbf{V}; \mathbf{X}, t)\mu$;
Sample a 3D standard Gaussian random vector η;
put $\mathbf{X} := \mathbf{X} - \mathbf{V}\Delta t$, $\mathbf{V} = \mathbf{V} - \mathbf{a}(t, \mathbf{X}, \mathbf{V})\Delta t + \sqrt{C_0 \bar{\varepsilon}(\mathbf{X}, t)\Delta t}\, \eta$;
$\mu := \mu \exp(-\rho(\mathbf{X}, \mathbf{V}, t)\Delta t)$; $t := t - \Delta t$;
if $t < 0$, then go to step 2. Otherwise go back to start the step 1.

Step 2.
put $\hat{\xi}_i := Q[S + q_0(\mathbf{X}) p_E(\mathbf{V}; \mathbf{X}, 0)\mu]$.

9.4 Impenetrable boundary

Note that to complete the description of the Lagrangian stochastic model we need to define the behavior of $\mathbf{X}(t), \mathbf{V}(t)$, the solution to (9.5) in the neighborhood of the boundary $\Gamma = \{\mathbf{x} : z = x_3 = 0\}$. We assume that the boundary is impenetrable,

i. e., that $w = u_3 = 0$ at the boundary of Γ. This implies that the true Lagrangian trajectories never reach Γ. Therefore it is reasonable to require that the same property holds for $\mathbf{X}(t)$, the solutions to (9.5). This can be done by the special choice of the function $\bar{\varepsilon}(z,t)$. Indeed, in the neighborhood of Γ it is reasonable to consider the flow as neutrally stratified. Therefore, $p_E(w)$ is Gaussian, with constant σ_w, and the vertical profile of $\bar{\varepsilon}(z)$ is given by [146]

$$\bar{\varepsilon}(z) = \frac{u_*^3}{\kappa z}, \quad \kappa \simeq 0.4. \tag{9.9}$$

Here κ is the Karman constant, and z_0 is the roughness height.

The equation of vertical motion $Z(t) = X_3(t)$, $W(t) = V_3(t)$ is then

$$dZ = W\,dt, \quad dW(t) = -\frac{a}{Z}W(t)dt + \frac{b}{\sqrt{Z}}\,dB(t), \tag{9.10}$$

where

$$a = \frac{C_0 u_*^3}{2\kappa\sigma_w^2}, \quad b = \left(\frac{C_0 u_*^3}{\kappa}\right)^{\frac{1}{2}}.$$

If we assume that the formula (9.9) is true for all $z > 0$, then all the solutions to (9.10) do not reach the boundary Γ. Indeed, let τ be a random time (which depends on the trajectory $Z(t)$) defined by

$$\tau(t) = \int_0^t \frac{ds}{Z(s)}.$$

Then, the vertical velocity in the new time variable $W(\tau)$ satisfies the equation

$$dW(\tau) = -aW(\tau)d\tau + b\,dB(\tau).$$

Therefore, from

$$\frac{dZ}{d\tau} = \frac{dZ}{dt}\frac{dt}{d\tau} = W(\tau)Z(\tau)$$

we have

$$Z(\tau) = Z(0) = \exp\{S(\tau)\}, \quad S(\tau) = \int_0^\tau W(\tau')\,d\tau'.$$

The function $W(\tau)$ is an Uhlenbeck–Ornstein process with continuous samples. Therefore, $|S(\tau)| < \infty$ with probability 1 for arbitrary $\tau > 0$. This implies that $Z(\tau) > 0$ provided that $Z(0) > 0$. Thus the function $Z(\tau)$ never reaches the boundary Γ. The same is true for $Z(t)$. To show this, it is sufficient to note that $t(\tau) \to \infty$ as $\tau \to \infty$. Let us show this property. We have

$$t(\tau) = \int_0^\tau \frac{dt}{d\tau'}\,d\tau' = \int_0^\tau Z(\tau')\,d\tau' = Z(0)\int_0^\tau \exp\{S(\tau')\}\,d\tau'.$$

In [108] it is shown that with probablity 1

$$\int_0^\infty \exp\{S(\tau)\}\,d\tau = \infty.$$

This implies that with probability 1 $t(\tau) \to \infty$ as $\tau \to \infty$.

In numerical implementation, it is convenient to simulate the trajectory in the neighborhood of Γ by numerical solution to

$$\begin{aligned}\frac{d\ln Z(\tau)}{d\tau} &= W(\tau),\\ dW(\tau) &= -aW(\tau) + b\,dB(\tau),\\ \frac{dt(\tau)}{d\tau} &= Z(\tau).\end{aligned}$$

The Euler scheme reads

$$\begin{aligned}Z(\tau+\Delta\tau) &= Z(\tau)\exp\{W(\tau)\Delta\tau\},\\ W(\tau+\Delta\tau) &= W(\tau) - aW(\tau)\Delta\tau + b\sqrt{\Delta\tau}\,\eta,\\ t(\tau+\Delta\tau) &= t(\tau) + Z(\tau)\Delta\tau,\end{aligned}$$

where $\Delta\tau$ is the discretization step, and η is a standard normal random number.

Note that by construction, this scheme ensures that the boundary is impenetrable, i. e., $Z(\tau) > 0$ for all τ.

9.5 Reacting species

In this section we show that the backward trajectory technique can be extanded to the case where a set of reacting species is in play. Assume that in the domain D with a boundary Γ which is impenetrable there are K reacting species governed in Lagrangian formulation by the system

$$\begin{aligned}d\mathbf{X}(t) &= \mathbf{V}(t)dt,\\ d\mathbf{V}(t) &= \mathbf{a}(t,\mathbf{X}(t),\mathbf{V}(t))dt + \sqrt{C_0\bar{\varepsilon}(\mathbf{X}(t),t)}\,d\mathbf{W}(t),\\ \frac{dN_k}{dt} &= f_k(t,\mathbf{X},\mathbf{N}), \quad k = 1,2.\ldots,K.\end{aligned} \tag{9.11}$$

with initial conditions

$$\mathbf{X}(0;\mathbf{x}_0) = \mathbf{x}_0, \quad \mathbf{V}(0;\mathbf{x}_0) = \mathbf{V}_0, \quad N_k(0;\mathbf{x}_0) = q_k(\mathbf{x}_0),\ k = 1,\ldots,K,$$

where $\mathbf{V}_0$ is the initial random velocity whose p. d. f. is $p_E(\mathbf{V};\mathbf{x}_0)$, and $q_k(\mathbf{x}_0)$ is the initial spatial distribution of the species, $\mathbf{N} = (N_1,\ldots,N_K)$. The Eulerian concentration of k-th specy is given by

$$n_k(\mathbf{x},t) = \int_D d\mathbf{x}_0 N_k(t;\mathbf{x}_0)\delta(\mathbf{x} - \mathbf{X}(t;\mathbf{x}_0)), \quad k = 1,\ldots,K.$$

Then the mean is

$$\langle n_k(\mathbf{x},t)\rangle = \int_{\mathrm{IR}^3} d\mathbf{V} \int d\mathbf{N} \int_D d\,\mathbf{x}_0 N_k p_L(\mathbf{x},\mathbf{V},\mathbf{N},t;\mathbf{x}_0), \quad k=1,\dots,K,$$

where

$$p_L(\mathbf{x},\mathbf{V},\mathbf{N},t;\mathbf{x}_0) = \langle \delta(\mathbf{x}-X(t;\mathbf{x}_0))\delta(\mathbf{V}-\mathbf{V}(t;\mathbf{x}_0))\delta(\mathbf{N}-\mathbf{N}(t;\mathbf{x}_0))\rangle.$$

This p. d. f. satisfies the Kolmogorov–Fokker–Planck equation

$$\frac{\partial p_L}{\partial t} + \frac{\partial}{\partial x_i}(\mathbf{V}_i p_L) + \frac{\partial}{\partial \mathbf{V}_i}(a_i p_L) + \frac{\partial}{\partial N_k}(f_k p_L) = \frac{C_0\bar{\varepsilon}}{2}\frac{\partial^2 p_L}{\partial V_i \partial V_j}$$

with the initial condition

$$p_L(\mathbf{x},\mathbf{V},\mathbf{N},0;\mathbf{x}_0) = \delta(\mathbf{x}-\mathbf{x}_0)\, p_E(\mathbf{V};\mathbf{x}_0)\delta(\mathbf{N}-\mathbf{q}(\mathbf{x}_0)).$$

Here $\mathbf{q}(\mathbf{x}_0) = (q_1(\mathbf{x}_0),\dots,q_K(\mathbf{x}_0))$. The flux of k-th specy in the i-th direction is given by

$$\langle u_i(\mathbf{x},t)n_k(\mathbf{x},t)\rangle = \int_{\mathrm{IR}^3} d\mathbf{V} \int d\mathbf{N} \int_D d\,\mathbf{x}_0 V_i N_k p_L(\mathbf{x},\mathbf{V},\mathbf{N},t;\mathbf{x}_0), \quad k=1,\dots,K.$$

Generally, the integral

$$I_h(\mathbf{x},t) = \int_{\mathrm{IR}^3} d\mathbf{V} \int d\mathbf{N} \int_D d\,\mathbf{x}_0 h(\mathbf{x},\mathbf{V},\mathbf{N},t) p_L(\mathbf{x},\mathbf{V},\mathbf{N},t;\mathbf{x}_0), \quad k=1,\dots,K \tag{9.12}$$

is to be evaluated for a given function h. Let

$$\theta(\mathbf{x},\mathbf{V},\mathbf{N},t) = \int_D d\,\mathbf{x}_0 p_L(\mathbf{x},\mathbf{V},\mathbf{N},t;\mathbf{x}_0).$$

Then this function satisfies

$$\frac{\partial \theta}{\partial t} + \frac{\partial}{\partial x_i}(\mathbf{V}_i \theta) + \frac{\partial}{\partial \mathbf{V}_i}(a_i \theta) + \frac{\partial}{\partial N_k}(f_k \theta) = \frac{C_0\bar{\varepsilon}}{2}\frac{\partial^2 \theta}{\partial V_i \partial V_j}$$

and the initial condition

$$\theta(\mathbf{x},\mathbf{V},\mathbf{N},0) = p_E(\mathbf{V};\mathbf{x})\delta(\mathbf{N}-\mathbf{q}(\mathbf{x})).$$

The probabilistic representation given in Appendix B in this case yields

$$\theta(\mathbf{x},\mathbf{V},\mathbf{N},t) = \mathbf{E} p_E(\mathbf{V}_0^{\mathbf{x},\mathbf{V},t};\mathbf{X}_0^{\mathbf{x},\mathbf{V},t}) \exp\left\{-\int_0^t \frac{\partial a_i}{\partial V_i}(s,\mathbf{X}_s^{\mathbf{x},\mathbf{V},t},\mathbf{V}_s^{\mathbf{x},\mathbf{V},t})ds\right\} \tag{9.13}$$
$$\times \exp\left\{-\int_0^t \frac{\partial f_k}{\partial N_k}(s,\mathbf{X}_s^{\mathbf{x},\mathbf{V},t},\mathbf{N}_s^{\mathbf{x},\mathbf{V},\mathbf{N},t})ds\right\}\delta\big(\mathbf{N}_0^{\mathbf{x},\mathbf{V},\mathbf{N},t} - \mathbf{q}(\mathbf{X}_0^{\mathbf{x},\mathbf{V},t})\big),$$

where $\mathbf{X}(s) = \mathbf{X}_s^{\mathbf{x},\mathbf{V},t}$, $\mathbf{V}(s) = \mathbf{V}_s^{\mathbf{x},\mathbf{V},t}$ and $\mathbf{N}(s) = \mathbf{N}_s^{\mathbf{x},\mathbf{V},\mathbf{N},t}$, $0 \le s \le t$ are the adjoint Lagrangian processes defined as the solutions to

$$
\begin{aligned}
d\mathbf{X}(s) &= \mathbf{V}(s)ds, \\
d\mathbf{V}(s) &= \mathbf{a}(s, \mathbf{X}(s), \mathbf{V}(s))ds + \sqrt{C_0\bar{\varepsilon}(\mathbf{X}(s), s)}\, d\mathbf{W}(s), \\
\frac{dN_k}{ds} &= f_k(s, \mathbf{X}, \mathbf{N}), \quad k = 1, 2 \ldots, K,
\end{aligned}
$$

with the terminal conditions

$$
\mathbf{X}(t) = \mathbf{x}, \quad \mathbf{V}(t) = \mathbf{V}, \quad N_k(t) = \mathbf{N}_k,\ k = 1, \ldots, K.
$$

The expectation $\mathbf{E}$ is taken over all backward Lagrangian trajectories. From (9.12) and (9.13) we find

$$
\begin{aligned}
I_h(\mathbf{x}, t) &= \int_{\mathrm{IR}^3} d\mathbf{V} \int d\mathbf{N}\, h(\mathbf{x}, \mathbf{V}, \mathbf{N}, t) \\
&\times \mathbf{E} p_E(\mathbf{V}_0^{\mathbf{x},\mathbf{V},t}; \mathbf{X}_0^{\mathbf{x},\mathbf{V},t}) Q_a(\mathbf{x}, \mathbf{V}, t) Q_f(\mathbf{x}, \mathbf{V}, \mathbf{N}, t) \delta(\mathbf{N}_0^{\mathbf{x},\mathbf{V},\mathbf{N},t} - \mathbf{q}(\mathbf{X}_0^{\mathbf{x},\mathbf{V},t})),
\end{aligned}
$$

where

$$
\begin{aligned}
Q_a(\mathbf{x}, \mathbf{V}, t) &= \exp\left\{ -\int_0^t \frac{\partial a_i}{\partial V_i}(s, \mathbf{X}_s^{\mathbf{x},\mathbf{V},t}, \mathbf{V}_s^{\mathbf{x},\mathbf{V},t}) ds \right\}, \\
Q_f(\mathbf{x}, \mathbf{V}, \mathbf{N}, t) &= \exp\left\{ -\int_0^t \frac{\partial f_k}{\partial N_k}(s, \mathbf{X}_s^{\mathbf{x},\mathbf{V},t}, \mathbf{N}_s^{\mathbf{x},\mathbf{V},\mathbf{N},t}) ds \right\}.
\end{aligned}
$$

Hence

$$
\begin{aligned}
I_h(\mathbf{x}, t) = \mathbf{E} &\int_{\mathrm{IR}^3} d\mathbf{V} p_E(\mathbf{V}_0^{\mathbf{x},\mathbf{V},t}; \mathbf{X}_0^{\mathbf{x},\mathbf{V},t}) Q_a(\mathbf{x}, \mathbf{V}, t) \\
&\times \int d\mathbf{N}\, Q_f(\mathbf{x}, \mathbf{V}, \mathbf{N}, t) h(\mathbf{x}, \mathbf{V}, \mathbf{N}, t) \delta(\mathbf{N}_0^{\mathbf{x},\mathbf{V},\mathbf{N},t} - \mathbf{q}(\mathbf{X}_0^{\mathbf{x},\mathbf{V},t})).
\end{aligned} \tag{9.14}
$$

To evaluate the last integral in (9.14) we use a known formula. Let F be a function of $\mathbf{x} = (x_1, \ldots, x_m)$, and g be a vector function $g(\mathbf{x}) = (g_1(\mathbf{x}), \ldots, g_m(\mathbf{x}))$, whose inverse g^{-1} exists, and let $\mathbf{b} = (b_1, \ldots, b_m)$ be a fixed vector. Then

$$
\int F(\mathbf{x}) \delta(g(\mathbf{x}) - \mathbf{b})\, d\mathbf{x} = \frac{F(\mathbf{x}_b)}{J(\mathbf{x}_b)}, \tag{9.15}
$$

where $\mathbf{x}_b = g^{-1}(\mathbf{b})$, and

$$
J(\mathbf{x}) = Det \left\| \frac{\partial g_i(\mathbf{x})}{\partial x_j} \right\|
$$

is the Jacobian of the transformation $\mathbf{x} \to g(\mathbf{x})$. Applying (9.15), we can evaluate the last integral in (9.14). Indeed, the transformation $g : \mathbf{N} \to \mathbf{N}_0^{\mathbf{x},\mathbf{V},\mathbf{N},t}$ has the Jacobian

$Q_f(\mathbf{x}, \mathbf{V}, \mathbf{N}, t)$. Choosing $\mathbf{b} = \mathbf{q}(\mathbf{X}_0^{\mathbf{x},\mathbf{V},t})$ we find that $g^{-1}(\mathbf{b}) = \mathbf{N}(t; \mathbf{b})$ where $\mathbf{N}(s; \mathbf{b})$ is the solution to

$$\frac{d\mathbf{N}}{ds} = f(\mathbf{N}, \mathbf{X}_s^{\mathbf{x},\mathbf{V},t}, s), \quad \mathbf{N}(0; \mathbf{b}) = \mathbf{b}.$$

Thus we have

$$I_h(\mathbf{x}, t) = \mathbf{E} \int_{\mathrm{IR}^3} d\mathbf{V} p_E(\mathbf{V}_0^{\mathbf{x},\mathbf{V},t}; \mathbf{X}_0^{\mathbf{x},\mathbf{V},t}) Q_a(\mathbf{x}, \mathbf{V}, t) h(\mathbf{x}, \mathbf{V}, \mathbf{N}(t; \mathbf{q}(\mathbf{X}_0^{\mathbf{x},\mathbf{V},t})), t).$$

Now we can formulate the backward algorithm for solving problem (9.11). Introduce a probability density function $r(\mathbf{V}) > 0$ in IR^3, and use the notation $\tilde{\mathbf{V}}$ for a sample from this density. We write for simplicity $\tilde{\mathbf{X}}_{t_0} = \mathbf{X}_{t_0}^{\mathbf{x},\tilde{\mathbf{V}},t}$ and $\tilde{\mathbf{V}}_{t_0} = \mathbf{V}_{t_0}^{\mathbf{x},\tilde{\mathbf{V}},t}$. We denote by $\tilde{\mathbf{N}}(s)$ the solution to

$$\frac{d\tilde{\mathbf{N}}}{ds} = f(\tilde{\mathbf{N}}, \tilde{\mathbf{X}}_s, s), \quad \tilde{\mathbf{N}}(0) = \mathbf{q}(\tilde{\mathbf{X}}_0).$$

The random estimator has the form:

$$\xi(\mathbf{x}, \tilde{\mathbf{V}}, t) = \frac{1}{r(\tilde{\mathbf{V}})} p_E(\tilde{\mathbf{V}}_0; \tilde{\mathbf{X}}_0)\, Q_a(\mathbf{x}, \tilde{\mathbf{V}}, t) h(\mathbf{x}, \tilde{\mathbf{V}}, \tilde{\mathbf{N}}(t), t),$$

and hence

$$I_h(\mathbf{x}, t) = \mathbf{E}_{\tilde{\mathbf{V}}} \mathbf{E} \xi(\mathbf{x}, \tilde{\mathbf{V}}, t).$$

Here $\mathbf{E}_{\tilde{\mathbf{V}}}$ means averaging over the samples $\tilde{\mathbf{V}}$, and $\mathbf{E}$ is the averaging over the trajectories $\tilde{\mathbf{X}}_s, \tilde{\mathbf{V}}_s, 0 \le s \le t$.

9.6 Numerical simulations

Numerical simulations are carried out for the following problem. A horizontally homogeneous stationary neutrally-stratified surface layer is considered. The source starts to generate particles at time $t = 0$. It is uniformly distributed over the plane at height $z = z_s$. In this section we use the SI metric system of units. We calculate the mean concentration and its vertical flux at a fixed point at height $z = z_d$ by three methods: the direct Monte Carlo described in Section 9.3.1, and two backward in time Monte Carlo algorithms presented in Section 9.3.2 and in Appendix C: the adjoint algorithm based on the probabilistic representation, and the backward method due to Thomson [237]. Comparison of the cost of these three methods is given.

Since our problem is described by horizontally homogeneous parameters, it is governed by a 1-dimensional stochastic model (9.9), (9.10).

First, let us present the details of random estimators for the direct estimator of the type (9.6). In our case it takes the form

$$\bar{c}(z_d, t) = Q\, \mathbf{E} \sum_{j=1}^{\nu_t(z_d)} \frac{1}{|W(\tau_j)|} \tag{9.16}$$

for the mean concentration and

$$F(z_d,t) = Q\,\mathbf{E}\sum_{j=1}^{\nu_t(z_d)} \frac{W(\tau_j)}{|W(\tau_j)|} \tag{9.17}$$

for the vertical concentration flux. Here $(Z(\tau), W(\tau))$ is the solution to (9.10) with the initial condition $Z(0) = z_s$, $W(0) = w_0$, where w_0 is a random velocity distributed with the Eulerian Gaussian p. d. f. $p_E(w) = \exp\{-w^2/2\sigma_w^2\}/\sqrt{2\pi\sigma_w^2}$. The values τ_j are random times at which the process $Z(s)$ intersects the level $z = z_d$, and $\nu_t(z_d)$ is the total number of such events in the interval $0 \le s \le t$. The constant Q is the strength of the surface area source.

The derivation of the estimator (9.16) can be obtained as follows.

Since

$$\begin{aligned}
\bar{c}(z_d,t) &= \int_{-\infty}^{\infty} dw \int_0^t dt_0 \int_0^{\infty} dz_0 \delta(z_0 - z_s) \\
&\quad \times \int_{-\infty}^{\infty} dw_0 p_E(w_0) p_L(z_d, w, t; z_0, w_0, t_0) \\
&= \int_{-\infty}^{\infty} dw \int_0^t dt_0 \int_0^{\infty} dz_0 \delta(z_0 - z_s) \int_{-\infty}^{\infty} dw_0 p_E(w_0) \\
&\quad \times \int_0^{\infty} dz_0 \delta(z - z_d) p_L(z, w, \tau; z_0, w_0, 0) \\
&= \mathbf{E} \int_0^t \delta(Z_\tau^{z_s,\tilde{w}_0} - z_d)\, d\tau.
\end{aligned} \tag{9.18}$$

Here $\tilde{w}_0$ is a random number distributed with $p_E(w_0)$, $Z_t^{z,w}$ is the solution to the system (9.10) with the initial deterministic condition $Z_0^{z,w} = z$, and $W_0^{z,w} = w$. It remains to note that for arbitrary continuously differentiable function $Z(\tau)$,

$$\int_0^t \delta(Z(\tau) - z)\, d\tau = \sum_{j=1}^{\nu_t(z)} \frac{1}{\left|\frac{dZ(\tau_j)}{d\tau}\right|},$$

where $\nu_t(z)$ is the number of intersections of the level z by the trajectory $Z(\tau)$ in the interval $0 \le \tau \le t$, and τ_j are the intersection times.

The random estimator of Thomson's backward algorithm used in [57] is constructed on the solutions to the backward-time stochastic differential equation

$$d\hat{Z}(t_0) = \hat{W}(t_0)dt_0, \quad d\hat{W}(t_0) = \frac{a}{\hat{Z}}\hat{W}(t_0)dt_0 + \frac{b}{\sqrt{\hat{Z}}}\, dB(t_0),$$

with terminal condition $\hat{Z}(t) = z_d$, $\hat{W}(t) = \hat{w}$, where $\hat{w}$ is a random number distributed with Gaussian $p_E(w)$ given above. Here

$$a = \frac{C_0 u_*^3}{2\kappa\sigma_w^2}, \quad b = \left(\frac{C_0 u_*^3}{\kappa}\right)^{\frac{1}{2}}.$$

The random estimators read

$$\bar{c}(z_d, t) = Q\,\mathbf{E} \sum_{j=1}^{\nu_t(z_s)} \frac{1}{|\hat{W}(\tau_j)|} \tag{9.19}$$

for the mean concentration and

$$F(z_d, t) = Q\,\mathbf{E} \sum_{j=1}^{\nu_t(z_s)} \frac{\hat{W}(t)}{|\hat{W}(\tau_j)|} \tag{9.20}$$

for the vertical concentration flux.

Here the values τ_j are random times at which the process $\hat{Z}(t_0)$ $(0 \le t_0 \le t)$ intersects the level $z = z_s$, and $\nu_t(z_s)$ is the total number of such events in the interval $0 \le t_0 \le t$.

Finally, let us consider the adjoint estimator based on probabilistic representation. The probabilistic representation (9.8) in our case (1D) reads

$$p(z, w, t) = \mathbf{E}_{z,w} \int_0^t \delta(\tilde{Z}(t_0) - z_s) p_E(\tilde{W}(t_0)) \exp\left\{\int_{t_0}^t \frac{b\,ds}{\tilde{Z}(s)}\right\} dt_0, \tag{9.21}$$

where $\tilde{Z}(t_0) = Z_{t_0}^{z,w}$, $\tilde{W}(t_0)) = W_{t_0}^{z,w}$, $0 \le t_0 \le t$ is the adjoint trajectory defined by

$$d\tilde{Z}(t_0) = \tilde{W}(t_0)\,dt, \quad d\tilde{W}(t_0) = -\frac{a}{\tilde{Z}}\tilde{W}(t_0)dt_0 + \frac{b}{\sqrt{\tilde{Z}}}\,dB(t_0),$$

with terminal condition $\tilde{Z}(t) = z$, $\tilde{W}(t) = w$.

Now, substituting (9.21) in

$$\bar{c}(z_d, t) = \int_{-\infty}^{+\infty} p(z_d, w, t)\,dw$$

we get

$$\bar{c}(z_d, t) = \int_{-\infty}^{+\infty} dw \mathbf{E}_{z_d,w} \sum_{j=1}^{\tilde{\nu}_t(z_s)} \frac{p_E(\tilde{W}(\tau_j))}{|\tilde{W}(\tau_j)|} \exp\left\{\int_{\tau_j}^t \frac{b\,ds}{\tilde{Z}(s)}\right\}. \tag{9.22}$$

Here we used the relation

$$\int_0^t \delta(\tilde{Z}(t_0) - z_s) p_E(\tilde{W}(t_0)) \exp\left\{\int_{t_0}^t \frac{b\,ds}{\tilde{Z}(s)}\right\} dt_0$$
$$= \sum_{j=1}^{\tilde{\nu}_t(z_s)} \frac{p_E(\tilde{W}(\tau_j))}{|\tilde{W}(\tau_j)|} \exp\left\{\int_{\tau_j}^t \frac{b\,ds}{\tilde{Z}(s)}\right\}.$$

where $\tilde{\nu}_{t(z_s)}$ is the number of intersections of the level $z = z_s$ by the trajectory $Z_{t_0}^{z_d, w}$, $0 \le t_o \le t$, and $\tau_1, \ldots, \tau_j, \ldots$ are the intersection times.

The resulting estimator is

$$\bar{c}(z_d, t) = \mathbf{E}\left\{\frac{1}{r(\tilde{w})} \sum_{j=1}^{\tilde{\nu}_t(z_s)} \frac{p_E(W_{\tau_j}^{z_d, \tilde{w}})}{|W_{\tau_j}^{z_d, \tilde{w}}|} \exp\left\{\int_{\tau_j}^t \frac{b\,ds}{Z_\tau^{z_d, \tilde{w}}}\right\}\right\}. \tag{9.23}$$

where $r(w)$ is a p. d. f. which is positive in $(-\infty, +\infty)$, and $\tilde{w}$ is a random number distributed with $r(w)$.

Analogous arguments lead to the random estimator for the vertical concentration flux:

$$F(z_d, t) = \mathbf{E}\left\{\frac{\tilde{w}}{r(\tilde{w})} \sum_{j=1}^{\tilde{\nu}_t(z_s)} \frac{p_E(W_{\tau_j}^{z_d, \tilde{w}})}{|W_{\tau_j}^{z_d, \tilde{w}}|} \exp\left\{\int_{\tau_j}^t \frac{b\,ds}{Z_\tau^{z_d, \tilde{w}}}\right\}\right\}. \tag{9.24}$$

In numerical calculations, one takes a cut-off in the integral over w in (9.22), and integrates from, say, $-A$ to A, A being sufficiently large; in our case we have chosen $A = 5\sigma_w$. Further parameters in calculations are $\sigma_w = 1.25\,u_*$, $u_* = 0.4$, and the function $\bar{\varepsilon}$ is defined in (9.9).

As to the penetrable boundary conditions, in calculations it cannot be strictly satisfied. In the numerical scheme it is convenient to follow the trajectories until some reflection height $z = z_r < z_0$, and then reflect them according to some reflection law. In calculations we found that beginning from $z_r < z_s/5$, the results are stable with respect to further decreasing of the reflection height z_r. The perfect reflection (symmetric to the plane $z = z_r$) was used.

It should be noted that usually (e. g., see [55]) one reflects the trajectories at the height $z = z_0$, which does not affect the calculations at large (compared to z_0) heights, but at a height of several z_0 the error may be about 10–30 %. In Table 9.1 we present the mean concentration and its vertical flux obtained by the direct (9.16), (9.17), the adjoint (9.23), (9.24), and the backward method (9.19), (9.20). The calculations are made for four time instances $t = T_L(z_s), 2T_L(z_s), 4T_L(z_s)$, and $8T_L(z_s)$, for $z_d = 1, z_s = 0.5$. The unit source strength uniformly distributed over the plane $z = z_s$ was taken. The Lagrangian time scale $T_L = T_L(z_s)$ is given by [215] $T_L(z_s) = 2\sigma_w^2/C_0\bar{\varepsilon}(z_s) = z_s/a = 0.39$. The error of the results presented in the table was less than 1 % calculated

Table 9.1. Comparison of three different methods.

Method		$t = T_L$	$t = 2T_L$	$t = 4T_L$	$t = 8T_L$
direct	$\bar{c}$	$2.08e-3$	$8.94e-2$	$4.67e-1$	1.26
	F	$2.71e-3$	$6.48e-2$	$2.13e-1$	0.4
	cost	200 [s]	200 [s]	85 [s]	60 [s]
adjoint	$\bar{c}$	$1.95e-3$	$8.42e-2$	$4.71e-1$	1.19
	F	$2.51e-3$	$6.10e-2$	$2.16e-1$	0.34
	cost	400 [s]	1,800 [s]	2,000 [s]	2,200 [s]
backw.	$\bar{c}$	$1.95e-3$	$9.03e-2$	$4.68e-1$	1.21
	F	$2.50e-3$	$6.58e-2$	$2.05e-1$	0.37
	cost	180 [s]	300 [s]	300 [s]	40 [s]

as the statistical error measured as $3 \times$ standard deviation/$\sqrt{\text{number of samples}}$. The *cost* means the computer time of a 233 MHz PC computer.

The results presented in the table show that in this special case of horizontally homogeneous problem the direct and backward methods have approximately equal cost and are both much more efficient than the adjoint method. It should be stressed, however, that this is because our model problem is actually 1-dimensional and the source is stationary. In the general case of 3D problems with a source generating particles during a short period of time, the backward algorithm is much more efficient.

As to the adjoint method, although our calculations show that it requires a lot of computer time, it has the following important advantages. The method allows us to solve problems of transport of reacting species as described in Section 9.5. Another advantage of the adjoint method is in treating the problems with boundary conditions. It should be noted that the boundary value problems cause difficulties for the backward method.

9.7 Conclusion

A generalized footprint problem is treated as a calculation of a linear integral over space, velocity and time of the space-velocity distribution of ensemble of particles in a turbulent flow. The Lagrangian stochastic description is used to solve this problem. As important particular cases the mean concentration and its flux are analysed in details.

Three different algorithms are presented: (1) direct Monte Carlo, (2) adjoint Monte Carlo, and (3) backward Monte Carlo algorithms. The direct Monte Carlo is quite general but it is not efficient in estimation of local functions like, e.g, the concentration and its flux at a fixed point.

The adjoint method is also general and is especially convenient for evaluation of local functionals. The method is based on the well developed probabilistic representations for the boundary value problems. Therefore, it allows to solve problems with quite general boundary conditions. Unfortunately, the method requires a lot of computer time because the variance increases with time instance under consideration.

The backward algorithm originally presented by Thomson is extended to more general case when the transport in the phase space is described by general stochastic differential equation. This extension allows to treat problems with absorption, which is of our current interest.

9.8 Appendices

9.8.1 Appendix A. Flux representation

Here we derive the representation (9.2). By the definition, the instant concentration is

$$c(\mathbf{x},t) = \int_0^t dt_0 \int_D d\mathbf{x}_0\, q(\mathbf{x}_0,t_0)\delta(\mathbf{x}-\mathbf{X}(t;\mathbf{x}_0,t_0)) + \int_D d\mathbf{x}_0\, q_0(\mathbf{x}_0)\delta(\mathbf{x}-\mathbf{X}(t;\mathbf{x}_0,0)).$$

From this,

$$\begin{aligned} F_i(\mathbf{x},t) &= \langle u_i(\mathbf{x},t)c(\mathbf{x},t)\rangle \\ &= \int_0^t dt_0 \int_D d\mathbf{x}_0\, q(\mathbf{x}_0,t_0)\langle u_i(\mathbf{x},t)\delta(\mathbf{x}-\mathbf{X}(t;\mathbf{x}_0,t_0))\rangle \\ &\quad + \int_D d\mathbf{x}_0\, q_0(\mathbf{x}_0)\langle u_i(\mathbf{x},t)\delta(\mathbf{x}-\mathbf{X}(t;\mathbf{x}_0,0))\rangle. \end{aligned} \tag{A1}$$

Since

$$\begin{aligned} u_i(\mathbf{x},t)\delta(\mathbf{x}-\mathbf{X}(t;\mathbf{x}_0,t_0)) &= u_i(\mathbf{X}(t;\mathbf{x}_0,t_0),t)\delta(\mathbf{x}-\mathbf{X}(t;\mathbf{x}_0,t_0)) \\ &= V_i(t;\mathbf{x}_0,t_0)\delta(\mathbf{x}-\mathbf{X}(t;\mathbf{x}_0,t_0)), \end{aligned}$$

then

$$\begin{aligned} \langle u_i(\mathbf{x},t)\delta(\mathbf{x}-\mathbf{X}(t;\mathbf{x}_0,t_0))\rangle &= \langle V_i(t;\mathbf{x}_0,t_0)\delta(\mathbf{x}-\mathbf{X}(t;\mathbf{x}_0,t_0))\rangle \\ &= \int_{\mathbb{R}^3} d\mathbf{V} \int_D d\mathbf{X}\, V_i\delta(\mathbf{x}-\mathbf{X})p_L(\mathbf{X},\mathbf{V},t;\mathbf{x}_0,t_0) \\ &= \int_{\mathbb{R}^3} V_i\, p_L(\mathbf{x},\mathbf{V},t;\mathbf{x}_0,t_0)\, d\mathbf{V}. \end{aligned}$$

From this we get in view of (A1) the desired representation (9.2).

9.8.2 Appendix B. Probabilistic representation

Let us write down the probabilistic representation of the solution $\theta(\mathbf{y},t) = \theta(y_1,\ldots, y_n,t)$, to the following general parabolic equation:

$$\frac{\partial\theta}{\partial t} + A_i(\mathbf{y},t)\frac{\partial\theta}{\partial y_i} + \frac{1}{2}B_{ij}(\mathbf{y},t)\frac{\partial^2\theta}{\partial y_i\partial y_j} + \beta(\mathbf{y},t)\theta + f(\mathbf{y},t) = 0, \quad t\in[0,T),$$

with the terminal condition

$$\theta(\mathbf{y},t)|_{t=T} = f_0(\mathbf{y}).$$

The probabilistic representation to this problem has the form [59]:

$$\theta(\mathbf{y},t) = \mathbf{E}_{(\mathbf{y},t)}\left\{ \int_t^T f(\mathbf{Y}_s^{\mathbf{y},t}, s) \exp\left(\int_t^s \beta(\mathbf{Y}_\tau^{\mathbf{y},t}, \tau) d\tau \right) ds + f_0(\mathbf{Y}_T^{\mathbf{y},t}) \exp\left(\int_t^T \beta(\mathbf{Y}_\tau^{\mathbf{y},t}, \tau) d\tau \right) \right\}, \quad \text{(B1)}$$

where $\mathbf{Y}_s^{\mathbf{y},t} \equiv \mathbf{Y}(s)$, $s \geq t$ solves the problem

$$dY_i(s) = A_i(\mathbf{Y}(s), s)ds + \sigma_{ij}(\mathbf{Y}(s), s)dW_j(s), \quad s > t, \quad \mathbf{Y}(s)|_{s=t} = \mathbf{y}. \quad \text{(B2)}$$

Here $\sigma_{ik}\sigma_{jk} = B_{ij}$. In (B1), $\mathbf{E}_{(\mathbf{y},t)}$ stands for the expectation taken over all trajectories (solutions to (B2)) starting from $\mathbf{y}$ at time t.

Note that the solution to the equation

$$\frac{\partial \theta}{\partial t} + A_i(\mathbf{y},t)\frac{\partial \theta}{\partial y_i} + \beta(\mathbf{y},t)\theta = \frac{1}{2}B_{ij}(\mathbf{y},t)\frac{\partial^2 \theta}{\partial y_i \partial y_j} + f(\mathbf{y},t), \qquad t \in [0,T),$$

satisfying $\theta(y,0) = f_0(y)$, has the probabilistic representation

$$\theta(\mathbf{y},t) = \mathbf{E}_{(\mathbf{y},t)}\left\{ \int_0^t f(\mathbf{X}_{t_0}^{\mathbf{y},t}, t_0) \exp\left(-\int_{t_0}^t \beta(\mathbf{X}_{t_0'}^{\mathbf{y},t}, t_0')dt_0' \right) dt_0 + f_0(\mathbf{X}_0^{\mathbf{y},t}) \exp\left(-\int_0^t \beta(\mathbf{X}_{t_0}^{\mathbf{y},t}, t_0)dt_0 \right) \right\},$$

where $\mathbf{X}_{t_0}^{\mathbf{y},t} \equiv \mathbf{X}(t_0)$, $0 \leq t_0 \leq t$ solves the problem

$$dX_i(t_0) = A_i(\mathbf{X}(t_0), t_0)dt_0 + \sigma_{ij}(\mathbf{X}(t_0), t_0)dW_j(t_0),$$
$$0 \leq t_0 \leq t, \quad \mathbf{X}(t_0)|_{t_0=t} = \mathbf{y}.$$

9.8.3 Appendix C. Forward and backward trajectory estimators

In this Appendix we treat the evaluation of the integral:

$$I_{h,q} = \int_D d\mathbf{y} \int_0^T dt \int_D d\mathbf{y}_0 \int_0^t dt_0 h(\mathbf{y},t)q(\mathbf{y}_0,t_0)p^f(\mathbf{y},t;\mathbf{y}_0,t_0), \qquad \text{(C1)}$$

where D is a domain in IR^n, $T > 0$, h and q are functions defined in $D \times [0, T]$, and $p^f(\mathbf{y}, t; \mathbf{y}_0, t_0) = \langle \delta(\mathbf{y} - \mathbf{Y}_t^{\mathbf{y}_0, t_0}) \rangle$ is the transition density of the n-dimensional diffusion process $\mathbf{Y}_t^{\mathbf{y}_0, t_0}$, the solution to

$$dY_i(t) = A_i(\mathbf{Y}(t), t)dt + \sigma_{ij}(\mathbf{Y}(t), t)dW_j(t), \quad s > t, \quad \mathbf{Y}(t)|_{t=t_0} = \mathbf{y}_0. \quad \text{(C2)}$$

We assume that the boundary of D is penetrable, i. e., the trajectories determined by (C2) do not reach the boundary. The direct Monte Carlo estimator for evaluating the integral (C1) is straightforward:

$$\begin{aligned} I_{h,q} &= \int_D d\mathbf{y}_0 \int_0^T dt_0 \int_D d\mathbf{y} \int_{t_0}^T dt h(\mathbf{y}, t) q(\mathbf{y}_0, t_0) p^f(\mathbf{y}, t; \mathbf{y}_0, t_0) \\ &= \mathbf{E}\left\{ \frac{q(\tilde{\mathbf{y}}_0, \tilde{t}_0)}{p_0(\tilde{\mathbf{y}}_0, \tilde{t}_0)} \int_{\tilde{t}_0}^T h(\mathbf{Y}_t^{\tilde{\mathbf{y}}_0, \tilde{t}_0}, t)\, dt \right\}. \end{aligned}$$

Here $p_0(\mathbf{y}_0, t_0)$ is an arbitrary p. d. f. in $D \times [0, T]$ consistent with the function $q(\mathbf{y}_0, t_0)$ in the sense that $p_0(\mathbf{y}_0, t_0) > 0$ if $q_0(\mathbf{y}_0, t_0) \neq 0$, and the expectation is taken over all sample points $(\tilde{\mathbf{y}}_0, \tilde{t}_0)$ and sample trajectory $\mathbf{Y}_t^{\tilde{\mathbf{y}}_0, \tilde{t}_0}$, $\tilde{t}_0 \le t \le T$; the random points $\tilde{\mathbf{y}}_0, \tilde{t}_0$ are distributed with $p_0(\mathbf{y}_0, t_0)$. A backward estimator can be obtained by a technique suggested by Thomson [237].

Assume that we have a positive function $\rho(\mathbf{y}, t)$ defined on $D \times [0, T]$ as a solution to the equation

$$\frac{\partial \rho}{\partial t} + \frac{\partial}{\partial y_i}(A_i \rho) = \frac{1}{2} \frac{\partial^2 (B_{ij} \rho)}{\partial y_i \partial y_j}, \quad \text{(C3)}$$

where $\sigma_{ik}\sigma_{jk} = B_{ij}$. Let $p^b(\mathbf{y}_0, t_0; \mathbf{y}, t) = \langle \delta(\mathbf{y}_0 - \mathbf{Z}_{t_0}^{\mathbf{y}, t}) \rangle$ is the transition density of the diffusion process $Z_{t_0}^{\mathbf{y}, t}$, $0 \le t_0 \le t$ which is defined by

$$dZ_i = A_i^*(\mathbf{Z}, t_0)\, dt_0 + B_{ij}(\mathbf{Z}, t_0), t_0)\, dW_j(t_0), \quad t_0 < t, \quad \mathbf{Z}(t) = \mathbf{y}. \quad \text{(C4)}$$

Here

$$A_i^*(\mathbf{y}, t) = A_i(\mathbf{y}, t) - \frac{1}{\rho(\mathbf{y}, t)} \frac{\partial}{\partial y_j}(B_{ij}(\mathbf{y}, t)\rho(\mathbf{y}, t)).$$

We again assume that the solutions to (C4) never reach the boundary of D. Then the following relation is true:

$$\rho(\mathbf{y}_0, t_0) p^f(\mathbf{y}, t; \mathbf{y}_0, t_0) = \rho(\mathbf{y}, t) p^b(\mathbf{y}_0, t_0; \mathbf{y}, t). \quad \text{(C5)}$$

To prove this, we first remark that the function p^b and

$$F(\mathbf{y}_0, t_0; \mathbf{y}, t) = \frac{\rho(\mathbf{y}_0, t_0)}{\rho(\mathbf{y}, t)} p^f(\mathbf{y}, t; \mathbf{y}_0, t_0)$$

satisfy the equations

$$\mathcal{L}_{\mathbf{y}_0,t_0} F = 0, \quad \mathcal{L}_{\mathbf{y}_0,t_0} p^b = 0,$$

where the operator $\mathcal{L}_{\mathbf{y}_0,t_0}$ acts on a function $g(\mathbf{y}_0, t_0)$ as follows:

$$\mathcal{L}_{\mathbf{y}_0,t_0} g = \frac{\partial g}{\partial t} + \frac{\partial}{\partial y_{0i}}(A_i^* g) + \frac{1}{2}\frac{\partial^2 (B_{ij} g)}{\partial y_{0i} \partial y_{0j}}.$$

Since the values of the functions F and p^b at $t_0 = t$ coincide,

$$p^b(\mathbf{y}_0, t; \mathbf{y}, t) = F(\mathbf{y}_0, t; \mathbf{y}, t) = \delta(\mathbf{y} - \mathbf{y}_0), \tag{C6}$$

we conclude that $F \equiv p^b$, provided that the equation $\mathcal{L}_{\mathbf{y}_0,t_0} g = 0$ with the initial condition $g(t, \mathbf{y}_0) = \delta(\mathbf{y}_0 - \mathbf{y})$ has a unique solution. This implies that (C5) is true. Now we note that $\mathcal{L}_{\mathbf{y}_0,t_0} p^b = 0$ is true indeed, since it is the first Kolmogorov equation for p^b. The equality $\mathcal{L}_{\mathbf{y}_0,t_0} F = 0$ then follows from (C3), the inverse Kolmogorov equation

$$\frac{\partial p^f(\mathbf{y}, t; \mathbf{y}_0, t_0)}{\partial t_0} + A_i(\mathbf{y}_0, t_0)\frac{\partial p^f}{\partial y_{0i}} + \frac{1}{2} B_{ij}(\mathbf{y}_0, t_0)\frac{\partial^2 p^f}{\partial y_{0i} \partial y_{0j}} = 0$$

and the expression for $A_i^*(\mathbf{y}_0, t_0)$ through A_i and B_{ij} given above. Now, we present the backward Monte Carlo algorithm based on the (C5). We proceed as follows. Substituting the expression for p^f from (C5) into (C1) we get

$$I_{h,q} = \int_D d\mathbf{y} \int_0^T dt \int_D d\mathbf{y}_0 \int_0^t dt_0 h(\mathbf{y}, t) q(\mathbf{y}_0, t_0) \frac{\rho(\mathbf{y}, t)}{\rho(\mathbf{y}_0, t_0)} p^b(\mathbf{y}_0, t_0; \mathbf{y}, t). \tag{C7}$$

Let $r(\mathbf{y}, t)$ be a probability density in $D \times [0, T]$ consistent with $h\rho$, i. e., $r > 0$ if $h\rho \neq 0$. Then from (C7) we get

$$I_{h,q} = \mathbf{E}\left\{ \frac{h(\tilde{\mathbf{y}}, \tilde{t})\rho(\tilde{\mathbf{y}}, \tilde{t})}{r(\tilde{\mathbf{y}}, \tilde{t})} \int_0^{\tilde{t}} \frac{q(\mathbf{Z}_{t_0}^{\tilde{\mathbf{y}},\tilde{t}}, t_0)}{\rho(\mathbf{Z}_{t_0}^{\tilde{\mathbf{y}},\tilde{t}}, t_0)} dt_0 \right\}. \tag{C8}$$

Here the expectation is taken over the random points $(\tilde{\mathbf{y}}, \tilde{t})$ distributed in $D \times [0, T]$ with density $r(\mathbf{y}, t)$, and backward trajectories $Z_{t_0}^{\tilde{\mathbf{y}},\tilde{t}}$, $0 \le t_0 \le \tilde{t}$. Another backward trajectory estimator which generalizes the estimator presented in Section 3.2 can be obtained as follows. Let

$$\theta(\mathbf{y}, t) = \int_D d\mathbf{y}_0 \int_0^t dt_0 q(\mathbf{y}_0, t_0) p^f(\mathbf{y}, t; \mathbf{y}_0, t_0).$$

This function solves the problem

$$\frac{\partial \theta}{\partial t} + \frac{\partial (A_i \theta)}{\partial y_i} = \frac{1}{2}\frac{\partial^2 (B_{ij}\theta)}{\partial y_i \partial y_j} + q(\mathbf{y}, t), \quad \theta(\mathbf{y}, 0) = 0. \tag{C9}$$

From the probabilistic representation given in Appendix B we get

$$\theta(\mathbf{y},t) = \mathbf{E}\left\{ \int_0^t q(\mathbf{X}_{t_0}^{\mathbf{y},t}, t_0) \exp\left\{ -\int_{t_0}^t R(\mathbf{X}_s^{\mathbf{y},t}, s)\, ds \right\} dt_0 \right\},$$

where the expectation is taken over the backward trajectories $\mathbf{X}_{t_0}^{\mathbf{y},t}$, $0 \le t_0 \le t$ determined from

$$dX_i(t_0) = \hat{A}_i(\mathbf{X}(t_0), t_0)\, dt_0 + C_{ij}(\mathbf{X}(t_0), t_0)\, dW_j(t_0), \quad t_0 \le t, \quad \mathbf{X}(t) = \mathbf{y}.$$

Here

$$\hat{A}_i(\mathbf{x},t) = A_i(\mathbf{x},t) + 2\frac{\partial B_{ik}(\mathbf{x},t)}{\partial x_k}$$

and

$$R(\mathbf{x},t) = \frac{\partial A_i(\mathbf{x},t)}{\partial x_i} + \frac{\partial^2 B_{ij}(\mathbf{x},t)}{\partial x_i \partial x_j}.$$

Thus we have

$$\begin{aligned} I_{h,q} &= \int_D d\mathbf{y} \int_0^T dt h(\mathbf{y},t)\theta(\mathbf{y},t) \\ &= \int_D d\mathbf{y} \int_0^T dt h(\mathbf{y},t) \mathbf{E}_{\mathbf{X}} \left\{ \int_0^t q(\mathbf{X}_{t_0}^{\mathbf{y},t}, t_0) \exp\left\{ -\int_{t_0}^t R(\mathbf{X}_s^{\mathbf{y},t}, s)\, ds \right\} dt_0 \right\}. \end{aligned}$$

Here $\mathbf{E}_{\mathbf{X}}$ stands for the averaging over trajectories $\mathbf{X}_{t_0}^{\mathbf{y},t}$, $0 \le t_0 \le t$. From this we find

$$I_{h,q} = \mathbf{E}\left(\frac{h(\tilde{\mathbf{y}},\tilde{t})}{r(\tilde{\mathbf{y}},\tilde{t})} \left\{ \int_0^{\tilde{t}} q(\mathbf{X}_{t_0}^{\tilde{\mathbf{y}},\tilde{t}}, t_0) \exp\left\{ -\int_{t_0}^{\tilde{t}} R(\mathbf{X}_s^{\tilde{\mathbf{y}},\tilde{t}}, s)\, ds \right\} dt_0 \right\} \right), \tag{C10}$$

where $\tilde{\mathbf{y}}, \tilde{t}$ is a random point distributed in $D \times [0,T]$ with density $r(\mathbf{y},t)$ consistent with $h(\mathbf{y},t)$. The notation $\mathbf{E}$ means the expectation over the random points $\tilde{\mathbf{y}}, \tilde{t}$ and trajectories $\mathbf{X}_{t_0}^{\tilde{\mathbf{y}},\tilde{t}}$, $0 \le t_0 \le \tilde{t}$.

Remark 9.2. In this appendix we assumed that the boundary is impenetrable. However, in practice one also treats situations where a part of boundary (say, the upper bound of a layer) can be reached by the Lagrangian trajectories. In this case the boundary conditions should be given. For instance, an absorption, reflection or other behavior at the boundary can be considered. For the direct algorithm this can be taken into account by simulating the relevant behavior of the trajectories of (C2) at the boundary (e. g., the trajectories are absorbing at the absorbing boundary, reflecting at the reflection boundary, etc.). In the adjoint algorithm the situation is more complicated. Indeed, it is not clear how to arrange the behavior of trajectories, the solutions to (C4), to guarantee that (C5) is fulfilled. Note that in the approach based on (C10) there is a need for the generalization of probabilistic representation of the solution to (C9) with the relevant boundary conditions. Such representation can be derived from the known probabilistic representations [59].

Chapter 10

Lagrangian stochastic models for turbulent dispersion in an atmospheric boundary layer

A 1-particle 3-dimensional stochastic Lagrangian model for transport of particles in a horizontally homogeneous atmospheric surface layer with arbitrary 1-point probability density function of Eulerian velocity fluctuations is suggested. A uniquely-defined Lagrangian stochastic model in the class of well-mixed models is constructed from physically plausible assumptions. These assumptions are (i) in the neutrally-stratified horizontally homogeneous surface layer, the vertical motion is mainly controlled by eddies whose size is of order of the current height, and (ii) the streamwise drift term is independent of the crosswind velocity. Numerical simulations for neutral stratification have shown a good agreement of our model with the well-known Thomson model, with Flesch and Wilson's model, and with experimental measurements as well. However there is a discrepancy in these results to the results obtained by Reynolds' model.

10.1 Introduction

This chapter deals with 1-particle Lagrangian stochastic models for 2-dimensional (2D) and 3-dimensional (3D) turbulent transport. Here we treat the flow in the atmospheric boundary layer as a fully developed turbulence (i. e., a flow with very high Reynolds number) and consider it to be a random velocity field (u, v, w) assumed to be incompressible. Therefore, the trajectories of particles in such flows are stochastic processes. To simulate these stochastic processes, two different approaches are known in the literature. The first one is based on the numerical solution of the system of random equations

$$\frac{\partial X}{\partial t} = u(X, Y, Z, t), \quad \frac{\partial Y}{\partial t} = v(X, Y, Z, t), \quad \frac{\partial Z}{\partial t} = w(X, Y, Z, t). \qquad (10.1)$$

Here $X(t), Y(t), Z(t)$ are the coordinates of the Lagrangian trajectory at the time t. The random fields u, v, w are simulated by Monte Carlo methods (e. g., see [42, 63, 102, 191, 198, 244], and the random trajectories are then obtained by the numerical solution of (10.1) with the relevant initial data.

In the second approach the true trajectory $X(t), Y(t), Z(t)$ is assumed to be approximated by a model trajectory $\hat{X}(t), \hat{Y}(t), \hat{Z}(t)$, a solution to a stochastic differential equation of Ito type (e. g., see [215,237,256] and the list of references in these papers):

$$d\hat{X} = \hat{U}\,dt, \quad d\hat{Y} = \hat{V}\,dt, \quad d\hat{Z} = \hat{W}\,dt,$$
$$d\hat{U} = a_u\,dt + b_u\,dB_u(t), \quad d\hat{V} = a_v\,dt + b_v\,dB_v(t),$$
$$d\hat{W} = a_w\,dt + b_w\,dB_w(t). \tag{10.2}$$

Here by $\hat{U}, \hat{V}, \hat{W}$ we denote the components of the model Lagrangian velocity, $B_u(t)$, $B_v(t)$, $B_w(t)$ are three standard independent Wiener processes; a_u, a_v, a_w and b_u, b_v, b_w are functions of $(t, \hat{X}, \hat{Y}, \hat{Z}, \hat{U}, \hat{V}, \hat{W})$, in general.

Ideally, one would have an approximation such that the true and the model Lagrangian velocities coincide:

$$\hat{U}(t) = u(\hat{X}(t), \hat{Y}(t), \hat{Z}(t), t),$$
$$\hat{V}(t) = v(\hat{X}(t), \hat{Y}(t), \hat{Z}(t), t), \quad \hat{W}(t) = w(\hat{X}(t), \hat{Y}(t), \hat{Z}(t), t), \tag{10.3}$$

which would assure that the true and the model trajectories are the same. However it is unrealistic to satisfy (10.1), and therefore one uses different consistency principles. The general *consistency principle* says that the statistics of the model process $\hat{X}(t), \hat{Y}(t), \hat{Z}(t)$, $\hat{U}(t), \hat{V}(t), \hat{W}(t)$ satisfies the same relations satisfied by the true process $X(t)$, $Y(t), Z(t)$, $U(t)$, $V(t)$, $W(t)$, where

$$U(t) = u(X(t), Y(t), Z(t), t), \quad V(t) = v(X(t), Y(t), Z(t), t),$$
$$W(t) = w(X(t), Y(t), Z(t), t)$$

are the components of the true Lagrangian velocity.

Two consistency principles used in the literature are:

(a) consistency with the Kolmogorov similarity theory;
(b) consistency with Thomson's well-mixed condition.

Here (a) reads

$$\langle (dU)^2 \rangle = \langle (dV)^2 \rangle = \langle (dW)^2 \rangle = C_0 \varepsilon dt,$$

and

$$\langle dU\, dV \rangle = \langle dU\, dW \rangle = \langle dW\, dV \rangle = 0,$$

where dU, dV, dW are the components of the increments of the Lagrangian velocity, ε is the mean rate of the dissipation of turbulence energy, C_0 is the universal constant (e. g., [145, 215, 237]); here and in what follows, the angle brackets stand for the ensemble average.

Note that (a) implies (e. g., see [237]) that in (10.1) all the terms b_u, b_v, b_w are equal to $\sqrt{C_0\varepsilon}$:

$$b_u = b_v = b_w = \sqrt{C_0\varepsilon}. \tag{10.4}$$

The Thomson well-mixed condition can be rigorously derived from Novikov's integral relation [153]

$$p_E(u,v,w;x,y,z,t) = \int_{R^3} p_L(x,y,z,u,v,w;x_0,y_0,z_0,t)dx_0dy_0dz_0. \quad (10.5)$$

Here p_E is the probability density function (p. d. f.) of the Eulerian velocity u, v, w, in the fixed point x, y, z, at the time t, and p_L is the joint p. d. f. of the true Lagrangian phase point X, Y, Z, U, V, W defined by the trajectory started at x_0, y_0, z_0.

It is natural to require that the p. d. f. of the model phase point governed by (10.1), say $\hat{p}_L$, satisfies

$$p_E(u,v,w;x,y,z,t) = \int_{R^3} \hat{p}_L(x,y,z,u,v,w;x_0,y_0,z_0,t)dx_0dy_0dz_0. \quad (10.6)$$

Note that (10.6), the Focker–Planck–Kolmogorov equation for $\hat{p}_L$ and (10.4) lead to the well-mixed condition due to Thomson [237]:

$$\frac{\partial p_E}{\partial t} + u\frac{\partial p_E}{\partial x} + v\frac{\partial p_E}{\partial y} + w\frac{\partial p_E}{\partial z} + \frac{\partial}{\partial u}(a_u p_E) + \frac{\partial}{\partial v}(a_v p_E) + \frac{\partial}{\partial w}(a_w p_E)$$
$$= \frac{C_0\varepsilon}{2}\left\{\frac{\partial^2 p_E}{\partial u^2} + \frac{\partial^2 p_E}{\partial v^2} + \frac{\partial^2 p_E}{\partial w^2}\right\}. \quad (10.7)$$

It is convenient to rewrite this equation in the form given by Thomson [237],

$$\frac{\partial p_E}{\partial t} + u\frac{\partial p_E}{\partial x} + v\frac{\partial p_E}{\partial y} + w\frac{\partial p_E}{\partial z} + \frac{\partial}{\partial u}(\phi_u) + \frac{\partial}{\partial v}(\phi_v) + \frac{\partial}{\partial w}(\phi_w) = 0, \quad (10.8)$$

where

$$\phi_u = a_u p_E - \frac{C_0\varepsilon}{2}\frac{\partial p_E}{\partial u}, \quad \phi_v = a_v p_E - \frac{C_0\varepsilon}{2}\frac{\partial p_E}{\partial v}, \quad \phi_w = a_w p_E - \frac{C_0\varepsilon}{2}\frac{\partial p_E}{\partial w}.$$

The vector function $\boldsymbol{\phi} = (\phi_u, \phi_v, \phi_w)$ is not uniquely defined from (10.8). Indeed, a series of solutions can be obtained by adding to $\boldsymbol{\phi}$ an arbitrary vector function whose divergence in velocity space is zero.

It should be noted that in the 1-dimensional case, the well-mixed condition uniquely defines the LS model even for non-Gaussian p_E [237]. In the multidimensional case, the *uniqueness problem* can be formulated as follows: give physically plausible assumptions which define uniquely the function $\boldsymbol{\phi}$ in (10.8).

The first 3D LS model satisfying (10.8) was suggested by Thomson [237]. The p. d. f. p_E in his model has a Gaussian form, and the drift terms a_u, a_v, a_w have a quadratic dependence on the velocity. Another example of a model (suggested by

Borgas; see, e. g., [255]) with a quadratic drift term and Gaussian p_E was studied in [217]. They found that the Borgas model gives slightly different results compared to Thomson's model. Reynolds [177] has constructed a 2-parameter class of well-mixed models, also quadratic and with Gaussian p_E, which includes Thomson's and Borgas' models. He demonstrated that two different models from his class produce essentially different predictions of the turbulent dispersion.

The non-Gaussian form of p_E in the 2D case was treated in [56]. To extract a unique model in the class of well-mixed models, they suggested the following ad hoc assumption: the term $(\phi_u/p_E, \phi_w/p_E)$ accelerates particles directly towards (or away from) the origin of (u, w) space. A 3D generalization of this model is given by Monti and Leuzzi [147]. A further generalization of the approach of Flesch and Wilson [56] was given in [255]: the vector $(\phi_u/p_E, \phi_w/p_E)$ is chosen so that there is no preferred direction of rotation of the velocity fluctuation vector ("zero-spin" models). However as shown by Reynolds [178], this approach does not solve the uniqueness problem.

It should be emphasized that all the above-mentioned LS models deal with quite general inhomogeneous turbulence flows. It is therefore difficult to formulate physically motivated assumptions which, together with the well-mixed condition, uniquely define the LS model. Therefore it is reasonable to consider special classes of turbulent flows (e. g., horizontally homogeneous) whose specific features can be used to construct uniquely the LS models under assumptions with a credible physical basis.

In the present chapter we treat a 3D horizontally homogeneous surface layer with a general form of p_E and formulate a physically plausible assumption about the structure of the drift terms a_u, a_v, a_w. This assumption uniquely defines our model in the class of well-mixed models. The model proposed is essentially different from all the models cited above, in particular, for Gaussian p_E, our model, being in this case also quadratic in velocities, is generally not in the class of models given by Reynolds [177]; an exception concerns the case of ideally neutral stratification (i. e., the Obukhov–Monin length scale is infinite: $L = \infty$): in this case our model belongs to Reynolds' class if the Reynolds parameters C_1 and C_2 are chosen as $C_1 = C_0 u_*^4/2\sigma_w^4$, and $C_2 = 0$ (see Appendix B of this chapter).

In Section 10.2 we formulate the assumption which ensures the unique definition of our model for the horizontally homogeneous neutrally-stratified surface layer. Comparison with other Lagrangian stochastic models and experimental measurements is given in Section 10.3. The convective case is treated in Section 10.4 . The behavior of trajectories of our model near the boundary is analyzed in Section 10.5. In Appendix A the drift terms are derived in the Gaussian case. In Appendix B we analyze how our model relates to Reynolds' and "zero-spin" classes of models.

10.2 Neutrally stratified boundary layer

10.2.1 General case of Eulerian p. d. f.

We consider a horizontally homogeneous incompressible boundary layer in the half-space $R^3_+ = \{(x,y,z) : z \geq 0\}$, where x, y are the horizontal coordinates and z is the vertical coordinate. Thus it is assumed that the mean velocity has no vertical component. It is supposed in this section that the mean velocity vector does not change its direction with height; it is directed along the X-axis, and the crosswind velocity fluctuations are symmetric with respect to the plane XZ. Thus the mean velocity vector is $(\bar{u}(z,t), 0, 0)$, while p_E does not depend on x, y.

We will write the p. d. f. p_E in the form

$$p_E(u, v, w; z, t) = p'_E(u', v', w'; z, t),$$

where $u' = u - \bar{u}(z,t)$, $v' = v$ and $w' = w$.

By (10.4), equation (10.1) in these variables has the form

$$\begin{aligned} dX &= (U' + \bar{u}(Z,t))dt,\ dY = V'dt,\ dZ = W'dt, \\ dU' &= a'_u(t, Z, U', V', W')dt + \sqrt{C_0\varepsilon}\, dB_u(t), \\ dV' &= a'_v(t, Z, U', V', W')dt + \sqrt{C_0\varepsilon}\, dB_v(t), \\ dW' &= a'_w(t, Z, U', V', W')dt + \sqrt{C_0\varepsilon}\, dB_w(t). \end{aligned} \tag{10.9}$$

To simplify the notation, here and in what follows we omit the hat sign introduced in Section 10.1 to denote the model trajectory.

The well-mixed condition in new variables is

$$\begin{aligned} \frac{\partial p'_E}{\partial t} + w'\frac{\partial p'_E}{\partial z} + \frac{\partial}{\partial u'}(a'_u p'_E) + \frac{\partial}{\partial v'}(a'_v p'_E) + \frac{\partial}{\partial w'}(a'_w p'_E) \\ = \frac{C_0\varepsilon}{2}\left\{\frac{\partial^2 p'_E}{\partial u'^2} + \frac{\partial^2 p'_E}{\partial v'^2} + \frac{\partial^2 p'_E}{\partial w'^2}\right\}. \end{aligned} \tag{10.10}$$

Now we give our main assumption about the structure of the Lagrangian model (10.9).

Assumption 10.1. *We assume in addition to the well-mixed condition that*

(i) *the vertical drift term does not depend on the horizontal velocity components:* $a'_w = a'_w(t, z, w')$*;*

(ii) *the streamwise term* a'_u *does not depend on the crosswind velocity* v'*:* $a'_u = a'_u(t, z, u', w')$.

This assumption meets the conditions of a surface layer with neutral (or close to neutral) stratification. Indeed, all the contributions to the vertical motions can be divided into two parts: the first comes from the vortices whose sizes are smaller or close to the current height z, and the second is due to the large horizontally stretched vortices. The second part of the contribution is much smaller than the first one, since the vertical velocities in such horizontal stretched vortices are much smaller than those of the small vortices whose sizes are of the order of the current height. The first part comes mainly from vortices which are isotropic or at least not very anisotropic. But in the isotropic case, the well-mixed condition leads to the dependence $a'_w = a'_w(t,z,w')$ (e. g., see [256]) which gives us the motivation of point (i) in our assumption. As for point (ii), we note that the coordinate system is chosen so that $\langle u'v'\rangle = 0$, $\langle v'w'\rangle = 0$, but $\langle u'w'\rangle \neq 0$, which suggests the approximation $a'_u = a'_u(t,z,u',w')$.

The approximation formulated in point (ii) is reasonable if the mean velocity is dominating over the fluctuated part. Otherwise, for instance in the convective case, this approximation fails, and the velocity components u' and v' must enter the drift terms symmetrically. In Section 10.3 we will treat this case separately.

Note that the dependence $a'_u = a'_u(t,z,u',w')$ holds both for Thomson's and Reynolds' models; see Appendix B of this chapter.

Thus the model (10.9), in view of the assumption reads

$$\begin{aligned} dX &= (U' + \bar{u}(Z,t))dt,\ dY = V'dt,\ dZ = W'dt, \\ dU' &= a'_u(t,Z,U',W')dt + \sqrt{C_0\varepsilon}\, dB_u(t), \\ dV' &= a'_v(t,Z,U',V',W')dt + \sqrt{C_0\varepsilon}\, dB_v(t), \\ dW' &= a'_w(t,Z,W')dt + \sqrt{C_0\varepsilon}\, dB_w(t). \end{aligned} \tag{10.11}$$

Integrating (10.10) over u' and v' yields

$$\frac{\partial p'_{1E}}{\partial t} + w'\frac{\partial p'_{1E}}{\partial z} + \frac{\partial}{\partial w'}(a'_w(t,z,w')p'_{1E}) = \frac{C_0\varepsilon}{2}\frac{\partial^2 p'_{1E}}{\partial (w')^2}, \tag{10.12}$$

where

$$p'_{1E} = p'_{1E}(w';z,t) = \int_{-\infty}^{\infty}\int_{-\infty}^{\infty} p'_E(u',v',w';z,t)\, du'dv'. \tag{10.13}$$

Here we have assumed that

$$a'_u p'_E,\ a'_v p'_E,\ \frac{\partial p'_E}{\partial u'},\ \frac{\partial p'_E}{\partial v'} \quad \text{all tend to zero as} \quad (u')^2 + (v')^2 \to \infty.$$

Similarly, the integration of (10.10) over v' leads to

$$\frac{\partial p'_{2E}}{\partial t} + w'\frac{\partial p'_{2E}}{\partial z} + \frac{\partial}{\partial u'}(a'_u(t,z,u',w')p'_{2E}) + \frac{\partial}{\partial w'}(a'_w(t,z,w')p'_{2E}) = \frac{C_0\varepsilon}{2}\left(\frac{\partial^2 p'_{2E}}{\partial (u')^2} + \frac{\partial^2 p'_{2E}}{\partial (w')^2}\right), \quad (10.14)$$

where

$$p'_{2E} = p'_{2E}(u',w';z,t) = \int_{-\infty}^{\infty} p'_E(u',v',w';z,t)\,dv'. \quad (10.15)$$

Now, under the assumption about the behavior in infinity, it is possible to uniquely define the coefficients a'_u, a'_v and a'_w. Indeed, from (10.12) one gets a'_w, then from (10.14) one finds a'_u, and from (10.10) one obtains a'_v. This yields

$$a'_w(t,z,w) = \frac{1}{p'_{1E}(w;z,t)}\left\{\frac{C_0\varepsilon}{2}\frac{\partial p'_{1E}}{\partial w} - \left(\frac{\partial f_{1E}}{\partial t} + \frac{\partial F_{1E}}{\partial z}\right)\right\}, \quad (10.16)$$

where

$$f_{1E}(w;z,t) = \int_{-\infty}^{w} p'_{1E}(w';z,t)\,dw',$$

$$F_{1E}(w;z,t) = \int_{-\infty}^{w} w' p'_{1E}(w';z,t)\,dw',$$

and

$$a'_u(t,z,u,w) = \frac{1}{p'_{2E}}\left\{\frac{C_0\varepsilon}{2}\left(\frac{\partial p'_{2E}}{\partial u} + \frac{\partial^2 f_{2E}}{\partial w^2}\right) - \left(\frac{\partial f_{2E}}{\partial t} + w\frac{\partial f_{2E}}{\partial z}\right) - \frac{\partial}{\partial w}(a'_w f_{2E})\right\}, \quad (10.17)$$

where

$$f_{2E}(u,w;z,t) = \int_{-\infty}^{u} p'_{2E}(u',w;z,t)\,du'.$$

Finally,

$$a'_v(t,z,u,w) = \frac{1}{p'_E}\left\{\frac{C_0\varepsilon}{2}\left(\frac{\partial^2 f_E}{\partial u^2} + \frac{\partial p'_E}{\partial v} + \frac{\partial^2 f_E}{\partial w^2}\right) - \left(\frac{\partial f_E}{\partial t} + w\frac{\partial f_E}{\partial z}\right) - \frac{\partial}{\partial u}(a'_u f_E) - \frac{\partial}{\partial w}(a'_w f_E)\right\}, \quad (10.18)$$

where

$$f_E(u,v,w;z,t) = \int_{-\infty}^{v} p'_E(u,v',w;z,t)\,dv'.$$

Thus the coefficients (10.16)–(10.18) define a unique stochastic model (10.11) through the p. d. f. p'_E.

In the case when the crosswind velocity fluctuations are independent of the streamwise and vertical fluctuations, i. e., if

$$p'_E(u,v,w;z,t) = p'_{2E}(u,w;z,t)\,p_{vE}(v;z,t), \tag{10.19}$$

then the expression (10.18) for the crosswind drift term can be simplified:

$$a'_v(t,z,u,v,w) = \frac{C_0\varepsilon}{2p_{vE}}\frac{\partial p_{vE}}{\partial v} - \frac{1}{p_{vE}}\frac{\partial f_{vE}}{\partial t} - \frac{w}{p_{vE}}\frac{\partial f_{vE}}{\partial z}, \tag{10.20}$$

where

$$f_{vE}(v) = \int_{-\infty}^{v} p_{vE}(v')\,dv'.$$

10.2.2 Gaussian p. d. f.

We present here expressions for the coefficients to (10.11) for Gaussian p. d. f. p_E. Recall that we deal here with horizontally homogeneous turbulence, and use a coordinate system where the direction of the mean velocity coincides with the X-axes, and the crosswind velocity fluctuations are symmetric relative to the plane XZ. Therefore, the Gaussian p. d. f. p_E has the form

$$\begin{aligned} p'_E(u,v,w;z,t) = {} & \frac{1}{2\pi\sigma_{u/w}\sigma_w}\exp\left\{-\frac{1}{2\sigma_{u/w}^2}(u-\rho w)^2 - \frac{w^2}{2\sigma_w^2}\right\} \\ & \times\frac{1}{\sqrt{2\pi}\sigma_v}\exp\left\{-\frac{v^2}{2\sigma_v^2}\right\}, \end{aligned} \tag{10.21}$$

where

$$\sigma_{u/w} = \frac{\Delta^{1/2}}{\sigma_w}, \qquad \rho = \frac{\overline{uw}}{\sigma_w^2}, \qquad \Delta = \sigma_u^2\sigma_w^2 - (\overline{uw})^2,$$

and $\sigma_u^2, \sigma_v^2, \sigma_w^2$ are the variances of the x-, y- and z-velocity components, respectively. Note that by definition, $\sigma_{u/w}$ is the standard deviation of u conditional on the value of w.

Using the result given in Section 2.1 we obtain (see Appendix A of this chapter) the following expressions:

$$a'_w(t,z,w) = -\left(\frac{C_0\varepsilon}{2\sigma_w^2} - \frac{1}{\sigma_w}\frac{\partial\sigma_w}{\partial t}\right)w + \frac{1}{2}\frac{\partial\sigma_w^2}{\partial z}\left(\frac{w^2}{\sigma_w^2}+1\right), \tag{10.22}$$

$$a'_u(t,z,u,w) = -\frac{C_0\varepsilon(1+\rho^2)}{2\sigma^2_{u/w}}(u-\rho w) + \frac{\rho}{2\sigma^2_w}\left(C_0\varepsilon + \frac{\partial\sigma^2_w}{\partial t}\right)w$$
$$+\frac{\rho}{2}\frac{\partial\sigma^2_w}{\partial z}\left(\frac{w^2}{\sigma^2_w}+1\right)\left(\frac{\partial\rho}{\partial t} + w\frac{\partial\rho}{\partial z}\right)w$$
$$-\frac{1}{\sigma_{u/w}}\left(\frac{\partial\sigma_{u/w}}{\partial t} + w\frac{\partial\sigma_{u/w}}{\partial z}\right)(u-\rho w), \tag{10.23}$$

and

$$a'_v(t,z,u,v,w) = -\left(\frac{C_0\varepsilon}{2\sigma^2_v} - \frac{1}{\sigma_v}\frac{\partial\sigma_v}{\partial t}\right)v + \frac{1}{2}\frac{\partial\sigma^2_v}{\partial z}\frac{vw}{\sigma^2_v}. \tag{10.24}$$

Note that in the stationary case these expressions can be simplified to

$$a'_u(t,z,u,w) = -\frac{C_0\varepsilon(1+\rho^2)}{2\sigma^2_{u/w}}(u-\rho w) + \frac{\rho C_0\varepsilon}{2\sigma^2_w}w + \frac{\rho}{2}\frac{\partial\sigma^2_w}{\partial z}\left(\frac{w^2}{\sigma^2_w}+1\right)$$
$$-\frac{\partial\rho}{\partial z}w^2 - \frac{1}{\sigma_{u/w}}\frac{\partial\sigma_{u/w}}{\partial z}(u-\rho w)w,$$

$$a'_v(t,z,u,v,w) = -\frac{C_0\varepsilon}{2\sigma^2_v}v + \frac{1}{2}\frac{\partial\sigma^2_v}{\partial z}\frac{vw}{\sigma^2_v}, \tag{10.25}$$

$$a'_w(t,z,w) = -\frac{C_0\varepsilon}{2\sigma^2_w}w + \frac{1}{2}\frac{\partial\sigma^2_w}{\partial z}\left(\frac{w^2}{\sigma^2_w}+1\right).$$

10.3 Comparison with other models and measurements

10.3.1 Comparison with measurements in an ideally-neutral surface layer (INSL)

In this section we analyse some quantities in the case of turbulent dispersion in a stationary, horizontally homogeneous, ideally-neutral surface layer (i. e., the Obukhov–Monin length L equals infinity).

We have calculated the following dimensionless Lagrangian characteristics:

$$A(t) = \frac{\langle Z^2(t)\rangle^{1/2}}{u_* t}, \quad B(t) = \frac{\langle Z(t)\rangle}{u_* t}, \quad C(t) = \frac{z_0}{u_* t}\exp\left\{\frac{\kappa\langle X(t)\rangle}{u_* t}+1\right\}, \tag{10.26}$$

and the ratio $pr(z) = k_\mu(z)/k_z(z)$, where κ is the von Karman constant, $k_\mu = \kappa u_* z$ is the vertical eddy viscosity, and k_z is the vertical eddy diffusivity coefficient defined through the Boussinesque hypothesis

$$\overline{c'w'}(z) = -k_z(z)\frac{\partial\bar{c}(z)}{\partial z}. \tag{10.27}$$

The importance of the characteristics $A(t)$, $B(t)$, and $C(t)$ is that these functions tend, as $t \to \infty$, to some universal constant values a, b, and c, respectively, provided that h_s and z_0 are much less than $u_* t$ (e. g., see [19, p. 77]). Here h_s is the height at which the Lagrangian trajectory starts. As to the ratio $pr(z)$, for values z much larger than the source height, it tends to the Prandtl constant Pr; its universal character is well known and is in the literature often approximately taken equal to unity (e. g., see [145, Sect. 8.2]).

Since all the four quantities $A(t)$, $B(t)$, $C(t)$, and $pr(z)$ do not depend on the cross-wind dispersion, we use the 2D stochastic models to simulate the dispersion:

$$\begin{aligned} dX &= (U' + \bar{u}(Z,t))dt, \; dZ = W'dt, \\ dU' &= a'_u(Z, U', W')dt + \sqrt{C_0 \varepsilon}\, dB_u(t), \\ dW' &= a'_w(Z, U', W')dt + \sqrt{C_0 \varepsilon}\, dB_w(t), \end{aligned} \tag{10.28}$$

where for the ISNL, the vertical profiles of ε and $\bar{u}$ can be taken as (e. g., see [145])

$$\varepsilon(z) = \frac{u_*^3}{\kappa z}, \quad \bar{u}(z) = \frac{u_*}{\kappa} \ln\big(z/z_0)\big),$$

and z_0 is the roughness height. The calculations were carried out by Thomson's, Reynolds', Flesch and Wilson's, and our models. Thomson's 2D model in this case is specified by

$$a'_u(z,u,w) = -\frac{C_0\varepsilon(z)}{2\Delta}(\sigma_w^2 u + u_*^2 w),$$

$$a'_w(z,u,w) = -\frac{C_0\varepsilon(z)}{2\Delta}(\sigma_u^2 w + u_*^2 u),$$

where σ_u and σ_w are given by $\sigma_u = b_u u_*$, $\sigma_w = b_w u_*$ with $u_*^2 = -\overline{uw} = \text{const}$; $\Delta = \sigma_u^2 \sigma_w^2 - \overline{uw}^2$, b_u and b_w are universal constants. Following Panofsky and Dutton [161], and Stull [232] we have taken $b_u = 2.5$, $b_w = 1.25$. These parameters enter all the models specified below.

The drift terms of the model due to Flesch and Wilson [56] can be written in the case of INSL as

$$a'_u(z,u,w) = -\frac{C_0\varepsilon(z)}{2}\frac{\partial \ln p'_E}{\partial u} = -\frac{C_0\varepsilon(z)}{2\sigma_{u/w}^2}(u - \rho w),$$

$$a'_w(z,u,w) = -\frac{C_0\varepsilon(z)}{2}\frac{\partial \ln p'_E}{\partial w} = \frac{C_0\varepsilon(z)\rho}{2\sigma_{u/w}^2}(u - \rho w) - \frac{C_0\varepsilon(z)}{2\sigma_w^2}\, w,$$

where $\sigma_{u/w} = \sqrt{\Delta}/\sigma_w$, $\rho = \overline{uw}/\sigma_w^2$. It turns out that the Thomson and the Flesch and Wilson models are identical for the case of INSL. This can be seen directly by comparing of the coefficients.

The model of Reynolds [178] in our case of ideally-neutral surface layer is specified by

$$a'_u(z,u,w) = -\left(\frac{C_0\varepsilon}{2} + C_1 u_*^2 \frac{d\bar{u}}{dz}\right)\frac{\sigma_w^2 u + u_*^2 w}{\Delta},$$

$$a'_w(z,u,w) = -\frac{C_0\varepsilon(z)}{2\Delta}\left(\sigma_u^2 w + u_*^2 u\right) + C_1\sigma_w^2 \frac{d\bar{u}}{dz}\frac{\sigma_w^2 u + u_*^2 w}{\Delta}, \quad (10.29)$$

with $C_1 = 3$ chosen by Reynolds [178] to fit the experimental results of Legg [127].

Our model (10.22)–(10.23) (in the following we call it a KS model) in the case of INSL is specified by

$$a'_u(z,u,w) = -\frac{C_0\varepsilon(z)(1+\rho^2)}{2\sigma^2_{u/w}}(u - \rho w) + \frac{\rho\, C_0\varepsilon(z)}{2\sigma_w^2} w,$$

$$a'_w(z,w) = -\frac{C_0\varepsilon(z)}{2\sigma_w^2} w. \quad (10.30)$$

Note that if we choose the parameter C_1 in (10.29) as $C_1 = C_0/2b_w^4$, then it reduces to our model (10.30) (see Appendix B of this chapter). Reynolds however suggests in his model $C_1 = 3$, and in all comparisons below, when referring to Reynolds' model, we take $C_1 = 3$.

In all models, the calculations were carried out for $z_0 = 0.01$ m, the trajectories started at $h_s = 0.02$ m, $u_* = 0.4\,\mathrm{m\,s^{-1}}$, the number of trajectories was $N = 10^5$ in the case of a, b, and c calculations, and $N = 10^6$ for the constant Pr. The vertical eddy diffusivity k_z was calculated from the relation (10.27) where a finite-difference approximation of the calculated mean concentration was used to find the mean concentration derivative. A stationary source was uniformly distributed on the plane $z = z_s = 0.02$ m.

The stochastic differential equations were solved by the explicit Euler scheme, with the varying time step $\Delta t = \alpha\tau_L(z)$, where $\tau_L(z) = 2\sigma_w/C_0\varepsilon(z)$ is the Lagrangian time scale; to reach stable numerical results, we found that $\alpha = 0.02$ was sufficient. At the boundary, a perfect reflection is made after the trajectory hits the layer $\{z < z_0\}$, z_0 being the roughness height.

The results of calculations and experimental data are presented in Table 10.1. The calculations were carried out for different values of C_0, since the constant C_0 is known to be scattered in the interval $(2, 8)$ (e. g., see [170]). The results for all four constants a, b, c, and Pr show that our model is in a good agreement with Thomson's model and experimental results, but in poor agreement with the results obtained by Reynolds' model. We recall that a, b, and c, as defined above, are universal constants determined as limits, as $t \to \infty$, of the functions $A(t)$, $B(t)$, and $C(t)$. In calculations, t was taken sufficiently large. Note that the Thomson and the Flesch and Wilson models (which are identical in this case) give slightly different values of pr, while the constants a, b, and c practically coincide. This can be explained by the fact that in the case of pr,

Table 10.1. Universal constants a, b, c and Pr calculated by different Lagrangian models, compared against experimental results.

Model	C_0	a	b	c	Pr
Thomson 1987 [237]	3.	0.85	0.65	0.25	0.47
	4.	0.71	0.54	0.22	0.6
	5.	0.61	0.46	0.2	0.74
	7.	0.48	0.35	0.16	1.
Flesch and Wilson 1992 [56]	3.	0.85	0.65	0.26	0.45
	5.	0.61	0.46	0.2	0.8
	7.	0.48	0.35	0.16	0.9
KS (see (6.14), Section 3.1)	3.	0.73	0.55	0.17	0.64
	4.	0.59	0.44	0.15	0.82
	5.	0.5	0.36	0.14	1.
	7.	0.37	0.27	0.11	1.43
Reynolds 1997 [177]	3.	0.13	0.09	0.04	5.26
	5.	0.18	0.13	0.06	3.3
	7.	0.21	0.15	0.07	2.86
Measurements					
Garger and Zhukov 1986 [64]		0.58	0.44	0.19	
Chaudhry and Meroney 1973 [21]			0.4		
Rider 1954 [179]					0.83
Gurvich 1965 [73]					1.25

additional error of the finite-difference approximation to the concentration derivative in (10.27) is involved.

As to the best fit to the experimental data, our model reaches it at $C_0 = 4$, while Thomson's and Flesch and Wilson's models fit best at $C_0 = 5$. Regarding the Reynolds model, it should be mentioned that as C_0 becomes larger, the discrepancy between the results obtained by his model and measurements slightly decreases, but even for $C_0 = 7$ it remains too large. Calculations for $C_0 = 10$ (not shown in the table) gave almost the same results as for $C_0 = 7$.

10.3.2 Comparison with the wind tunnel experiment by Raupach and Legg (1983)

In this section we present a comparison of the same models analysed in the previous section against the data of the wind tunnel experiment by Raupach and Legg [176]. The vertical profiles of the mean concentration $\bar{c}$, the streamwise and vertical fluxes of concentration $\overline{c'u'}$, $\overline{c'w'}$, were analysed. A stationary line source at the height $h_s = 0.06\ m$ directed along the y-axis was considered, and all the profiles were calculated

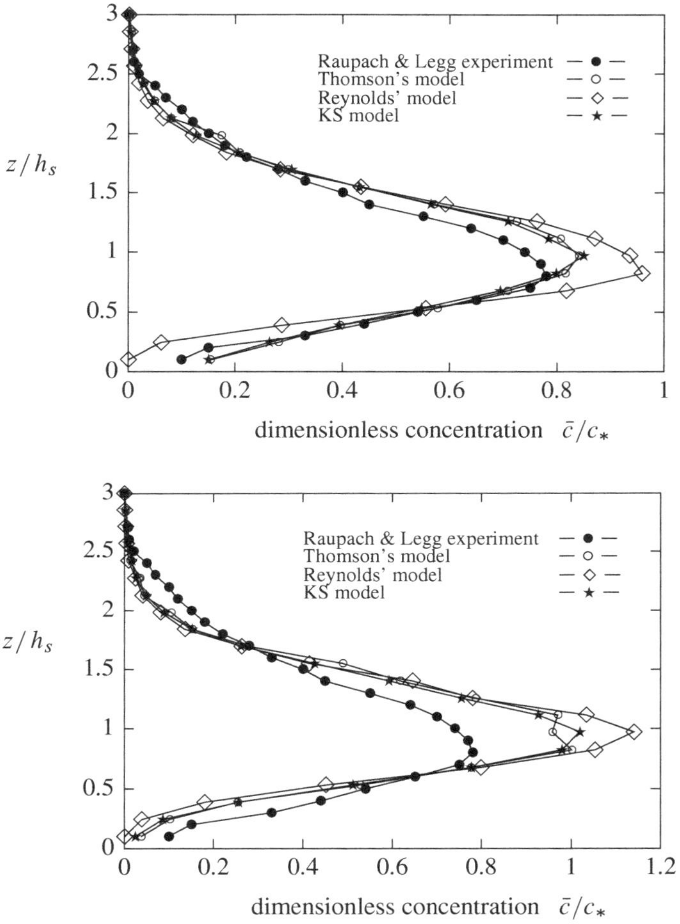

Figure 10.1. A comparison of three model predictions of vertical profile of mean concentrtaion with Raupach and Legg's measurement, for $C_0 = 3$ (upper panel) and $C_0 = 7$ (lower panel).

at the downwind distance $x = 7.5\,h_s$. The problem is governed by the 2D equations used in the previous subsection.

In Figure 10.1, upper panel, the scaled mean concentration $\bar{c}(x,z)/c_*$ and temperature $\bar{\theta}(x,z)/\theta_*$ profiles are plotted for $C_0 = 3$, where

$$c_* = Q/(h_s \bar{u}(h_s)), \quad \theta_* = Q/(\rho c_p h_s \bar{u}(h_s)).$$

Here Q is the line source strength per unit length, ρ the air density and c_p the specific heat of air at constant pressure. The temperature profiles were taken from the mea-

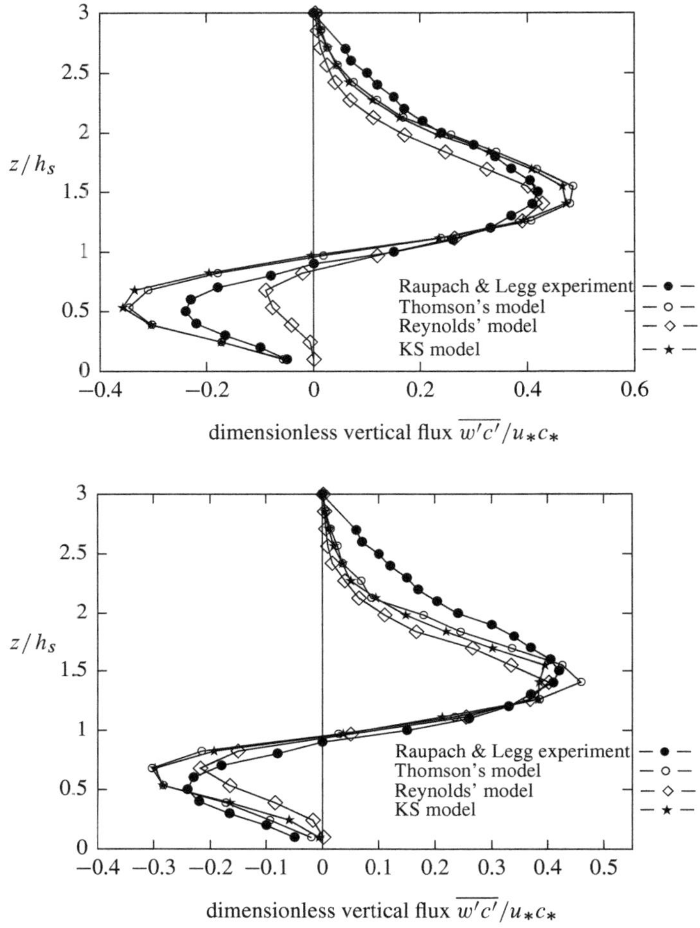

Figure 10.2. A comparison of three model predictions of the vertical profile of mean vertical flux with Raupach and Legg's measurement, for $C_0 = 3$ (upper panel) and $C_0 = 7$ (lower panel).

surements by Raupach and Legg [176] where the temperature was the tracer in the experiments.

All the models predict the experimental results qualitatively well. We mention that the results obtained by Flesch and Wilson's model are very close to the results obtained by Thomson's model, and therefore we do not plot them in our figures. Above the height $z = 1.75h_s$ all three models agree well with the experimental results. Below the height $1.5h_s$ the Thomson and the KS models give results close to the measurements, while Reynolds's model overestimates the maximum concentration and underestimates the concentration near the ground. As to the sensitivity to the constant C_0, we have also made calculations for $C_0 = 2, 4, 5$, and 7. The best fit of Thomson's and KS models was found at $C_0 = 3$. For larger values of C_0 (see Figure 10.1, lower panel for $C_0 = 7$) all the models overestimate the values at the maximum, and underestimate them at small and large heights.

In Figure 10.2 the vertical profile of the vertical flux of concentration is shown for $C_0 = 3$. From these curves, it is clearly seen that at the height $z < z_s$ the Thomson and the KS models underestimates, and Reynolds' model overestimates the experimental results. Above the height $z = 1.5z_s$ all three models are in a good agreement with the measurements. Note that for $C_0 = 7$ the picture is different (see the lower panel of Figure 10.2): the models give slightly better predictions for heights $z < 1.5z_s$

In Figure 10.3 the vertical profiles of the streamwise flux of concentration are presented for $C_0 = 3$ (upper panel) and $C_0 = 7$ (lower panel). Here the Reynolds model significantly overpredicts the maximum and underpredicts the minimum values. the Thomson and the KS models show better agreement with the measurements. Note, however, that the agreement between the Thomson and KS models in this case is not so good as in Figures 10.1 and 10.2. Calculations of the vertical and horizontal fluxes by our model with different values of C_0 have shown that the best fit with the Raupach and Legg experimental results was around $C_0 = 3.5 \pm 0.5$. This value is in good agreement with the value $C_0 = 3$ suggested in [43].

10.4 Convective case

In this section we consider a horizontally homogeneous boundary layer under strong convective conditions at sufficiently large heights compared to $|L|$. In this case, the velocity fluctuations can be considered as horizontally isotropic (e. g., see [145]). Therefore, the mean velocity is zero, and the correlation between the vertical and horizontal velocities is zero.

In this section we show that the horizontal isotropy and the dependence supposed in (i) of the assumption ensure the unique choice of the Lagrangian stochastic model for the convective layer.

To construct the Lagrangian 1-particle model in the convective case, we have to specify the Eulerian velocity p. d. f.. For simplicity, we will treat the case when the Eulerian p. d. f. has the form

$$p_E(u, v, w; z, t) = p_E^{\|}(w; z, t) p_E^{\perp}(u_\perp; z, t), \tag{10.31}$$

where $u_\perp = \sqrt{u^2 + v^2}$, $p_E^{\|}$ is the p. d. f. of the vertical velocity component, and $p_E^{\perp}$ is the p. d. f. of the horizontal velocity components satisfying the relation

$$\int_{-\infty}^{\infty} du \int_{-\infty}^{\infty} dv p_E^{\perp}(\sqrt{u^2 + v^2}; z, t) = 2\pi \int_0^{\infty} du_\perp u_\perp p_E^{\perp}(u_\perp; z, t) = 1.$$

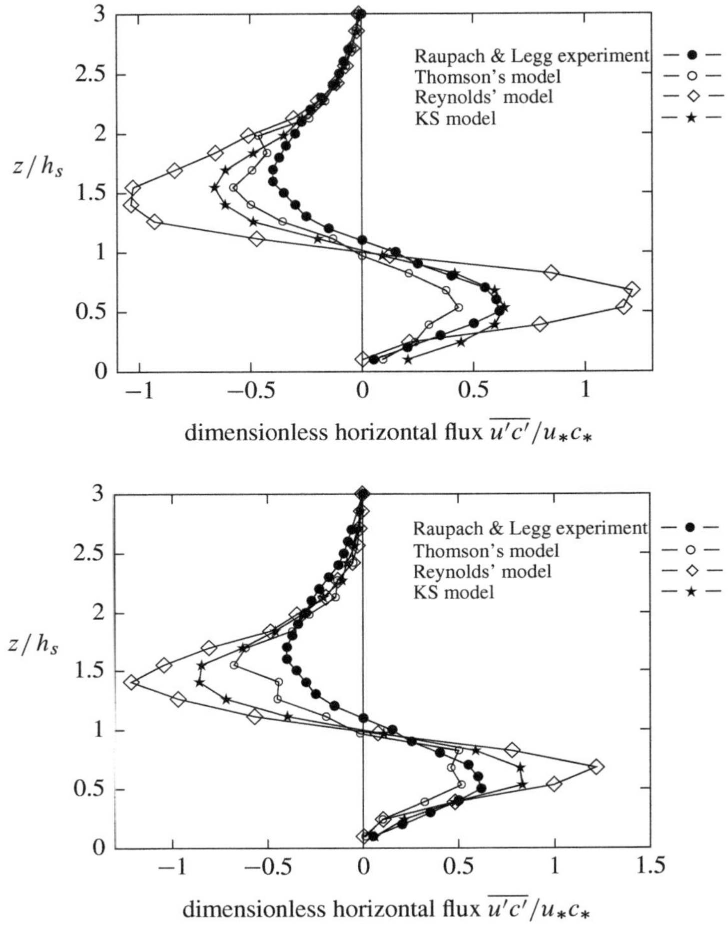

Figure 10.3. A comparison of three model predictions of the vertical profile of mean streamwise flux with Raupach and Legg's measurement, for $C_0 = 3$ (upper panel) and $C_0 = 7$ (lower panel).

Note that in the convective case, the assumption (10.31) is quite reasonable, because the vertical and horizontal velocity fluctuations can be considered as approximately independent.

Under the horizontally isotropy and assuming that the vertical velocity component is governed as assumed in the assumption, point (i), the model (10.9) takes the form

$$\begin{aligned} dX &= Udt, \quad dY = Vdt, \quad dZ = Wdt, \\ dU &= Ug(t, Z, U_\perp, W)dt + \sqrt{C_0\varepsilon}dB_u(t), \\ dV &= Vg(t, Z, U_\perp, W)dt + \sqrt{C_0\varepsilon}dB_v(t), \\ dW &= a_w(t, Z, W)dt + \sqrt{C_0\varepsilon}dB_w(t). \end{aligned} \tag{10.32}$$

Thomson's well-mixed condition implies in our case that

$$\begin{aligned} u_\perp \frac{\partial p_E(u_\perp, w; z, t)}{\partial t} &+ \frac{\partial}{\partial z}(w u_\perp p_E) + \frac{\partial}{\partial u_\perp}(u_\perp^2 g\, p_E) + \frac{\partial}{\partial w}(u_\perp a_w\, p_E) \\ &= \frac{C_0\varepsilon}{2}\left\{\frac{\partial}{\partial u_\perp}\left(u_\perp \frac{\partial p_E}{\partial u_\perp}\right) + \frac{\partial^2}{\partial w^2}(u_\perp p_E)\right\}. \end{aligned} \tag{10.33}$$

This relation follows from (10.7) and from the following simple equalities:

$$\frac{\partial}{\partial u}(ugp_E) + \frac{\partial}{\partial v}(vgp_E) = \frac{1}{u_\perp}\frac{\partial}{\partial u_\perp}(u_\perp^2 g p_E),$$

$$\frac{\partial^2 p_E}{\partial u^2} + \frac{\partial^2 p_E}{\partial v^2} + \frac{\partial^2 p_E}{\partial w^2} = \frac{1}{u_\perp}\frac{\partial}{\partial u_\perp}\left(u_\perp \frac{\partial p_E}{\partial u_\perp}\right) + \frac{\partial^2 p_E}{\partial w^2}.$$

The well-mixed condition (10.33) can be simplified as follows. Integrate (10.33) over $u_\perp$ and use the relation $2\pi \int_0^\infty p_E(u_\perp, w; z, t) u_\perp du_\perp = p_E^{\|}(w; z, t)$. Assuming $u_\perp^2 p_E$ and $u_\perp \frac{\partial p_E}{\partial u_\perp}$ tend to zero as $u_\perp \to \infty$, this yields

$$\frac{\partial p_E^{\|}}{\partial t} + \frac{\partial}{\partial z}(w p_E^{\|}) + \frac{\partial}{\partial w}(a_w p_E^{\|}) = \frac{C_0\varepsilon}{2}\frac{\partial^2 p_E^{\|}}{\partial w^2}. \tag{10.34}$$

This is the 1-dimensional well-mixed condition (10.12). As in the case (10.12), we can find from (10.34) the coefficient $a_w(t, z, w)$

$$a_w(t, z, w) = \frac{1}{p_E^{\|}(w; z, t)}\left\{\frac{C_0\varepsilon}{2}\frac{\partial p_E^{\|}}{\partial w} - \left(\frac{\partial f_{1E}}{\partial t} + \frac{\partial F_{1E}}{\partial z}\right)\right\}, \tag{10.35}$$

where

$$f_{1E}(w; z, t) = \int_{-\infty}^{w} p_E^{\|}(w'; z, t)\, dw',$$

$$F_{1E}(w; z, t) = \int_{-\infty}^{w} w' p_E^{\|}(w'; z, t)\, dw'.$$

This form agrees with that used by Luhar and Britter [133], who explored the value of a_w in detail for a particular choice of p_E.

To find the function g, we substitute (10.31) in (10.33), which in view of (10.34) yields

$$u_\perp \frac{\partial p_E^\perp}{\partial t} + \frac{\partial}{\partial z}(w u_\perp p_E^\perp) + \frac{\partial}{\partial u_\perp}(u_\perp^2 g\, p_E^\perp) = \frac{C_0 \varepsilon}{2} \frac{\partial}{\partial u_\perp}\left(u_\perp \frac{\partial p_E^\perp}{\partial u_\perp}\right). \quad (10.36)$$

Integrating (10.36) over $u_\perp$ from 0 to ∞ we get, under the condition $u_\perp^2 g p_E^\perp \to 0$ as $u_\perp \to 0$, that

$$\frac{\partial P_E}{\partial t} + w \frac{\partial P_E}{\partial z} + u_\perp^2 g(t, z, u_\perp, w) p_E^\perp = \frac{C_0 \varepsilon}{2} u_\perp \frac{\partial p_E^\perp}{\partial u_\perp}, \quad (10.37)$$

where

$$P_E(u_\perp; z, t) = \int_0^{u_\perp} u p_E^\perp(u; z, t) du.$$

This defines the function g if $p_E^\perp$ is given.

For example, if

$$p_E^\perp(u_\perp) = \frac{1}{2\pi\sigma^2(t,z)} \exp\left\{-\frac{u_\perp^2}{2\sigma^2(t,z)}\right\},$$

$$P_E(u_\perp) = \frac{1}{2\pi}\left(1 - \exp\left\{-\frac{u_\perp^2}{2\sigma^2(t,z)}\right\}\right),$$

then

$$\frac{\partial P_E}{\partial z} = \frac{\partial \ln \sigma_\perp}{\partial z}(u_\perp^2 p_E^\perp), \quad \frac{\partial P_E}{\partial t} = \frac{\partial \ln \sigma_\perp}{\partial t}(u_\perp^2 p_E^\perp), \quad \frac{\partial p_E^\perp}{\partial u_\perp} = -\frac{u_\perp}{\sigma_\perp^2} p_E^\perp,$$

and we get

$$\begin{aligned} g &= \frac{1}{u_\perp^2 p_E^\perp}\left\{\frac{C_0 \varepsilon}{2} u_\perp \frac{\partial p_E^\perp}{\partial u_\perp} - \frac{\partial P_E}{\partial t} - w \frac{\partial P_E}{\partial z}\right\} \\ &= -\frac{C_0 \varepsilon}{2\sigma_\perp^2} - \frac{\partial \ln \sigma_\perp}{\partial t} - w \frac{\partial \ln \sigma_\perp}{\partial z}. \end{aligned}$$

Remark 10.1. We have assumed here the factorization (10.31), which simplifies the form of g. Generally, when (10.31) is not true, the function g can be found analogously, but its structure is more complicated.

10.5 Boundary conditions

To complete the description of the Lagrangian stochastic model, we need to define the behavior of $(X(t), Y(t), Z(t), U(t), V(t), W(t))$, the solution to (10.1) in the neighborhood of the boundary $\Gamma = \{(x, y, z) : z = 0\}$. We assume that the boundary is impenetrable, i. e., that $w = 0$ at the boundary of Γ. This implies that the true Lagrangian trajectories never reach Γ. Therefore it is reasonable to require that the same property holds for $X(t), Y(t), Z(t)$, the solutions to (10.1). This can be done by special choice of the function $\varepsilon(z, t)$. Indeed, in the neighborhood of Γ, it is reasonable to consider the flow as ideally neutral stratified. Therefore, $p_E(w)$ is Gaussian, with constant σ_w, and the vertical profile of $\varepsilon(z)$ is given by (e. g., see [145])

$$\varepsilon(z) = \frac{u_*^3}{\kappa z}, \quad \kappa \simeq 0.4, \quad z > z_0. \tag{10.38}$$

Here κ is the Karman constant, and z_0 is the roughness height.

The equation of vertical motion then is

$$dZ = W\,dt, \quad dW(t) = -\frac{a}{Z} W(t)dt + \frac{b}{\sqrt{Z}}\,dB(t), \tag{10.39}$$

where

$$a = \frac{u_*^3}{2\kappa\sigma_w^2}, \quad b = \left(\frac{C_0 u_*^3}{\kappa}\right)^{\frac{1}{2}}.$$

If we assume that the formula (10.38) is true for all $z > 0$, then all the solutions to (10.39) do not reach the boundary Γ. Indeed, let τ be a random variable (which depends on the trajectory $Z(t)$) defined by

$$\tau(t) = \int_0^t \frac{ds}{Z(s)}.$$

Then, the vertical velocity in new variable $W(\tau)$ satisfies the equation

$$dW(\tau) = -aW(\tau)d\tau + b\,dB(\tau).$$

Therefore, from

$$\frac{dZ}{d\tau} = \frac{dZ}{dt}\frac{dt}{d\tau} = W(\tau)Z(\tau)$$

we have

$$Z(\tau) = Z(0)\exp\{S(\tau)\}, \quad S(\tau) = \int_0^\tau W(\tau')\,d\tau'.$$

The function $W(\tau)$ is an Uhlenbeck–Ornstein process with continuous samples. Therefore, $|S(\tau)| < \infty$ for all $\tau > 0$ with probability 1. This implies that $Z(\tau) > 0$, provided that $Z(0) > 0$. Thus the function $Z(\tau)$ never reaches the boundary Γ. The same is true for $Z(t)$. To show this, it is sufficient to note that $t(\tau) \to \infty$ as $\tau \to \infty$. Let us show this property. We have

$$t(\tau) = \int_0^\tau \frac{dt}{d\tau'}\,d\tau' = \int_0^\tau Z(\tau')\,d\tau' = Z(0)\int_0^\tau \exp\{S(\tau')\}\,d\tau'.$$

In [108] it is shown that with probability 1

$$\int_0^\infty \exp\{S(\tau)\}\,d\tau = \infty.$$

This implies that with probability 1 $t(\tau) \to \infty$ as $\tau \to \infty$.

10.6 Conclusion

A uniquely-defined Lagrangian stochastic model in the class of well-mixed models was constructed from physically plausible assumptions: (i) in the neutrally-stratified horizontally homogeneous surface layer, the vertical motion is mainly controlled by eddies whose size is of the order of the current height, and (ii), the streamwise drift term is independent of the crosswind velocity fluctuations. Supposition (i) is motivated by the well-known property that the vertical motion of vortices whose size is much larger than the current height is damped by the ground surface. Therefore, it is reasonable to assume that the vertical drift term is the same as in the isotropic case: $a'_w = a'_w(t, z, w)$. With regard to point (ii), it comes from the assumption that in the special coordinate system where the X-axis is oriented along the mean velocity vector, the crosswind velocity fluctuations are symmetrically distributed with respect to the plane XZ.

In the free convective layer the mean velocity vector vanishes, and the horizontal motion is isotropic. This property is used to define uniquely the model using only e point (i) of the assumption.

In the model presented the Eulerian p. d. f., p_E may be not Gaussian, as, for instance, in the forest canopy [255]. The Gaussian case was analyzed in detail. The model was compared against the wind tunnel experiment of Raupach and Legg [176] and models due to Thomson [237], Flesch and Wislon [56], and Reynolds [178]. Numerical experiments have shown good agreement of our model with the models of Thomson [237], Flesch and Wilson [56] , and with experimental measurements as well. However there is a large discrepancy of these results with the results obtained by Reynolds' model. Our model shows the best fit to the measurements for $C_0 = 3.5 \pm 0.5$; namely, at $C = 4$, we found the best agreement between the calculated and measured values of the universal constants a, b, c, and Pr; at $C_0 = 3$, the best agreement with the wind-tunnel experiments by Raupach and Legg [176] was achieved. It is interesting to note that our model, also being quadratic in velocity (in the Gaussian case), does not belong to the general 2-parameter class of models suggested by Reynolds (1997); it is also not in the family of "zero-spin" models introduced by Wilson and Flesch [255].

It is believed that the model proposed is well suited for the case of a neutrally (or close to neutrally) stratified surface layer. This generalization might be possible for the whole boundary layer with the mean velocity vector varying with the height, but this needs special study. The same is true for the generalization to the stably stratified surface layer where buoyancy is important.

10.7 Appendices

10.7.1 Appendix A. Derivation of the coefficients in the Gaussian case

Here we derive the coefficients (10.22)–(10.24) from (10.16)–(10.18) in the case of Gaussian p. d. f. (10.21). First we remark that from (10.13) and (10.15) it follows

$$p'_{1E}(w;z,t) = \frac{1}{\sqrt{2\pi}\sigma_w} \exp\left\{-\frac{w^2}{2\sigma_w^2}\right\},$$
$$p'_{2E}(u,w;z,t) = \frac{1}{2\pi\sigma_{u/w}\sigma_w} \exp\left\{-\frac{1}{2\sigma_{u/w}^2}(u-\rho w)^2 - \frac{w^2}{2\sigma_w^2}\right\}.$$

Consequently,

$$f_{1E}(w;z,t) = \int_{-\infty}^{\frac{w}{\sigma_w}} \frac{1}{\sqrt{2\pi}} \exp(-t^2/2)\,dt = \Phi\left(\frac{w}{\sigma_w}\right),$$
$$F_{1E}(w;z,t) = -\sigma_w^2 p'_{1E}(w;z,t).$$

Note that

$$\frac{1}{p'_{1E}}\frac{\partial p'_{1E}}{\partial w} = -\frac{w}{\sigma_w^2}, \qquad -\frac{1}{p'_{1E}}\frac{\partial F_{1E}}{\partial z} = \frac{1}{2}(w^2+1)\frac{\partial \sigma_w^2}{\partial z},$$

and

$$\frac{\partial f_{1E}}{\partial t} = -\frac{w}{\sigma_w^2}\frac{\partial \sigma_w}{\partial t}\dot{\Phi}(w/\sigma_w),$$

where

$$\Phi(\tau) = \int_{-\infty}^{\tau} \frac{1}{\sqrt{2\pi}} \exp(-t^2/2)\,dt, \quad \dot{\Phi}(\tau) = \frac{d\Phi}{d\tau}.$$

From (10.16) we find

$$a'_w(t,z,w) = -\left(\frac{C_0\varepsilon}{2\sigma_w^2} - \frac{1}{\sigma_w}\frac{\partial \sigma_w}{\partial t}\right) w + \frac{1}{2}\frac{\partial \sigma_w^2}{\partial z}\left(\frac{w^2}{\sigma_w^2}+1\right).$$

Note that this coincides with Thomson's relevant expression in his 1D model.

By the definition,

$$f_{2E}(u,w,z,t) = p'_{1E}(w;z,t)\,\Phi\left(\frac{u-\rho w}{\sigma_{u/w}}\right).$$

To find a'_u from (10.17) we need the expressions for

$$\frac{\partial f_{2E}}{\partial t},\quad \frac{\partial f_{2E}}{\partial z},\quad \frac{\partial f_{2E}}{\partial w},\quad \frac{\partial p'_{2E}}{\partial u},\quad \frac{\partial^2 f_{2E}}{\partial w^2}.$$

By definition we get

$$\frac{\partial p'_{2E}}{\partial u} = -\frac{(u-\rho w)}{\sigma^2_{u/w}} p'_{2E},$$

$$\frac{\partial f_{2E}}{\partial t} = f_{2E}\left\{\frac{1}{2\sigma_w^2}\frac{\partial \sigma_w^2}{\partial t}\left(\frac{w^2}{\sigma_w^2}-1\right) + \Psi(\xi)\frac{\partial \xi}{\partial t}\right\},$$

$$\frac{\partial f_{2E}}{\partial z} = f_{2E}\left\{\frac{1}{2\sigma_w^2}\frac{\partial \sigma_w^2}{\partial z}\left(\frac{w^2}{\sigma_w^2}-1\right) + \Psi(\xi)\frac{\partial \xi}{\partial z}\right\},$$

$$\frac{\partial f_{2E}}{\partial w} = f_{2E}\left\{-\frac{w}{\sigma_w^2} - \Psi(\xi)\eta\right\},$$

$$\frac{\partial^2 f_{2E}}{\partial w^2} = f_{2E}\left\{\left[-\frac{w}{\sigma_w^2} - \Psi(\xi)\eta\right]^2 - \frac{1}{\sigma_w^2} + \eta^2\dot{\Psi}(\xi)\right\},$$

where

$$\Psi(\tau) = \frac{d}{d\tau}\ln\Phi(\tau),\ \dot{\Psi}(\tau) = \frac{d\Psi(\tau)}{d\tau},\ \xi = \frac{u-\rho w}{\sigma_{u/w}},\ \eta = \frac{\overline{uw}}{\sigma_{u/w}\sigma_w^2}.$$

Substituting these expressions in (10.17) yields

$$\begin{aligned}
a'_u &= -\frac{C_0\varepsilon}{2\sigma^2_{u/w}}(u-\rho w) \\
&\quad + \frac{1}{p'_{2E}}\left\{-\frac{\partial f_{2E}}{\partial t} - w\frac{\partial f_{2E}}{\partial z} - f_{2E}\frac{\partial a'_w}{\partial w} - a'_w\frac{\partial f_{2E}}{\partial w} + \frac{C_0\varepsilon}{2}\frac{\partial^2 f_{2E}}{\partial w^2}\right\} \\
&= -\frac{C_0\varepsilon}{2\sigma^2_{u/w}}(u-\rho w) + \frac{f_{2E}}{p'_{2E}}\left\{-\frac{1}{2\sigma_w^2}\frac{\partial \sigma_w^2}{\partial t}\left(\frac{w^2}{\sigma_w^2}-1\right) - \Psi(\xi)\frac{\partial \xi}{\partial t}\right. \\
&\quad - w\left[\frac{1}{2\sigma_w^2}\frac{\partial \sigma_w^2}{\partial z}\left(\frac{w^2}{\sigma_w^2}-1\right) + \Psi(\xi)\frac{\partial \xi}{\partial z}\right] - \frac{\partial a'_w}{\partial w} - a'_w\left[-\frac{w}{\sigma_w^2} - \Psi(\xi)\eta\right] \\
&\quad \left. + \frac{C_0\varepsilon}{2}\left[\left(-\frac{w}{\sigma_w^2} - \Psi(\xi)\eta\right)^2 - \frac{1}{\sigma_w^2} + \eta^2\dot{\Psi}(\xi)\right]\right\}. \qquad \text{(A1)}
\end{aligned}$$

Since

$$\frac{f_{2E}\Psi}{p'_{2E}} = \sigma_{u/w}, \quad \dot{\Psi}(\xi) = -\Psi(\xi)(\xi + \Psi(\xi)),$$

we find from (A1)

$$\begin{aligned}
a'_u(t,z,u,w) &= -\frac{C_0\varepsilon(1+\rho^2)}{2\sigma^2_{u/w}}(u-\rho w) + \frac{\rho}{2\sigma^2_w}\left(C_0\varepsilon + \frac{\partial\sigma^2_w}{\partial t}\right) w \\
&\quad + \frac{\rho}{2}\frac{\partial\sigma^2_w}{\partial z}\left(\frac{w^2}{\sigma^2_w}+1\right) - \sigma_{u/w}\left(\frac{\partial\xi}{\partial t} + w\frac{\partial\xi}{\partial z}\right) \\
&= -\frac{C_0\varepsilon(1+\rho^2)}{2\sigma^2_{u/w}}(u-\rho w) + \frac{\rho}{2\sigma^2_w}\left(C_0\varepsilon + \frac{\partial\sigma^2_w}{\partial t}\right) w + \frac{\rho}{2}\frac{\partial\sigma^2_w}{\partial z}\left(\frac{w^2}{\sigma^2_w}+1\right) \\
&\quad - \left(\frac{\partial\rho}{\partial t} + w\frac{\partial\rho}{\partial z}\right) w - \frac{1}{\sigma_{u/w}}\left(\frac{\partial\sigma_{u/w}}{\partial t} + w\frac{\partial\sigma_{u/w}}{\partial z}\right)(u-\rho w),
\end{aligned}$$

Since for the case considered the condition (10.19) is satisfied, we use here the expression (10.20). Substituting

$$\frac{1}{p_{vE}}\frac{\partial p_{vE}}{\partial v} = -\frac{v}{\sigma^2_v}, \ \frac{1}{p_{vE}}\frac{\partial f_{vE}}{\partial t} = -\frac{v}{\sigma_v}\frac{\partial\sigma_v}{\partial t}, \ \frac{1}{p_{vE}}\frac{\partial f_{vE}}{\partial z} = -\frac{v}{\sigma_v}\frac{\partial\sigma_v}{\partial z}$$

into (10.20), we get

$$a'_v(t,z,u,v,w) = -\left(\frac{C_0\varepsilon}{2\sigma^2_v} - \frac{1}{\sigma_v}\frac{\partial\sigma_v}{\partial t}\right) v + \frac{1}{2}\frac{\partial\sigma^2_v}{\partial z}\frac{vw}{\sigma^2_v}.$$

10.7.2 Appendix B. Relation to other models

2-parameter class of models due to Reynolds

Here we analyse Reynolds' 2-parameter class of models in the case of horizontally homogeneous turbulence with the mean velocity direction not varying with height. It is also assumed that the X-axis is oriented along the mean velocity vector, and that the crosswind velocity fluctuations are symmetric with respect to the plane XZ. Then the 2-parameter class of models quadratic in velocity, which satisfies the well-mixed condition for Gaussian p_E, considered by Reynolds (1997), reads

$$\begin{aligned}
&dX_1 = (U'_1 + \bar{u}(X_3,t))dt, \ dX_2 = U'_2 dt, \ dX_3 = U'_3 dt, \\
&dU'_1 = a'_1(t, X_3, U'_1, U'_2, U'_3)dt + \sqrt{C_0\varepsilon}\, dB_1(t), \\
&dU'_2 = a'_2(t, X_3, U'_1, U'_2, U'_3)dt + \sqrt{C_0\varepsilon}\, dB_2(t), \qquad \text{(B1)} \\
&dU'_3 = a'_3(t, X_3, U'_1, U'_2, U'_3)dt + \sqrt{C_0\varepsilon}\, dB_3(t),
\end{aligned}$$

where

$$
\begin{aligned}
a_i'(t,z,u_1,u_2,u_3) = & -\frac{C_0\varepsilon}{2}\lambda_{ij}u_j + \frac{1}{2}\frac{\partial\sigma_{i3}}{\partial z} + \frac{1}{2}C_2\Big(\frac{\partial\sigma_{i3}}{\partial z} + \sigma_{i3}\sigma_{km}\frac{\partial\lambda_{km}}{\partial z}\Big) \\
& - \frac{1}{2}\sigma_{im}\frac{\partial\lambda_{jm}}{\partial t}u_j + C_1\Big(\sigma_{i3}\frac{\partial\bar{u}}{\partial z}\lambda_{j1} - \delta_{1i}\delta_{3j}\frac{\partial\bar{u}}{\partial z}\Big)u_j \\
& - \frac{1}{2}C_2\sigma_{i3}\frac{\partial\lambda_{jk}}{\partial z}u_ju_k - \frac{1}{2}(1-C_2)\sigma_{im}\frac{\partial\lambda_{km}}{\partial z}u_3u_k, \\
& i = 1,2,3. \qquad \text{(B2)}
\end{aligned}
$$

Here we adopt the summation convention, therefore the notation $(X,Y,Z) = (X_1,X_2,X_3)$ and $(U,V,W) = (U_1,U_2,U_3)$ is used; δ_{ij} is the Kronecker symbol, $\sigma_{ij} = (\lambda^{-1})_{ij}$ are the velocity covariances which in the case considered have the form

$$\sigma_{11} = \sigma_u^2, \quad \sigma_{22} = \sigma_v^2, \quad \sigma_3 = \sigma_w^2, \quad \sigma_{12} = \sigma_{21} = \sigma_{23} = \sigma_{32} = 0,$$

$$\sigma_{13} = \sigma_{31} = \overline{uw}; \quad \lambda_{11} = \frac{\sigma_w^2}{\Delta}, \quad \lambda_{22} = \frac{1}{\sigma_v^2}, \quad \lambda_{33} = \frac{\sigma_u^2}{\Delta},$$

$$\lambda_{12} = \lambda_{21} = \lambda_{23} = \lambda_{32} = 0, \quad \lambda_{13} = \lambda_{31} = -\frac{\overline{uw}}{\Delta},$$

where σ_u^2, σ_v^2, and σ_w^2 are the variances of velocity components, $\Delta = \sigma_u^2\sigma_w^2 - (\overline{uw})^2$.

Thus the model includes two free parameters C_1 and C_2. Reynolds (1998) has suggested $C_1 = \pm 3$, and $C_2 = 0$ to fit the experimental results for the wind tunnel boundary layer by Legg (1983).

It is interesting to see whether or not there are some values of C_1, C_2 such that the model (B2) reduces to our model (10.22)–(10.24). To this end, it is sufficient to check if the model (B2) satisfies the assumption of our model (see Section 2.1). It is clear that point (ii) of the assumption is satisfied iff $C_2 = 0$, since in the expression for a_1', the dependence on u_2 can be eliminated only if $C_2 = 0$. Thus, taking $C_2 = 0$, we analyse point (i) of the assumption. In the expression for $a_3'(t,z,u_1,u_2,u_3)$ we are interested in the terms depending on u_1 and u_2, which ee therefore we write as

$$a_3' = \{\cdots\} + C_1\sigma_{33}\frac{\partial\bar{u}}{\partial z}\lambda_{11}u_1 - \frac{1}{2}C_0\varepsilon\lambda_{31}u_1 - \frac{1}{2}\sigma_{3m}\frac{\partial\lambda_{1m}}{\partial z}u_3u_1,$$

where $\{\cdots\}$ stands for the terms not depending on u_1 and u_2. From this we see that if the term $\frac{\partial\lambda_{1m}}{\partial z}$ is not equal to zero, then point (i) cannot be satisfied. Note that this term is zero in the ideally-neutral stratification ($L = \infty$). In this case (i) is satisfied, iff $C_1 = C_0u_*^4/(2\sigma_w^4)$.

From this we conclude that only in the case of ideally-neutral stratification does our model belong to the class of models (B1)-(B2), and, for this case, the values of C_1 and C_2 corresponding to our model are $C_2 = 0$, and $C_1 = C_0u_*^4/(2\sigma_w^4)$.

The "zero-spin" property

Here we show that in our model (for simplicity we consider the stationary turbulence) the average increment $\langle d\theta; z\rangle$ to the orientation $\theta = \arctan(w'/u')$ of the Lagrangian velocity fluctuation vector in 2D case is negative, and hence it does not belong to the "zero-spin" class of models.

Flesch and Wilson [56] show that

$$\langle d\theta; z\rangle = dt \int_{-\infty}^{\infty}\int_{-\infty}^{\infty} \frac{u\phi'_w - w\phi'_u}{u^2 + w^2}\, du\, dw,$$

where

$$\frac{\phi'_u}{p'_E} = a'_u - \frac{C_0\varepsilon}{2}\frac{\partial \ln p'_E}{\partial u}, \qquad \frac{\phi'_w}{p'_E} = a'_w - \frac{C_0\varepsilon}{2}\frac{\partial \ln p'_E}{\partial w}.$$

For our model (see (10.25)) we find

$$\begin{aligned}\frac{\phi'_u}{p'_E} &= w^2\frac{d\rho}{dz} + \frac{1}{\sigma}\frac{d\sigma}{dz}w(u - \rho w) - \frac{C_0\varepsilon\rho^2}{2\sigma^2}(u - \rho w)\\ &\quad + \frac{C_0\varepsilon\rho}{2\sigma_w^2}w + \frac{\rho}{2}\frac{d\sigma_w^2}{dz}\left(1 + \frac{w^2}{\sigma_w^2}\right),\\ \frac{\phi'_w}{p'_E} &= \frac{1}{2}\frac{d\sigma_w^2}{dz}\left(1 + \frac{w^2}{\sigma_w^2}\right) - \frac{C_0\varepsilon\rho}{2\sigma^2}(u - \rho w),\end{aligned}$$

where we use the notation $\sigma = \sigma_{u/w}$. After some algebra we can find that

$$\langle d\theta; z\rangle = -\,dt\,\frac{C_0\varepsilon\rho}{2}\int_{-\infty}^{\infty}\int_{-\infty}^{\infty}\frac{p'_E}{u^2 + w^2}\left(\frac{u^2}{\sigma^2} + \frac{\rho^2 w^2}{\sigma^2} + \frac{w^2}{\sigma_w^2}\right) du\, dw > 0.$$

Chapter 11

Analysis of the relative dispersion of two particles by Lagrangian stochastic models and DNS methods

Stochastic Lagrangian models for the relative dispersion of two particles in a stationary, spatially isotropic incompressible, fully developed turbulent flow are studied. Along with a review of existing models, we suggest new models and compare the results with the DNS data.

11.1 Introduction

Turbulent dispersion of a contaminant, for example pollutant dispersion in the atmosphere, is conveniently described in terms of Lagrangian statistics sampled along the paths of fluid particles. In practice, however, the Eulerian statistics sampled at fixed points in space are better known from experiments. Hence the basic problem of turbulent dispersion is to calculate Lagrangian statistics from given Eulerian statistics. Lagrangian stochastic (LS) models of turbulent dispersion address the problem by statistically characterizing particle paths from an Eulerian input. LS models formulate the time evolution of the particle coordinate and velocity in terms of stochastic differential equations. The LS models are best understood for the description of 1-particle statistics, which contain only 1-point statistical information and hence lack any information about the different scales of turbulent eddies. A modeled ensemble of single particles allows the calculation of mean concentrations, whereas an ensemble of particle pairs allows the calculation of concentration fluctuations. When we consider the motion of a pair particles, the modeling can be seen as the superposition of a relative motion and the motion of single particle, or particle centroid (e. g., see [44, 199, 238]). The relative motion reflects more directly the internal turbulent structure because of the appearance of an internal lengths (particle distance), and its description permits the introduction of concepts developed within the theory of turbulence [146].

To be more specific, we introduce the following notations. Throughout the chapter, we use the summation convention. The Eulerian velocity field is considered as a 3D random field denoted by $\mathbf{U}_E(\mathbf{x}, t) = (U_{E1}(\mathbf{x}, t), U_{E2}(\mathbf{x}, t), U_{E3}(\mathbf{x}, t))$, whose samples are incompressible : $\frac{\partial}{\partial x_i} U_{Ei}(\mathbf{x}, t) = 0$. The concentration of a conservative passive scalar scattered by this field is governed by the transport equation

$$\frac{\partial c(\mathbf{x}, t)}{\partial t} + U_{Ei}(\mathbf{x}, t) \frac{\partial c}{\partial x_i} = 0, \quad t \geq 0,$$

$$c(\mathbf{x}, 0) = S(\mathbf{x})$$

where $S(\mathbf{x})$ is the initial distribution of the concentration.

In practice, the following quantities are of special interest: $\langle c(\mathbf{x}, t)\rangle$, the mean concentration, $\langle U_{Ei}(\mathbf{x}, t)c(\mathbf{x}, t)\rangle$, the mean fluxes of concentration, and $\langle c(\mathbf{x}, t)c(\mathbf{x}', t)\rangle$, the concentration covariance.

There are two main approaches to evaluating these quantities. The first one is based on the averaging of the transport equation to extract closed equations for the quantities in question. This is the so-called closure problem, which faces well-known difficulties [145].

The second approach is based on the Lagrangian description, where the following representations are used:

$$\begin{aligned}
\langle c(\mathbf{x}, t)\rangle &= \int d\mathbf{x}_0\, S(\mathbf{x}_0)\, p_{1L}(\mathbf{x}, t; \mathbf{x}_0), \\
\langle U_{Ei}(\mathbf{x}, t)c(\mathbf{x}, t)\rangle &= \int d\vec{v} \int d\mathbf{x}_0\, v_i\, S(\mathbf{x}_0)\, p_{1L}(\vec{v}, \mathbf{x}, t; \mathbf{x}_0), \\
\langle c(\mathbf{x}, t)c(\mathbf{x}', t)\rangle &= \int d\mathbf{x}_0 \int d\mathbf{x}_0'\, S(\mathbf{x}_0)\, S(\mathbf{x}_0')\, p_{2L}(\mathbf{x}, \mathbf{x}', t; \mathbf{x}_0, \mathbf{x}_0').
\end{aligned}$$

Here p_{1L} and p_{2L} are the Lagrangian transition densities:

$$\begin{aligned}
p_{1L}(\mathbf{x}, t; \mathbf{x}_0) &= \langle \delta(\mathbf{x} - \mathbf{X}(t; \mathbf{x}_0))\rangle, \\
p_{1L}(\vec{v}, \mathbf{x}, t; \mathbf{x}_0) &= \langle \delta(\mathbf{x} - \mathbf{X}(t; \mathbf{x}_0))\delta(\vec{v} - \mathbf{V}(t; \mathbf{x}_0))\rangle, \\
p_{2L}(\mathbf{x}, \mathbf{x}', t; \mathbf{x}_0, \mathbf{x}_0') &= \langle \delta(\mathbf{x} - \mathbf{X}(t; \mathbf{x}_0))\delta(\mathbf{x}' - \mathbf{X}(t; \mathbf{x}_0'))\rangle.
\end{aligned}$$

In these formula, the Lagrangian variables $\mathbf{X}, \mathbf{V}$ are defined through

$$\frac{\partial \mathbf{X}}{\partial t} = \mathbf{V}(t; \mathbf{x}_0) = \mathbf{U}_E(t, \mathbf{X}(t; \mathbf{x}_0)), \quad \mathbf{X}(0; \mathbf{x}_0) = \mathbf{x}_0.$$

In this chapter we focus on the 2-particle models which describe the motion of two fluid particles. It is governed by

$$\frac{d\mathbf{X}_1(t)}{dt} = \mathbf{U}_E(t, \mathbf{X}_1), \qquad \frac{d\mathbf{X}_2(t)}{dt} = \mathbf{U}_E(t, \mathbf{X}_2),$$

where $\mathbf{X}_1(t) = \mathbf{X}(t; \mathbf{x}_0)$ and $\mathbf{X}_2(t) = \mathbf{X}(t; \mathbf{x}_0')$.

It is convenient to rewrite this system as

$$\begin{aligned}
\frac{d\mathbf{R}}{dt} &= \frac{1}{2}\left\{\mathbf{U}_E\left(\mathbf{R} + \frac{\mathbf{r}}{2}, t\right) + \mathbf{U}_E\left(\mathbf{R} - \frac{\mathbf{r}}{2}, t\right)\right\}, \\
\frac{d\mathbf{r}}{dt} &= \mathbf{U}_E\left(\mathbf{R} + \frac{\mathbf{r}}{2}, t\right) - \mathbf{U}_E\left(\mathbf{R} - \frac{\mathbf{r}}{2}, t\right),
\end{aligned} \tag{11.1}$$

where

$$\mathbf{R}(t) = \frac{1}{2}(\mathbf{X}_1(t) + \mathbf{X}_2(t)), \quad \mathbf{r}(t) = \mathbf{X}_1(t) - \mathbf{X}_2(t).$$

This form clearly illustrates the division of large and small scales of turbulence. Indeed if the distance between the two particles is much less than the external scale ($r \ll L$), then the terms

$$\mathbf{U}_E\left(\mathbf{R} + \frac{\mathbf{r}}{2}, t\right) + \mathbf{U}_E\left(\mathbf{R} - \frac{\mathbf{r}}{2}, t\right)$$

and

$$\mathbf{U}_E\left(\mathbf{R} + \frac{\mathbf{r}}{2}, t\right) - \mathbf{U}_E\left(\mathbf{R} - \frac{\mathbf{r}}{2}, t\right)$$

are approximately statistically independent.

This property can be used to simulate the motion of two particles in small scales according to

$$\frac{d\mathbf{R}}{dt} = \frac{1}{2}\left\{\tilde{\mathbf{U}}\left(\mathbf{R} + \frac{\mathbf{r}}{2}, t\right) + \tilde{\mathbf{U}}\left(\mathbf{R} - \frac{\mathbf{r}}{2}, t\right)\right\},$$

$$dr_i(t) = v_i(t)\,dt, \quad dv_i(t) = a_i(\mathbf{r}, \vec{v}, t)\,dt + b_{ij}(\mathbf{r}, \vec{v}, t)\,dW_j(t), \quad i = 1, 2, 3.$$

Here the large scale velocity field is approximated by a field marked by the tilde (e. g., extracted from DNS or LES methods), while the small scale motion is described as a diffusion process governed by a Langevin type equation.

Two alternative modelling approaches include Eulerian statistics in LS models: the *well-mixed* approach of Thomson [237] and the *moments approximation* method of Novikov [155]. The importance of Thomson's approach is that when the material distribution is uniform, the model does not artificially unmix the material.

A one dimensional well-mixed LS model of relative dispersion of two particles has been proposed by Thomson [236]. 3-dimensional models based on well-mixed criteria have been considered in Thomson [15, 238]. Gaussian Eulerian statistics are used in these articles. Well-mixed quasi-1-dimensional (Q1D) models of relative dispersion of two particles in the case of arbitrary Eulerian statistics were considered in our papers [108, 114]. Recall that the Q1D model of relative dispersion is aimed at describing time evolution of the distance and longitudinal component of relative velocity between two particles. A 3-dimensional well-mixed model of relative dispersion consistent with arbitrary Eulerian statistics was proposed in [110].

LS models of relative dispersion based on the moments approximation approach have been proposed by Novikov [155], and have been developed in [77, 163, 164].

11.2 Basic assumptions

We deal here with the process of relative dispersion of a pair of fluid particles in a stationary, spatially isotropic incompressible fully developed turbulent flow.

We introduce the Eulerian velocity difference $\vec{u}_E(\mathbf{r}) = \mathbf{U}_E(\mathbf{x} + \mathbf{r}, t) - \mathbf{U}_E(\mathbf{x}, t)$ considered at two fixed points separated by vector $\mathbf{r}$.

11.2.1 Markov assumption

Let $(\mathbf{r}(t), \vec{v}(t))$ be the Lagrangian variables for the separation vector and the relative velocity between two fluid particles. It is usually assumed that $(\mathbf{r}(t), \vec{v}(t))$ is a 6D (continuous) Markov process (i. e., given the values of $\mathbf{r}(t)$ and $\vec{v}(t)$ at time t, the values at time greater than t are independent of the values at times less than t). Under the Markov assumption for $(\mathbf{r}(t), \vec{v}(t))$ the most general equation used to describe the time evolution of $(\mathbf{r}(t), \vec{v}(t))$ is the Ito-type stochastic differential equation:

$$dr_i = v_i\, dt, \quad dv_i = a_i(\mathbf{r}, \vec{v})\, dt + b_{ij}(\mathbf{r}, \vec{v})\, dW_j(t), \quad i = 1, 2, 3. \tag{11.2}$$

The main problem here is the following: how can we determine the functions $a_i(\mathbf{r}, \vec{v})$ and $b_{ij}(\mathbf{r}, \vec{v})$, (called *the drift* and *diffusion* terms, respectively) so that the model random process (11.2) is in a sense close to the true process (11.1)? To this end, one uses two consistency principles: (i) consistency with Kolmogorov's similarity theory and (ii) consistency with Thomson's well-mixed condition.

11.2.2 Consistency with the second Kolmogorov similarity hypothesis

Given r, assume that there exists τ such that

$$\tau_\eta \ll \tau \ll \frac{r^{2/3}}{\bar{\varepsilon}^{1/3}}. \tag{11.3}$$

The consistency with the second Kolmogorov similarity hypothesis requires that

$$b_{ij}(\mathbf{r}, \vec{v}) = (2\, C_0\, \bar{\varepsilon})^{1/2}\, \delta_{ij}. \tag{11.4}$$

Here ε is the mean rate of the dissipation of turbulence energy, and C_0 is a universal constant in the Lagrangian structure function.

Some comments: let us denote by $\mathbf{V}_1(t) = (V_{11}(t), V_{12}(t), V_{13}(t))$ and $\mathbf{V}_2(t) = (V_{21}(t), V_{22}(t), V_{23}(t))$ the Lagrangian velocity of the first and the second particles, respectively. Then,

- $$\begin{aligned}\Delta_\tau v_i(t) \equiv v_i(t + \tau) - v_i(t) &= V_{2i}(t + \tau) - V_{2i}(t) - (V_{1i}(t + \tau) - V_{1i}(t)) \\ &= \Delta_\tau V_{2i}(t) - \Delta_\tau V_{1i}(t).\end{aligned}$$

In view of (11.3), the quantities

$$\vec{v}(t) = \mathbf{V}_2(t) - \mathbf{V}_1(t), \quad \Delta_\tau \mathbf{V}_2(t), \quad \Delta_\tau \mathbf{V}_1(t)$$

are approximately mutually independent. Therefore

$$\begin{aligned}
&\langle \Delta_\tau v_i(t)\Delta_\tau v_j(t) \mid \vec{v}(t) = \vec{v},\ \mathbf{r}(t) = \mathbf{r}\rangle \\
&\quad = \langle \{\Delta_\tau V_{2i}(t) - \Delta_\tau V_{1i}(t)\}\{\Delta_\tau V_{2j}(t) - \Delta_\tau V_{1j}(t)\} \mid \vec{v}(t) = \vec{v},\ \mathbf{r}(t) = \mathbf{r}\rangle \\
&\qquad\qquad \simeq \langle \Delta_\tau V_{2i}(t)\Delta_\tau V_{2j}(t)\rangle + \langle \Delta_\tau V_{1i}(t)\Delta_\tau V_{1j}(t)\rangle \simeq 2\, C_0\, \bar{\varepsilon}\delta_{ij}\, \tau,
\end{aligned}$$

which implies (11.4).

11.2.3 Thomson's well-mixed condition

The following relation between the true Eulerian and Lagrangian p. d. f.'s are known [153]:

$$p_E(\vec{v};\mathbf{r},t) = \int p_L(\mathbf{r},\vec{v};t,\mathbf{r}_0)\, d\mathbf{r}_0,$$

where

$$\begin{aligned}
p_E(\vec{v};\mathbf{r},t) &= \langle \delta(\vec{v} - \vec{u}_E(\mathbf{r}))\rangle, \\
p_L(\mathbf{r},\vec{v};t,\mathbf{r}_0) &= \langle \delta(\mathbf{r} - \mathbf{r}(t))\delta(\vec{v} - \vec{v}(t))\rangle.
\end{aligned}$$

The model is considered consistent with Novikov's integral relation if its p. d. f. also satisfies such a relation. It is well known that this leads to Thomson's well-mixed condition written in the form [237]

$$v_i \frac{\partial p_E}{\partial r_i} + \frac{\partial}{\partial v_i}(a_i p_E) = C_0 \bar{\varepsilon} \frac{\partial^2 p_E}{\partial v_i \partial v_i}.$$

It should be noted that all this does not define the model uniquely.

11.3 Well-mixed Lagrangian stochastic models

Note that in the case of isotropic turbulence the structure of the drift term is defined by two scalar functions. Indeed,

$$a_i(\mathbf{r},\vec{v}) = \varphi(r, v_\parallel, v_\perp)\frac{r_i}{r} + \psi(r, v_\parallel, v_\perp)\frac{v_i}{v},$$

where

$$r = (r_i r_i)^{1/2}, \quad v_\parallel = v_i r_i / r, \quad v_\perp = (v_i\, v_i - v_\parallel^2)^{1/2}.$$

It should be noted that if we could find an additional relation between the functions φ and ψ, then the well-mixed condition would provide a unique choice of the drift term. For instance, this is the case when $\varphi \equiv 0$ (e. g., see the 1-particle model treated in [147], or $\psi \equiv 0$ [114]. But generally, since such relations are not known, different approaches can be used to extract the unique model. We present below two such approaches. In the first one, the drift term is assumed to be quadratic in velocity, and the Eulerian p. d. f. p_E is Gaussian ([15, 238]. The second approach is based on a Markovian character of the evolution of the 2D process $r(t), u_r(t)$ where $r(t)$ is the distance between the two particles, and $u_r(t)$ is the longitudinal component of the relative velocity [108, 114].

11.3.1 Quadratic-form models

Following [15] let us assume that p. d. f. p_E is Gaussian:

$$p_E(\vec{u}, \mathbf{r}) = \frac{\lambda^{1/2}}{(2\pi)^{3/2}} \exp\left(-\frac{1}{2}\lambda_{ij} u_i u_j \right),$$

where

$$\lambda_{ij}^{-1}(\mathbf{r}) = \langle u_{Ei}(\mathbf{r}) u_{Ej}(\mathbf{r}) \rangle, \quad \lambda = det(\lambda_{ij}), \quad \lambda_{ik}\lambda_{kj}^{-1} = \delta_{ij},$$
$$u_{Ei}(\mathbf{r}) = U_{Ei}(\mathbf{x} + \mathbf{r}) - U_{Ei}(\mathbf{x}).$$

The drift term is constructed with the form

$$a_i = \varphi_i + C_0\, \bar{\varepsilon}\, \frac{1}{p_E} \frac{\partial p_E}{\partial u_i},$$

with

$$\varphi_i = \Gamma_i + \gamma_{ijk}\, u_j u_k.$$

As shown in [15], the well-mixed condition is satisfied for the following three cases:

$$\gamma_{ijk} = -\frac{1}{2}\lambda_{il}^{-1} \frac{\partial \lambda_{jk}}{\partial r_l},$$
$$\gamma_{ijk} = -\frac{1}{2}\lambda_{il}^{-1} \frac{\partial \lambda_{ik}}{\partial r_j} = \frac{1}{2}\lambda_{kl} \frac{\partial \lambda_{il}^{-1}}{\partial r_j},$$
$$\gamma_{ijk} = -\frac{1}{2}\lambda_{il}^{-1} \frac{\partial \lambda_{lj}}{\partial r_k} = \frac{1}{2}\lambda_{jl} \frac{\partial \lambda_{il}^{-1}}{\partial r_k},$$

where

$$\Gamma_i = \lambda_{ij}^{-1} \left(-\frac{1}{2\lambda} \frac{\partial \lambda}{\partial r_j} + \gamma_{kkj} + \gamma_{kjk} \right).$$

Thus this approach gives a specific structure of the drift term, but it also does not provide a unique solution. Let us now consider the second approach.

11.3.2 Quasi-1-dimensional models

The motion of two particles is described in the Lagrangian framework by the distance $r(t)$ and the longitudinal velocity component defined here by $u_r(t)$:

$$r(t) = (r_i(t)r_i(t))^{1/2}, \quad u_r(t) = v_i(t)r_i(t)/r(t).$$

Here $v_i(t)$ is the i-th component of the relative velocity $\vec{v}(t)$, and $\mathbf{r}(t)$ is the separation vector.

Now we formulate the main assumption:
Assume that $(r(t), u_r(t))$ is a continuous 2D Markov process:

$$dr = u_r\, dt, \quad du_r = X(r, u_r)\, dt + (2C_0\bar{\varepsilon})^{1/2} dW(t). \tag{11.5}$$

where $X(r, u_r)$ is an unknown drift term. In the case of quasi-1-dimensional model the well-mixed condition reads [114]

$$v\frac{\partial r^2 p_E^{\|}(v,r)}{\partial r} + r^2\frac{\partial (X(r,v)p_E^{\|})}{\partial v} = C_0\bar{\varepsilon}r^2\frac{\partial^2 p_E^{\|}}{\partial v^2}, \tag{11.6}$$

where $p_E^{\|}(v,r) = \langle \delta(v - u_{Er})\rangle$ is the p. d. f. of u_{Er}, and the longitudinal component of the Eulerian velocity difference $\vec{u}_E(\mathbf{r}) = \mathbf{U}_E(\mathbf{x}+\mathbf{r}) - \mathbf{U}_E(\mathbf{x})$, i. e.,

$$u_{Er} = (U_{Ei}(\mathbf{x}+\mathbf{r}) - U_{Ei}(\mathbf{x}))\,\frac{r_i}{r}.$$

Assuming $Xp_E^{\|}\,|_{|v|\to\infty} = 0$, it is easy get

$$X(r,v) = C_0\bar{\varepsilon}\frac{\partial}{\partial v}\ln p_E^{\|}(v,r) - \frac{1}{p_E^{\|}(v,r)}\int_{-\infty}^{v}\frac{v'}{r^2}\frac{\partial}{\partial r}\left(r^2 p_E^{\|}(v',r)\right)dv'.$$

In the inertial subrange ($\eta \ll r \ll L$) this expression can be considerably simplified. In the inertial subrange the unique external parameter is $\bar{\varepsilon}$, so it is possible to turn to a nondimensional density $f_E(\xi)$:

$$p_E^{\|}(v,r) = \frac{1}{(\bar{\varepsilon}r)^{1/3}} f_E\left(\frac{v}{(\bar{\varepsilon}r)^{1/3}}\right).$$

Consequently,

$$X(r,v) = \frac{\bar{\varepsilon}^{2/3}}{r^{1/3}}\tilde{X}\left(\frac{v}{(\bar{\varepsilon}r)^{1/3}}\right),$$

$$\tilde{X}(\xi) = C_0\frac{d\ln f_E}{d\xi} + \frac{\xi^2}{3} - \frac{\frac{7}{3}\int_{-\infty}^{\xi}\xi' f_E(\xi')\,d\xi'}{f_E(\xi)}.$$

It is convenient to use the equation in the nondimensional form $\xi_r(t) = u_r(t)/(\bar{\varepsilon}r)^{1/3}$:

$$dr = (\bar{\varepsilon}r)^{1/3}\xi_r\,dt, \quad d\xi_r = \left(\frac{\bar{\varepsilon}}{r^2}\right)^{1/3} a_0(\xi_r)dt + \left(\frac{\bar{\varepsilon}}{r^2}\right)^{1/6}\sqrt{2C_0}dW(t),$$

where

$$a_0(\xi) = \tilde{X}(\xi) - \frac{\xi^2}{3} = C_0\frac{d\ln f_E}{d\xi} - \frac{\frac{7}{3}\int_{-\infty}^{\xi}\xi' f_E(\xi')\,d\xi'}{f_E(\xi)}.$$

In the case of Gaussian p. d. f. f_E

$$f_E(\xi) = \frac{1}{(2\pi C)^{1/2}}\exp\left(-\frac{\xi^2}{2C}\right),$$

the coefficient a_0 has the most simple form:

$$a_0(\xi) = -\frac{C_0}{C}\xi + \frac{7}{3}C.$$

Here C is the Kolmogorov universal constant defined by $\langle u_{Er}^2\rangle = C(\bar{\varepsilon}\,r)^{2/3}$.

11.3.3 3-dimensional extension of Q1D models

Assume that the turbulence is isotropic and stationary. Let us consider a 3D model of relative dispersion in the subrange $\eta \ll r$:

$$dr_i = v_i\,dt, \quad dv_i = a_i(\mathbf{r},\vec{v})\,dt + (C_0\bar{\varepsilon})^{1/2}\,dW_i(t), \quad i = 1,2,3,$$

where

$$a_i(\mathbf{r},\vec{v}) = \varphi(r, v_\parallel, v_\perp)\frac{r_i}{r} + \psi(r, v_\parallel, v_\perp)\frac{v_i}{v}$$

with unknown φ and ψ.

We derive from it the Q1D model [110]:

$$dr = u_r\,dt, \quad du_r = \left(\frac{a_i r_i}{r} + \frac{u_\perp^2}{r}\right)dt + (C_0\bar{\varepsilon})^{1/2}\,dW(t),$$

where $u_\perp^2(t) = u^2(t) - u_r^2(t)$. Assuming that Q1D process $(r(t), v_r(t))$ is Markovian, we can write

$$\frac{a_i r_i}{r} + \frac{v_\perp^2}{r} = \varphi(r, v_\parallel, v_\perp) + \frac{v_\parallel}{v}\psi(r, v_\parallel, v_\perp) + \frac{v_\perp^2}{r} = X(r, v_\parallel).$$

From the last relation and the 3D well-mixed condition it follows that

$$\psi(r, v_\parallel, v_\perp) = C_0\bar{\varepsilon}\frac{v}{v_\perp}\frac{\partial}{\partial v_\perp}\ln p_E(v_\parallel, v_\perp; r) + \frac{v_\parallel v}{r}$$
$$- \frac{v}{v_\perp^2 p_E(v_\parallel, v_\perp; r)}\int_0^{v_\perp}\left\{\frac{\partial}{\partial v_\parallel}(Xp_E) + \frac{v_\parallel}{r^2}\frac{\partial}{\partial r}r^2 p_E(v_\parallel, v_\perp'; r)\right\}v_\perp'\,dv_\perp',$$

$$\varphi(r, v_\parallel, v_\perp) = X(r, v_\parallel) - \frac{v_\perp^2}{r} - \frac{v_\parallel}{v}\psi(r, v_\parallel, v_\perp).$$

This yields in the case of Gaussian p. d. f. p_E, for the inertial subrange $\eta \ll r \ll L$

$$\begin{aligned} a_i(\mathbf{r}, \vec{v}) = {} & \frac{\bar{\varepsilon}^{2/3}}{r^{1/3}}\Big[\frac{7}{3}C - \frac{C_0}{4C}\frac{v_\parallel}{(\bar{\varepsilon}r)^{1/3}} - \frac{v^2}{(\bar{\varepsilon}r)^{2/3}}\Big]\frac{r_i}{r} \\ & + \frac{\bar{\varepsilon}^{2/3}}{r^{1/3}}\Big[\frac{4}{3}\frac{v_\parallel}{(\bar{\varepsilon}r)^{1/3}} - \frac{C_0}{C'}\Big]\frac{v_i}{(\bar{\varepsilon}r)^{1/3}}, \end{aligned}$$

where $C' = \frac{4}{3}C$.

11.4 Stochastic Lagrangian models based on the moments approximation method

The evaluation of the p. d. f. $p_E(\vec{v}; \mathbf{r})$ is a very difficult problem, because it needs to construct a family of solutions to Navier–Stokes equations. Therefore, in practice one uses the method of moments: one constructs an approximation to the Eulerian p. d. f. under the condition that its first several moments coincide with those of the true velocity moments.

The true moments can be found via DNS method solving the Navier–Stokes equation, or extracted from experiments.

11.4.1 Moments approximation conditions

In a more general case, when the intermittency is taken into account, the model has the form

$$dr_i = v_i\,dt, \quad dv_i = a_i(\mathbf{r}, \vec{v})\,dt + b_{ij}(\mathbf{r}, \vec{v})\,dW_j(t), \quad i = 1, 2, 3.$$

The well-mixed condition reads in this case (e. g., see [155, 237])

$$v_i\frac{\partial p_E}{\partial r_i} + \frac{\partial}{\partial v_i}(a_i p_E) = \frac{1}{2}\frac{\partial^2 m_{ij} p_E}{\partial v_i \partial v_j}, \tag{11.7}$$

where $m_{ij} = b_{ik}b_{jk}$. Multiplying this equation by v_j and integrating over $\vec{v}$, we get [164]

$$\frac{\partial}{\partial r_i}\langle u_{Ei}(\mathbf{r})u_{Ej}(\mathbf{r})\rangle = \langle a_j(\mathbf{r}, \vec{u}_E(\mathbf{r}))\rangle,$$

i. e., (due to incompressibility)

$$\langle a_j(\mathbf{r}, \vec{u}_E(\mathbf{r}))\rangle = 0, \quad j = 1, 2, 3.$$

Multiplying (11.7) by $v_j v_k$ and integrating over $\vec{v}$ yields

$$\frac{\partial}{\partial r_i}\langle u_{Ei}(\mathbf{r})u_{Ej}(\mathbf{r})u_{Ek}(\mathbf{r})\rangle$$
$$= \langle a_j(\mathbf{r},\vec{u}_E(\mathbf{r}))\,u_{Ek}(\mathbf{r}) + a_k(\mathbf{r},\vec{u}_E(\mathbf{r}))\,u_{Ej}(\mathbf{r})\rangle + \langle m_{jk}(\mathbf{r},\vec{u}_E(\mathbf{r}))\rangle$$

for $j,k = 1,2,3$.

Kolmogorov's relation for third-order moments in the inertial subrange (e. g., see [146, 155]) is

$$\langle u_{Ei}(\mathbf{r})u_{Ej}(\mathbf{r})u_{Ek}(\mathbf{r})\rangle = -\frac{4}{15}\bar{\varepsilon}(r_i\delta_{jk} + r_j\delta_{ik} + r_k\delta_{ij}), \quad \eta \ll r \ll L.$$

Hence

$$\frac{\partial}{\partial r_i}\langle u_{Ei}(\mathbf{r})u_{Ej}(\mathbf{r})u_{Ek}(\mathbf{r})\rangle = -\frac{4}{3}\bar{\varepsilon}\delta_{jk}, \quad (\eta \ll r \ll L).$$

Therefore, in the inertial subrange the moments approximation conditions have the form

$$\begin{aligned}&\langle a_j(\mathbf{r},\vec{u}_E(\mathbf{r}))\,u_{Ek}(\mathbf{r}) + a_k(\mathbf{r},\vec{u}_E(\mathbf{r}))\,u_{Ej}(\mathbf{r})\rangle + \langle m_{jk}(\mathbf{r},\vec{u}_E(\mathbf{r}))\rangle = -\frac{4}{3}\,\bar{\varepsilon}\delta_{jk},\\ &\langle a_j(\mathbf{r},\vec{u}_E(\mathbf{r}))\rangle = 0, \qquad j,k = 1,2,3; \quad (\eta \ll r \ll L).\end{aligned} \tag{11.8}$$

11.4.2 Realizability of LS models based on the moments approximation method

Here we have to analyse whether or not the scheme presented in the previous subsection can indeed be realized.

The input parameters of the moments approximation method are

$$\langle u_{Ei}(\mathbf{r})u_{Ej}(\mathbf{r})\rangle, \quad \langle u_{Ei}(\mathbf{r})u_{Ej}(\mathbf{r})u_{Ek}(\mathbf{r})\rangle, \quad i,j,k = 1,2,3.$$

Let us assume that the functions a_i and b_{ij} satisfying the moments approximation conditions (11.8) are found.

Now we have to check whether there exists a *positive* solution $p_E(\vec{v},\mathbf{r})$ to equation (11.7) which satisfies the conditions

$$\int_{\mathrm{IR}^3} p_E(\vec{v},\mathbf{r})\,d\vec{v} = 1, \qquad \int_{\mathrm{IR}^3} v_i\,p_E(\vec{v},\mathbf{r})\,d\vec{v} = 0,$$
$$\int_{\mathrm{IR}^3} v_i v_j\,p_E(\vec{v},\mathbf{r})\,d\vec{v} = \langle u_{Ei}(\mathbf{r})u_{Ej}(\mathbf{r})\rangle,$$
$$\int_{\mathrm{IR}^3} v_i v_j v_k\,p_E(\vec{v},\mathbf{r})\,d\vec{v} = \langle u_{Ei}(\mathbf{r})u_{Ej}(\mathbf{r})u_{Ek}(\mathbf{r})\rangle, \quad i,j,k = 1,2,3.$$

Example ([164]).

$$a_i = -k\frac{(\bar{\varepsilon}r)^{1/3}}{r}(v_i + \gamma v_j n_j n_i), \quad n_i = r_i/r, \tag{11.9}$$

$$m_{ij} = \bar{\varepsilon}\Big\{\beta\delta_{ij} + (\alpha - \beta)n_i n_j\Big\}, \quad (\alpha \geq 0, \beta \geq 0), \tag{11.10}$$

where α, β, γ and k are dimensionless coefficients of the model ($\alpha \geq 0$, $\beta \geq 0$, $k \geq 0$), and C is Kolmogorov's constant in the law of two-thirds ($C \simeq 2$). The moments approximation conditions (11.8) imply the following relations between these coefficients [164]:

$$8Ck = 3\beta + 4, \quad (6\gamma - 2)Ck = (3\alpha - \beta).$$

From $\alpha \geq 0, \beta \geq 0$ it follows that

$$Ck \geq 0.5, \quad (9\gamma + 1)Ck \geq 2.$$

Thus for any γ and $k \geq 0$ satisfying these inequalities the model (11.9)–(11.10) satisfies the moments approximation conditions (11.8). However as shown in [164], the realizability conditions are satisfied only for a special subrange of the pairs (γ, k). Some examples of such pairs are: (1/3,3), (1,2.5), (3,1.3), (5,0.9).

A generalization of the model (11.9)–(11.10) is given in [163], where the drift term is defined by

$$a_i = -\frac{k}{r}\left((\bar{\varepsilon}r)^{1/3}(v_i + \gamma v_j n_j n_i) + \nu\, v_j n_j \tilde{v}_i\right), \quad \tilde{v}_i = v_i - v_j n_j n_i \tag{11.11}$$

and with diffusion term (11.10). In absence of intermittency, the moments approximation conditions (11.8) imply [163]

$$\alpha = -\frac{4}{3} + 2k(\gamma + 1), \quad \beta = -\frac{4}{3} + \frac{8}{15}k(5C - \nu).$$

Realizability conditions imply that the resting parameters k, γ and ν are not completely free. For example, as shown in [163], in the case of isotropic forcing ($\alpha = \beta$) which implies $\gamma = 1/3 - 4\nu/15C$, and assuming $\nu = 0, 0.25, 1$, the corresponding values of parameter k, for which the realizability conditions are satisfied, are 3.8, 4.2, and 6, respectively.

The examples cited will be used below in the next section when comparing different models of relative dispersion in the inertial subrange (see Table 11.1).

11.5 Comparison of different models of relative dispersion for the inertial subrange of a fully developed turbulence

11.5.1 Q1D quadratic-form model of Borgas and Yeung

The exact Eulerian transport equation for the Eulerian velocity p. d. f. $p_E(\vec{v}, \mathbf{r})$ is (e. g., see [77])

$$v_i \frac{\partial p_E}{\partial r_i} + \frac{\partial}{\partial v_i}(\langle a_i | \vec{v}, \mathbf{r}\rangle p_E) = 0,$$

where $a_i(t) = \frac{du_i}{dt}$ is the relative acceleration between two fluid particles, and $\langle a_i | \vec{v}, \mathbf{r}\rangle$ is its conditionally averaged value. The Q1D analog of this equation is the following exact transport equation for $p_E^{\|}$, e. g., see [17]):

$$\frac{u_r}{r^2} \frac{\partial r^2 p_E^{\|}}{\partial r} + \frac{\partial}{\partial u_r}(\langle a_r | u_r, r\rangle p_E^{\|}) = 0, \tag{11.12}$$

where

$$a_r(t) = \frac{du_r(t)}{dt} = \frac{a_i(t) r_i(t)}{r(t)} + \frac{v_i(t) v_i(t) - u_r^2(t)}{r}$$

is the longitudinal component of the relative acceleration, and $\langle a_r | u_r, r\rangle$ is the conditional acceleration.

From (11.6) and (11.12) it follows that

$$X(r, u_r) = C_0 \bar{\varepsilon} \frac{1}{p_E^{\|}} \frac{\partial p_E^{\|}(u_r, r)}{\partial u_r} + \langle a_r | u_r, r\rangle. \tag{11.13}$$

The quadratic-form assumption for the conditional acceleration is

$$\langle a_r | u_r, r\rangle = \alpha(r) + \beta(r) u_r + \gamma(r) u_r^2 = \frac{\sigma^2}{r}(\alpha_0 + \beta_0 \xi + \gamma_0 \xi^2), \tag{11.14}$$

where $\sigma^2 = \langle u_{Er}^2\rangle$, $\xi = u_r/\sigma$. In order to satisfy the moment constraints

$$\langle u_{Er}^2\rangle = \sigma^2, \quad \langle u_{Er}^3\rangle = S\sigma^3, \quad \langle u_{Er}^4\rangle = F\sigma^4$$

the coefficients α, β, and γ should be chosen from the following equations [17]:

$$\frac{1}{r^2} \frac{d(r^2 \sigma^2)}{dr} = \alpha + \gamma \sigma^2,$$

$$\frac{1}{r^2} \frac{d(r^2 S \sigma^3)}{dr} = 2\beta \sigma^2 + 2\gamma S \sigma^3,$$

$$\frac{1}{r^2} \frac{d(r^2 F \sigma^4)}{dr} = 3\alpha \sigma^2 + 3\beta S \sigma^3 + 3\gamma F \sigma^4,$$

which can be solved for parameters α, β and γ to give

$$\alpha = \alpha_0 \frac{\sigma^2}{r}, \quad \beta = \beta_0 \frac{\sigma}{r}, \quad \gamma = \gamma_0 \frac{1}{r}, \tag{11.15}$$

where the dimensionless parameters α_0, β_0, and γ_0 are

$$\alpha_0 = -\gamma_0 - 2\psi_0 + 2, \quad \beta_0 = -\gamma_0 - \frac{3}{2} S\psi_0 + S + \frac{r}{2}\frac{dS}{dr}, \quad \psi_0 = -r\frac{d\ln\sigma}{dr},$$

$$\gamma_0 = \frac{\left(-6 - 3S^2 + 2F + \psi_0(6 + \frac{9}{2}S^2 - 4F) - \frac{3}{2}Sr\frac{dS}{dr} + r\frac{dF}{dr}\right)}{3(F - S^2 - 1)}. \tag{11.16}$$

In the inertial subrange $\eta \ll r \ll L$ we have $\sigma^2(r) = C(\bar{\varepsilon}r)^{2/3}$, $S = \text{const}$, $F = \text{const}$, due to Kolmogorov's second similarity hypothesis (e. g., see [146]) which yields $\psi_0 = -1/3$. Therefore (11.16) yields

$$\gamma_0 = \frac{\frac{10}{9}F - \frac{3}{2}S^2 - \frac{8}{3}}{F - S^2 - 1}, \quad \alpha_0 = \frac{8}{3} - \gamma_0, \quad \beta_0 = \frac{3}{2}S - \gamma_0.$$

In the inertial subrange there exists a dimensionless p. d. f. $f(\xi)$ such that

$$p_E^{\|}(u_r, r) = \frac{1}{\sigma(r)} f\left(\frac{u_r}{\sigma(r)}\right).$$

Then in this case equation (11.12) can be rewritten as the following ODE:

$$(b_0 + b_1\xi + b_2\xi^2)\frac{df(\xi)}{d\xi} + (c_0 + c_1\xi)f(\xi) = 0, \tag{11.17}$$

where b_0, b_1, b_2, c_0, c_1 are the dimensionless constants

$$b_0 = \alpha_0, \quad b_1 = \beta_0, \quad b_2 = \gamma_0 - \frac{1}{3},$$
$$c_0 = b_1, \quad c_1 = 2 + 2\gamma_0 - \frac{1}{3}.$$

From (11.13), (11.14), and (11.17) we get

$$X(r, u_r) = \frac{\bar{\varepsilon}^{2/3}}{r^{1/3}}\tilde{X}\left(\frac{u_r}{\sigma(r)}\right), \quad \sigma(r) = \sqrt{C}(\bar{\varepsilon}r)^{1/3},$$

where

$$\tilde{X}(\xi) = -\frac{C_0}{\sqrt{C}}\frac{c_0 + c_1\xi}{b_0 + b_1\xi + b_2\xi^2} + C(\alpha_0 + \beta_0\xi + \gamma_0\xi^2).$$

Table 11.1. The universal constant g in the Richardson law $\langle r^2(t)\rangle = g\varepsilon t^3$ calculated by different well-mixed Lagrangian stochastic models.

Well-mixed models	$C_0 = 4$	$C_0 = 5$	$C_0 = 6$	$C_0 = 7$	$C_0 = 10$
Borgas and Sawford (1994)	0.75	0.7	0.6	0.5	0.3
Thomson (1990)	1.8	1.4	1.1	0.85	0.45
Q1D, p. d. f. of Borgas and Yeung (1998)	4.8	2.6	1.7	1.1	0.45
Q1D, bi-Gaussian p. d. f. $\sigma_1 = \sigma_2$.	4.4	2.55	1.67	1.15	0.47
Q1D, bi-Gaussian p. d. f. $\mu_1/\sigma_1 = -\mu_2/\sigma_2$.	8.25	4.	2.27	1.4	0.51
Q1D, Gaussian p. d. f.	7.	4.7	2.8	1.9	0.75

11.5.2 Comparison of different models in the inertial subrange

In this subsection we study the process of the relative dispersion of two particles in the inertial subrange (ISR) $\eta \ll r \ll L$. We first consider Q1D models which are determined by the Eulerian p. d. f. $f_E(\xi)$ of the dimensionless velocity difference $\xi_r = u_{Er}/(\bar{\varepsilon}r)^{1/3}$. Suppose

$$f_E(\xi) = \frac{p}{\sqrt{2\pi}\sigma_1}\exp\left\{-\frac{(\xi-\mu_1)^2}{2\sigma_1^2}\right\} + \frac{1-p}{\sqrt{2\pi}\sigma_2}\exp\left\{-\frac{(\xi-\mu_2)^2}{2\sigma_2^2}\right\},$$

where the unknown parameters $p,\ \mu_1,\ \mu_2,\ \sigma_1,\ \sigma_2$ should be chosen to fit the first four moments of ξ_r:

$$\begin{aligned}
\langle \xi_r\rangle &= p\mu_1 + (1-p)\mu_2 = 0,\\
\langle \xi_r^2\rangle &= p(\mu_1^2+\sigma_1^2) + (1-p)(\mu_2^2+\sigma_2^2) = C,\\
\langle \xi_r^3\rangle &= p(\mu_1^3+3\mu_1\sigma_1^2) + (1-p)(\mu_2^3+3\mu_2\sigma_2^2) = -\frac{4}{5},\\
\langle \xi_r^4\rangle &= p(\mu_1^4+6\mu_1^2\sigma_1^2+3\sigma_1^4) + (1-p)(\mu_2^4+6\mu_2^2\sigma_2^2+3\sigma_2^4) = 3.4\,C^2.
\end{aligned}$$

Thus we have four equations for five unknown parameters $p,\ \mu_1,\ \mu_2,\ \sigma_1,\ \sigma_2$. To find these parameters, we need an additional (closure) assumption. Further we will use two kinds of closure assumptions: (i) $\sigma_1 = \sigma_2$, and (ii) $\mu_1/\sigma_1 = -\mu_2/\sigma_2$.

In Table 11.1 we present the Richardson constant g obtained by different models. In the models [163, 164], the isotropic forcing was considered ($\alpha = \beta$; see Section 4.2)

Table 11.2. The universal constant g evaluated by different moments approximation models.

Model	*realizability*	g	
Pedrizzetti & Novikov	+	0.50	$(k = 3.8,\ C_0 = 9.47)$
Pedrizzetti	+	0.35	$(\nu = 0.25,\ k = 4.2,\ C_0 = 10.25)$
	+	0.20	$(\nu = 1,\ k = 6,\ C_0 = 13.73)$
Heppe	unknown	0.44	$(C_0 = 5.3)$
	unknown	0.31	$(C_0 = 9.47)$

Table 11.3. The universal constant g evaluated by other methods.

Method	g
LHDI, Kraichnan 1966 [101]	2.42
Modified LHDI, Lundgren 1981 [135]	3.0
EDQNM, Larcheveque & Lesieur 1981 [125]	3.5
Thomson's corrected EDQNM, Thomson 1996 [239]	1.4
Effective Hamiltonian method, Nakao 1991 [149]	3.5

11.6 Comparison of different Q1D models of relative dispersion for modestly large Reynolds number turbulence ($Re_\lambda \simeq 240$)

The models discussed in the previous section were developed for the case of a turbulence with a reach inertial subrange. In this case the DNS methods cannot be used, since today the DNS data is available around $Re_\lambda \simeq 240$. Therefore, to make the validation through comparison with the DNS method, we can do this only for models with modestly large Reynolds-number turbulence.

11.6.1 Parametrization of Eulerian statistics

Here we turn to a turbulent flow with $L/\eta \simeq 500$. In this case the scales go from the viscous ones, pass through a transitional region, and goes on to the external interval. But it should be noted that the transitional range can be considered to be an analog of the inertial subrange only in terms of the Eulerian statistical characteristics. For the Lagrangian statistics however this interval does not show the inertial subrange behavior.

Thus we have to make the assumption that $\mathbf{r}(t), \vec{v}(t)$ is a Markov random process governed by (11.5) with the value C_0 depending on the distance r such that C_0 tends to a constant value as r/η is increases.

Dimensional arguments show that in the case of stationary and isotropic turbulence

$$p_E^{\|}(v,r) = \frac{1}{\sigma(r)} f_E(\frac{v}{\sigma(r)};r),$$

where $\sigma(r) = \langle u_{Er}^2 \rangle^{1/2}$ is r. m. s of the longitudinal component of the Eulerian velocity difference $\vec{u}_E(\mathbf{r}) = \mathbf{U}_E(\mathbf{x}+\mathbf{r}) - \mathbf{U}_E(\mathbf{x})$:

$$u_{Er} = (U_{Ei}(\mathbf{x}+\mathbf{r}) - U_{Ei}(\mathbf{x}))\,\frac{r_i}{r},$$

and $f_E(\xi;r)$ is a dimensionless p. d. f. of dimensionless velocity $\xi_r = u_{Er}/\sigma(r)$. It follows from (11.13) that

$$X(r,v) = \frac{\bar{\varepsilon}^{2/3}}{r^{1/3}} \tilde{X}(r,\frac{v}{\sigma(r)}),$$

$$\tilde{X}(r,\xi) = C_0(r)\frac{\bar{\varepsilon} r}{\sigma^3}\frac{\partial \ln f_E}{\partial \xi} - \psi_0(r)\xi^2 - (2-\psi_0)\,\frac{\int_{-\infty}^{\xi} \xi' f_E(\xi';r)\,d\xi'}{f_E(\xi;r)},$$

where

$$\psi_0 = \psi_0(r) = -r\,\frac{d\ln\sigma}{dr}.$$

If we pass to the dimensionless velocity $\xi_r(t) = u_r(t)/\sigma(r(t))$, then the Q1D model reads

$$dr = \sigma(r)\xi_r\,dt, \quad d\xi_r = \frac{\sigma(r)}{r}\left\{\tilde{X}(r,\xi_r) + \psi_0(r)\xi_r^2\right\} + \frac{(C_0(r)\bar{\varepsilon})^{1/2}}{\sigma(r)} dW(t). \tag{11.18}$$

Now we have to specify the Eulerian velocity $f_E(\xi;r)$ and the function $\sigma(r)$.

Concerning the density $f_E(\xi;r)$, we assume that the first four moments are given in [17]:

$$\langle \xi_r \rangle = 0, \quad \langle \xi_r^2 \rangle = 1, \quad \langle \xi^3 \rangle = \frac{\langle u_{Er}^3 \rangle}{\langle u_{Er}^2 \rangle^{3/2}}, \quad \langle \xi^4 \rangle = \frac{\langle u_{Er}^4 \rangle}{\langle u_{Er}^2 \rangle^2},$$

where

$$u_{Er}^2(r) = 2\sigma_1^2\left(\frac{r^2}{\omega_2\eta^2 + r^2}\right)^{2/3}\left(\frac{r^2}{\Lambda_2 L^2 + r^2}\right)^{1/3},$$

$$\langle u_{Er}^3 \rangle = \theta\sigma_1^3\left(\frac{r}{L}\right)^3\left(\frac{L^2}{\omega_3\eta^2 + r^2}\right)\left(\frac{L^2}{\Lambda_3 L^2 + r^2}\right)^4,$$

$$\langle u_{Er}^4 \rangle = 4k\sigma_1^4\left(\frac{r^2}{\omega_4\eta^2 + r^2}\right)^{4/3}\left(\frac{r^2}{\Lambda_4 L^2 + r^2}\right)^{2/3}.$$

Here $\sigma_1 = [U_{Ei}U_{Ei}]^{1/2}$ is the one-point root-mean-square (rms) velocity fluctuations, $L = \sigma_1^3/\bar{\varepsilon}$, the dimensionless constants $\theta, k, \omega_2, \omega_3, \omega_4, \Lambda_2, \Lambda_3, \Lambda_4$ in this parametrization are

$$\Lambda_2 = \left(\frac{2}{C}\right)^3, \quad \Lambda_3 = 1., \quad \Lambda_4 = \Lambda_2\left(\frac{k}{K_i}\right)^{3/2}, \quad k = 3., \quad \theta = -\frac{4}{5}(\Lambda_3)^4,$$

$$\omega_2 = \frac{164.32}{\sqrt{\Lambda_2}}, \quad \omega_3 = -\frac{4}{5}\frac{\omega_2\Lambda_2^{1/2}}{2^{3/2}S_0}, \quad \omega_4 = \omega_2\left(\frac{K_i}{K_0}\right)^{3/4},$$

$$K_0 = 7.5, \quad S_0 = -0.5, \quad K_i = 3.4\,.$$

11.6.2 Bi-Gaussian p. d. f.

Here we compare the Lagrangian statistical characteristics obtained by the model (11.18) with the results obtained by DNS. The p. d. f. $f_E(\xi; r)$ is chosen as the following bi-Gaussian density with unknown parameters $p, \mu_1, \mu_2, \sigma_1, \sigma_2$:

$$f_E(\xi, r) = \frac{p}{\sqrt{2\pi}\sigma_1}\exp\left\{-\frac{(\xi-\mu_1)^2}{2\sigma_1^2}\right\} + \frac{1-p}{\sqrt{2\pi}\sigma_2}\exp\left\{-\frac{(\xi-\mu_2)^2}{2\sigma_2^2}\right\},$$

$$p\mu_1 + (1-p)\mu_2 = 0, \quad p(\mu_1^2+\sigma_1^2) + (1-p)(\mu_2^2+\sigma_2^2) = 1,$$

$$p(\mu_1^3 + 3\mu_1\sigma_1^2) + (1-p)(\mu_2^3 + 3\mu_2\sigma_2^2) = S,$$

$$p(\mu_1^4 + 6\mu_1^2\sigma_1^2 + 3\sigma_1^4) + (1-p)(\mu_2^4 + 6\mu_2^2\sigma_2^2 + 3\sigma_2^4) = F,$$

where $S = S(r) = \frac{\langle u_{Er}^3\rangle}{\langle u_{Er}^2\rangle^{3/2}}$ and $F = F(r) = \frac{\langle u_{Er}^4\rangle}{\langle u_{Er}^2\rangle^2}$ are the skewness and the flatness of u_{Er}, respectively.

To determine uniquely the five unknown parameters, we use the closure assumption [125] which has proven to be reasonable:

$$\frac{\mu_1}{\sigma_1} = -\frac{\mu_2}{\sigma_2} = m, \tag{11.19}$$

where $m = m(r)$ is the solution to the equation

$$F = \left(1 + \frac{(1+m^2)^3 S^2}{(3+m^2)^2 m^2}\right)\frac{3+6m^2+m^4}{(1+m^2)^2}.$$

The assumption (11.19) allows us to obtain the unknown parameters σ_1 and σ_2 explicitly:

$$\sigma_1 = \left(\frac{1-p}{p(1+m^2)}\right)^{1/2}, \quad \sigma_2 = \left(\frac{p}{(1-p)(1+m^2)}\right)^{1/2},$$

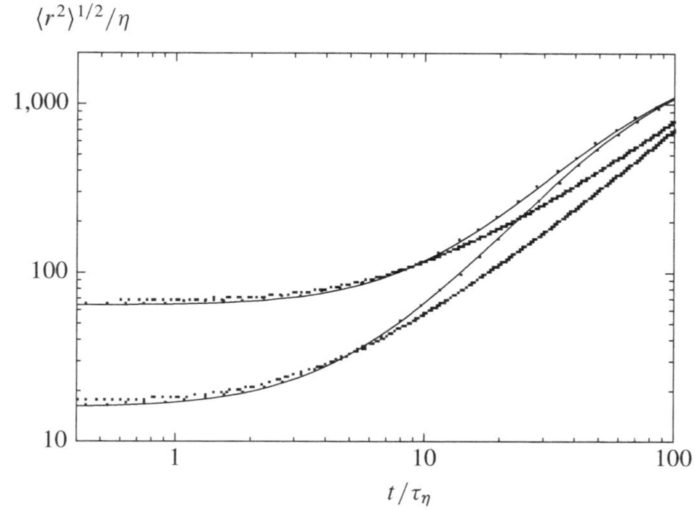

Figure 11.1. The dimensionless rms of the relative separation; Data obtained by the Lagrangian stochastic model: $r_0 = 16\eta$ (lower solid curve) and $r_0 = 64\eta$ (upper solid curve). The relevant DNS data are shown as the dotted curves: $r_0 = 16\eta$ (lower dotted curve) and $r_0 = 64\eta$ (upper dotted curve).

where

$$p = \frac{1}{2}\left[1 - \left(\frac{a}{4+a}\right)^{1/2}\right], \quad a = \frac{(1+m^2)^3 S^2}{(3+m^2)^2 m^2}.$$

In Figure 11.1, the dimensionless rms of the relative separation as a function of dimensionless time is presented. Calculations by the model (11.18) were carried out for two initial separations: $r_0 = 16\eta$ (lower solid curve) and $r_0 = 64\eta$ (upper solid curve). The relevant DNS data are shown as dotted curves.

In Figure 11.2 the skewness factor is shown as a function of dimensionless time. In this curve, the upper solid lines correspond to the initial separation $r_0 = 16\eta$, and the lower solid lines to $r_0 = 64\eta$. The DNS data are also shown as dotted curves, where $r_0 = 16\eta$ and $r_0 = 64\eta$ correspond to the upper and lower lines, respectively. As the results of Figure 11.1 show, the model describes the rms of the separation well for dimensionalless times from 0 to 10. There is some discrepancy between the model and DNS data for larger times. As to the skewness and flatness factors, we can only state a qualitative agreement with the DNS data: the model overestimates these factors, compared to the DNS results, in the time interval intermediate between the viscous and external ranges.

It should be noted that our calculations with the bi-Gaussian p. d. f. with other types of closure assumptions qualitatively show the same picture. We conclude thus that the bi-Gaussian p. d. f. is not the right choice. Therefore we try another class of densities,

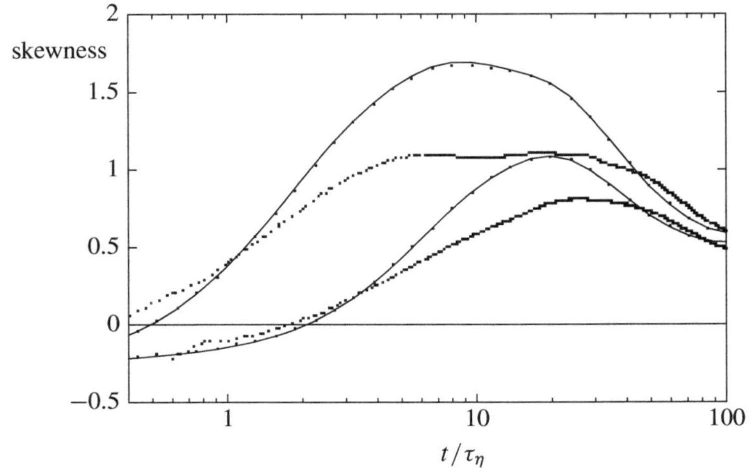

Figure 11.2. The skewness factor of the relative separation; Data obtained by the Lagrangian stochastic model: $r_0 = 64\eta$ (lower solid curve) and $r_0 = 16\eta$ (upper solid curve). The relevant DNS data are shown as the dotted curves: $r_0 = 64\eta$ (lower dotted curve) and $r_0 = 16\eta$ (upper dotted curve).

based on the quadratic-form approximation of the conditional acceleration. These densities have a different tail behavior, which might have a large impact on the skewness and flatness factors. Such indications were reported by M. Borgas [14].

11.6.3 Q1D quadratic-form model

Let us show that the realizability condition will be satisfied if there exists a positive solution $p_E^{\|}$ to equation (11.12) with the quadratic-form of the conditional acceleration (11.14) which is a p. d. f. of the variable u_{Er}:

$$\int_{-\infty}^{\infty} p_E^{\|}(u_r;r)\,du_r = 1, \quad \forall r \geq 0, \tag{11.20}$$

and there exists a point r_0 such that

$$\overline{u}_{Er}(r_0) = 0, \quad \overline{u^2_{Er}}(r_0) = \sigma^2(r_0), \quad \overline{u^3_{Er}}(r_0) = \kappa(r_0), \quad \overline{u^4_{Er}}(r_0) = \vartheta(r_0). \tag{11.21}$$

Here and below we denote by an overbar the average over this p. d. f.. In order to prove this assertion let us integrate equation (11.12) on u_r to get $r^2\overline{u_{Er}} = \text{const}$ which yields $\overline{u_{Er}}(r) = 0$, since $\overline{u_{Er}}(r_0) = 0$.

Further multiplying equation (11.12) by u_r and then integrating over u_r, we get

$$\frac{1}{r^2}\frac{dr^2\overline{u^2_{Er}}}{dr} = \alpha + \gamma\overline{u^2_{Er}}.$$

This yields $\overline{u_{Er}^2}(r) = \sigma^2(r)$ since the functions $\overline{u_{Er}^2}$ and $\sigma^2(r)$ satisfy the same equations and the common initial conditions at $r = r_0$ (see (11.21)).

Multiplying equation (11.12) by u_r^2 and u_r^3 and integrating over u_r we find that the functions $\overline{u_r^3}$ and $\overline{u_{Er}^4}$ satisfy the same equations as the functions $\kappa(r)$ and $\vartheta(r)$, respectively. From these and the conditions (11.21) it follows that

$$\overline{u_r^3}(r) = \kappa(r), \quad \overline{u_r^4}(r) = \vartheta(r),$$

i. e., the realizability condition is established.

Remark 11.1. If we assume that for some α, r_1, and r_2 the conditions

$$\frac{\langle u_{Er}^2 \rangle}{r^\alpha} = \text{const}, \quad S(r) = \text{const}, \quad F(r) = \text{const}, \quad r_1 \ll r \ll r_2. \tag{11.22}$$

are valid, then from (11.16) and (11.22) it follows that the coefficients b_0, b_1, b_2, c_0, and c_1 are constants, say, $\tilde{b}_0$, $\tilde{b}_1$, $\tilde{b}_2$, $\tilde{c}_0$, $\tilde{c}_1$ in the subrange $r_1 \ll r \ll r_2$. For example in the far-viscous subrange $r \ll \eta$, in the inertial subrange $\eta \ll r \ll L$, and in the external subrange $r \gg L$ these conditions are fulfilled.

This property can be used to define the boundary conditions to the transport equation for the function $f(\xi, r)$:

$$\xi r \frac{\partial f(\xi, r)}{\partial r} + (b_0 + b_1\xi + b_2\xi^2)\frac{\partial f(\xi, r)}{\partial \xi} + (c_0 + c_1\xi) f(\xi, r) = 0, \tag{11.23}$$

where b_0, b_1, b_2, c_0, c_1 are dimensionless functions depending on r:

$$b_0(r) = \alpha_0(r), \quad b_1(r) = \beta_0(r), \quad b_2(r) = \gamma_0(r) + \psi_0(r),$$

$$c_0(r) = b_1(r), \quad c_1(r) = 2 + 2\gamma_0(r) + \psi_0(r),$$

and $\alpha_0(r)$, $\beta_0(r)$, $\gamma_0(r)$ and $\psi_0(r)$ are determined in (11.16). Indeed, let $f_0(\xi)$ be a solution to the ODE

$$(\tilde{b}_0 + \tilde{b}_1\xi + \tilde{b}_2\xi^2)\frac{df_0(\xi)}{d\xi} + (\tilde{c}_0 + \tilde{c}_1\xi) f_0(\xi) = 0,$$

satisfying the condition

$$\int_{-\infty}^{\infty} f_0(\xi)\, d\xi = 1.$$

We can consider the p. d. f. $f_0(\xi)$ as an approximation to the p. d. f. $f(\xi, r)$ in the subrange $r_1 \ll r \ll r_2$.

Choosing r_0 in the interval $r_1 \ll r_0 \ll r_2$, we can put the following boundary condition for equation (11.23):

$$f(\xi, r = r_0) = f_0(\xi). \tag{11.24}$$

This brief remark is just to recall that there are many possible choices of the Eulerian p. d. f..

Chapter 12

Evaluation of mean concentration and fluxes in turbulent flows by Lagrangian stochastic models

Forward and backward stochastic Lagrangian trajectory simulation methods for calculation of the mean concentration of scalars and their fluxes for sources arbitrarily distributed in space and time are constructed and justified. Generally, absorption of scalars by medium is taken into account. A special case of the source structure, when the scalar is generated by a plane source, say, located close to the ground, is treated. This practically interesting particular case is known in the literature as the *footprint problem*.

12.1 Introduction

The turbulent dispersion of particles in the framework of statistical fluid mechanics is described as the transport of particles in a random velocity field (e. g., [146]). In particular, the concentration of scalars and their fluxes are random fields. There are mainly two different approaches for calculation of the mean values of these fields: (i) the conventional deterministic methods, based on the semiempirical turbulent diffusion equation and closure assumptions (e. g., see [79, 162, 225]), and (ii) the stochastic approach, which utilizes trajectory simulations (e. g., see [62, 102, 113, 191, 237, 244, 256]).

The deterministic approach directly deals with the equation governing the mean concentration, and relies on the Boussinesq hypothesis, whose applicability is restricted (e. g., see [19, 175]). For instance, this hypothesis cannot be true if the concentration is calculated close to the sources [19, 146]. More generally, the high-order closure methods are developed, but a different closure hypothesis also should be made (see e. g., [95, 146]).

The stochastic approach based on the modeling of stochastic Lagrangian trajectories in principle does not require a closure hypothesis. Two main issues in this approach are (i) the development of adequate Lagrangian stochastic models governed by generalized Langevin-type equations and (ii) construction of Monte Carlo random estimators for the evaluation of desired statistical characteristics (for instance, the mean concentration, the mean height of a cloud of particles, etc.).

It should be noted that in the Monte Carlo methods, when using the random estimators the results are obtained with statistical errors. Recall that a random variable ξ is said to be a Monte Carlo estimator for a quantity a if the mathematical expectation of ξ is equal to a: $\mathbf{E}\xi = a$. If $\xi_1, \xi_2, \dots, \xi_N$ are N-independent samples of the random variable ξ, then the average $S_N = \frac{1}{N}\sum_{i=1}^{N} \xi_i$ almost surely tends to a

(i. e., with probability 1) as N tends to infinity, and the error in using S_N to approximately $a = \mathbf{E}\xi$ (for sufficiently large N) is proportional to the standard deviation of ξ. As N increases, this statistical error decreases as $N^{-1/2}$. The well-known "law of three sigmas" gives the rate of convergence: $\mathbf{P}(|S_N - a| < 3\sigma_\xi/\sqrt{N}) \approx 0.997$. Here $\sigma_\xi = (\mathbf{E}\xi^2 - \mathbf{E}^2\xi)^{1/2}$ is the standard deviation of ξ. The larger N is, the closer the distribution of S_N to the Gaussian one, and the better the approximation.

Issue (i) attracted attention in many publications (e. g., see [115,178,200,255,256]). In this chapter we concentrate on issue (ii). It should be noted that this field is not well developed, and we can give only a few references [57, 111, 215, 237].

In this chapter we treat simulation methods based on the forward and backward Lagrangian trajectories. The general principle is quite clear: one uses the backward trajectories originating at the detector if it is a point detector in space (or the detector occupies a small volume); the forward trajectories are used if the detector is quite extended in space.

The chapter is organized as follows. In Section 12.2 one relates the calculation of the mean concentration and its flux with the averages over Lagrangian trajectories governed by generalized Langevin-type equations. The forward and backward estimators are presented in Section 12.3. Applications of these estimators to the footprint problem are given in Section 12.4. Some technical details are included in Appendices A–C at the end of the chapter.

12.2 Formulation of the problem

Let us assume that a passive but generally nonconservative scalar is dispersed by a turbulent velocity field $\vec{u}(\mathbf{x}, t)$ in the half-space $D = \{\mathbf{x} = (x_1, x_2, x_3) : x_3 \geq 0\}$, for example in the surface layer of the atmosphere. Throughout this chapter the following notation of spatial and velocity co-ordinates is used: $\mathbf{x} = (x_1, x_2, x_3) = (x, y, z)$ and $\vec{u} = (u_1, u_2, u_3) = (u, v, w)$; and analogously $\mathbf{X} = (X_1, X_2, X_3) = (X, Y, Z)$ and $\mathbf{V} = (V_1, V_2, V_3) = (U, V, W)$ for the Lagrangian coordinates.

The passive scalar is assumed to be uninertial, i. e., it follows the streamlines of the flow.

The evolution of scalar concentration field from a source of intensity $q(\mathbf{x}, t)$ (the amount of emitted scalar per unit volume in a unit time interval at the phase point $(\mathbf{x}, t)$) is controlled by the turbulent transport and absorption by a medium:

$$\frac{\partial c(\mathbf{x}, t)}{\partial t} + u_i(\mathbf{x}, t)\frac{\partial c}{\partial x_i} + \gamma(\mathbf{x}, t)c(\mathbf{x}, t) = q(\mathbf{x}, t), t > 0; \quad c(\mathbf{x}, 0) = q_0(\mathbf{x}),$$

where $\gamma(\mathbf{x}, t)$ ($\gamma \geq 0$) denotes the coefficient of absorption, the initial spatial distribution of concentration is given by $q_0(\mathbf{x})$, and the molecular diffusion is neglected. Here and in the following the summation convention is assumed over repeated indices.

The turbulent velocity field $\vec{u}(\mathbf{x},t)$ is assumed to be an incompressible 3-dimensional (3D) random field. Accordingly the concentration $c(\mathbf{x},t)$ is also a scalar random field. We consider the simplest statistical characteristics of this field, the mean concentration $\langle c(\mathbf{x},t)\rangle$, the mean flux of scalar concentration $\langle u_i(\mathbf{x},t)c(\mathbf{x},t)\rangle$, and the spatial-temporal average of these statistical characteristics. Here and below the angle brackets denote the average over samples of turbulent velocity fluctuations.

The above-mentioned means are calculated by simulation of the Lagrangian trajectories $\mathbf{X}(t) = \mathbf{X}(t;\mathbf{x}_0,t_0)$, $t \geq t_0$, determined by

$$\frac{dX_i(t)}{dt} = u_i(\mathbf{X}(t),t) = V_i(t), \quad \mathbf{X}(t_0) = \mathbf{x}_0,$$

where $\mathbf{V}(t) = \mathbf{V}(t;\mathbf{x}_0,t_0)$ is the Lagrangian velocity.

The instantaneous concentration can be expressed as (see Appendix A at the end of the chapter)

$$\begin{aligned} c(\mathbf{x},t) = &\int_0^t dt_0 \int_D d\mathbf{x}_0\, \mu(t;\mathbf{x}_0,t_0) q(\mathbf{x}_0,t_0)\delta(\mathbf{x}-\mathbf{X}(t;\mathbf{x}_0,t_0)) \\ &+ \int_D d\mathbf{x}_0\, \mu(t;\mathbf{x}_0,0) q_0(\mathbf{x}_0)\delta(\mathbf{x}-\mathbf{X}(t;\mathbf{x}_0,0)), \end{aligned} \tag{12.1}$$

where $\delta(\cdot)$ is the Dirac delta function, and $\mu(t) = \mu(t;\mathbf{x}_0,t_0)$ is defined by

$$\frac{d\mu(t)}{dt} + \gamma(\mathbf{X}(t;\mathbf{x}_0,t_0),t)\mu(t) = 0, \quad \mu(t_0) = 1. \tag{12.2}$$

The expression for instantaneous concentration can be rewritten as

$$\begin{aligned} c(\mathbf{x},t) = &\int_{\mathrm{IR}^3} d\vec{u} \int_0^\infty d\mu \int_0^t dt_0 \int_D d\mathbf{x}_0\, \mu\, Q(\mathbf{x}_0,t_0) \\ &\times \delta(\mathbf{x}-\mathbf{X}(t;\mathbf{x}_0,t_0))\delta(\vec{u}-\mathbf{V}(t;\mathbf{x}_0,t_0))\delta(\mu-\mu(t;\mathbf{x}_0,t_0)), \end{aligned}$$

where $Q(\mathbf{x}_0,t_0) = q(\mathbf{x}_0,t_0)+q_0(\mathbf{x}_0)\delta(t_0)$. Averaging the last equation yields the mean concentration

$$\langle c(\mathbf{x},t)\rangle = \int_{\mathrm{IR}^3} d\vec{u} \int_0^\infty d\mu \int_0^t dt_0 \int_D d\mathbf{x}_0\, \mu\, Q(\mathbf{x}_0,t_0) p_L(\mathbf{x},\vec{u},\mu,t;\mathbf{x}_0,t_0), \tag{12.3}$$

where

$$p_L(\mathbf{x},\vec{u},\mu,t;\mathbf{x}_0,t_0) = \langle \delta(\mathbf{x}-\mathbf{X}(t;\mathbf{x}_0,t_0))\delta(\vec{u}-\mathbf{V}(t;\mathbf{x}_0,t_0))\delta(\mu-\mu(t;\mathbf{x}_0,t_0))\rangle$$

is the joint probability density function (p. d. f.) of Lagrangian characteristics $\mathbf{X}(t;\mathbf{x}_0,t_0)$, $\mathbf{V}(t;\mathbf{x}_0,t_0)$, and $\mu(t;\mathbf{x}_0,t_0)$.

Analogously, the mean flux of concentration can be represented as

$$\langle u_i(\mathbf{x},t)c(\mathbf{x},t)\rangle = \int_{\mathrm{IR}^3} d\vec{u} \int_0^\infty d\mu \int_0^t dt_0 \int_D \times d\mathbf{x}_0\, u_i \mu\, Q(\mathbf{x}_0,t_0) p_L(\mathbf{x},\vec{u},\mu,t;\mathbf{x}_0,t_0), \quad i = 1,2,3. \tag{12.4}$$

For convenience, the mean characteristics (12.3) and (12.4) will be written in the general form

$$\langle g(\vec{u}(\mathbf{x},t))c(\mathbf{x},t)\rangle = \int_{\mathrm{IR}^3} d\vec{u} \int_0^\infty d\mu \int_0^t dt_0 \int_D d\mathbf{x}_0\, g(\vec{u})\mu\, Q(\mathbf{x}_0,t_0) p_L(\mathbf{x},\vec{u},\mu,t;\mathbf{x}_0,t_0), \tag{12.5}$$

where $g(\vec{u})$ equals 1 and u_i for the mean concentration and fluxes, respectively.

Our problem can be formulated as follows. It is necessary to represent the integral (12.5) as an expectation of a random estimator defined on Lagrangian trajectories. But since the exact form of $p_L(\mathbf{x},\vec{u},\mu,t;\mathbf{x}_0,t_0)$ is not known, we have to use some approximation, usually taken as a p. d. f. of the solution to the following generalized Langevin-type equation (e. g., see [237]):

$$\begin{aligned} d\mathbf{X}(t) &= \mathbf{V}(t)dt, \\ d\mathbf{V}(t) &= \mathbf{a}(t,\mathbf{X}(t),\mathbf{V}(t))dt + \sqrt{C_0\bar{\varepsilon}(\mathbf{X}(t),t)}\, d\mathbf{W}(t), \end{aligned} \tag{12.6}$$

where C_0 is the universal Kolmogorov constant ($C_0 \approx 46$), $\bar{\varepsilon}(\mathbf{x},t)$ is the mean dissipation rate of the kinetic energy of turbulence, and $\mathbf{W}(t) = (W_1(t), W_2(t), W_3(t))$ is the standard 3D Wiener process. The function $\mathbf{a}$ is to be specified in each specific situation (e. g., [56, 133, 237, 256]). We mention only that in all these models Thomson's well-mixed condition should be satisfied [237].

We will deal in this chapter with two different types of random estimators, namely, with forward estimators, which are defined on forward Lagrangian trajectories. which emanate from the source and move toward the detector, and with backward estimators, which are defined on backward trajectories starting at the detector and moving toward the source.

More exactly, the solution to (12.6) with the initial conditions

$$\mathbf{X}(t_0) = \mathbf{x}_0, \quad \mathbf{V}(t_0) = \vec{u}_0$$

is called the *forward Lagrangian trajectory*. We denote it by $\mathbf{X}_t^{\mathbf{x}_0,\vec{u}_0,t_0}$ and $\mathbf{V}_t^{\mathbf{x}_0,\vec{u}_0,t_0}$. Then the true Lagrangian trajectory $\mathbf{X}(t;\mathbf{x}_0,t_0)$, $\mathbf{V}(t;\mathbf{x}_0,t_0)$ can be approximated by the model trajectory $\mathbf{X}_t^{\mathbf{x}_0,\vec{u}_0,t_0}$ and $\mathbf{V}_t^{\mathbf{x}_0,\vec{u}_0,t_0}$ with the random initial velocity $\vec{u}_0$ chosen according to the Eulerian p. d. f. p_E, which is defined by $p_E(\vec{u};\mathbf{x}_0,t_0) = \langle\delta(\vec{u} - \vec{u}(\mathbf{x}_0,t_0))\rangle$.

Let $p_L(\mathbf{x}, \vec{u}, \mu, t; \mathbf{x}_0, \vec{u}_0, t_0)$ be the conditional p. d. f. under the condition that $\mathbf{V}(t; \mathbf{x}_0, t_0) = u_0$:

$$\begin{aligned} &p_L(\mathbf{x}, \vec{u}, \mu, t; \mathbf{x}_0, \vec{u}_0, t_0) \\ &\quad = \big\langle \delta(\mathbf{x} - \mathbf{X}(t; \mathbf{x}_0, t_0))\delta(\vec{u} - \mathbf{V}(t; \mathbf{x}_0, t_0))\delta(\mu - \mu(t; \mathbf{x}_0, t_0)) \big| \mathbf{V}(t; \mathbf{x}_0, t_0) = u_0 \big\rangle. \end{aligned}$$

By the theorem on conditional probability we get

$$p_L(\mathbf{x}, \vec{u}, \mu, t; \mathbf{x}_0, t_0) = \int_{\mathrm{IR}^3} d\vec{u}_0 \; p_E(\vec{u}_0; \mathbf{x}_0, t_0) p_L(\mathbf{x}, \vec{u}, \mu, t; \mathbf{x}_0, \vec{u}_0, t_0).$$

Let us introduce the model conditional p. d. f. as

$$\begin{aligned} &p_L^f(\mathbf{x}, \vec{u}, \mu, t; \mathbf{x}_0, \vec{u}_0, t_0) \\ &\qquad = \mathbf{E}_{\mathbf{x}_0, \vec{u}_0, t_0} \{\delta(\mathbf{x} - \mathbf{X}_t^{\mathbf{x}_0, \vec{u}_0, t_0})\delta(\vec{u} - \mathbf{V}_t^{\mathbf{x}_0, \vec{u}_0, t_0})\delta(\mu - \mu_t^{\mathbf{x}_0, \vec{u}_0, t_0})\}, \end{aligned} \tag{12.7}$$

where $\mu(t) = \mu_t^{\mathbf{x}_0, \vec{u}_0, t_0}$ is defined by

$$\frac{d\mu(t)}{dt} + \gamma(\mathbf{X}_t^{\mathbf{x}_0, \vec{u}_0, t_0}, t)\mu(t) = 0, \quad \mu(t_0) = 1.$$

Here $\mathbf{E}_{\mathbf{x}_0, \vec{u}_0, t_0}$ means the expectation over samples of stochastic processes $\mathbf{X}_t^{\mathbf{x}_0, \vec{u}_0, t_0}$, $\mathbf{V}_t^{\mathbf{x}_0, \vec{u}_0, t_0}$, and $\mu_t^{\mathbf{x}_0, \vec{u}_0, t_0}$, starting at time $t = t_0$ from the point $\mathbf{x}_0, \vec{u}_0, 1$.

Taking the model transition density $p_L^f(\mathbf{x}, \vec{u}, \mu, t; \mathbf{x}_0, \vec{u}_0, t_0)$ as an approximation to $p_L(\mathbf{x}, \vec{u}, \mu, t; \mathbf{x}_0, \vec{u}_0, t_0)$, the true Lagrangian p. d. f. $p_L(\mathbf{x}, \vec{u}, \mu, t; \mathbf{x}_0, t_0)$ is approximated as

$$p_L(\mathbf{x}, \vec{u}, \mu, t; \mathbf{x}_0, t_0) \approx \int_{\mathrm{IR}^3} d\vec{u}_0 \; p_E(\vec{u}_0; \mathbf{x}_0, t_0) p_L^f(\mathbf{x}, \vec{u}, \mu, t; \mathbf{x}_0, \vec{u}_0, t_0). \tag{12.8}$$

Substituting the approximation (12.8) into the integral (12.5), we come to the approximate equality

$$\begin{aligned} \langle g(\vec{u}(\mathbf{x}, t))c(\mathbf{x}, t)\rangle = &\int_{\mathrm{IR}^3} d\vec{u} \int_0^\infty d\mu \int_0^t dt_0 \int_D d\mathbf{x}_0 \, g(\vec{u})\mu \, Q(\mathbf{x}_0, t_0) \\ &\times \int_{\mathrm{IR}^3} d\vec{u}_0 \, p_E(\vec{u}_0; \mathbf{x}_0, t_0) p_L^f(\mathbf{x}, \vec{u}, \mu, t; \mathbf{x}_0, \vec{u}_0, t_0). \end{aligned} \tag{12.9}$$

The *backward Lagrangian trajectory*, denoted in the following by $\hat{\mathbf{X}}(t_0) = \hat{\mathbf{X}}_{t_0}^{\mathbf{x}, \vec{u}, t}$, $\hat{\mathbf{V}}(t_0) = \hat{\mathbf{V}}_{t_0}^{\mathbf{x}, \vec{u}, t}$, $t_0 \le t$, is defined as the solution to (e.g, [57, 237])

$$\begin{aligned} d\hat{\mathbf{X}}(t_0) &= \hat{\mathbf{V}}(t_0) dt_0, \\ d\hat{\mathbf{V}}(t_0) &= \hat{\mathbf{a}}(t_0, \hat{\mathbf{X}}(t_0), \hat{\mathbf{V}}(t_0)) dt_0 + \sqrt{C_0 \bar{\varepsilon}(\hat{\mathbf{X}}(t_0), t_0)} \; \overleftarrow{d} \, \mathbf{W}(t_0), \end{aligned} \tag{12.10}$$

with the terminal condition $\hat{\mathbf{X}}(t) = \mathbf{x}$, $\hat{\mathbf{V}}(t) = \vec{u}$, where the drift term $\hat{\mathbf{a}} = (\hat{a}_1, \hat{a}_2, \hat{a}_3)$ of the backward model (12.10) is related to the drift term $\mathbf{a} = (a_1, a_2, a_3)$ of the forward model (12.6) via

$$\hat{a}_i(t, \mathbf{x}, \vec{u}) = a_i(t, \mathbf{x}, \vec{u}) - C_0 \bar{\varepsilon}(\mathbf{x}, t) \frac{\partial}{\partial u_i} \ln p_E(\vec{u}; \mathbf{x}, t). \tag{12.11}$$

This form of the drift term is the consequence of Thomson's well-mixed condition (see [237]). It ensures the relation between the forward and backward p. d. f.'s used in the construction of backward algorithms in Section 3.3. Note that in Appenix B at the end of the chapter such a relation is given for a more general case.

Remark 12.1. In (12.10), the differential $\overleftarrow{d} W$ means that here the backward Ito integral is taken[1]. From this, the finite-difference form of the backward Ito equation (12.10) reads

$$\begin{aligned}
&\hat{\mathbf{X}}(t_0) - \hat{\mathbf{X}}(t_0 - \Delta t_0) = \hat{\mathbf{V}}(t_0)\Delta t_0, \\
&\hat{\mathbf{V}}(t_0) - \hat{\mathbf{V}}(t_0 - \Delta t_0) \\
&\quad = \hat{\mathbf{a}}(t_0, \hat{\mathbf{X}}(t_0), \hat{\mathbf{V}}(t_0))\Delta t_0 + \sqrt{C_0 \bar{\varepsilon}(\hat{\mathbf{X}}(t_0), t_0)} \left[W(t_0) - W(t_0 - \Delta t_0)\right],
\end{aligned}$$

where the integration step Δt_0 is positive.

Thus in this chapter we will deal with the construction of Monte Carlo estimators for the integral (12.9) based on the simulation of forward and backward Lagrangian trajectories.

12.3 Monte Carlo estimators for the mean concentration and fluxes

In this section we construct Monte Carlo estimators for the mean concentration and fluxes at a fixed point and for integrals over space and time of these mean fields. In Section 12.3.1 we deal with forward estimators for the general case of nonstationary, possibly horizontally inhomogeneous turbulence. In Section 12.3.2 we modify these estimators to the horizontally homogeneous turbulence. Backward estimators are suggested in Section 12.3.3.

---

[1] The backward Ito integral is defined by

$$\int_s^t \xi(\tau) \overleftarrow{d} W(\tau) := \int_{T-t}^{T-s} \xi(T - \tau) \, d W_T(\tau),$$

$s \le t \le T$, $W_T(\tau) := W(T) - W(T - \tau)$ is a standard Wiener process. This integral does not depend on the choice of T. For details see, e. g., [105].

12.3.1 Forward estimator

Calculation of the mean concentration and fluxes at a fixed point by forward simulation is generally not possible (e. g., [57]).

However if calculation of an integral of $\langle g(\vec{u}(\mathbf{x},t))c(\mathbf{x},t)\rangle$ over space and time is desired,

$$I_H = \int_D d\mathbf{x} \int_0^T dt \, \langle g(\vec{u}(\mathbf{x},t))c(\mathbf{x},t)\rangle \, H(\mathbf{x},t), \tag{12.12}$$

where $T > 0$ and $H(\mathbf{x},t)$ is a weight function defined on $D \times [0,T]$, then the forward estimator can be successfully used. As an example we mention the problem of evaluation of the center and size of a cloud.

Let us give now a forward Monte Carlo estimator for the integral (12.12) with arbitrary function $H(\mathbf{x},t)$. Substituting (12.9) into the right-hand side of (12.12), we get

$$\begin{aligned}
&\int_D d\mathbf{x} \int_0^T dt \, \langle g(\vec{u}(\mathbf{x},t))c(\mathbf{x},t)\rangle \, H(\mathbf{x},t) \\
&= \int_0^T dt_0 \int_D d\mathbf{x}_0 \, g(\vec{u})\mu \, Q(\mathbf{x}_0,t_0) \int_{\mathrm{IR}^3} d\vec{u}_0 p_E(\vec{u}_0;\mathbf{x}_0,t_0) \int_D d\mathbf{x} \int_{t_0}^T dt \int_{\mathrm{IR}^3} \\
&\quad \times d\vec{u} \int_0^\infty d\mu \, H(\mathbf{x},t) p_L^f(\mathbf{x},\vec{u},\mu,t;\mathbf{x}_0,\vec{u}_0,t_0) \\
&= \int_0^T dt_0 \int_D d\mathbf{x}_0 \, Q(\mathbf{x}_0,t_0) \int_{\mathrm{IR}^3} d\vec{u}_0 \, p_E(\vec{u}_0;\mathbf{x}_0,t_0) \mathbf{E}_{\mathbf{x}_0,\vec{u}_0,t_0} \int_{t_0}^T \\
&\quad \times dt \, \mu_t^{\mathbf{x}_0,\vec{u}_0,t_0} g(\mathbf{V}_t^{\mathbf{x}_0,\vec{u}_0,t_0}) H(\mathbf{X}_t^{\mathbf{x}_0,\vec{u}_0,t_0},t),
\end{aligned}$$

where $\mathbf{E}_{\mathbf{x}_0,\vec{u}_0,t_0}$ is the expectation over samples of stochastic processes $\mathbf{X}_t^{\mathbf{x}_0,\vec{u}_0,t_0}$, $\mathbf{V}_t^{\mathbf{x}_0,\vec{u}_0,t_0}$, $\mu_t^{\mathbf{x}_0,\vec{u}_0,t_0}$ (for fixed $\mathbf{x}_0$, $\vec{u}_0$, t_0).

The forward estimator can be obtained by applying the randomization procedure (see e. g., [143, 191]) to the integrals (over t_0, $\mathbf{x}_0$ and $\vec{u}_0$) in the second line of the last equality. Randomization can be done by choosing an arbitrary p. d. f. $r(\mathbf{x},t)$ defined in $D \times [0,T]$, which is consistent with $Q(\mathbf{x},t)$ in the sense that $r(\mathbf{x},t) \neq 0$ if $Q(\mathbf{x},t) \neq 0$. If $(\mathbf{x}_0,t_0)$ is a random point in $D \times [0,T]$ with the p. d. f. $r(\mathbf{x},t)$, and $\vec{u}_0$ is a 3D random variable with the p. d. f. $p_E(\vec{u};\mathbf{x}_0,t_0)$, then

$$\int_D d\mathbf{x} \int_0^T dt \, < g(\vec{u}(\mathbf{x},t))c(\mathbf{x},t) > H(\mathbf{x},t) = \mathbf{E}\xi_H,$$

where

$$\xi_H = \frac{Q(\mathbf{x}_0,t_0)}{r(\mathbf{x}_0,t_0)} \int_{t_0}^T dt \mu_t^{\mathbf{x}_0,\vec{u}_0,t_0} g(\mathbf{V}_t^{\mathbf{x}_0,\vec{u}_0,t_0}) H(\mathbf{X}_t^{\mathbf{x}_0,\vec{u}_0,t_0},t), \tag{12.13}$$

and $\mathbf{E}$ stands for the expectation over ensemble of trajectories $\mathbf{X}_t^{\mathbf{x}_0,\vec{u}_0,t_0}$, $\mathbf{V}_t^{\mathbf{x}_0,\vec{u}_0,t_0}$, $\mu_t^{\mathbf{x}_0,\vec{u}_0,t_0}$ with random initial points $\mathbf{x}_0, \vec{u}_0, t_0$.

It is reasonable to choose $r(\mathbf{x},t)$ proportional to $Q(\mathbf{x},t)$. In this case the factor Q/r in (12.13) is a constant, and this might result in a variance reduction.

12.3.2 Modified forward estimators in case of horizontally homogeneous turbulence

Time averaged mean characteristics

In this subsection the turbulence is assumed to be horizontally homogeneous, generally nonstationary, and the coefficient of absorption does not depend on the horizontal coordinates: $\gamma(\mathbf{x},t) = \gamma(z,t)$.

We use the horizontal homogeneity to calculate the time averaged mean $\int_0^T dt$ $\langle g(\vec{u}(\mathbf{x},t))c(\mathbf{x},t)\rangle h(t)$ at a fixed point $\mathbf{x} = (x,y,z)$. Here $h(t)$ is a weight function defined on $[0,T]$.

From the horizontal homogeneity it follows that

$$p_L^f(x,y,z,\vec{u},\mu,t;x_0,y_0,z_0,\vec{u}_0,t_0) = p_L^f(x-x_0,y-y_0,z,\vec{u},\mu,t;0,0,z_0,\vec{u}_0,t_0).$$

Taking into account that $p_E(\vec{u}_0;\mathbf{x}_0,t_0) = p_E(\vec{u}_0;z_0,t_0)$, we get

$$\begin{aligned}
&\int_0^T dt\, \langle g(\vec{u}(\mathbf{x},t))c(\mathbf{x},t)\rangle\, h(t) \qquad (12.14)\\
&= \int_0^T dt_0 \int_0^\infty dz_0 \int_{\mathrm{IR}^3} d\vec{u}_0\, p_E(\vec{u}_0;z_0,t_0) \int_{t_0}^T dt\, h(t) \int_D d\mathbf{x}' \int_{\mathrm{IR}^3} d\vec{u}\\
&\quad \times \int_0^\infty d\mu \delta(z-z')g(\vec{u})\,\mu\, Q(x-x',y-y',z_0,t_0)p_L^f(\mathbf{x}',\vec{u},\mu,t;0,0,z_0,\vec{u}_0,t_0)\\
&= \int_0^T dt_0 \int_0^\infty dz_0 \int_{\mathrm{IR}^3} d\vec{u}_0\, p_E(\vec{u}_0;z_0,t_0)\mathbf{E}_{z_0,\vec{u}_0,t_0} \int_{t_0}^T dt\, h(t)\delta(z-Z_t^{z_0,\vec{u}_0,t_0})\\
&\quad \times g(\mathbf{V}_t^{z_0,\vec{u}_0,t_0})\mu_t^{z_0,\vec{u}_0,t_0} Q(x-X_t^{z_0,\vec{u}_0,t_0}, y-Y_t^{z_0,\vec{u}_0,t_0}, z_0,t_0),
\end{aligned}$$

where $\mathbf{E}_{z_0,\vec{u}_0,t_0}$ is the expectation over ensemble of trajectories $\mathbf{X}_t^{z_0,\vec{u}_0,t_0}$, $\mathbf{V}_t^{z_0,\vec{u}_0,t_0}$, $\mu_t^{z_0,\vec{u}_0,t_0}$, $t \geq t_0$, which are defined as $\mathbf{X}_t^{z_0,\vec{u}_0,t_0} = (X_t^{z_0,\vec{u}_0,t_0}, Y_t^{z_0,\vec{u}_0,t_0}, Z_t^{z_0,\vec{u}_0,t_0}) = \mathbf{X}_t^{\mathbf{x}_0,\vec{u}_0,t_0}$, $\mathbf{V}_t^{z_0,\vec{u}_0,t_0} = (U_t^{z_0,\vec{u}_0,t_0}, V_t^{z_0,\vec{u}_0,t_0}, W_t^{z_0,\vec{u}_0,t_0}) = \mathbf{V}_t^{\mathbf{x}_0,\vec{u}_0,t_0}$, $\mu_t^{z_0,\vec{u}_0,t_0} = \mu(t;\mathbf{x}_0,t_0)$, with $\mathbf{x}_0 = (0,0,z_0)$.

Now we will use the following property of the Dirac delta function (e. g., see [252, p. 36, eq. (9.3)]): for arbitrary continuous function $f(\tau)$ and continuously differentiable function $Z(\tau)$

$$\int_0^t f(\tau)\delta(Z(\tau)-z)\,d\tau = \sum_{j=1}^{\nu_t(z)} \frac{f(\tau_j)}{\left|\frac{dZ(\tau_j)}{d\tau}\right|}, \tag{12.15}$$

where $\nu_t(z)$ is the number of intersections of the level z by the trajectory $Z(\tau)$ in the interval $0 \le \tau \le t$, and τ_j are the intersection times.

Thus, from (12.14), taking into account (12.15) we find

$$\begin{aligned}\int_0^T dt\,\langle g(\vec{u}(\mathbf{x},t))c(\mathbf{x},t)\rangle\,h(t) \qquad (12.16)\\ = \int_0^T dt_0 \int_0^\infty dz_0 \int_{\mathrm{IR}^3} d\vec{u}_0\; p_E(\vec{u}_0; z_0, t_0)\mathbf{E}_{z_0,\vec{u}_0,t_0}\\ \times \sum_{j=1}^{\nu_{t_0,T}(z)} h(\tau_j)\frac{g(\mathbf{V}_{\tau_j}^{z_0,\vec{u}_0,t_0})}{|W_{\tau_j}^{z_0,\vec{u}_0,t_0}|}\mu_{\tau_j}^{z_0,\vec{u}_0,t_0} Q(x - X_{\tau_j}^{z_0,\vec{u}_0,t_0}, y - Y_{\tau_j}^{z_0,\vec{u}_0,t_0}, z_0, t_0),\end{aligned}$$

where $\nu_{t_0,T}(z)$ is the number of intersections of the level z by the trajectory $Z_t^{z_0,\vec{u}_0,t_0}$ in the interval $t_0 \le t \le T$, and τ_j are the intersection times. Now the randomization of integrals over t_0, z_0 and $\vec{u}_0$ in the right-hand side of the last equality enables us to obtain the final estimator. For this, consider a p. d. f. $r(z_0, t_0)$ on $[0,\infty)\times[0,T]$ which is consistent with the source $Q(x,y,z,t)$ in the sense that $r(z_0,t_0) \ne 0$ if there exist x, y such that $Q(x,y,z_0,t_0) \ne 0$. Then,

$$\int_0^T dt\,\langle g(\vec{u}(\mathbf{x},t))c(\mathbf{x},t)\rangle h(t) = \mathbf{E}\xi_1(\mathbf{x}),$$

where

$$\begin{aligned}\xi_1(\mathbf{x}) = \frac{1}{r(z_0,t_0)} \sum_{j=1}^{\nu_{t_0,T}(z)} h(\tau_j)\frac{g(\mathbf{V}_{\tau_j}^{z_0,\vec{u}_0,t_0})}{|W_{\tau_j}^{z_0,\vec{u}_0,t_0}|}\mu_{\tau_j}^{z_0,\vec{u}_0,t_0}\\ \times Q(x - X_{\tau_j}^{z_0,\vec{u}_0,t_0}, y - Y_{\tau_j}^{z_0,\vec{u}_0,t_0}, z_0, t_0).\end{aligned}$$

Here z_0, t_0 is a 2D random variable chosen from $[0,\infty)\times[0,T]$ with the p. d. f. $r(z_0,t_0)$, and $\vec{u}_0$ is a 3D random variable with the p. d. f. $p_E(\vec{u}_0; z_0, t_0)$.

Crosswind and time-averaged mean characteristics

In this subsection the turbulence is assumed to be horizontally homogeneous (generally nonstationary), and the coefficient of absorption does not depend on the horizontal coordinates : $\gamma(\mathbf{x},t) = \gamma(z,t)$. Let us estimate the crosswind and time-averaged mean characteristic

$$I_h = \int_0^T dt \int_{-\infty}^{\infty} dy \, \langle g(\vec{u}(x,y,z,t))c(x,y,z,t)\rangle \, h(y,t)$$

at a fixed point (x,z). Here $h(y,t)$ is a weight function defined on $(-\infty,\infty) \times [0,T]$. Using the same arguments as in the previous subsection we get

$$\begin{aligned} I_h &= \int_0^T dt \int_{-\infty}^{\infty} dy \, \langle g(\vec{u}(x,y,z,t))c(x,y,z,t)\rangle \, h(y,t) \qquad (12.17) \\ &= \int_0^T dt_0 \int_{-\infty}^{\infty} dy_0 \int_0^{\infty} dz_0 \int_{\mathbb{R}^3} d\vec{u}_0 \, p_E(\vec{u}_0; z_0, t_0) \mathbf{E}_{y_0,z_0,\vec{u}_0,t_0} \\ &\times \int_{t_0}^T dt \, h(Y_t^{y_0,z_0,\vec{u}_0,t_0}, t)\delta(z - Z_t^{y_0,z_0,\vec{u}_0,t_0}) g(\mathbf{V}_t^{y_0,z_0,\vec{u}_0,t_0}) \mu_t^{y_0,z_0,\vec{u}_0,t_0} \\ &\times Q(x - X_t^{y_0,z_0,\vec{u}_0,t_0}, y_0, z_0, t_0), \end{aligned}$$

where $\mathbf{E}_{y_0,z_0,\vec{u}_0,t_0}$ is the expectation over the ensemble of trajectories $\mathbf{X}_t^{y_0,z_0,\vec{u}_0,t_0}$, $\mathbf{V}_t^{y_0,z_0,\vec{u}_0,t_0}$, $\mu_t^{y_0,z_0,\vec{u}_0,t_0}$, $t \geq t_0$, which are defined as

$$\begin{aligned} \mathbf{X}_t^{y_0,z_0,\vec{u}_0,t_0} &= (X_t^{y_0,z_0,\vec{u}_0,t_0}, Y_t^{y_0,z_0,\vec{u}_0,t_0}, Z_t^{y_0,z_0,\vec{u}_0,t_0}) = \mathbf{X}_t^{\mathbf{x}_0,\vec{u}_0,t_0}, \\ \mathbf{V}_t^{y_0,z_0,\vec{u}_0,t_0} &= (U_t^{y_0,z_0,\vec{u}_0,t_0}, V_t^{y_0,z_0,\vec{u}_0,t_0}, W_t^{y_0,z_0,\vec{u}_0,t_0}) = \mathbf{V}_t^{\mathbf{x}_0,\vec{u}_0,t_0}, \mu_t^{y_0,z_0,\vec{u}_0,t_0} \\ &= \mu(t; \mathbf{x}_0, t_0), \end{aligned}$$

with $\mathbf{x}_0 = (0, y_0, z_0)$.

From (12.17) we get by the property (12.15) that

$$\begin{aligned} I_h &= \int_0^T dt_0 \int_{-\infty}^{\infty} dy_0 \int_0^{\infty} dz_0 \int_{\mathbb{R}^3} d\vec{u}_0 \, p_E(\vec{u}_0, z_0, t_0) \, \mathbf{E}_{y_0,z_0,\vec{u}_0,t_0} \qquad (12.18) \\ &\times \sum_{j=1}^{\nu_{t_0,T}(z)} h(Y_{\tau_j}^{y_0,z_0,\vec{u}_0,t_0}, \tau_j) \frac{g(\mathbf{V}_{\tau_j}^{y_0,z_0,\vec{u}_0,t_0})}{|W_{\tau_j}^{y_0,z_0,\vec{u}_0,t_0}|} \mu_{\tau_j}^{y_0,z_0,\vec{u}_0,t_0} Q(x - X_{\tau_j}^{y_0,z_0,\vec{u}_0,t_0}, y_0, z_0, t_0), \end{aligned}$$

where $\nu_{t_0,T}(z)$ is the number of intersections of the level z by the trajectory $Z_t^{y_0,z_0,\vec{u}_0,t_0}$ during the interval $t_0 \leq t \leq T$, and τ_j are the intersection times.

Now it is not difficult to construct a random estimator for I_h by applying a standard Monte Carlo randomization procedure for evaluation of integrals. In our case we apply it to the multiple integral in (12.18) over t_0, y_0, z_0 and $\vec{u}_0$. To this end, we consider a probability density $r(y_0, z_0, t_0)$ on $(-\infty,\infty) \times [0,\infty) \times [0,T]$ which is consistent with the source $Q(x,y,z,t)$ in the sense that $r(y_0,z_0,t_0) \neq 0$ if there exists x such that $Q(x,y_0,z_0,t_0) \neq 0$. Then

$$\mathbf{E}\xi_2(x,z) = \int_0^T dt \int_{-\infty}^{\infty} dy \, \langle g(\vec{u}(x,y,z,t))c(x,y,z,t)\rangle \, h(y,t),$$

where

$$\xi_2(x,z) = \frac{1}{r(y_0,z_0,t_0)} \sum_{j=1}^{\nu_{t_0,T}(z)} h(Y_{\tau_j}^{y_0,z_0,\vec{u}_0,t_0}, \tau_j) \frac{g(\mathbf{V}_{\tau_j}^{y_0,z_0,\vec{u}_0,t_0})}{|W_{\tau_j}^{y_0,z_0,\vec{u}_0,t_0}|} \mu_{\tau_j}^{y_0,z_0,\vec{u}_0,t_0}$$
$$\times Q(x - X_{\tau_j}^{y_0,z_0,\vec{u}_0,t_0}, y_0, z_0, t_0).$$

Here y_0, z_0, t_0 is a random variable chosen in $(-\infty,\infty) \times [0,\infty) \times [0,T]$ with the density $r(y_0,z_0,t_0)$, and $\vec{u}_0$ is a 3D random variable with the p. d. f. $p_E(\vec{u}_0; z_0, t_0)$.

Stationary turbulence

In this subsection the turbulence is assumed to be horizontally homogeneous and stationary, and the coefficient of absorption depends only on height: $\gamma(\mathbf{x},t) = \gamma(z)$. In addition, the initial concentration is assumed to be zero: $q_0(\mathbf{x}) = 0$. These assumptions allow us to estimate $\langle g(\vec{u}(\mathbf{x},t))c(\mathbf{x},t)\rangle$ directly at a fixed point $(\mathbf{x},t)$. Indeed, under the assumed conditions

$$p_L^f(x,y,z,\vec{u},\mu,t; x_0,y_0,z_0,\vec{u}_0,t_0) = p_L^f(x-x_0, y-y_0, z, \vec{u}, \mu, t-t_0; 0,0,z_0,\vec{u}_0,0).$$

Therefore, by $p_E(\vec{u}_0; \mathbf{x}_0, t_0) = p_E(\vec{u}_0; z_0)$ we get

$$\begin{aligned} &\langle g(\vec{u}(\mathbf{x},t))c(\mathbf{x},t)\rangle \qquad (12.19)\\ &= \int_0^{\infty} dz_0 \int_{\mathrm{IR}^3} d\vec{u}_0 \, p_E(\vec{u}_0; z_0) \int_0^t d\tau \int_D d\mathbf{x}' \int_{\mathrm{IR}^3} d\vec{u} \int_0^{\infty} d\mu \delta(z-z') g(\vec{u}) \\ &\quad \times \mu \, q(x-x', y-y', z_0, t-\tau) p_L^f(\mathbf{x}', \vec{u}, \mu, \tau; 0,0,z_0,\vec{u}_0,0) \\ &= \int_0^{\infty} dz_0 \int_{\mathrm{IR}^3} d\vec{u}_0 \, p_E(\vec{u}_0; z_0) \mathbf{E}_{z_0,\vec{u}_0} \int_0^t d\tau \delta(z - Z_\tau^{z_0,\vec{u}_0}) g(\mathbf{V}_\tau^{z_0,\vec{u}_0}) \mu_\tau^{z_0,\vec{u}_0} \\ &\quad \times q(x - X_\tau^{z_0,\vec{u}_0}, y - Y_\tau^{z_0,\vec{u}_0}, z_0, t-\tau), \end{aligned}$$

where $\mathbf{E}_{z_0,\vec{u}_0}$ is the expectation over the ensemble of trajectories $\mathbf{X}_\tau^{z_0,\vec{u}_0}$, $\mathbf{V}_\tau^{z_0,\vec{u}_0}$, $\mu_\tau^{z_0,\vec{u}_0}$, $\tau \geq 0$, which are defined as $\mathbf{X}_\tau^{z_0,\vec{u}_0} = (X_\tau^{z_0,\vec{u}_0}, Y_\tau^{z_0,\vec{u}_0}, Z_\tau^{z_0,\vec{u}_0}) = \mathbf{X}_\tau^{\mathbf{x}_0,\vec{u}_0,0}$, $\mathbf{V}_\tau^{z_0,\vec{u}_0} = (U_\tau^{z_0,\vec{u}_0}, V_\tau^{z_0,\vec{u}_0}, W_\tau^{z_0,\vec{u}_0}) = \mathbf{V}_\tau^{\mathbf{x}_0,\vec{u}_0,0}$, $\mu_\tau^{z_0,\vec{u}_0} = \mu(\tau; \mathbf{x}_0, 0)$, with $\mathbf{x}_0 = (0,0,z_0)$.

Now, from (12.19) we get by (12.15)

$$\langle g(\vec{u}(\mathbf{x},t))c(\mathbf{x},t)\rangle = \int_0^\infty dz_0 \int_{\mathrm{IR}^3} d\vec{u}_0\, p_E(\vec{u}_0,z_0)\mathbf{E}_{z_0,\vec{u}_0} \sum_{j=1}^{\nu_t(z)} \frac{g(\mathbf{V}_{\tau_j}^{z_0,\vec{u}_0})}{|W_{\tau_j}^{z_0,\vec{u}_0}|}\mu_{\tau_j}^{z_0,\vec{u}_0}$$
$$\times\, q(x - X_{\tau_j}^{z_0,\vec{u}_0}, y - Y_{\tau_j}^{z_0,\vec{u}_0}, z_0, t-\tau_j),$$

where $\nu_t(z)$ is the number of intersections of the level z by the trajectory $Z_\tau^{z_0,\vec{u}_0}$ during the interval $0 \le \tau \le t$, and τ_j are the intersection times.

We apply here the standard Monte Carlo randomization procedure to evaluate the integrals over z_0 and $\vec{u}_0$. To this end, we consider a probability density $r(z_0)$ on $[0,\infty)$ which is consistent with the source $q(x,y,z,t)$ in the sense that $r(z_0) \ne 0$ if there exist x, y, t such that $q(x,y,z_0,t) \ne 0$. Then

$$\langle g(\vec{u}(\mathbf{x},t))c(\mathbf{x},t)\rangle = \mathbf{E}\xi_3(\mathbf{x},t),$$

where

$$\xi_3(\mathbf{x},t) = \frac{1}{r(z_0)} \sum_{j=1}^{\nu_t(z)} \frac{g(\mathbf{V}_{\tau_j}^{z_0,\vec{u}_0})}{|W_{\tau_j}^{z_0,\vec{u}_0}|}\mu_{\tau_j}^{z_0,\vec{u}_0} q(x - X_{\tau_j}^{z_0,\vec{u}_0}, y - Y_{\tau_j}^{z_0,\vec{u}_0}, z_0, t-\tau_j). \quad (12.20)$$

Here z_0 is a random variable chosen in $[0,\infty)$ with the density $r(z_0)$, and $\vec{u}_0$ is a 3D random variable with the p. d. f. $p_E(\vec{u}_0; z_0)$.

Analogously the crosswind averaged mean can be evaluated at a fixed point (x,z,t): $\int_{-\infty}^{\infty} dy\, \langle g(\vec{u}(x,y,z,t))c(x,y,z,t)\rangle\, h(y)$. Here $h(y)$ is a weight function defined on $(-\infty,\infty)$. Let $r(y_0,z_0)$ be a probability density defined on $(-\infty,\infty)\times[0,\infty)$ which is consistent with the source $q(x,y,z,t)$ in the sense that $r(y_0,z_0) \ne 0$ if there exist x, t such that $q(x,y_0,z_0,t) \ne 0$. Then

$$\int_{-\infty}^{\infty} dy\, \big\langle g(\vec{u}(x,y,z,t))c(x,y,z,t)\big\rangle\, h(y) = \mathbf{E}\xi_4(x,z,t),$$

where

$$\xi_4(x,z,t) = \frac{1}{r(y_0,z_0)} \sum_{j=1}^{\nu_t(z)} h(Y_{\tau_j}^{y_0,z_0,\vec{u}_0}) \frac{g(\mathbf{V}_{\tau_j}^{y_0,z_0,\vec{u}_0})}{|W_{\tau_j}^{y_0,z_0,\vec{u}_0}|}\mu_{\tau_j}^{y_0,z_0,\vec{u}_0}$$
$$\times\, q(x - X_{\tau_j}^{y_0,z_0,\vec{u}_0}, y_0, z_0, t-\tau_j).$$

Here (y_0,z_0) is a random variable chosen in $(-\infty,\infty)\times[0,\infty)$ with the density $r(y_0,z_0)$, $\vec{u}_0$ is a 3D random variable with the p. d. f. $p_E(\vec{u}_0; z_0)$ and $\mathbf{X}_\tau^{y_0,z_0,\vec{u}_0}$, $\mathbf{W}_\tau^{y_0,z_0,\vec{u}_0}$, $\mu_\tau^{y_0,z_0,\vec{u}_0}$ $(\tau \ge 0)$ are stochastic processes defined as $\mathbf{X}_\tau^{y_0,z_0,\vec{u}_0} = (X_\tau^{y_0,z_0,\vec{u}_0}, Y_\tau^{y_0,z_0,\vec{u}_0}, Z_\tau^{y_0,z_0,\vec{u}_0}) = \mathbf{X}_\tau^{\mathbf{x}_0,\vec{u}_0,0}$, $\mathbf{V}_\tau^{y_0,z_0,\vec{u}_0} = (U_\tau^{y_0,z_0,\vec{u}_0}, V_\tau^{y_0,z_0,\vec{u}_0}, W_\tau^{y_0,z_0,\vec{u}_0}) = \mathbf{V}_\tau^{\mathbf{x}_0,\vec{u}_0,0}$, $\mu_\tau^{y_0,z_0,\vec{u}_0} = \mu(\tau; \mathbf{x}_0, 0)$, with $\mathbf{x}_0 = (0, y_0, z_0)$.

12.3.3 Backward estimator

Unlike the forward algorithm, the backward technique enables us to estimate the mean concentration and fluxes at a fixed point in space and time, even in the general case of nonstationary turbulence. Therefore, the estimation can be done directly for $\langle g(\vec{u})c\rangle$. Note that taking $g(\vec{u})$ equal to 1 or to u_i, we get $\langle g(\vec{u})c\rangle = \langle c\rangle$ or $\langle g(\vec{u})c\rangle = \langle u_i c\rangle$, respectively.

Analogously to forward Lagrangian p. d. f. (12.7), the backward Lagrangian p. d. f. can be defined as

$$p_L^b(\mathbf{x}_0, \vec{u}_0, \mu_0, t_0; \mathbf{x}, \vec{u}, t) = \mathbf{E}_{\mathbf{x},\vec{u},t}\{\delta(\mathbf{x}_0 - \hat{\mathbf{X}}_{t_0}^{\mathbf{x},\vec{u},t})\delta(\vec{u}_0 - \hat{\mathbf{V}}_{t_0}^{\mathbf{x},\vec{u},t})\delta(\mu_0 - \hat{\mu}_{t_0}^{\mathbf{x},\vec{u},t})\},$$

where $\hat{\mu}(t_0) = \hat{\mu}_{t_0}^{\mathbf{x},\vec{u},t}$ is defined by

$$\frac{d\hat{\mu}(t_0)}{dt_0} = \gamma(\hat{\mathbf{X}}_{t_0}^{\mathbf{x},\vec{u},t}, t_0)\hat{\mu}(t_0), \quad \hat{\mu}(t) = 1,$$

and $\mathbf{E}_{\mathbf{x},\vec{u},t}$ means the expectation over samples of stochastic processes $\hat{\mathbf{X}}_{t_0}^{\mathbf{x},\vec{u},t}$, $\hat{\mathbf{V}}_{t_0}^{\mathbf{x},\vec{u},t}$, and $\hat{\mu}_{t_0}^{\mathbf{x},\vec{u},t}$, $t_0 \le t$, starting at final time $t_0 = t$ at point $\mathbf{x}, \vec{u}, 1$. In appendix C it is shown that

$$p_E(\vec{u}_0; \mathbf{x}_0, t_0)p_L^f(\mathbf{x}, \vec{u}, \mu, t; \mathbf{x}_0, \vec{u}_0, t_0) = p_E(\vec{u}; \mathbf{x}, t)p_L^b(\mathbf{x}_0, \vec{u}_0, \mu, t_0; \mathbf{x}, \vec{u}, t). \tag{12.21}$$

Substituting the right-hand side of this equality to the right-hand side of (12.9), we get

$$\begin{aligned}
&\langle g(\vec{u}(\mathbf{x}, t))c(\mathbf{x}, t)\rangle \\
&= \int_{\mathrm{IR}^3} d\vec{u} \int_0^\infty d\mu \int_0^t dt_0 \int_D d\mathbf{x}_0\, g(\vec{u})\mu\, Q(\mathbf{x}_0, t_0) \int_{\mathrm{IR}^3} d\vec{u}_0\, p_E(\vec{u}; \mathbf{x}, t) \\
&\quad \times p_L^b(\mathbf{x}_0, \vec{u}_0, \mu, t_0; \mathbf{x}, \vec{u}, t) \\
&= \int_{\mathrm{IR}^3} d\vec{u}\, p_E(\vec{u}; \mathbf{x}, t)g(\vec{u})\mathbf{E}_{\mathbf{x},\vec{u},t} \int_0^t dt_0\, \hat{\mu}_{t_0}^{\mathbf{x},\vec{u},t} Q(\hat{\mathbf{X}}_{t_0}^{\mathbf{x},\vec{u},t}, t_0).
\end{aligned}$$

From the last expression, using the standard Monte Carlo arguments, one gets

$$\langle g(\vec{u}(\mathbf{x}, t))c(\mathbf{x}, t)\rangle = \mathbf{E}\hat{\xi}(\mathbf{x}, t), \tag{12.22}$$

where

$$\begin{aligned}
\hat{\xi}(\mathbf{x}, t) &= g(\vec{u}) \int_0^t dt_0\, \hat{\mu}_{t_0}^{\mathbf{x},\vec{u},t} Q(\hat{\mathbf{X}}_{t_0}^{\mathbf{x},\vec{u},t}, t_0) \\
&= g(\vec{u}) \left(\hat{\mu}_0^{\mathbf{x},\vec{u},t} q_0(\hat{\mathbf{X}}_0^{\mathbf{x},\vec{u},t}) + \int_0^t dt_0\, \hat{\mu}_{t_0}^{\mathbf{x},\vec{u},t} q(\hat{\mathbf{X}}_{t_0}^{\mathbf{x},\vec{u},t}, t_0) \right).
\end{aligned} \tag{12.23}$$

Here $\vec{u}$ is $3D$ random variable with the p. d. f. $p_E(\vec{u}; \mathbf{x}, t)$.

Now we are in a position to construct a Monte Carlo estimator for the integral (12.12) from $\langle g(\vec{u})c\rangle$ over space and time with an averaging function $H(\mathbf{x},t)$. For this, we consider an arbitrary p. d. f. $p(\mathbf{x},t)$ defined in $D \times [0,T]$ which is consistent with $H(\mathbf{x},t)$ in the sense that $p(\mathbf{x},t) \neq 0$ if $H(\mathbf{x},t) \neq 0$. Let $(\mathbf{x},t)$ be a random point in $D \times [0,T]$ with the p. d. f. $p(\mathbf{x},t)$, and $\vec{u}$ be a 3D random variable with the p. d. f. $p_E(\vec{u};\mathbf{x},t)$. Then from (12.22)–(12.23) and the standard Monte Carlo arguments it follows that the random variable

$$\begin{aligned}\hat{\xi}_H &= \frac{H(\mathbf{x},t)}{p(\mathbf{x},t)} g(\vec{u}) \int_0^t dt_0\, \hat{\mu}_{t_0}^{\mathbf{x},\vec{u},t}\, Q(\hat{\mathbf{X}}_{t_0}^{\mathbf{x},\vec{u},t}, t_0)\\ &= \frac{H(\mathbf{x},t)}{p(\mathbf{x},t)}\, g(\vec{u}) \left(\hat{\mu}_0^{\mathbf{x},\vec{u},t} q_0(\hat{\mathbf{X}}_0^{\mathbf{x},\vec{u},t}) + \int_0^t dt_0\, \hat{\mu}_{t_0}^{\mathbf{x},\vec{u},t} q(\hat{\mathbf{X}}_{t_0}^{\mathbf{x},\vec{u},t}, t_0)\right)\end{aligned}$$

is a Monte Carlo estimator for the integral (12.12):

$$\int_D d\mathbf{x} \int_0^T dt\, \langle g(\vec{u}(\mathbf{x},t))c(\mathbf{x},t)\rangle H(\mathbf{x},t) = \mathbf{E}\hat{\xi}_H\ .$$

12.4 Application to the footprint problem

The footprint problem as formulated in the literature (e. g., see [55, 221]) essentially deals with the calculation of the contribution to the mean concentration and its flux at a fixed point from a surface source of a scalar.

Let us consider a surface source at a height z_s, and let $F(x,y,t)$ be an amount of emitted scalar per unite time and area (at time t near the surface point (x,y)). Then the distribution function $q(\mathbf{x},t)$ has the form

$$q(\mathbf{x},t) = q(x,y,z,t) = F(x,y,t)\,\delta(z-z_s). \tag{12.24}$$

We assume that the turbulence is horizontally homogeneous and stationary. The coefficient of absorption is assumed to depend only on height: $\gamma(\mathbf{x},t) = \gamma(z)$. The initial concentration distribution is assumed to be zero: $q_0(\mathbf{x}) = 0$. The Lagrangian trajectories are perfectly reflected at roughness height z_*. Therefore we will naturally assume that $z_s \geq z_*$.

First, let us construct a Monte Carlo estimator for $\langle g(\vec{u}(\mathbf{x},t))c(\mathbf{x},t)\rangle$ based on the *forward Lagrangian trajectory* $\mathbf{X}_\tau^{z_s,\vec{u}_0}$, $\mathbf{V}_\tau^{z_s,\vec{u}_0}$, $\mu_\tau^{z_s,\vec{u}_0}$, $\tau \geq 0$ (see Section 3.2.3). Indeed, choosing in (12.20) $r(z_0) = \delta(z_0 - z_s)$ and taking into account (12.24), we have

$$\begin{aligned}&\langle g(\vec{u}(\mathbf{x},t))c(\mathbf{x},t)\rangle\\ &= \mathbf{E}_{z_s,\vec{u}_0}\left(\sum_{j=1}^{\nu_t(z)} \frac{g(\mathbf{V}_{\tau_j}^{z_s,\vec{u}_0})}{|W_{\tau_j}^{z_s,\vec{u}_0}|} \mu_{\tau_j}^{z_s,\vec{u}_0} F(x - X_{\tau_j}^{z_s,\vec{u}_0}, y - Y_{\tau_j}^{z_s,\vec{u}_0}, t-\tau_j)\right),\end{aligned} \tag{12.25}$$

where $\vec{u}_0$ is a 3D random variable with the p. d. f. $p_E(\vec{u}_0; z_s)$ and $\mathbf{E}_{z_s,\vec{u}_0}$ means an expectation over samples of the stochastic processes $\mathbf{X}_\tau^{z_s,\vec{u}_0}$, $\mathbf{V}_\tau^{z_s,\vec{u}_0}$, $\mu_\tau^{z_s,\vec{u}_0}$, $\tau \geq 0$. In practical implementation the mathematical expectation on the right-hand side of (12.25) is approximately calculated as

$$\mathbf{E}_{z_s,\vec{u}_0}\Bigg(\sum_{j=1}^{\nu_t(z)} \frac{g(\mathbf{V}_{\tau_j}^{z_s,\vec{u}_0})}{|W_{\tau_j}^{z_s,\vec{u}_0}|}\mu_{\tau_j}^{z_s,\vec{u}_0} F(x - X_{\tau_j}^{z_s,\vec{u}_0}, y - Y_{\tau_j}^{z_s,\vec{u}_0}, t - \tau_j)\Bigg)$$
$$\simeq \frac{1}{N}\sum_{i=1}^{N}\sum_{j=1}^{\nu_i} \frac{g(\mathbf{V}_{\tau_{ij}}^i)}{|W_{\tau_{ij}}^i|}\mu_{\tau_{ij}}^i F(x - X_{\tau_{ij}}^i, y - Y_{\tau_{ij}}^i, t - \tau_{ij}), \tag{12.26}$$

where i denotes the trajectory starting with the initial velocity $\vec{u}_{0i}$ (which is random with the p. d. f. $p_E(\vec{u}_0; z_s)$ and independent for different i), N is the number of trajectories, ν_i is the number of intersections of the level z by i-th trajectory, and τ_{ij} are the intersection times.

Now let us construct a Monte Carlo estimator for $\langle g(\vec{u}(\mathbf{x},t))c(\mathbf{x},t)\rangle$ based on the *backward Lagrangian trajectory* $\hat{\mathbf{X}}_{t_0}^{\mathbf{x},\vec{u},t}$, $\hat{\mathbf{V}}_{t_0}^{\mathbf{x},\vec{u},t}$, and $\hat{\mu}_{t_0}^{\mathbf{x},\vec{u},t}$, $t_0 \leq t$ (see Section 3.3). First we will assume that $z_s > z_*$. Taking into account (12.24) and using the property (12.15) of the Dirac delta function, from (12.22)–(12.23) we find

$$\langle g(\vec{u}(\mathbf{x},t))c(\mathbf{x},t)\rangle = \mathbf{E}_{\mathbf{x},\vec{u},t}\left(g(\vec{u})\int_0^t dt_0\, \hat{\mu}_{t_0}^{\mathbf{x},\vec{u},t} q(\hat{\mathbf{X}}_{t_0}^{\mathbf{x},\vec{u},t}, t_0)\right)$$
$$= \mathbf{E}_{\mathbf{x},\vec{u},t}\left(g(\vec{u})\sum_{j=1}^{\hat{\nu}_t(z_s)} \frac{\hat{\mu}_{\tau_j}^{\mathbf{x},\vec{u},t}}{|\hat{W}_{\tau_j}^{\mathbf{x},\vec{u},t}|} F(\hat{X}_{\tau_j}^{\mathbf{x},\vec{u},t}, \hat{Y}_{\tau_j}^{\mathbf{x},\vec{u},t}, \tau_j)\right), \tag{12.27}$$

where $\hat{\nu}_t(z_s)$ is the number of intersections of the level z_s by the backward trajectory $\hat{Z}_\tau^{\mathbf{x},\vec{u},t}$ in the interval $0 \leq \tau \leq t$; τ_j are the intersection times; $\vec{u}$ is $3D$ random variable with the p. d. f. $p_E(\vec{u}; \mathbf{x}, t)$, and $\mathbf{E}_{\mathbf{x},\vec{u},t}$ is the expectation taken over samples of the stochastic processes $\hat{\mathbf{X}}_{t_0}^{\mathbf{x},\vec{u},t}$, $\hat{\mathbf{V}}_{t_0}^{\mathbf{x},\vec{u},t}$, and $\hat{\mu}_{t_0}^{\mathbf{x},\vec{u},t}$, $t_0 \leq t$. The surface emission at the height where the trajectories are reflected, (the case $z_s = z_*$) can be handled by letting $z_s \to z_*$ ($z_s > z_*$). Taking into account that for each time τ_j the trajectory $\hat{Z}_\tau^{\mathbf{x},\vec{u},t}$ will simultaneously pass twice (first in dawnward direction and, then, in upward one) the level z_s, it is easy to establish that

$$\langle g(\vec{u}(\mathbf{x},t))c(\mathbf{x},t)\rangle = \mathbf{E}\left(g(\vec{u})\int_0^t dt_0\, \hat{\mu}_{t_0}^{\mathbf{x},\vec{u},t} q(\hat{\mathbf{X}}_{t_0}^{\mathbf{x},\vec{u},t}, t_0)\right)$$
$$= 2\,\mathbf{E}\left(g(\vec{u})\sum_{j=1}^{\hat{\nu}_t(z_*)} \frac{\hat{\mu}_{\tau_j}^{\mathbf{x},\vec{u},t}}{|\hat{W}_{\tau_j}^{\mathbf{x},\vec{u},t}|} F(\hat{X}_{\tau_j}^{\mathbf{x},\vec{u},t}, \hat{Y}_{\tau_j}^{\mathbf{x},\vec{u},t}, \tau_j)\right). \tag{12.28}$$

In practice, the approximate calculation of mathematical expectations in the right-hand sides of (12.27)–(12.28) is carried out by similiar technique as in (12.26). For example, in the case $z_s = z_*$ we have

$$\mathbf{E}_{\mathbf{x},\vec{u},t}\left(g(\vec{u})\sum_{j=1}^{\hat{\nu}_t(z_*)}\frac{\hat{\mu}_{\tau_j}^{\mathbf{x},\vec{u},t}}{|\hat{W}_{\tau_j}^{\mathbf{x},\vec{u},t}|}q(\hat{\mathbf{X}}_{\tau_j}^{\mathbf{x},\vec{u},t},\tau_j)\right)$$

$$\simeq 2\frac{1}{N}\sum_{i=1}^{N}\sum_{j=1}^{\hat{\nu}_i} g(\vec{u})\frac{\hat{\mu}_{\tau_{ij}}^{i}}{|\hat{W}_{\tau_{ij}}^{i}|}F(\hat{X}_{\tau_{ij}}^{i},\hat{X}_{\tau_{ij}}^{\mathbf{x},\vec{u},t},\tau_{ij}). \qquad (12.29)$$

12.5 Conclusion

Direct and backward Lagrangian stochastic algorithms for the numerical evaluation of the mean concentration of scalars and their fluxes were suggested and justified. The random estimators were constructed in the form of expectations over stochastic Lagrangian trajectories governed by Langevin-type equations derived from Thomson's well-mixed condition. The transported scalar may be absorbed. Detailed expressions for random estimators for the mean characteristics (concentration, fluxes, time and space averages of concentration and fluxes) for quite general cases of sources were given. A practically important case of a plane source (related to the so-called "footprint problem") was treated in detail. Advantages of the methods developed were that they are flexible to the structure of the source and the measured statistical characteristics.

12.6 Appendices

12.6.1 Appendix A. Representation of concentration in Lagrangian description

Here we show that the equality (12.1) is true. The total instantaneous concentration $c(\mathbf{x},t)$ can be represented as the sum of $c_0(\mathbf{x},t)$ and $c_1(\mathbf{x},t)$, defined by

$$\frac{\partial c_0(\mathbf{x},t)}{\partial t} + u_i(\mathbf{x},t)\frac{\partial c_0}{\partial x_i} + \gamma(\mathbf{x},t)c_0(\mathbf{x},t) = 0, t > 0; \quad c_0(\mathbf{x},0) = q_0(\mathbf{x}), \qquad \text{(A1)}$$

and

$$\frac{\partial c_1(\mathbf{x},t)}{\partial t} + u_i(\mathbf{x},t)\frac{\partial c_1}{\partial x_i} + \gamma(\mathbf{x},t)c_1(\mathbf{x},t) = q(\mathbf{x},t), t > 0; \quad c_1(\mathbf{x},0) = 0,$$

respectively. First we show that

$$c_0(\mathbf{x},t) = \int_D d\mathbf{x}_0\, \mu(t;\mathbf{x}_0,0)q_0(\mathbf{x}_0)\delta(\mathbf{x}-\mathbf{X}(t;\mathbf{x}_0,0)). \qquad \text{(A2)}$$

According to (A1) the function $C_0(t) = C_0(t;\mathbf{x}_0) = c_0(\mathbf{X}(t;\mathbf{x}_0,0),t)$ satisfies the equation

$$\frac{dC_0(t)}{dt} + \gamma(\mathbf{X}(t;\mathbf{x}_0,0),t)C_0(t) = 0, \quad C_0(0) = q_0(\mathbf{x}_0).$$

Therefore from the definition of $\mu(t;\mathbf{x}_0,t_0)$ given by (2) it follows that $C_0(t;\mathbf{x}_0) = \mu(t;\mathbf{x}_0,0)q_0(\mathbf{x}_0)$ and

$$\begin{aligned}\int_D & d\mathbf{x}_0\, \mu(t;\mathbf{x}_0,0)q_0(\mathbf{x}_0)\delta(\mathbf{x}-\mathbf{X}(t;\mathbf{x}_0,0)) \\ &= \int_D d\mathbf{x}_0\, c_0(\mathbf{X}(t;\mathbf{x}_0,0),t)\delta(\mathbf{x}-\mathbf{X}(t;\mathbf{x}_0,0)) \\ &= \int_D d\mathbf{y}_0 c_0(\mathbf{y}_0,t)\delta(\mathbf{x}-\mathbf{y}_0) = c_0(\mathbf{x},t).\end{aligned}$$

Here in the last integral the substitution of variables $\mathbf{x}_0 \to \mathbf{y}_0 = \mathbf{X}(t;\mathbf{x}_0,0)$ was performed, and it was taken into account that the Jacobian of this transformation equals unity due to incompressibility of the velocity field $\vec{u}(\mathbf{x},t)$ ([145]). With this, (A2) is established.

Now the following equality will be shown:

$$c_1(\mathbf{x},t) = \int_0^t dt_0 \int_D d\mathbf{x}_0\, \mu(t;\mathbf{x}_0,t_0)q(\mathbf{x}_0,t_0)\delta(\mathbf{x}-\mathbf{X}(t;\mathbf{x}_0,t_0)). \tag{A3}$$

Indeed, it can be easily shown (by taking suitable derivatives) that

$$c_1(\mathbf{x},t) = \int_0^t dt_0\, g_{t_0}(\mathbf{x},t), \tag{A4}$$

where $g_{t_0}(\mathbf{x},t)$, $(t_0 > 0)$ is defined by

$$\frac{\partial g_{t_0}(\mathbf{x},t)}{\partial t} + u_i(\mathbf{x},t)\frac{\partial g_{t_0}}{\partial x_i} + \gamma(\mathbf{x},t)g_{t_0}(\mathbf{x},t) = 0, t > t_0; \quad g_{t_0}(\mathbf{x},t_0) = q(\mathbf{x},t_0). \tag{A5}$$

From (A5) and by the definition of the function $\mu(t;\mathbf{x}_0,t_0)$ given by (12.2), it follows that

$$g_{t_0}(\mathbf{X}(t;\mathbf{x}_0,t_0),t) = \mu(t;\mathbf{x}_0,t_0)q(\mathbf{x}_0,t_0).$$

From this equality and since the Jacobian of the transformation $\mathbf{x}_0 \to \mathbf{y}_0 = \mathbf{X}(t;\mathbf{x}_0,t_0)$ is equal to unity, we obtain

$$\begin{aligned}\int_D & d\mathbf{x}_0\, \mu(t;\mathbf{x}_0,t_0)q(\mathbf{x}_0,t_0)\delta(\mathbf{x}-\mathbf{X}(t;\mathbf{x}_0,t_0)) \\ &= \int_D d\mathbf{x}_0\, g_{t_0}(\mathbf{X}(t;\mathbf{x}_0,t_0),t)\delta(\mathbf{x}-\mathbf{X}(t;\mathbf{x}_0,t_0)) \\ &= \int_D d\mathbf{y}_0 g_{t_0}(\mathbf{y}_0,t)\delta(\mathbf{x}-\mathbf{y}_0) = g_{t_0}(\mathbf{x},t).\end{aligned}$$

The last equality and (A4) yields (A3). Since $c(\mathbf{x}, t) = c_0(\mathbf{x}, t) + c_1(\mathbf{x}, t)$, from (A2) and (A3) it follows that the representation (12.1) holds.

12.6.2 Appendix B. Relation between forward and backward transition density functions

Here we present the relation between the forward and backward p. d. f.'s used further in Appendix C. Let $p^f(\mathbf{y}, t; \mathbf{y}_0, t_0) = \langle \delta(\mathbf{y} - \mathbf{Y}_t^{\mathbf{y}_0, t_0}) \rangle$ be the transition density function of the n-dimensional diffusion process $\mathbf{Y}_t^{\mathbf{y}_0, t_0}$, the solution to

$$dY_i(t) = A_i(\mathbf{Y}(t), t)dt + \sigma_{ij}(\mathbf{Y}(t), t)dW_j(t),\ t > t_0,\ i = 1, \dots, n,\ \mathbf{Y}(t)|_{t=t_0} = \mathbf{y}_0, \tag{B1}$$

where $A_i(\mathbf{y}, t)$ and $\sigma_{ij}(\mathbf{y}, t)$ are functions defined in $D \times [0, T]$, $W_1(t), \dots, W_n(t)$ are independent standard Wiener processes, and D is a domain in $\mathbb{R}^n$, $T > 0$.

We assume that the boundary of D is impenetrable, i. e., the trajectories determined by (B1) do not reach the boundary. Assume that we have a positive function $\rho(\mathbf{y}, t)$ defined on $D \times [0, T]$ as a solution to the equation

$$\frac{\partial \rho}{\partial t} + \frac{\partial}{\partial y_i}(A_i \rho) = \frac{1}{2} \frac{\partial^2 (B_{ij} \rho)}{\partial y_i \partial y_j}, \tag{B2}$$

where $\sigma_{ik}\sigma_{jk} = B_{ij}$. Let $p^b(\mathbf{y}_0, t_0; \mathbf{y}, t) = \langle \delta(\mathbf{y}_0 - \mathbf{Z}_{t_0}^{\mathbf{y},t}) \rangle$ be the transition density of the diffusion process $Z_{t_0}^{\mathbf{y},t}$, $0 \le t_0 \le t$ which is defined by

$$dZ_i = A_i^*(\mathbf{Z}, t_0)\, dt_0 + \sigma_{ij}(\mathbf{Z}, t_0), t_0)\, \overleftarrow{d}\, W_j(t_0), \quad t_0 < t, \quad \mathbf{Z}(t) = \mathbf{y}. \tag{B3}$$

Here $\overleftarrow{d}\, W_j(t_0)$ is defined as in the footnote to (12.10) in Section 12.2, and

$$A_i^*(\mathbf{y}, t) = A_i(\mathbf{y}, t) - \frac{1}{\rho(\mathbf{y}, t)} \frac{\partial}{\partial y_j}(B_{ij}(\mathbf{y}, t)\rho(\mathbf{y}, t)).$$

We assume again, that the solutions to (B3) do never reach the boundary of D. Then the following relation is true (see [111], Appendix C):

$$\rho(\mathbf{y}_0, t_0) p^f(\mathbf{y}, t; \mathbf{y}_0, t_0) = \rho(\mathbf{y}, t) p^b(\mathbf{y}_0, t_0; \mathbf{y}, t). \tag{B4}$$

12.6.3 Appendix C. Derivation of the relation between the forward and backward densities

Here the derivation of the relation between forward and backward Lagrangian transition p. d. f.'s, equation (12.21), is presented. It is assumed that the boundary $z = 0$ is impenetrable, i. e., the trajectory, the solution to (12.6), will never reach this boundary.

Let us define $C(t) = C_t^{\mathbf{x}_0,\vec{u}_0,c_0,t_0}$ as the solution to

$$\frac{dC(t)}{dt} + \gamma(\mathbf{X}_t^{\mathbf{x}_0,\vec{u}_0,t_0}, t)C(t) = 0, \quad t > t_0; \quad C(t_0) = c_0,$$

and the extended forward Lagrangian transition p. d. f.

$$P_L^f(\mathbf{x}, \vec{u}, c, t; \mathbf{x}_0, \vec{u}_0, c_0, t_0) \\ = \mathbf{E}_{\mathbf{x}_0,\vec{u}_0,c_0,t_0}\big[\delta(\mathbf{x} - \mathbf{X}_t^{\mathbf{x}_0,\vec{u}_0,t_0})\delta(\vec{u} - \mathbf{V}_t^{\mathbf{x}_0,\vec{u}_0,t_0})\delta(c - C_t^{\mathbf{x}_0,\vec{u}_0,c_0,t_0})\big],$$

where the forward trajectory $\mathbf{X}_t^{\mathbf{x}_0,\vec{u}_0,t_0}$, $\mathbf{V}_t^{\mathbf{x}_0,\vec{u}_0,t_0}$ is defined in Section 12.2 and $\mathbf{E}_{\mathbf{x}_0,\vec{u}_0,c_0,t_0}$ means an expectation over samples of stochastic processes $\mathbf{X}_t^{\mathbf{x}_0,\vec{u}_0,t_0}$, $\mathbf{V}_t^{\mathbf{x}_0,\vec{u}_0,t_0}$, and $C_t^{\mathbf{x}_0,\vec{u}_0,c_0,t_0}$ starting at time $t = t_0$ from the point $\mathbf{x}_0, \vec{u}_0, c_0$. Analogously, we define $\hat{C}(t_0) = \hat{C}_{t_0}^{\mathbf{x},\vec{u},c,t}$ as the solution to

$$\frac{d\hat{C}(t_0)}{dt_0} + \gamma(\hat{\mathbf{X}}_{t_0}^{\mathbf{x},\vec{u},t}, t_0)\hat{C}(t_0) = 0, \quad t_0 < t; \quad \hat{C}(t) = c,$$

and the extended backward Lagrangian transition p. d. f.

$$P_L^b(\mathbf{x}_0, \vec{u}_0, c_0, t_0; \mathbf{x}, \vec{u}, c, t) = \mathbf{E}_{\mathbf{x},\vec{u},c,t}\{\delta(\mathbf{x} - \hat{\mathbf{X}}_{t_0}^{\mathbf{x},\vec{u},t})\delta(\vec{u} - \hat{\mathbf{V}}_{t_0}^{\mathbf{x},\vec{u},t})\delta(c - \hat{C}_{t_0}^{\mathbf{x},\vec{u},c,t})\},$$

where $\mathbf{E}_{\mathbf{x},\vec{u},c,t}$ means the expectation over samples of stochastic processes $\hat{\mathbf{X}}_{t_0}^{\mathbf{x},\vec{u},t}$, $\hat{\mathbf{V}}_{t_0}^{\mathbf{x},\vec{u},t}$, $\hat{C}_{t_0}^{\mathbf{x},\vec{u},c,t}$, $t_0 \le t$, starting at final time $t_0 = t$ at point $\mathbf{x}, \vec{u}, c$.

To derive the relation (12.21), first we establish the following equality:

$$\frac{p_E(\vec{u}_0; \mathbf{x}_0, t_0)}{c_0} P_L^f(\mathbf{x}, \vec{u}, c, t; \mathbf{x}_0, \vec{u}_0, c_0, t_0) = \frac{p_E(\vec{u}; \mathbf{x}, t)}{c} P_L^b(\mathbf{x}_0, \vec{u}_0, c_0, t_0; \mathbf{x}, \vec{u}, c, t). \tag{C1}$$

To this end, we use the well-mixed condition [237]:

$$\frac{\partial p_E}{\partial t} + \frac{\partial(u_i p_E)}{\partial x_i} + \frac{\partial(a_i p_E)}{\partial u_i} = \frac{1}{2}C_0\bar{\varepsilon}(\mathbf{x}, t)\frac{\partial^2 p_E}{\partial u_i \partial u_i}.$$

Denote

$$\rho(\mathbf{x}, \vec{u}, c, t) = \frac{p_E(\vec{u}; \mathbf{x}, t)}{c}.$$

From the well-mixed condition we get

$$\frac{\partial \rho}{\partial t} + \frac{\partial(u_i \rho)}{\partial x_i} + \frac{\partial(a_i \rho)}{\partial u_i} + \frac{\partial(-\gamma(\mathbf{x}, t)\, c\, \rho)}{\partial c} = \frac{1}{2}C_0\bar{\varepsilon}(\mathbf{x}, t)\frac{\partial^2 \rho}{\partial u_i \partial u_i}. \tag{C2}$$

Now (C1) follows from (C2) and from the result obtained in Appendix B.

Using

$$p_L^f(\mathbf{x}, \vec{u}, c, t; \mathbf{x}_0, \vec{u}_0, t_0) = P_L^f(\mathbf{x}, \vec{u}, c, t; \mathbf{x}_0, \vec{u}_0, 1, t_0)$$

and assuming in (C1) that $c_0 = 1$ and $c = \mu$, we get

$$p_E(\vec{u}_0; \mathbf{x}_0, t_0) p_L^f(\mathbf{x}, \vec{u}, \mu, t; \mathbf{x}_0, \vec{u}_0, t_0) = \frac{p_E(\vec{u}; \mathbf{x}, t)}{\mu} P_L^b(\mathbf{x}_0, \vec{u}_0, 1, t_0; \mathbf{x}, \vec{u}, \mu, t). \tag{C3}$$

Further taking into account that

$$\hat{C}_{t_0}^{\mathbf{x},\vec{u},\mu,t} = \frac{\mu}{\hat{\mu}_{t_0}^{\mathbf{x},\vec{u},t}},$$

and using the following property of the Dirac delta function

$$\delta(a\mu - b) = \frac{1}{a}\delta(\mu - \frac{b}{a}), \quad \mu, a, b \in (-\infty, \infty),$$

with $b = 1$ and $a = 1/\hat{\mu}_{t_0}^{\mathbf{x},\vec{u},t}$, we get

$$\delta(1 - \hat{C}_{t_0}^{\mathbf{x},\vec{u},\mu,t}) = \delta\left(\frac{\mu}{\hat{\mu}_{t_0}^{\mathbf{x},\vec{u},t}} - 1\right) = \hat{\mu}_{t_0}^{\mathbf{x},\vec{u},t}\delta(\mu - \hat{\mu}_{t_0}^{\mathbf{x},\vec{u},t}) = \mu\delta(\mu - \hat{\mu}_{t_0}^{\mathbf{x},\vec{u},t}).$$

From this and the definition of the function P_L^b it follows that

$$\begin{aligned} &P_L^b(\mathbf{x}_0, \vec{u}_0, 1, t_0; \mathbf{x}, \vec{u}, \mu, t) \\ &= \mu \mathbf{E}_{\mathbf{x},\vec{u},t}\{\delta(\mathbf{x}_0 - \hat{\mathbf{X}}_{t_0}^{\mathbf{x},\vec{u},t})\delta(\vec{u}_0 - \hat{\mathbf{V}}_{t_0}^{\mathbf{x},\vec{u},t})\delta(\mu_0 - \hat{\mu}_{t_0}^{\mathbf{x},\vec{u},t})\} \\ &= \mu p_L^b(\mathbf{x}_0, \vec{u}_0, \mu_0, t_0; \mathbf{x}, \vec{u}, t). \end{aligned}$$

Substitution of the right-hand side of the last equality into (C3) completes the proof of the relation (12.21).

Chapter 13

Stochastic Lagrangian footprint calculations over a surface with an abrupt change of roughness height

Forward and backward stochastic Lagrangian trajectory simulation methods are developed to calculate the footprint and cumulative footprint functions of concentration and fluxes in the case when the ground surface has an abrupt change of the roughness height. The statistical characteristics to the stochastic model are extracted numerically from a closure model which we developed for the atmospheric boundary layer. The flux footprint function is perturbed in comparison with the footprint function for surface without change in properties. The perturbation depends on the observation level as well as roughness change and distance from the observation point. It is concluded that the footprint function for horizontally homogeneous surface, widely used in estimating sufficient fetch for measurements, can be seriously biased in many cases of practical importance.

13.1 Introduction

Over a horizontally homogeneous surface the flux measured by the micrometeorological technique equals the surface flux. This principle is used to determine the surface exchange by the eddy covariance (EC) technique. The flux footprint function (e. g., see [221]) links the surface emissions to the observed fluxes above surface at EC measurement level. The footprint function is therefore used to estimate a distance required to make reliable EC measurements, i. e., if the horizontal extent of the underlying surface of interest is sufficient to determine its exchange rate. Extended tower measurements of fluxes over forests have been used during the past ten years to obtain detailed information on carbon and water exchanges between forest canopies and the atmosphere [90, 187, 246]). Large areas of forest are, however, common neither in Europe nor in the US. The footprint models based on analytical diffusion theory [79, 221, 225] as well as the Lagrangian stochastic simulation of the ensemble of fluid parcel trajectories [6, 55, 126] assume a horizontally homogeneous surface. For forest canopies the footprint models involve a number of uncertainties originating from the parametrization of the canopy turbulence features [174]. Such models are frequently applied in order to estimate the contribution of an area at a certain upwind distance, or to estimate the fetch to ensure that the given area contributes a certain percent to the observed flux, by vaguely assuming that the footprint function for a horizontally homogeneous surface is a good approximation for more complex situations with changes in surface properties. In reality, changes in surface roughness can be very drastic, for example in the case of forest and field interface. Also the thermal inhomogeneities induced by albedo and the repartitioning of available energy into sensible and latent heat fluxes can be significant; this will be analyzed, however, in the second part of this chapter.Here we will deal with pure mechanical turbulence caused by the surface roughness.

One additional remark here should be made. Over rough surfaces, such as tall vegetation, vertical displacement of surface layer profiles occurs relative to the ground surface. Height displacement is usually 2/3 to 3/4 of the height of the roughness elements, whereas the roughness length is usually 1/30 to 1/10 of the vegetation height. In this study the flow inside vegetation is not considered, and the observation level (detector height) is equivalent to that of relative to the displacement height in a real measurement setup. Displacement height would effectively elevate flow streamlines and not affect the results qualitatively, whereas roughness change induces transition in horizontal wind speed and also, via mass conservation, nonzero vertical winds.

Lagrangian trajectory simulation can be used in the inhomogeneous flow field (e. g., see [237], and [115]). However, to make the stochastic trajectory simulation possible, the mean flow and some other statistical moments have to be found. We will extract this data from a closure model, conventionally obtained from the Reynolds-averaging equations. The footprint function for inhomogeneous surface is estimated by backward Lagrangian trajectory simulation, and the perturbations relative to footprint function for horizontally homogeneous case are analyzed.

13.2 The governing equations

There exists a variety of closure models for turbulent mixing, ranging from constant eddy coefficient parametrization to detailed large eddy simulations and direct numerical simulation. As mentioned in [5], the performance of a k-model is almost identical to that of the k–ε-model. We assume that the mean profiles in the boundary layer of atmosphere are described by the following system [253]:

$$\begin{aligned} u\frac{\partial u}{\partial x} + w\frac{\partial u}{\partial z} &= \frac{\partial}{\partial z} k \frac{\partial u}{\partial z} + f(v - G\sin\alpha) \\ u\frac{\partial v}{\partial x} + w\frac{\partial v}{\partial z} &= \frac{\partial}{\partial z} k \frac{\partial v}{\partial z} - f(u - G\cos\alpha) \end{aligned} \tag{13.1}$$

is the momentum equation, where the Cariolis parameter is given by $f = 2\Omega\sin\varphi$, $\Omega = 7.29\,10^{-5}\mathrm{s}^{-1}$, G is the geostrophical wind, and φ is the angle between the x-axis and the isobare; α is the angle between the x-axis and the direction of the geostrophical wind. Further,

$$\frac{\partial u}{\partial x} + \frac{\partial w}{\partial z} = 0$$

is the continuity equation, where k is the turbulent exchange coefficient for the momentum.

The balance of the kinetic energy is written as

$$u\frac{\partial b}{\partial x} + w\frac{\partial b}{\partial z} = \alpha_b \frac{\partial}{\partial z} k \frac{\partial b}{\partial z} + k\left[\left(\frac{\partial u}{\partial z}\right)^2 + \left(\frac{\partial v}{\partial z}\right)^2\right] - \bar{\varepsilon}, \qquad (13.2)$$

where $\bar{\varepsilon} = \frac{cb^2}{k}$ is the mean rate of dissipation of the turbulent energy, $c = 0.0286$, $\alpha_b \approx 0.7$.

The functions k and b are related through

$$l = \left(\frac{1}{\kappa z} + \frac{1}{l_0}\right)^{-1}, \quad k = C_k l \sqrt{b}, \qquad (13.3)$$

with $\kappa \approx 0.4$, $l_0 = 100$ m, $C_k = 0.41$. We will deal in this first part of the chapter with purely mechanical turbulence, and our system of governing equations consists of (13.1), (13.2), and the relation (13.3).

The functions vary in the layer $z_0 \leq z \leq h$, h being the height of the boundary layer, and z_0 the roughness height. The system of equations is considered with the following boundary conditions:

$$u = 0, \quad v = 0, \quad w = 0, \quad \text{at} \quad z = z_0,$$
$$u = G\cos\alpha, \quad v = G\sin\alpha \quad \text{at} \quad z = h,\ x \leq 0,$$
$$\frac{\partial u}{\partial z} = \frac{\partial v}{\partial z} = 0 \quad \text{at} \quad z = h,\ x > 0,$$
$$\frac{\partial b}{\partial z} = 0 \quad \text{at} \quad z = z_0, \quad \text{and} \quad b = 0 \quad \text{at} \quad z = h,$$

At $z = z_0$ we take $l = \kappa z_0$.

13.2.1 Evaluation of footprint functions

We assume that our 3D flow is homogeneous in the y-direction and inhomogeneous in the z- and x-directions. Roughness inhomogeneity along the x-direction is only assumed in our numerical analysis.

The flow is considered in the boundary layer of height h, with a roughness height $z_0 = z_{01}$ which is a constant for $x < 0$, then in a small interval $0 < x < \Delta$ it linearly changes from z_{01} to z_{02}, and then it is again constant: for $x > \Delta$ it equals z_{02}. The detector is placed at a point (x_d, y_d, z_d). We will deal with the footprint function of concentration $c(x) = c(x; x_d, y_d, z_d)$ defined as the mean concentration at the detector point from a linear source with the coordinate x, placed at roughness height, directed along the axis Y. Analogously the footprint function of the vertical flux is defined as $q(x) = q(x; x_d, y_d, z_d) = \langle w(x_d, y_d, z_d)\, c'(x; x_d, y_d, z_d)\rangle$, where c' is the fluctuating part of the concentration.

The cumulative footprint functions of concentration and of vertical flux are defined as

$$C(x) = \int_x^{\infty} c(x')dx', \quad Q(x) = \int_x^{\infty} q(x')dx'.$$

The normalized footprint function of flux is defined as $Q_n(x) = Q(x)/Q(-\infty)$. In this section we present the cumulative footprint functions for smooth-to-rough and rough-to-smooth changes of the roughness height.

We calculate the mean concentration c_i and q_{ix}, the vertical concentration fluxes at the detector point, from a surface source uniformly distributed over strips of width $\Delta_i = b_i - a_i$: $D_i = \{x, z : a_i \le x \le b_i\ z = z_0,\}$, $i = 1, \dots, n_{sr}$. The corresponding cumulative and normalized cumulative footprint functions of flux are also calculated. Calculations are carried out for small values of Δ_i, and the result is then normalized by the strip width.

The forward simulation technique cannot be applied for estimation of concentration and flux exactly at a point in space in case of horizontally inhomogeneous turbulence, and/or exactly at moment t in case of non-stationary turbulence. Instead, one might consider averages over space and/or time localized near $\mathbf{x}$ and/or t (e.g., see [111]). However, such a simulation might be computationally inefficient if the extension of the source is much larger than that of the detector. For these cases, the backward trajectory simulation, starting from the space-time point of interest, is more efficient (in the case of Eulerian approach see [191], and in the Lagrangian framework see, e.g., [56]).

To be more specific, let us present now the backward estimators for the evaluation of footprint functions in the case of the boundary layer with the sources uniformly distributed over the strips D_i, $i = 1, \dots, n_{sr}$. For simplicity, we have taken the x-axis coincident with the direction of the geostrophical wind, i.e., $\alpha = 0$.

The backward trajectory starts at time t, at the detector point with the velocity sampled from the Eulerian velocity p.d.f. $p_E(u, x)$ which is assumed to be Gaussian, see Appendix B at the end of this chapter. We note only that below, we denote by $\bar{u}_{Ek}$ the k-th component ($k = 1, 2, 3$) of the mean Eulerian velocity vector, and the hat over the symbols x and u is to indicate that this is a finite-difference approximation to the true Lagrangian trajectory.

The backward trajectory simulation is conveniently carried out through the semi-implicit Euler scheme, which can be written for one time step as follows:

$$\begin{aligned}\hat{x}_k(t-\Delta t) &= \hat{x}_k(t) - (\hat{u}'_k(t) + \bar{u}_{Ek}(t,\hat{x}(t)))\,\Delta t,\\ \hat{u}'_k(t-\Delta t) &= \hat{u}'_k(t) - \hat{a}'_k(t,\hat{x}(t-\Delta t),\hat{u}(t))\,\Delta t\\ &\quad + \sqrt{C_0\bar{\varepsilon}(t-\Delta t,\hat{x}(t-\Delta t))\,\Delta t}\;\eta_{tk},\end{aligned}$$

where η_{tk}, $k = 1,2,3$ are independent standard Gaussian random variables. Here for a reason of practical convenience, we work in the "primed" velocity variables $\hat{u}'_k = \hat{u}_k - \bar{u}_{Ek}$, so that

$$\begin{aligned}d\hat{x}_k &= (\hat{u}'_k + \bar{u}_{Ek})\,ds,\\ d\hat{u}'_k &= \hat{a}'_k\,ds + \sqrt{C_0\bar{\varepsilon}}\;\overleftarrow{d}\;W_k(s), \quad s < t,\ k = 1,2,3,\end{aligned}$$

with the condition that the trajectory starts at the detector position with the velocity sampled from the Gaussian p. d. f. p_E. Here according the formula given in Appendix B of this chapter we have

$$\begin{aligned}\hat{a}'_k &= \hat{a}_k - \frac{\partial \bar{u}_{Ek}}{\partial x_j}\bar{u}_{Ej} - \frac{\partial \bar{u}_{Ek}}{\partial x_j}\hat{u}'_j\\ &= \frac{C_0\bar{\varepsilon}}{2}\lambda_{kj}\hat{u}'_j + \frac{1}{2}\frac{\partial \sigma_{kj}}{\partial x_j} + \frac{\lambda_{km}}{2}\bar{u}_{Ei}\frac{\partial \sigma_{jm}}{\partial x_i}\hat{u}'_j + \frac{\lambda_{km}}{2}\frac{\partial \sigma_{im}}{\partial x_j}\hat{u}'_j\hat{u}'_i\\ &= \hat{\alpha}_k + \hat{\beta}_{kj}\,\hat{u}'_j + \gamma_{kji}\,\hat{u}'_j\hat{u}'_i\end{aligned}$$

and

$$\hat{\alpha}_k = \frac{1}{2}\frac{\partial \sigma_{kj}}{\partial x_j}, \quad \hat{\beta}_{kj} = \frac{C_0\bar{\varepsilon}}{2}\lambda_{kj} + \frac{\lambda_{km}}{2}\bar{u}_{Ei}\frac{\partial \sigma_{jm}}{\partial x_i}, \quad \gamma_{kji} = \frac{\lambda_{km}}{2}\frac{\partial \sigma_{im}}{\partial x_j}.$$

All the expressions for the input functions are given in the Appendices of this chapter.

Recall that we use here the summation convention taking the sum over repeated indices $i, j, l = 1,2,3$, and $\overleftarrow{d}\;W_k(s)$ stands for the backward Wiener differential (see [115]) which implies for the Euler scheme that the increments are taken back in time.

Let us denote by τ_{ij} the time at which the trajectory $(\hat{\mathbf{x}}(s), \hat{\mathbf{u}}(s))$, $s \le t$ reaches the ground surface and touches the i-th strip: the first touchdown at τ_{i1}, the second (after a reflection from the boundary) at τ_{i2}. etc., and the last one at τ_{iN_i}. The random estimators have the following form (see Appendix B), for the concentration:

$$c_i = \left\langle \sum_{j=1}^{N_i} \frac{2}{\Delta_i}\frac{1}{|\hat{u}_3(\tau_{ij})|} \right\rangle,$$

and for the vertical flux:

$$q_{iz} = \left\langle \sum_{j=1}^{N_i} \frac{2}{\Delta_i} \frac{\hat{u}_3(t)}{|\hat{u}_3(\tau_{ij})|} \right\rangle, \quad i = 1, \dots, n_{sr}.$$

Here the angle brackets stands for the averaging over the ensemble of independent backward trajectories.

13.3 Results

In this section we will examine the impact of the roughness change on the footprint functions. The developed code calculates the footprint and cumulative footprint functions of concentration and flux for the horizontally inhomogeneous case when the roughness height is constant z_{01} for $x < 0$, then in a small interval $0 < x < \Delta$ it linearly changes from z_{01} to z_{02}, and then it is again constant: for $x > \Delta$ it equals z_{02}. In calculations, we have taken $\Delta = |z_{02} - z_{01}|$. Note that when we speak of the inhomogeneous case, two essentially different cases are considered: $z_{01} < z_{02}$ (smooth-to-rough) and $z_{01} > z_{02}$ (rough-to-smooth).

To be specific, in all calculations we have taken the geostrophical wind as $G = 10$ m/s, and the boundary layer height as $h = 1$ km. So the spatial scale in the figures is given in km. To give a sensitivity analysis to the change of these parameters, we present simultaneously the footprint functions of the inhomogeneous and homogeneous cases with the corresponding roughness height. These enables us to find the regions of applicability of the results obtained for the homogeneous case and, moreover, to conclude where the inhomogeneous case shows considerable differences compared to the homogeneous roughness. We also present some other footprints, in particular, the cumulative footprint functions of concentration and vertical flux. In all calculations we run $4\,10^5$ backward trajectories, the strip width was 2 m, the time step was varying according to $\Delta t = 0.025\tau$, where $\tau = 2\sigma_w^2/(C_0\bar{\varepsilon})$ is the Lagrangian time scale at the trajectory instantaneous position. Since the variance of the random estimators was large, we have made a Gaussian smoothing procedure with a band width equal to 4 strip widths.

13.3.1 Footprint functions of concentration and flux

Let us describe the results of numerical simulations. We have made the calculations for two different cases: (i) smooth-to-rough and (ii) rough-to-smooth change of the roughness height. The detector was placed at the height $z_d = 2$ m, at a distance x_d.

Smooth-to-rough case

In these calculations, along with the inhomogeneous case, we plot simultaneously the footprint functions for the homogeneous case with roughness height $z_0 = z_{02}$.

In Figure 13.1, left, we present the footprint function of concentration for the case of roughness change indicated above, for $x_d = 20$ m, $x_d = 50$ m and $x_d = 100$ m (the same curve is given for the homogeneous case at $z_0 = 5$ cm) as functions of the dimensionless upwind distance X/h, where the upwind distance X is defined as $X = -x + x_d$, the distance to the detector. The same curves are shown in the Figure 13.1, right, for the case $z_{01} = 1$ cm, $z_{02} = 25$ cm.

It is seen that the footprint functions of concentration for inhomogeneous cases are all smaller compared to the homogeneous case in the near-region which is $X/h < 0.04$ in the left panel, and $X/h < 0.08$ in the right panel. In the far-region the situation is reverse: all the inhomogeneous curves are over the homogeneous curve after $X/h > 0.15$.

The inhomogeneous curves have local minima at the position of roughness change; after this point they increase and become higher than the homogeneous curve. It is clearly seen that the minima are more pronounced in the case of larger roughness change. Note that this behavior explains why the homogeneous curve in the left panel of Figure 13.2 (the corresponding cumulative footprint function of concentration) is higher than all the curves for the small upwind distances, and then decreases down all the curves; here we give more detailed dependence on x_d starting from $x_d = 20$ m, with the last value $x_d = 200$ m, where the influence of the roughness change is expected have almost disappeared. The same is true for the right panel (note that we have shown the curves only for distances $X/h < 1$, and therefore we cannot see the position where the homogeneous curve is below all the inhomogeneous curves, but in calculations this took place).

In Figure 13.3 we present the corresponding footprint functions of flux, for the case of small (left panel) and larger roughness change (right panel). Note that here the difference between the homogeneous and inhomogeneous curves in the near-region is in the case of larger roughness change (right panel) much higher than that of smaller roughness change (left panel). In all cases the homogeneous curve is positioned almost everywhere below the inhomogeneous curves; one exception is in the neighborhood of the roughness change which is clearly seen in the right picture for the case $x_d = 20$ m.

In Figure 13.4 the cumulative footprint function of flux Q is presented. It is clearly seen that the difference between the homogeneous and inhomogeneous curves becomes less and less as the value of x_d increases. Note that the homogeneous curve tends to 1 as the upwind distance increases, while in the inhomogeneous case the curves tend to asymptotic values which are larger than 1. To find the footprint area, it is convenient to use the normalized cumulative footprint function of flux, which is defined as the cumulative footprint function of flux divided by the corresponding asymptotic value. These curves are shown in Figure 13.5. To illustrate this, here we show through a horizontal dashed line the level of 90 % contribution to the detector made by the surface around the detector position. Calculations show that the fetch in the homogeneous case is smaller than that of the inhomogeneous case; in Figure 13.5 it is seen that the corresponding fetch in the homogeneous case is about 200 m, while for $x_d = 20$ m it is about two times larger. From this figure we can conclude that the inhomogeneous case with $x_d = 200$ m is approximately coincident with the homogeneous case, and hence the analytical formulas known for the homogeneous case can be applied inside a region whose diameter is not less than 200 m. For $x_d < 200$ m, the change of roughness should be taken into account. This confirms the known practical recommendation which says that the inhomogeneity can be neglected if $x_d/z_d > 100$.

Rough-to-smooth case

Here we present the calculations for the rough-to-smooth case showing the same footprint functions plotted in Figures 13.1–13.5. These footprint functions, when compared to the corresponding footprint functions presented in Figures 13.1–13.5, have the following features: the local minima at the roughness change in Figures 1,3 correspond to the local maxima in Figures 13.6 and 13.8. Also, in Figures 13.6 and 13.8 the homogeneous curve is below the inhomogeneous curves in the near-region, while in the far-region it is above these curves. The same is true for the cumulative footprint functions plotted in Figures 13.7 and 13.9. Note that in the inhomogeneous case, the asymptotic value of cumulative footprint functions of flux at large distances is less than 1; it is seen that the smaller x_d is, the smaller is this asymptotic value. For larger roughness change it is becoming even less.

As to the fetch, we can conclude from Figure 13.10 that, in contrast to the smooth-to-rough case, here the fetch of homogeneous case is larger than that of the inhomogeneous cases; for instance, in the case of the roughness change $z_{01} = 25$ cm, $z_{02} = 1$ cm, the fetch is about 80 m for $x_d = 50$ m, while for the homogeneous case it is about 300 m.

Some features of the qualitative behavior of the footprint functions

Let us describe some features of the qualitative behavior of the footprint functions for the inhomogeneous case. In Figure 13.11 we plot the homogeneous curves for two cases of the roughness height: $z_0 = 1$ cm and $z_0 = 25$ cm, and the inhomogeneous curves for the case of roughness change from $z_{01} = 1$ cm to $z_{02} = 25$ cm (left panel: the footprint of concentration; right panel: the footprint function of the flux). The position of the roughness change is shown by the dashed vertical line.

First consider the results plotted in the left panel. In the near-region (left of the dashed vertical line) the inhomogeneous curve behaves qualitatively like the homogeneous curve for $z_0 = 25$ cm, lying however considerably below with its maximum position shifted to the left (closer to the detector) when compared with the maximum position of the homogeneous curve. In the region $X/h > 0.08$ we observe a qualitatively similar behavior of the inhomogeneous curve and the homogeneous curve, but for $z_0 = 1$ cm; the curves are converging in the far-region. The maximum position of the inhomogeneous curve is shifted to the right. Thus the qualitative behavior of the inhomogeneous curve is controlled by the two homogeneous curves – in the close-region by the case $z_0 = 25$ cm, and in the far-region by the case $z_0 = 1$ cm. This leads to the bimodal shape of the inhomogeneous curve. But from a simple superposition of the two homogeneous footprint functions, we would not expect such a deep drop between the two modes. This drop is caused by the flow structure around the change of roughness: in contrast to the homogeneous case here we have the positive vertical component of the mean velocity.

Generally, the same arguments are true for the footprint function of flux shown in the right panel of Figure 13.11, with a not so deep drop between the two modes. This can be explained by the fact that the maximum position of the homogeneous curve with $z_0 = 1$ cm is closer to the roughness jump.

Note that from this picture we can clearly see that the cumulative footprint function of flux Q for the inhomogeneous case is larger than 1, in contrast to the homogeneous case, where it is always less than 1. Indeed, note that the area below the homogeneous curve $z_0 = 25$ cm equals 1, so if we take the area below the homogeneous curve $z_0 = 25$ cm in the region $X < x_d$ and add the area below the homogeneous curve $z_0 = 1$ cm in the region $X > x_d$, we get a value which is larger than 1, which follows from a simple comparison of the behavior of the curve.

Let us turn to the rough-to-smooth case. In Figure 13.12 we plot the same curves as in Figure 13.11, but for the roughness change from $z_{01} = 25$ cm to $z_{01} = 1$ cm. At the roughness change position we observe a small jump in the inhomogeneous curve presenting the footprint function of concentration (left panel). Again, the general form of this curve can be deduced from the superposition of the two homogeneous curves, while the jump can be explained here by the negative values of the vertical component of the mean flow around the roughness change.

An analysis of the footprint function of flux (right panel) analogous to that made for the smooth-to-rough case above shows that the cumulative footprint function of flux can here be less than 1.

A larger change of the roughness height

Further, we have made calculations for the larger change in roughness height, namely, for $z_{01} = 1$ cm, $z_{02} = 100$ cm in the smooth-to-rough case, and $z_{01} = 100$ cm, $z_{02} = 1$ cm in the rough-to-smooth case. The detector height was taken at $z_d = 20$ m. The relevant footprint functions of concentration are shown in Figure 13.13, the footprint functions of flux q in Figure 13.14, and the cumulative footprint functions of flux Q in Figure 13.15, for different values of x_d. The results are in good agreement with the conclusions made for the smaller change of the roughness height, showing even more clearly the qualitative behavior of the curves discussed above. It should be noted that in this case we cannot expect a good quantitative prediction, because for the high roughness height we need to make a correction of the mean flow model which takes into account that the inhomogeneity affects the mean flow in a more complicated manner.

Dependence on the detector height z_d

Let us now consider the dependence of the footprint functions on the detector height z_d. In Figure 13.16 we present the footprint functions of concentration (left panel, in log-log scale) and flux (right panel, in log-line scale) for the smooth-to-rough case ($z_{01} = 1$ cm, $z_{01} = 25$ cm; the x-coordinate of the detector is fixed at $x_d = 100$ m, and its height is varying from $z_d = 2$ m to $z_d = 16$ m. In the left panel it is seen that the local minima can be observed in all curves, but the higher the detector is, the smaller is the drop whose width is becoming larger with the height. The influence of the roughness change on the footprint function of flux is observable (right panel) until the height of about $z_d = 16$ m.

The corresponding cumulative footprint functions of concentration C (left panel) and flux Q (right panel) are shown in Figure 13.17. Note that the function C is uniformly decreased (in the considered region) with the height z_d. For the function Q this is not the case: the curve for $z_d = 4$ m first rapidly increases, being larger than all the other curves, and then slows down and finally tends to its asymptotic value at distances (about $X/h \sim 1$) where the other curves are still increasing. Note that there is no monotonic behavior of the asymptotics $Q(X/h)$ at large values of X/h: first it increases with height z_d, for $z_d = 4$ m it is smaller than for $z_d = 8$, while for $z_d = 32$ m the value is the smallest compared to all the other values. This can be explained as follows. When the height z_d is small, the detector is well inside the inner boundary layer generated by the roughness change, and the situation is close to the homogeneous case (with the roughness height $z_0 = z_{02}$) where the asymptotic value is 1. For large heights z_d (larger than the height of the inner boundary layer) we are again in a situation close to the homogeneous case with the roughness height $z_0 = z_{01}$, and therefore the corresponding asymptotics is again close to 1. In between, for intermediate heights, the asymptotic value is larger than 1 (e. g., for $z_d = 4, 8, 16$ m). The additional contribution is coming from the trajectories with positive vertical velocities in the neighborhood of the position of the roughness change (the source is in a sense effectively lifted). The same arguments are true for the rough-to-smooth case, with the feature that in the intermediate heights the asymptotic values are less than 1, because in this case the trajectories get negative vertical velocities at the position of the roughness change. It should be noted that if we define the cumulative footprint function differently by omitting the convective part $\langle w \rangle \langle c \rangle$, i. e., as $Q_t = \langle w' c' \rangle$, then the above-mentioned asymptotics is almost always true for Q_t. This holds for the smooth-to-rough case while for the rough-to-smooth case it is true only for large values of x_d ($x_d > 20 z_d$); see Figure 13.18.

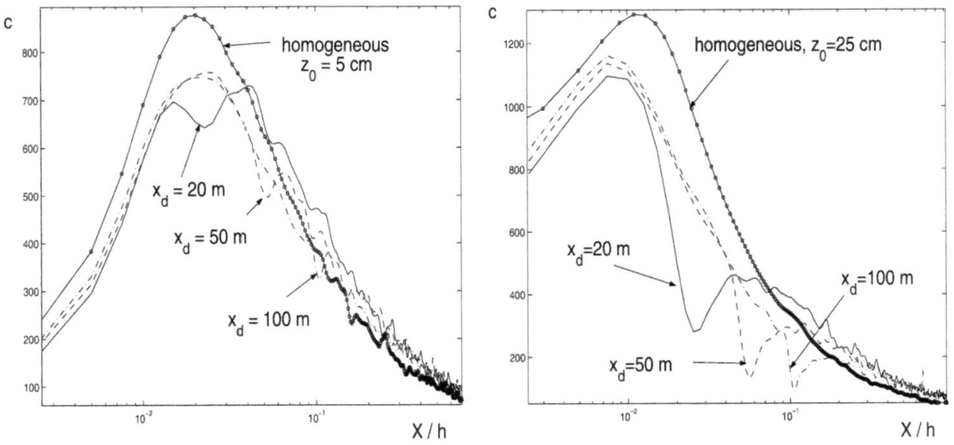

Figure 13.1. The footprint function of concentration c ($z_{01} = 1$ cm, $z_{02} = 5$ cm (left panel), and $z_{01} = 1$ cm, $z_{02} = 25$ cm (right panel)), vs. the dimensionless upwind distance X/h, for different values of x_d.

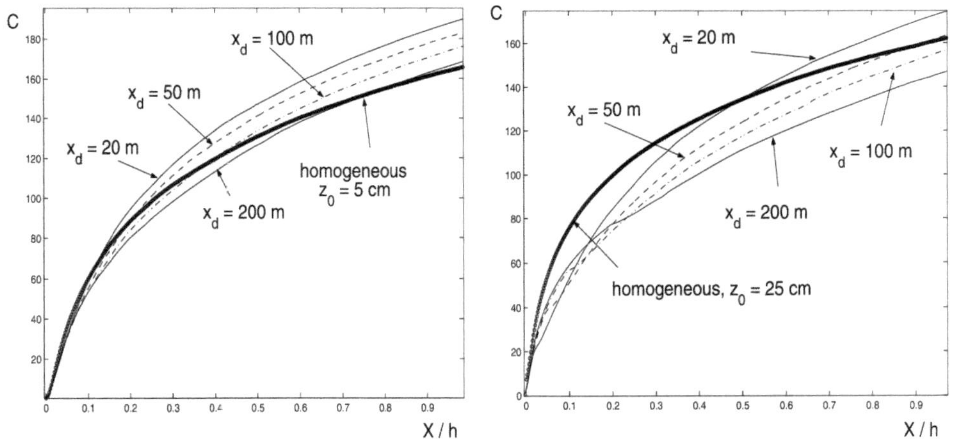

Figure 13.2. The cumulative footprint functions of concentration C, versus the dimensionless upwind distance X/h, for different values of x_d. The roughness change is the same as in Figure 13.1.

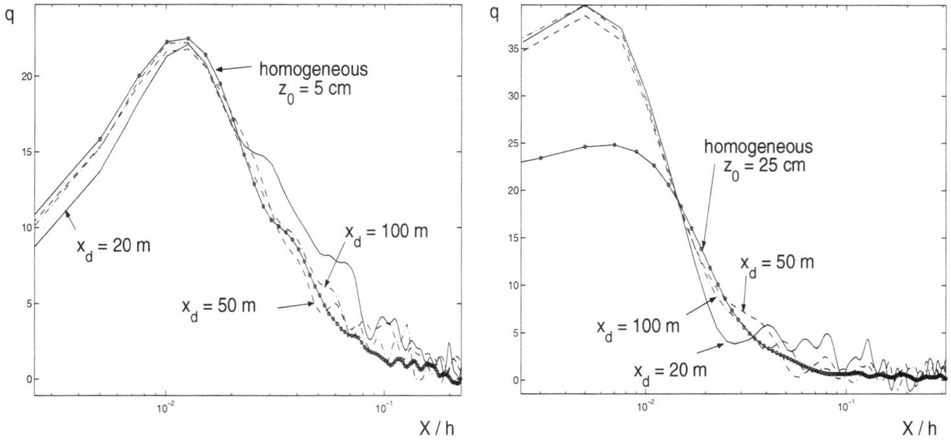

Figure 13.3. The footprint function of flux q (z_{01} = 1 cm, z_{02} = 5 cm (left panel), and z_{01} = 1 cm, z_{02} = 25 cm (right panel)), vs. the dimensionless upwind distance X/h, for different values of x_d.

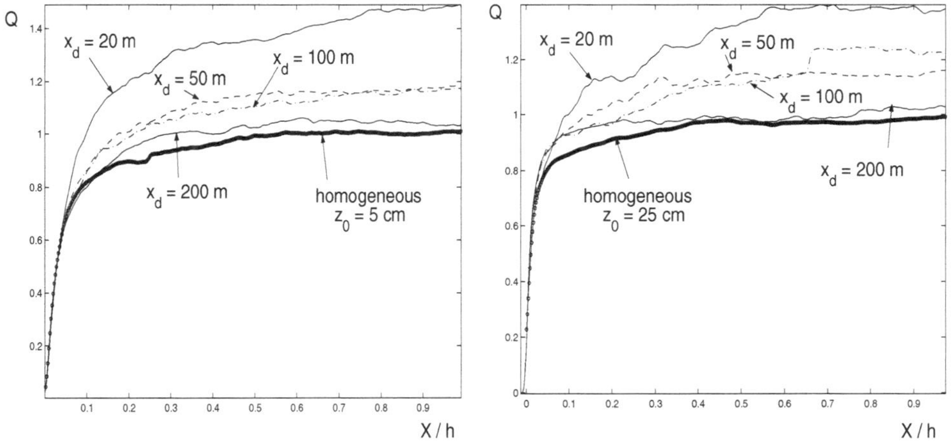

Figure 13.4. The cumulative footprint functions of flux Q, vs. the dimensionless upwind distance X/h, for different values of x_d. The roughness change is the same as in Figure 13.3.

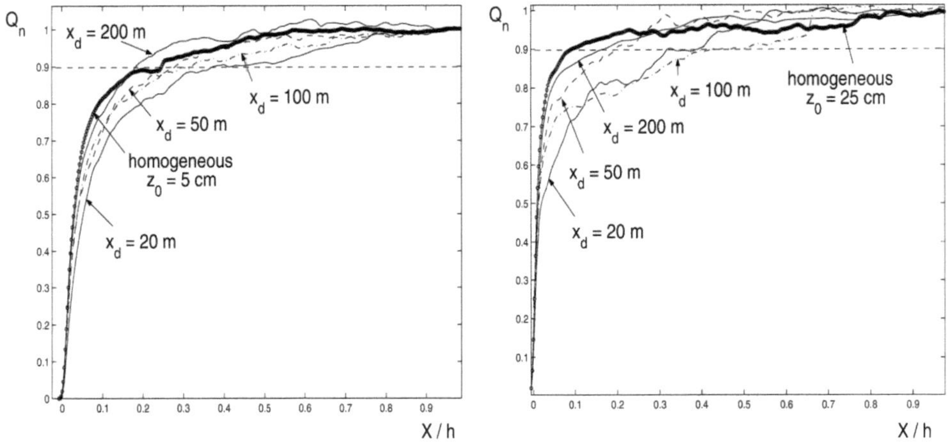

Figure 13.5. The normalized cumulative footprint functions of flux Q, vs. the dimensionless upwind distance X/h, for different values of x_d. The roughness change is the same as in Figure 13.3.

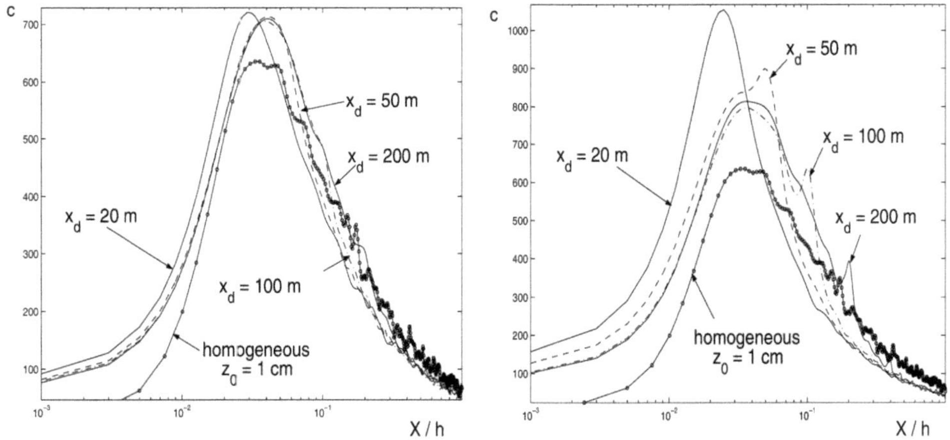

Figure 13.6. The footprint function of concentration c ($z_{01} = 5$ cm, $z_{02} = 1$ cm (left panel), and $z_{01} = 25$ cm, $z_{02} = 1$ cm (right panel)), vs. the dimensionless upwind distance X/h, for different values of x_d.

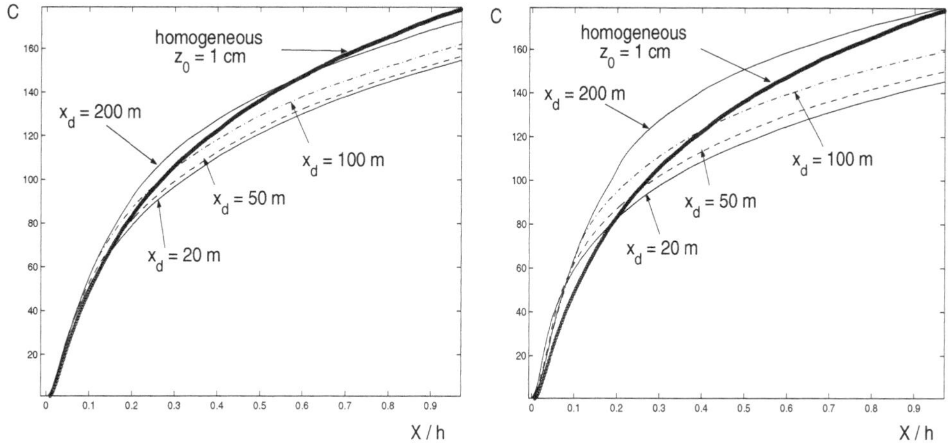

Figure 13.7. The cumulative footprint functions of concentration C, vs. the dimensionless upwind distance X/h, for different values of x_d. The roughness change is the same as in Figure 13.6.

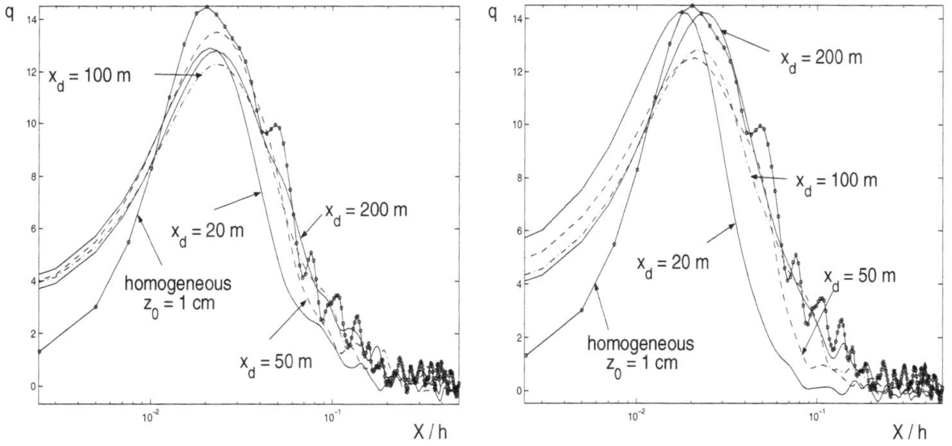

Figure 13.8. The footprint function of flux q ($z_{01} = 5$ cm, $z_{02} = 1$ cm (left panel), and $z_{01} = 25$ cm, $z_{02} = 1$ cm (right panel)), versus the dimensionless upwind distance X/h, for different values of x_d.

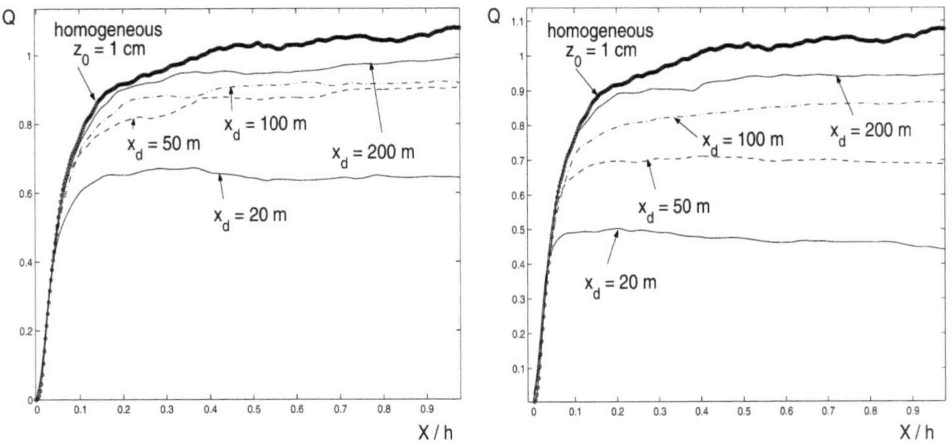

Figure 13.9. The cumulative footprint functions of flux Q, vs. the dimensionless upwind distance X/h, for different values of x_d. The roughness change is the same as in Figure 13 8.

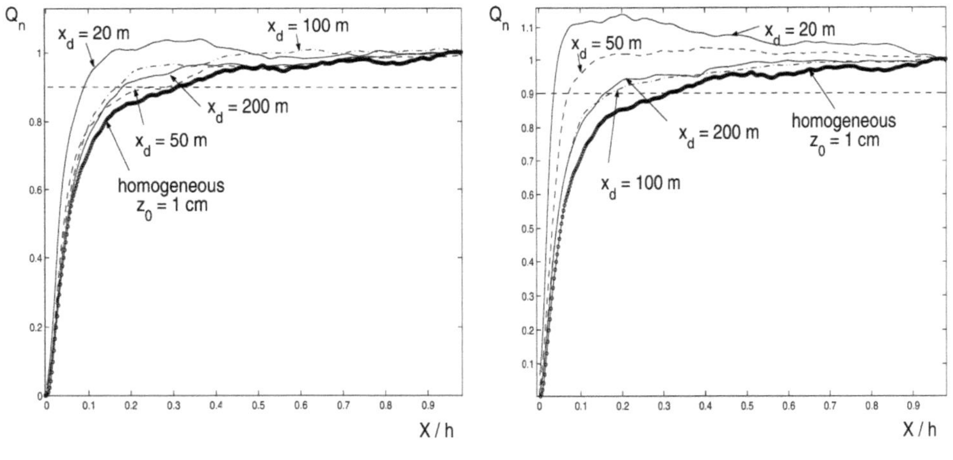

Figure 13.10. The normalized cumulative footprint functions of flux Q, vs. the dimensionless upwind distance X/h, for different values of x_d. The roughness change is the same as in Figure 13.8.

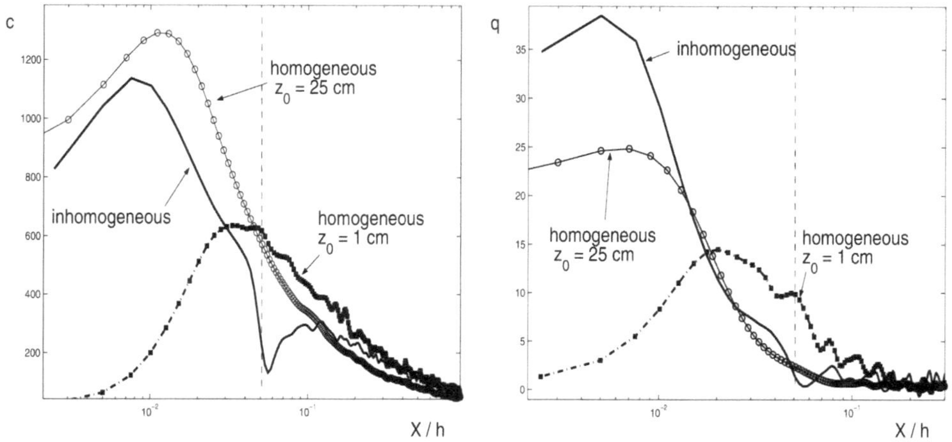

Figure 13.11. The footprint functions of concentration c (left panel) and flux q (right panel). For comparison, three curves are shown: the homogeneous curves for $z_0 = 1$ cm and $z_0 = 25$ cm, and the inhomogeneous curve for the roughness change from $z_0 = 1$ cm to $z_0 = 25$ cm, for $x_d = 50$ m.

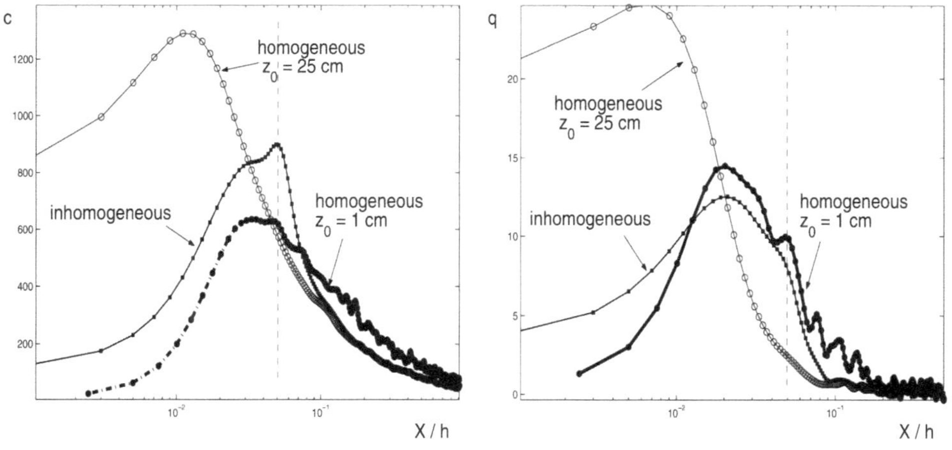

Figure 13.12. The same as in Figure 13.11, but for the rough-to-smooth case: the roughness change from $z_0 = 25$ cm to $z_0 = 1$ cm.

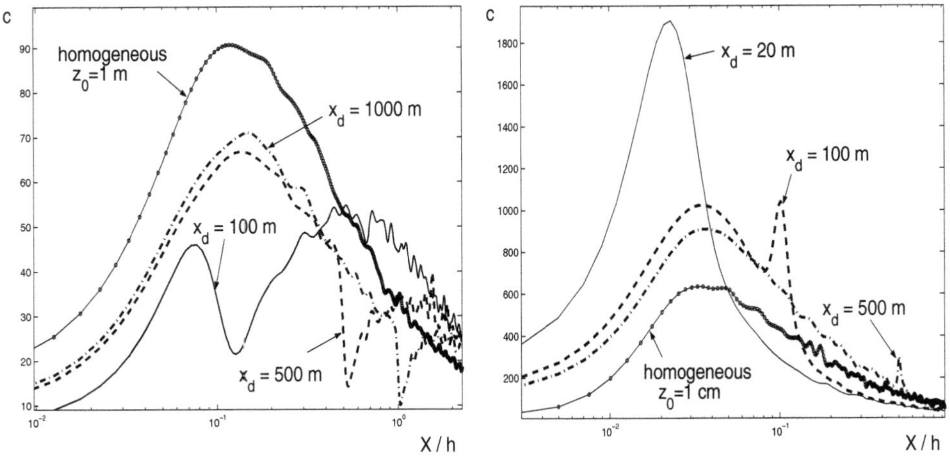

Figure 13.13. The footprint function of concentration for the smooth-to-rough case, the roughness height changes from $z_{01} = 1$ cm to $z_{02} = 100$ cm (left panel), and rough-to-smooth case with $z_{01} = 1$ cm, $z_{02} = 100$ cm (right panel). In all curves, $z_d = 20$ m.

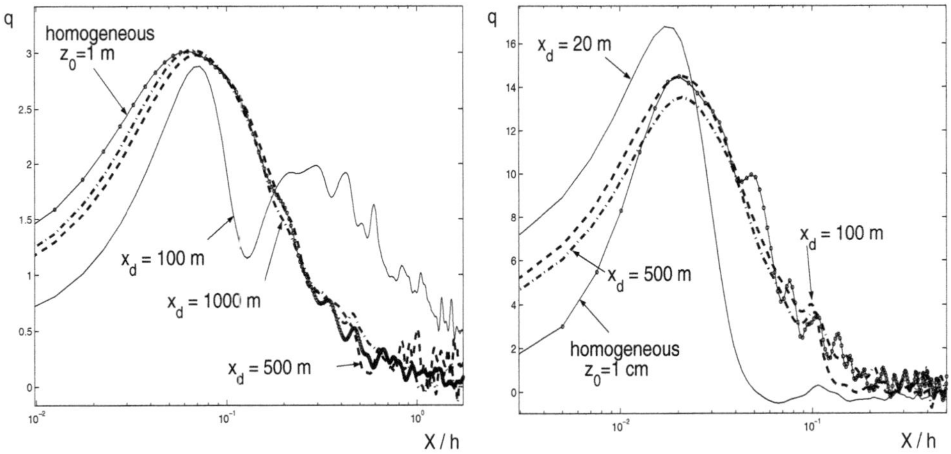

Figure 13.14. The same as in Figure 13.13, but for the footprint function of flux q.

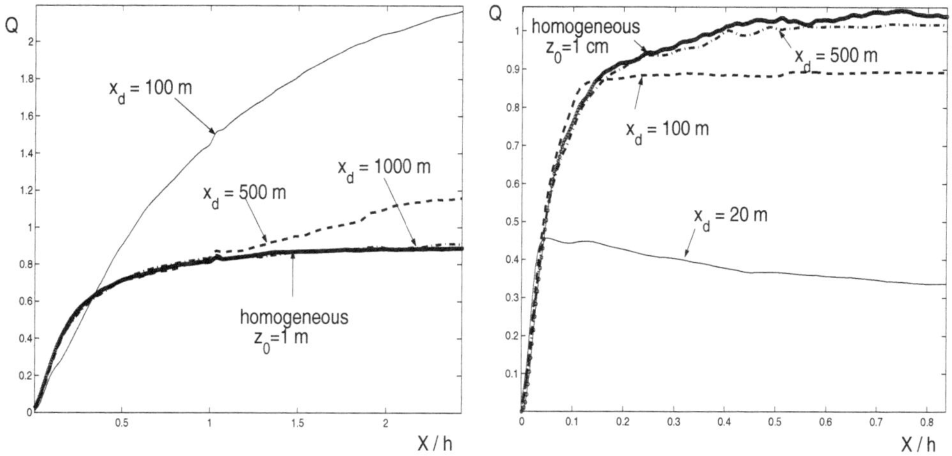

Figure 13.15. The same as in Figure 13.13, but for the cumulative footprint function of flux Q.

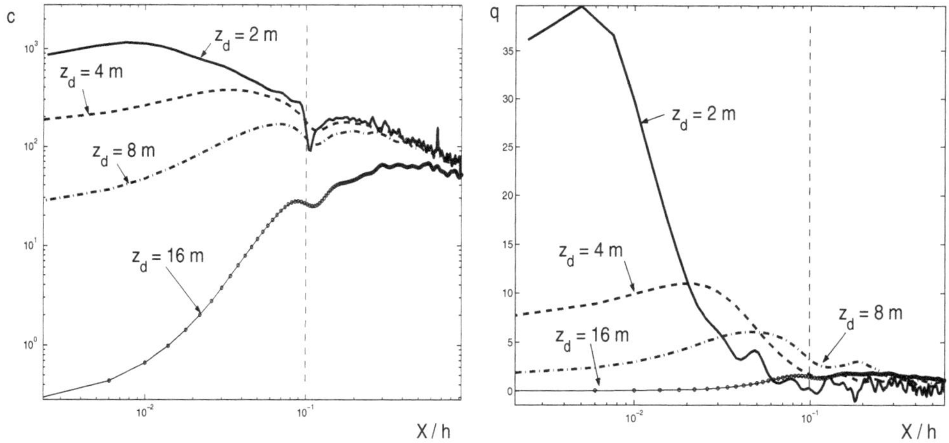

Figure 13.16. The footprint functions of concentration c (left panel) and flux q (right panel), for $x_d = 100$ m, and different values of the detector heights z_d.

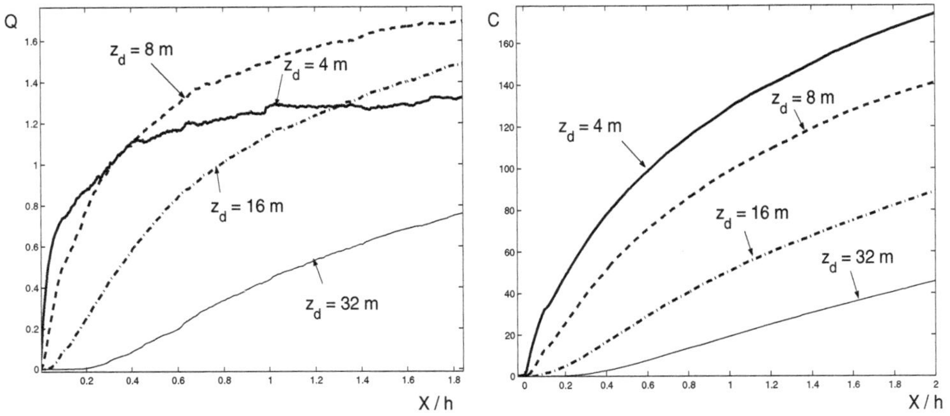

Figure 13.17. The cumulative footprint functions of concentration C (left panel) and flux Q (right panel), for $x_d = 100$ m, and different values of the detector heights z_d.

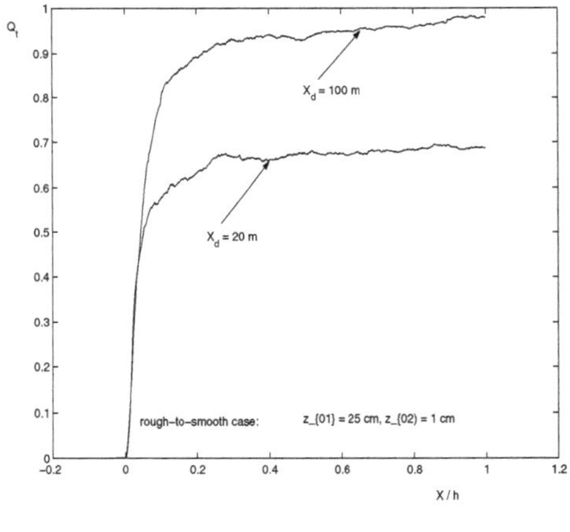

Figure 13.18. The cumulative footprint function of flux Q_t, the rough-to-smooth case ($z_{01} = 25$ cm, $z_{02} = 1$ cm), for $x_d = 20$ m and $x_d = 100$ m, and for the detector height $z_d = 2$ m.

13.4 Discussion and conclusions

A closure model was used to evaluate the mean flow and the Reynolds stress tensor required in the stochastic Lagrangian model we applied to calculate the footprint functions of concentration and its horizontal and vertical fluxes. This model provides the mean velocities and other characteristics of the flow over the roughness height.

A sensitivity analysis was made for the footprint functions under perturbation of the roughness height; two cases were considered: (i) smooth-to-rough and (ii) rough-to-smooth change of the roughness height. The calculations show that the footprint function of concentration is more sensitive than that of the vertical flux.

It iwa concluded that the footprint and cumulative footprint functions of concentration for horizontally homogeneous surface, widely used in for estimating the sufficient fetch for measurements, can be seriously biased in many cases of practical importance. The calculations show that the footprint area based on the cumulative concentrations, if estimated through the homogeneous case, can be essentially under- or overestimated, compared to the true inhomogeneous case. For instance, in the case when the detector is placed at $x_d = 100$ m from the roughness jump (from $z_{01} = 1$ cm to $z_{02} = 25$ cm), at the height $z_d = 2$ m, the fetch calculated from the homogeneous curve is about 100 m, while the inhomogeneous curve predicts the fetch of 400 m (see the right panel of Figure 13.5). In the smooth-to-rough case, the cumulative footprint function of flux for the inhomogeneous case is larger than 1, in contrast to the homogeneous case where it is always less than 1. In the rough-to-smooth case the situation is different: here the cumulative footprint function of flux can be considerably less than 1.

13.5 Appendices

13.5.1 Appendix A. Dimensionless mean-flow equations

It is convenient to work in dimensionless variables by introducing

$$\zeta = z/h, \quad \xi = x/h, \quad u_g = u/G, \quad v_g = v/G, \quad w_g = w/G,$$
$$b_g = b/G^2, \quad k_g = k/Gh, \quad l_g = l/h, \quad m = fh/G.$$

In the dimensionaless form, the systems of equations read

$$u_g \frac{\partial u_g}{\partial \xi} + w_g \frac{\partial u_g}{\partial \zeta} = \frac{\partial}{\partial \zeta} k_g \frac{\partial u_g}{\partial \zeta} + m v_g,$$
$$u_g \frac{\partial v_g}{\partial \xi} + w_g \frac{\partial v_g}{\partial \zeta} = \frac{\partial}{\partial \zeta} k_g \frac{\partial v_g}{\partial \zeta} - m(u_g - 1),$$
$$\frac{\partial u_g}{\partial \xi} + \frac{\partial w_g}{\partial \zeta} = 0,$$

and

$$u_g \frac{\partial b_g}{\partial \xi} + w_g \frac{\partial b_g}{\partial \zeta} = \alpha_b \frac{\partial}{\partial \zeta} k_g \frac{\partial b_g}{\partial \zeta} + k_g \left[\left(\frac{\partial u_g}{\partial \zeta} \right)^2 + \left(\frac{\partial v_g}{\partial \zeta} \right)^2 \right] - \frac{c b_g^2}{k_g},$$
$$l_g = \left(\frac{1}{\kappa \zeta} + \frac{h}{l_0} \right)^{-1}, \quad k_g = C_k \, l_g \sqrt{b_g}.$$

with the boundary conditions:

$$
\begin{aligned}
&u_g = 0, \quad v_g = 0, \quad w_g = 0, \quad \text{at} \quad \zeta = z_0/h, \\
&u_g = 1, \quad v_g = 0, \quad \text{at} \quad \zeta = 1, \quad x \le 0, \\
&\frac{\partial u_g}{\partial \zeta} = \frac{\partial v_g}{\partial \zeta} = 0 \quad \text{at} \quad \zeta = 1, \quad x > 0, \\
&\frac{\partial b_g}{\partial \zeta} = 0 \quad \text{at} \quad \zeta = z_0/h, \quad \text{and} \quad b_g = 0 \quad \text{at} \quad \zeta = 1, \\
&l_g = \kappa\zeta \quad \text{at} \quad \zeta = z_0/h.
\end{aligned}
$$

In the stochastic Lagrangian models, the following statistical characteristics of the flow are used: the tensor $\langle u'_\alpha u'_\beta \rangle$, and the energy dissipation rate $\bar{\varepsilon}$. We extract these functions from the following closure assumptions: $\bar{\varepsilon} = cb^2/k$, and

$$
\begin{aligned}
&\sigma_{11} = \langle (u'_1)^2 \rangle = b/C_u, \\
&\sigma_{12} = \langle u'_1 u'_2 \rangle = -k\frac{\partial \bar{u}_{E2}}{\partial x_1}, \qquad \sigma_{22} = \langle (u'_2)^2 \rangle = b/C_v, \\
&\sigma_{13} = \langle u'_1 u'_3 \rangle = -k\left(\frac{\partial \bar{u}_{E1}}{\partial x_3} + \frac{\partial \bar{u}_{E3}}{\partial x_1}\right), \\
&\sigma_{23} = \langle u'_2 u'_3 \rangle = -k\frac{\partial \bar{u}_{E2}}{\partial x_3}, \\
&\sigma_{33} = \langle (u'_3)^2 \rangle = b/C_w.
\end{aligned}
$$

where

$$
C_u = \frac{b_u^2 + b_v^2 + b_w^2}{2\,b_u^2}, \qquad C_v = \frac{b_u^2 + b_v^2 + b_w^2}{2\,b_v^2}, \qquad C_w = \frac{b_u^2 + b_v^2 + b_w^2}{2\,b_w^2},
$$

where b_u, b_v, b_w are the universal constants in the relations $\langle u'^2 \rangle = b_u^2 u_*^2$, $\langle v'^2 \rangle = b_v^2 u_*^2$, $\langle w'^2 \rangle = b_w^2 u_*^2$ which are true in the surface layer with a constant shear $\langle u'w' \rangle = -u_*^2$. In our calculations we took $b_u = 2.5, b_v = 2., b_w = 1.25$ (e. g., see [88]). All the parameters $C_k, c, C_u, C_v, C_w, etc.$ were chosen to fit the theory of the surface layer with neutral stratification.

13.5.2 Appendix B. Lagrangian stochastic trajectory model

The main input function of the Lagrangian stochastic models is the Eulerian p. d. f. which is in our case assumed to be Gaussian:

$$
p_E(u; x) = (2\pi)^{-3/2}(det\,\sigma)^{-1/2} \exp\left\{ -\frac{1}{2}(u_i - \bar{u}_{Ei})\lambda_{ij}(u_j - \bar{u}_{Ej}) \right\}.
$$

Here λ_{ij} are the elements of a matrix Λ which is the inverse to the matrix σ defined by the entries

$$\sigma_{ij} = \langle (u_{Ei} - \bar{u}_{Ei})(u_{Ej} - \bar{u}_{Ej}) \rangle,$$

i.e., $\sigma_{ik}\lambda_{kj} = \delta_{ij}$, or in matrix form, $\sigma\Lambda = I$, I being the identity matrix. The expressions for the entries of the matrix σ are given in Appendix A.

Forward Lagrangian trajectories

In the forward trajectory model, the governing equations are

$$\begin{aligned} dx_i &= u_i\, dt, \\ du_i &= a_i(x,u,t)\, dt + \sqrt{C_0\bar{\varepsilon}}\, dW_i(t), \end{aligned}$$

where

$$\begin{aligned} a_i = &-\left(\frac{C_0\bar{\varepsilon}}{2}\right)\lambda_{ik}(u_k - \bar{u}_{Ek}) + \bar{u}_{Ej}\frac{\partial \bar{u}_{Ei}}{\partial x_j} + \frac{1}{2}\frac{\partial \sigma_{ij}}{\partial x_j} \\ &+ \left[\frac{\partial \bar{u}_{Ei}}{\partial x_j} + \frac{\lambda_{im}}{2}\bar{u}_{Ek}\frac{\partial \sigma_{jm}}{\partial x_k}\right](u_j - \bar{u}_{Ej}) + \frac{\lambda_{im}}{2}\frac{\partial \sigma_{km}}{\partial x_j}(u_j - \bar{u}_{Ej})(u_k - \bar{u}_{Ek}), \end{aligned}$$

Backward Lagrangian trajectories

The backward trajectories are defined by

$$\begin{aligned} d\hat{x} &= \hat{u}\, ds. \\ d\hat{u}_i &= \hat{a}_i\, ds + \sqrt{C_0\bar{\varepsilon}}\, \overleftarrow{d}\, W_i(s), \quad s < t, \end{aligned}$$

where

$$\begin{aligned} \hat{a}_i &= a_i - C_0\bar{\varepsilon}\frac{\partial}{\partial u_i}\ln p_E = a_i + C_0\bar{\varepsilon}\lambda_{ij}(u_j - \bar{u}_{Ej}) \\ &= \left(\frac{C_0\bar{\varepsilon}}{2}\right)\lambda_{ik}(u_k - \bar{u}_{Ek}) + \bar{u}_{Ej}\frac{\partial \bar{u}_{Ei}}{\partial x_j} + \frac{1}{2}\frac{\partial \sigma_{ij}}{\partial x_j} \\ &+ \left[\frac{\partial \bar{u}_{Ei}}{\partial x_j} + \frac{\lambda_{im}}{2}\bar{u}_{Ek}\frac{\partial \sigma_{jm}}{\partial x_k}\right](u_j - \bar{u}_{Ej}) + \frac{\lambda_{im}}{2}\frac{\partial \sigma_{km}}{\partial x_j}(u_j - \bar{u}_{Ej})(u_k - \bar{u}_{Ek}). \end{aligned}$$

Chapter 14
Stochastic flow simulation in 3D porous media

Stochastic models and Monte Carlo algorithms for simulation of flow through porous media beyond the small hydraulic conductivity fluctuation assumptions are presented following our papers [98, 99, 196]. The hydraulic conductivity is modeled as an isotropic random field with a log-normal distribution and prescribed correlation or spectral functions. It is sampled by a Monte Carlo method based on a randomized spectral representation. The Darcy and continuity equations with the random hydraulic conductivity are solved numerically, using the successive overrelaxation method in order to extract statistical characteristics of the flow. Hybrid averaging is used: we combine spatial and ensemble averaging to get an efficient numerical procedure. We provide some conceptual and numerical comparisons of various stochastic simulation techniques and focus on the prediction of applicability of the randomized spectral models derived under the assumption of small hydraulic conductivity fluctuations.

14.1 Introduction

The porosity study has received renewed attention in recent years. Motivations for studying porosity came from many applied fields, in particular, from material science, biomedicine, geology, environment, etc. Simulating flows in natural porous media such as soils, aquifers, oil and gas reservoirs is drastically complicated by the extreme heterogeneities and with insufficient data characterizing the medium.

Generally the porous media is characterized by large irregularities in the sizes and forms of pores. To approximate the corresponding flow equations, a huge number of nodes is required to get practically relevant results. It should be noted that a reasonable description of the hydraulic conductivity behavior by a deterministic function meets with serious difficulties. Therefore, a natural choice used in this field is the statistical description of the hydraulic conductivity via a random field with a given statistical structure.

Freeze [58] has analyzed the available data and has found the field of hydraulic conductivity to be well described by the random log-normal distribution. In hydrogeology this approach is often used for the flow analysis in saturated zone, or for the transport of a dissolved pollutant in a saturated aquifer [1, 32, 66, 67, 137, 222, 223, 242]; see also the overview in the books [11, 34, 65].

Generally, when dealing with boundary value problems for PDEs with random parameters, one uses two main instruments to analyse the statistical characteristics of the solution: (i) the small perturbation method (based on first- or higher order approximation) applied in the case of small parameter fluctuations and (ii) the direct numerical solution of PDEs for the given samples of the random inputs.

The methods derived in the first-order approximation under small hydraulic conductivity fluctuation assumption are widely used [8, 66]. However, they have strong restrictions, and as a rule, the applicability conditions are uncertain. Therefore it is important to develop a general direct numerical method which is able to provide calculations beyond the small fluctuation assumptions; in addition, it could be used to validate the small perturbation results. On the other hand, the results obtained by the first approximation method are very useful, since they are explicit, and can therefore be used as a benchmark for testing the complicated direct numerical method.

In [196] we constructed a randomized spectral model (RSM) for the simulation of a steady flow in porous media in the 3D case under small fluctuation assumptions. The method follows the following scheme: first we explicitly derive the spectral tensor of the velocity from that of the hydraulic conductivity, then we construct a Monte Carlo simulation technique for the random velocity with the derived spectral tensor.

Note that a small perturbation analysis using a simulation formula inspired by [102], based on a numerical evaluation of the random field representation through the spectral measure, was also applied in Schwarze et al. [219]. In this approach, when constructing the realizations of a random field, the wave vectors are sampled in the whole space. This may cause a poor statistical representation for large wave vectors, which in turn may lead to large errors for small scale evaluations such as the mean squared separation of two particles. We suggest a different simulation technique which uses a stratified sampling of wave vectors described in [119] and further developed in [104].

Concerning the higher-order corrections of the spectral expansion method, we mention that Dagan [33] derived a second-order correction of the head covariances in the 3D case. He noted that the first-order approximation is very robust, even for a log-conductivity variance equal to unity, and the second-order correction of the head variance is smaller than 10 % of the first-order approximation. Thus, for small to moderate values of σ_f^2, it is suggested that the first-order approximation is accurate enough. Similar research on velocity covariance has been carried out in [38]. These authors explored the accuracy of the first-order approximation and reported that for $\sigma_f^2 \ll 1$, the second-order corrections to the velocity covariance are unimportant, but as σ_f^2 approaches unity they become significant.

We mention a high-order perturbation approach via the Karhunen–Loève decomposition reported in [265]. In this method, the log hydraulic conductivity Y is decomposed into an infinite series on the basis of a set of noncorrelated Gaussian standard random variables. The coefficients of the series are related to eigenvalues and eigenfunctions of the covariance function of the log hydraulic conductivity. The advantage of this method is that it suggests an approximation up to the fourth order in σ_Y. However its practical use is limited by the need for solving the complicated eigenvalue problem.

In the general case, when the fluctuations are not small, the only rigorous way to tackle the problem is direct numerical simulation, which allows us to analyze flows in complex domains, although it requires large computing resources.

There have been some attempts to develop direct numerical simulation for the problem of transport in porous media. In [222, 223] an analysis of one- and 2-dimensional steady groundwater flows in the bounded domain is carried out. The modeling domain has a block structure with a prescribed correlation of the hydraulic conductivity in the neighboring blocks.

Thus for such a simple piecewise-constant approximation of the hydraulic conductivity, the authors [222, 223] solved the Darcy equation by a finite-difference method to get samples of the hydraulic potential. It should be noted that their method neither guaranties homogeneity of the generated fields nor a specified correlation structure. An improved version based on a direct matrix inversion method is used in [12], which, however, is still time-consuming.

Another attempt to construct a model of a 3-dimensional stationary saturated flow beyond the small hydraulic conductivity fluctuations assumption was made in [1]. In this paper, the estimation of the head variance was calculated, and a comparison with the first order approximation results was carried out. However, the authors were faced with the demand fpr large computer resources: it was concluded that in oder to obtain reasonable computational results, the domain and the number of nodes should be increased up to an unrealistic level (about 10^6 nodes).

Statistical characteristics of the velocity field were also estimated by direct numerical simulation in [120], where the authors developed a stochastic Lagrangian model for the transport in a statistically isotropic porous medium. However the accuracy in these simulations also was not high enough to make the desired definite conclusions.

The applicability of the first-order approximation models for the velocity covariance in the mean flow direction in the 2-dimensional case was examined in [76]. It was found that these models give very accurate results for the longitudinal velocity covariance for the values of σ_f up to unity. However, the transversal velocity covariance deviates from the direct numerical simulation results as σ_f approaches unity.

Chin and Wang [23] used Monte Carlo simulation for a 3-dimensional flow to investigate the accuracy of the first-order approximation, in relation to the Eulerian–Lagrangian covariance relationship. They used the turning bands algorithm due to Thompson et al. [241]. This method superimposes independent random processes constructed along the lines; this is a kind of projection method constructed from the 1-dimensional spectral representations. It also cannot be considered to be an efficient simulation method, and in [23] the authors had to restrict the calculations to a crude mesh.

In this chapter we present a direct simulation model in three dimensions based on a numerical evaluation of the random realizations of the hydraulic conductivity by the successive overrelaxation method (DSM-SOR method). The samples of the hydraulic log-conductivity are constructed by a randomized spectral method. Since the

DSM-SOR method works for arbitrary large fluctuations, we are able to investigate the applicability of the models derived in the first-order approximation. The results extracted from the numerical simulations are also useful for the parametrization of the Lagrangian stochastic model developed in [120].

Note that both in the DSM-SOR method and in the first-order approximation approach we use the randomized spectral method to simulate the random fields with the desired spectral tensor. Hence, the construction of an efficient random field simulation method is a very important issue in this study. Interesting insights into the dynamics of transport in disordered media can be already achieved through relatively simple random models for the velocity field, such as finite superposition of Fourier modes, with each amplitude independently evolving according to an Ornstein–Uhlenbeck process [26, 220]. Instead, we will use randomized spectral methods for scalar and vector Gaussian fields described in [191] and further developed in [104].

This chapter is organized as follows. We start by formulating the stochastic boundary value problem in Section 14.2. The DSM-SOR method is described in Section 14.3. The first-order approximation and the relevant randomized spectral model are presented in Section 14.4. Section 14.5 includes calculations aimed at testing the DSM-SOR method by comparing with the results obtained by the randomized spectral model under small fluctuation assumptions. The main numerical simulation results obtained by the DSM-SOR method for the general case of hydraulic conductivity fluctuations are presented in Section 14.6.

14.2 Formulation of the problem

We consider a steady flow through saturated porous formation. For a stationary 3D flow, the specific discharge is determined by the Darcy law

$$\mathbf{q}(\mathbf{x}) = \theta(\mathbf{x})\mathbf{u}(\mathbf{x}) = -K(\mathbf{x})\nabla(\varphi(\mathbf{x})), \quad \mathbf{x} \in D \subset \mathrm{IR}^3, \tag{14.1}$$

where $\mathbf{q}$ is the so-called Darcy's velocity, or specific discharge, $\mathbf{u}$ is the pore velocity, θ is the porosity, φ is the hydraulic potential $\varphi = \frac{p}{\rho g} + z$, p is the fluid pressure, z is the height, and K is the hydraulic conductivity assumed to be a homogeneous log-normal random field with a given spectral density.

Thus $\mathbf{q}$ is a random field defined by (14.1) where φ is the solution of the following conservation of mass equation:

$$\sum_{j=1}^{3} \frac{\partial}{\partial x_j}\left(K(\mathbf{x})\frac{\partial \varphi}{\partial x_j}\right) = 0. \tag{14.2}$$

The functions K and θ are the key parameters of the flow. Experimental measurements show a high heterogeneous behavior of K in space with the following remarkable property: when considering K as a random field, its distribution is well approximated by a log-normal law (e. g., see [58]).

The porosity θ is also often considered in some models to be a random field. However, its variability is in the problems we tackle generally much smaller than that of K. We assume therefore $\theta(\mathbf{x}) = \theta = \text{const}$.

We will consider the hydraulic log-conductivity $\ln K = F + f$ as a statistically homogeneous random field with Gaussian distribution $N(m_f, \sigma_f)$. Here $m_f = F$ is the mean, and σ_f is the standard deviation. We denote by

$$C_{ff}(\mathbf{r}) = \langle f(\mathbf{x}) f(\mathbf{x} + \mathbf{r}) \rangle$$

the autocorrelation function, where $\mathbf{r}$ is the separation vector.

Moreover we assume first that f is statistically homogeneous and isotropic with the exponential autocorrelation function

$$C_{ff}(r) = \sigma_f^2 exp(-r/I_f), \tag{14.3}$$

where $r = |\mathbf{r}|$, I_f is a given correlation length.

The equation (14.2) will be solved in a finite domain, with the relevant boundary value conditions formulated in the next section.

14.3 Direct numerical simulation method: DSM-SOR

In this section we present the direct simulation method based on the successive over relaxation iterative solution of the relevant PDE. For brevity, we will call it the DSM-SOR method.

In numerical simulations we deal with the following boundary value problem in the domain $D = [0 : L_1] \times [0 : L_2] \times [0 : L_3]$:

$$\sum_{j=1}^{3} \frac{\partial}{\partial x_j} \left(K(\mathbf{x}) \frac{\partial \varphi}{\partial x_j} \right) = 0, \tag{14.4}$$

with the constant head (on the left and right bounds: $x_1 = 0$ and $x_1 = L_1$) and impervious (on other bounds of the domain D) boundary conditions:

$$\begin{cases} \varphi(\mathbf{x}) = \varphi_0 & , \quad x_1 = 0 \\ \varphi(\mathbf{x}) = \varphi_0 - \mathbf{J} L_1 & , \quad x_1 = L_1 \\ \dfrac{\partial \varphi(\mathbf{x})}{\partial n} = 0 & , \quad x_2 = 0, x_2 = L_2, x_3 = 0, x_3 = L_3 \end{cases} . \tag{14.5}$$

Here $J_j = -\partial \langle \varphi \rangle / \partial x_j$ is the mean hydraulic gradient in x_j-direction, $\mathbf{J} = (J_1, J_2, J_3)$ is taken in this chapter as a constant vector $\mathbf{J} = (J, 0, 0)$, and φ_0 being a constant. The hydraulic log-conductivity $\ln K$ is assumed to be a Gaussian isotropic random field with the mean $F = 3.4012$ and the autocovariance (14.3).

To construct the solution of the equation (14.4), for a chosen sample of $K(x)$ and satisfying the boundary conditions (14.5) we use the following finite-difference scheme in the interior nodes [86]:

$$\begin{aligned}
\big[K_{i-\frac{1}{2}jl} + K_{i+\frac{1}{2}jl} + K_{ij-\frac{1}{2}l} + K_{ij+\frac{1}{2}l} + K_{ijl-\frac{1}{2}} + K_{ijl+\frac{1}{2}}\big]\varphi_{ijl} \\
- K_{i-\frac{1}{2}jl}\varphi_{i-1jl} - K_{i+\frac{1}{2}jl}\varphi_{i+1jl} - K_{ij-\frac{1}{2}l}\varphi_{ij-1l} - K_{ij+\frac{1}{2}l}\varphi_{ij+1l} \\
- K_{ij-\frac{1}{2}l}\varphi_{ijl-1} - K_{ij+\frac{1}{2}l}\varphi_{ijl+1} = 0
\end{aligned}$$

with the uniform grid $h_1 = h_2 = h_3 = h$.

The normal derivative on the boundary was approximated by simply using the first-order approximation along the normal vector.

The finite-difference scheme can be written in a matrix form

$$A\varphi_h = (Diag(A) - L - U)\,\varphi_h = f_h, \tag{14.6}$$

where φ_h is the approximating solution vector which depends on the indexation, f_h is the vector in the right-hand side; Diag(A) is the diagonal matrix whose entries coincide with the diagonal elements of the matrix A; analogously, L and U are the left and right triangular matrices extracted from the matrix A. We use the successive over relaxation (SOR) iterative method for solving (14.6):

$$\varphi_h^n = w\,[\mathrm{Diag}(A)]^{-1}(f_h + L\varphi_h^n + U\varphi_h^{n-1} + (1-w)\varphi_h^{n-1}.$$

In [183, 184], the influence of boundary conditions in a 2-dimensional case was investigated. It was shown that beyond 3 or 4 correlation lengths (for constant head boundary conditions, or even less for impervious boundary conditions) the influence of boundary effects on the second moment of the hydraulic potential can be neglected. The influence of the impervious boundary conditions on the head covariances in 3-dimensional flow is quite small and restricted to the neighborhood of the boundary [34, 148]. The head increments are even less sensitive to the boundary conditions. In [120], all the statistical characteristics were calculated in a domain placed five correlation lengths from the head constant or impervious boundaries of the domain D. In our calculations, four correlation lengths were enough for correct the evaluation of the velocity correlations. Thus we consider two domains: the region $D = [0 : L_1] \times [0 : L_2] \times [0 : L_3]$, and $\tilde{D} = [4I_f : L_1 - 4I_f] \times [4I_f : L_2 - 4I_f] \times [4I_f : L_3 - 4I_f]$. In the region D the equations (14.1), (14.2) are solved numerically by the SOR method, and the desired statistical characteristics are calculated in the domain $\tilde{D}$.

The hydraulic potential is chosen as $\varphi_0 = 100$ m. To reproduce the field K, in [120] is recommended to choose the grid size h at least not larger than $I_f/4$; in [1] this value was recommended as $I_f/5$. We have taken the grid size equal to $h = I_f/5$ and $h = I_f/6$, and the time step was $\Delta t = 0.25h/\langle u\rangle$.

14.4 Randomized spectral model (RSM)

In this section we present the randomized spectral model (RSM) applied along with the first-order approximation expansion under assumption of small fluctuation. So let us first describe the first-order approximation model; small random perturbation about the mean values for the potential, specific discharge and pore velocity components are assumed:

$$\varphi = \langle \varphi \rangle + \varphi' = H + h, \quad q_j = \langle q_j \rangle + q_j', \quad u_j = \langle u_j \rangle + u_j', \quad j = 1, 2, 3.$$

The autocovariance (14.3) has the spectrum

$$S_{ff}(k) = I_f^3 \sigma_f^2 / [\pi^2 (1 + I_f^2 k^2)^2], \tag{14.7}$$

where $\mathbf{k} = (k_1, k_2, k_3)$ is the wave number vector, and $k = |\mathbf{k}|$.

Under the assumption of small hydraulic conductivity fluctuations the spectrum of specific discharge has the form [66]

$$S_{q_j q_l}(k) = K_G^2 J_m J_n \left(\delta_{jm} - \frac{k_j k_m}{k^2} \right) \left(\delta_{ln} - \frac{k_l k_n}{k^2} \right) S_{ff}(k). \tag{14.8}$$

In [98] the randomized simulation approach developed in [191] is used to construct a divergenceless vector field with a given spectral tensor. We have constructed Monte Carlo simulation formulas for the specific discharge perturbation $\mathbf{q}'$, and hence the velocity perturbation $\mathbf{u}'$. We simulate $i = 1, 2, \ldots, N$ independent random fields with $S(\mathbf{k})$, and then we set

$$\mathbf{u}'^{(N)}(\mathbf{x}) = \frac{1}{\sqrt{N}} \sum_{i=1}^{N} \left[\frac{1}{\sqrt{p(\mathbf{k}_i)}} \left(\vec{\xi}'_{ki}(\mathbf{a}) \cos(\mathbf{k}_i, \mathbf{x}) + \eta'_{ki}(\mathbf{a}) \sin(\mathbf{k}_i, \mathbf{x}) \right) \right],$$

where

$$\vec{\xi}'_{k_i}(\mathbf{a}) = \xi_{ki}\, \mathbf{a}(\mathbf{k}), \quad \eta'_{k_i}(\mathbf{a}) = \eta_{ki}\, \mathbf{a}(\mathbf{k}),$$

$$a_j(\mathbf{k}) = \frac{K_G}{\theta} \left(J_j - \frac{k_j J_m k_m}{k^2} \right) \left(S_{ff}(\mathbf{k}) \right)^{1/2}, \quad j = 1, 2, 3,$$

ξ_{ki} and η_{ki} being random variables with zero mean and unit variance, and $\mathbf{k}_i$, ξ_{ki}, η_{ki} are all sampled independently. Here $\mathbf{k}$ is sampled according to the density $p(\mathbf{k})$, which is generally an arbitrary density function that can be chosen from rather different arguments. For instance, it is recommended in [191] to use $p(\mathbf{k}) = a^2(\mathbf{k}) / \int_{R^3} a^2(\mathbf{k}) d\mathbf{k}$.

The central limit theorem ensures, under some general assumption [106], that $\mathbf{u}'^{(N)}(\mathbf{x})$ converges to an ergodic Gaussian random field with the spectral tensor $S(\mathbf{k})$, as $N \to \infty$.

Assuming small perturbations, $\langle \mathbf{q} \rangle = K_G \mathbf{J}$ (see [34]), and so the velocity is modeled as $\mathbf{u}(\mathbf{x}) = (K_G \mathbf{J})/\theta + \mathbf{u}'(\mathbf{x})$.

A more general randomized spectral simulation method is constructed by introducing a stratified sampling of the wave numbers (see [104, 119]). Let us present the simulation formula in its general form.

Let $\vec{u}(\mathbf{x}) = u_1(\mathbf{x}, \dots, u_l(\mathbf{x}))^T$, $\mathbf{x} \in I\!R^d$ be a homogeneous vector Gaussian random field with the given spectral tensor $F(\mathbf{k})$ related to the correlation tensor $B(r) = \langle \vec{u}(\mathbf{x} + \vec{r})\, \vec{u}^T(\mathbf{x}) \rangle$ by

$$B(r) = \int_{\mathbb{R}^d} \exp\{2\pi\, \mathrm{i}\, \mathbf{k}\vec{r}\} F(\mathbf{k})\, d\mathbf{k}, \quad F(\mathbf{k}) = \int_{\mathbb{R}^d} \exp\{-2\pi\, \mathrm{i}\, \mathbf{k}\vec{r}\} B(\vec{r})\, d\vec{r}. \tag{14.9}$$

Here $()^T$ is used for the notation of the transpose operation.

Let

$$p(\mathbf{k}) = \sum_{j=1}^{l} F_{jj}(\mathbf{k}),$$

and assume that $\sigma^2 = \int_{\mathbb{R}^d} p(\mathbf{k})\, d\mathbf{k} < \infty$.

We will use the Holeski decomposition

$$F(\mathbf{k}) = p(\mathbf{k})\, Q(\mathbf{k})\, Q^*(\mathbf{k}), \tag{14.10}$$

where the matrix Q^* is defined as a complex conjugate transposition $Q^* = \overline{Q^T}$.

We denote by Q' and Q'' as the real and imaginare parts of the tensor Q: $Q(\mathbf{k}) = Q'(\mathbf{k}) + \mathrm{i}\, Q''(\mathbf{k})$.

Let us denote by $\Delta = \operatorname{supp}(p)$ the support of the spectral density $p(\mathbf{k})$. We choose a subdivision of Δ: $\Delta = \sum_{i=1}^{n} \Delta_i$.

Let $\mathbf{k}_{i1}, \dots, \mathbf{k}_{in_0}$ be a family of mutually independent identically distributed random points lying in Δ_i sampled from the p. d. f.

$$f_i(\mathbf{k}) = \begin{cases} \frac{p(\mathbf{k})}{\sigma_i^2}, & \mathbf{k} \in \Delta_i, \\ 0, & \text{else}, \end{cases} \qquad \sigma_i^2 = \int_{\Delta_i} p(\mathbf{k})\, d\mathbf{k}. \tag{14.11}$$

The randomization spectral model can be written in the form

$$\begin{aligned} \vec{u}_{nn_0}(\mathbf{x}) = \sum_{i=1}^{n} \frac{\sigma_i}{\sqrt{n_0}} \sum_{j=1}^{n_0} \Big\{ &\xi_{ij} \big[Q'(\mathbf{k}_{ij}) \cos\, \theta_{ij} - Q''(\mathbf{k}_{ij}) \sin\, \theta_{ij} \big] \\ &+ \eta_{ij} \big[Q''(\mathbf{k}_{ij}) \cos\, \theta_{ij} + Q'(\mathbf{k}_{ij}) \sin\, \theta_{ij} \big] \Big\}, \end{aligned} \tag{14.12}$$

where $\theta_{ij} = 2\pi\, \mathbf{k}_{ij} \cdot \mathbf{x}$, and ξ_{ij}, η_{ij}, $i = 1, \dots, n$; $j = 1, \dots, n_0$ are mutually independent and independent of the family $\mathbf{k}_{ij}$ standard *l-dimensional* Gaussian random variables (with zero mean and unity covariance matrix): $\xi_{ij} = (\xi_{ij}^{(1)}, \dots, \xi_{ij}^{(l)})$, $\eta_{ij} = (\eta_{ij}^{(1)}, \dots, \eta_{ij}^{(l)})$.

14.5 Testing the simulation procedure

In this section we test the direct numerical technique by comparing the results against the calculations obtained by RSM. Obviously RSM is a reasonable approach for small fluctuations, so we compare the results mainly for $\sigma_f^2 = 0.01$, and fix $\theta = 1.0$ and $I_f = 1$. For testing the DNS-SOR method we calculate the correlation functions

$$C_{u_j u_l}(\mathbf{r}) = \langle u_j(\mathbf{x}) u_l(\mathbf{x} + \mathbf{r}) \rangle$$

by the direct numerical simulation based on the SOR iterative procedure and compare them with the results obtained by the randomized spectral model constructed in [196] for the spectrum (14.8). As concluded in [196], RSM has shown good agreement with the exact results presented through a numerical integration in the spectral representation

$$C_{u_j u_l}(\mathbf{r}) = \int_{R^3} S_{jl}(\mathbf{k}) e^{i(\mathbf{r},\mathbf{k})} d\mathbf{k}, \tag{14.13}$$

where Simpson's rule was used to evaluate the integral (14.13). The expectations in RSM were evaluated by averaging over $N = 10^5$ samples.

For evaluation of statistical characteristics of a stochastic flow one usually uses two different averaging procedures: (i) space averaging [1, 120, 242] and (ii) ensemble averaging [23, 76, 222, 223]. To use the space averaging, we have to be sure that our randomized spectral model has good ergodic properties. As reported in [104], this is the case if the number of harmonics is sufficiently large, say, more than 1,000.

A compromise which seeks to avoid the problems of both types of averaging is the hybrid method, or a combined averaging: first the space averaging is taken, and then the result is averaged over n independent samples, n being not so large as in the pure ensemble averaging.

In the calculations presented in Figures 14.1–14.5 of Section 14.5 we use the ensemble averaging, while in Section 14.6 we use the combined averaging.

In Figure 14.1 we plot the function $C_{u_1u_1}$ (left panel) and $C_{u_2u_2}$ (right panel) calculated by the DSM-SOR method (solid lines) and by RSM (dashed lines); both functions are presented for the longitudinal direction r_1, $J = 0.01$, and the expectation is calculated as an arithmetic mean over $N = 10^4$ samples. The maximal relative difference between the results of two methods for $r_1 = 1$ was 3 % for the curves presented in the left panel and 9 % in the right panel. The statistical error of the direct simulation results was about 7 %, and 12 %, respectively; the statistical error of RSM was less than 3 %.

The statistical error in calculating an ensemble average of a random estimator ξ was measured by $\varepsilon(\xi) = 3\sigma_\xi/\sqrt{N}$, where σ_ξ^2 is the variance of the random estimator ξ. From Figure 14.1 it can be seen that the difference between the solid and dashed curves is small everywhere except for small values of the separated vector. This is presumably caused by the limit space resolution of the DSM-SOR method.

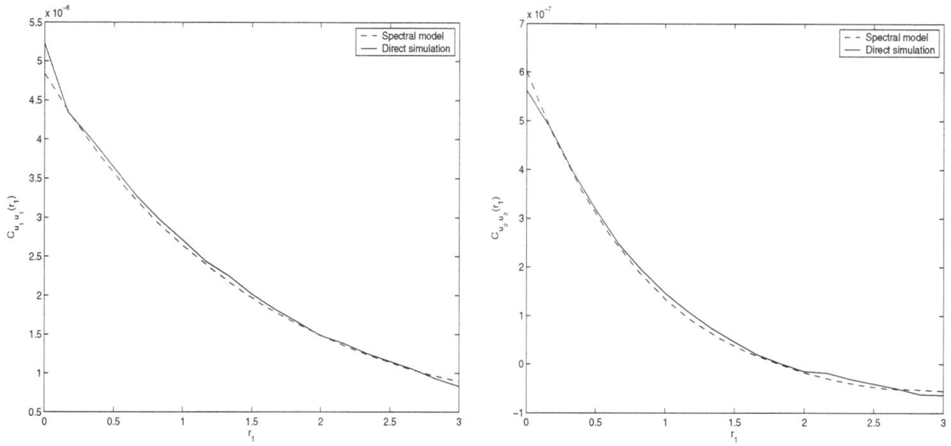

Figure 14.1. The auto-correlation function $C_{u_1u_1}(\mathbf{r})$ (left panel) and $C_{u_2u_2}(\mathbf{r})$ (right panel) in the longitudinal direction.

To control the space resolution, which is related to the large values of the wave number k in the log-conductivity power spectrum, we introduce a cut-off in the spectrum, so that the spectrum $S_{ff}(\mathbf{k})$ is defined on the interval $[0, k_{\max}]$. In Figure 14.2 we plot similar results for the spectrum (14.7) defined on the interval [0,15]. The total energy of this spectrum is more than 91.5 % of the total energy of the full power spectrum on $[0, \infty)$. The relative error (measured as the relative difference with the results obtained by RSM) in calculations of $C_{u_1u_1}(0)$, $C_{u_2u_2}(0)$ is decreased from 8 % (left panel) and 6.5 % (right panel) to 3 %. This indicates obviously that for spectra with rapidly decaying tails we can expect the accuracy of the direct numerical method to be higher.

Indeed, in [196] we used also a log-conductivity random field with the Gaussian form of the covariance

$$C_{ff}(r) = \sigma_f^2 \exp\left(-\frac{r^2}{l_f^2}\right)$$

whose spectrum has also a Gaussian form:

$$S_{ff}(\mathbf{k}) = \frac{\sigma_f^2 l_f^3}{\pi^{5/2}} \exp\left(-\frac{l_f^2 k^2}{4}\right), \tag{14.14}$$

where $I_f = l_f\sqrt{\pi}/2$. From Figure 14.3 it is seen that the velocity field corresponding to this log-conductivity spectrum with $l_f = 1$ is simulated more precisely than in the case of the exponential auto-correlation function (14.3).

As the fluctuations of the hydraulic conductivity get smaller ($\sigma_f \to 0$) then, analogous to the case analysed by RSM, the velocities tend to have Gaussian distributions.

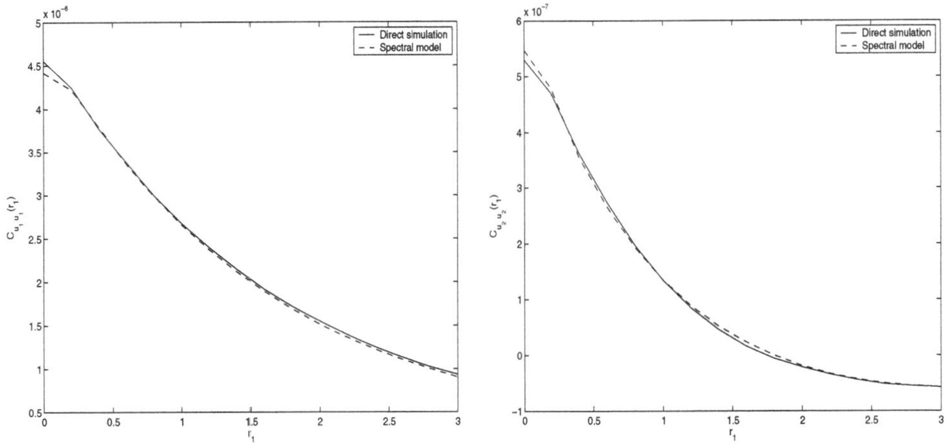

Figure 14.2. The auto-correlation functions $C_{u_1u_1}(\mathbf{r})$ (left panel) and $C_{u_2u_2}(\mathbf{r})$ (right panel) in longitudinal direction. Spectrum $S_{ff}(\mathbf{k})$ is defined on the interval $[0, 15]$.

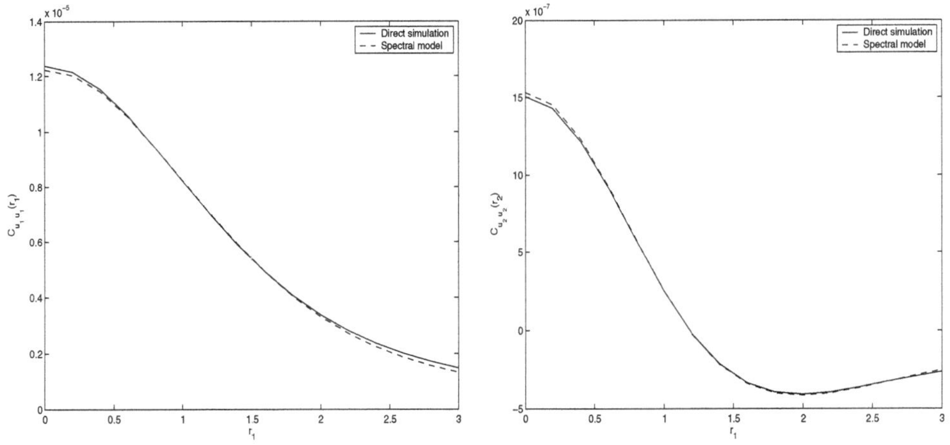

Figure 14.3. The auto-correlation functions $C_{u_1u_1}(\mathbf{r})$ (left panel) and $C_{u_2u_2}(\mathbf{r})$ (right panel) in the longitudinal direction; Gaussian spectrum $S_{ff}(\mathbf{k})$.

In Figures 14.4–14.5 the probability density functions of the longitudinal and transversal velocities are shown for two different (small and large) values of σ_f. As predicted, for $\sigma_f = 0.01$ the modeled densities (solid lines) are well approximated by Gaussian densities (dashed lines) with the mean $\langle u_1 \rangle = K_G J$ and $\sigma_{u_1} = 2.1991392\ E - 03$, $\sigma_{u_2} = 7.7482074E - 04$ (left panels). For large variance values ($\sigma_f = 1$) the velocity distributions are strongly non-Gaussian (right panels).

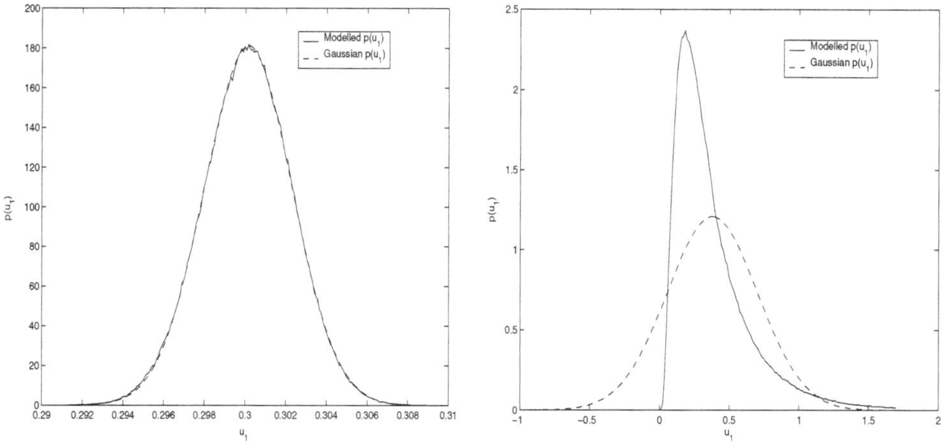

Figure 14.4. The probability density of the longitudinal velocity $p(u_1)$ for $\sigma_f = 0.01$ (left panel) and $\sigma_f = 1$ (right panel).

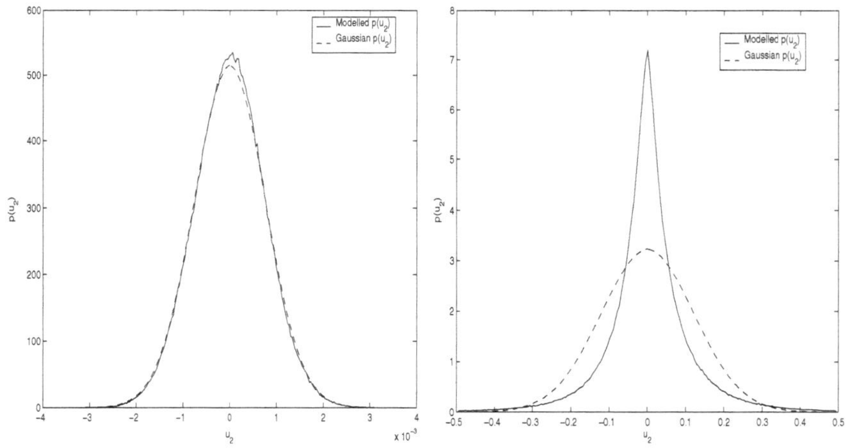

Figure 14.5. The probability density of the transversal velocity $p(u_2)$ for $\sigma_f = 0.01$ (left panel) and $\sigma_f = 1$ (right panel).

Note that the cross-correlation functions $C_{u_1 u_2}(\mathbf{r})$, $C_{u_2 u_3}(\mathbf{r})$ are identically equal to zero for arbitrary longitudinal and transversal separated vectors. This property was well confirmed by our calculations.

14.6 Evaluation of Eulerian and Lagrangian statistical characteristics by the DNS-SOR method

In this section we present the results of direct numerical simulation for some Eulerian and Lagrangian statistical characteristics of the flow. The key results are presented in Figure 14.7 where the Eulerian velocity autocorrelation functions are shown for different intensity fluctuations of the hydraulic conductivity, compared against the results obtained by the randomized spectral method under small fluctuation assumptions. From these curves one easily extracts the region of applicability of the small perturbation approach and RSM. Another important result is presented in Figure 14.12: here we show the behavior of the mean square separation of two particles. This complicated 2-particle Lagrangian statistical characteristic plays a crucial role in the turbulent diffusion study [145]. In the Kolmogorov inertial subrange the behavior of the mean square separation is described by the well-known Richardson's law, which predicts a cubic dependence on time. In our case, we cannot extract a universal structure for the function $\rho^2(t)$, not depending on the initial separation. However, it has shown an interesting subdiffusion behavior in transverse direction and a superdiffusion picture in longitudinal direction.

In the two subsections which follow, we calculate the expectations by the hybrid averaging: we combine spatial and ensemble averaging to get efficient numerical procedure. The Eulerian statistical characteristics were first calculated by spatial averaging over 21^3 nodes, and then by averaging over 300 independent samples of random velocities. The Lagrangian statistical characteristics were calculated by averaging over 252 trajectories per one sample of the random velocity, with subsequent averaging over 400 realizations of the velocity field. In all cases, the number of harmonics in the randomized spectral method was taken as $n_0 = 100$.

14.6.1 Eulerian statistical characteristics

Here we present the results of calculations of the mean Eulerian velocity and velocity autocorrelation functions. The spectrum $S_{ff}(\mathbf{k})$ is chosen in the form (14.7) which corresponds to the exponential decorrelation.

In Figure 14.6 we plot the mean velocity in a normalized form $\frac{\langle u_1\rangle\theta}{K_G J} - 1$, for different values of σ_f. This normalization is convenient, since the small perturbation method concludes that the mean velocity equals to $K_G\, J/\theta$.

Note that for $\sigma_f = 1$ our mean longitudinal velocity agrees well with that obtained in [120], being approximately 4 % larger. For values of σ_f up to 1.5 the mean velocity was also calculated in [23], where the relative difference with the result predicted by the first-order approximation was about 15 %. This difference in our calculations was 26 %, and 22 % in [120].

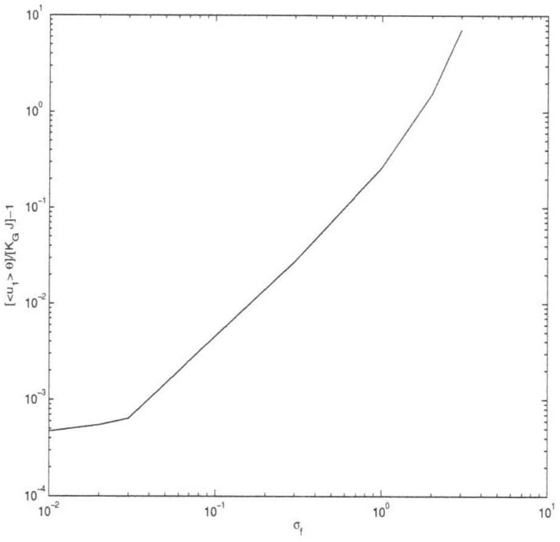

Figure 14.6. The behavior of $\frac{\langle u_1 \rangle \theta}{K_G J} - 1$, as a function of σ_f.

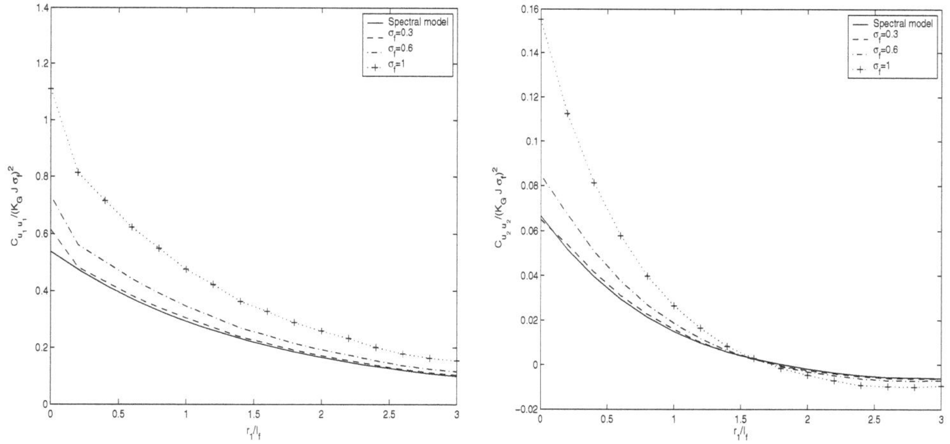

Figure 14.7. The dimensionless functions $C_{u_1u_1}(\mathbf{r}/I_f)$ (left panel) and $C_{u_2u_2}(\mathbf{r}/I_f)$ (right panel) in the longitudinal direction at different values σ_f in comparison against results of spectral model.

Figure 14.7 shows how good the small perturbation method and RSM work. This can be seen by comparing the RSM results with the data obtained by DSM-SOR method. Here we plot the dimensionless functions $C_{u_1u_1}$ (left panel) and $C_{u_2u_2}$ (right panel) in the longitudinal direction $r_1' = r_1/I_f$, for $\sigma_f = 0.3, 0.6$ and $\sigma_f = 1$. In the left panel: as expected, the relative difference between the RSM and DSM-SOR results is rapidly increasing with the growth of the fluctuation intensity, i. e., as σ_f increases. So, for $r_1' = 1$, this difference behaves like 9 %, 29 % and 84 % for $\sigma_f = 0.3, 0.6$ and 1, respectively. As to the statistical error of these calculations, it was less than 1 % for DSM-SOR method, and 2 % for RSM.

In the right panel: the relative difference between the two methods (again, for $r_1' = 1$) is less than 6 %, 31 %, and 108 % for $\sigma_f = 0.3, 0.6$, and 1, respectively. The statistical error: less than 2 % for DSM-SOR method and 3 % for RSM.

Thus the curves shown in Figure 14.7 present a clear picture of the region where the small perturbation approach and RSM can be applied, and of how fast this approximation fails as the fluctuation intensity increases.

14.6.2 Lagrangian statistical characteristics

In this section we present the results of numerical simulations for some Lagrangian statistical characteristics of the flow, $\sigma_f = 1$ is fixed. The calculations were carried out for the exponential correlations controlled by the spectrum (14.7) and by the Gaussian correlations with the spectrum (14.14).

The random Lagrangian trajectory $\mathbf{X}(t) = (X_1(t), X_2(t), X_3(t))$ starting at a point $\mathbf{x}_0$ is defined as a function satisfying the following stochastic equation:

$$\frac{d\mathbf{X}}{dt} = \mathbf{u}(\mathbf{X}), \quad \mathbf{X}(0) = \mathbf{x}_0. \tag{14.15}$$

The displacement covariances are defined by

$$D_{jl}(t) = \langle (X_j(t) - \langle X_j \rangle(t))(X_l(t) - \langle X_l \rangle(t)) \rangle.$$

In what follows we deal with the normalized quantities:

$$D_{jl}' = D_{jl}/I_f^2, \quad j, l = 1, 2, 3,$$

and dimensionless time $t' = t\,U/I_f$, where $U = \langle u_1 \rangle$, and $K_G = \exp(F)$.

In Figure 14.8, the dispersions $D_{11}'(t')/\sigma_f^2$ (left panel), and $D_{22}'(t')/\sigma_f^2$ (right panel) are shown. The curves seem to follow a linear law in time from, say, $t' = 3$, which would be in accordance with the classical Taylor's formula. To confirm this, we have extended these calculations to the times up to $t' = 30$ (see Figure 14.9). From these results we can conclude that the linear law happens to be true after the time $t' = 5$, both for the longitudinal and transverse dispersions.

Important Lagrangian statistical characteristic is the Lagrangian correlation tensor of velocity:

$$R_{jl}(\tau) = \langle \left[(u_j(X(t)) - \langle u_j(X(t)) \rangle \right] \left[(u_l(X(t+\tau)) - \langle u_l(X(t+\tau)) \rangle \right] \rangle,$$

where $X(t)$ is a Lagrangian trajectory started at the time t.

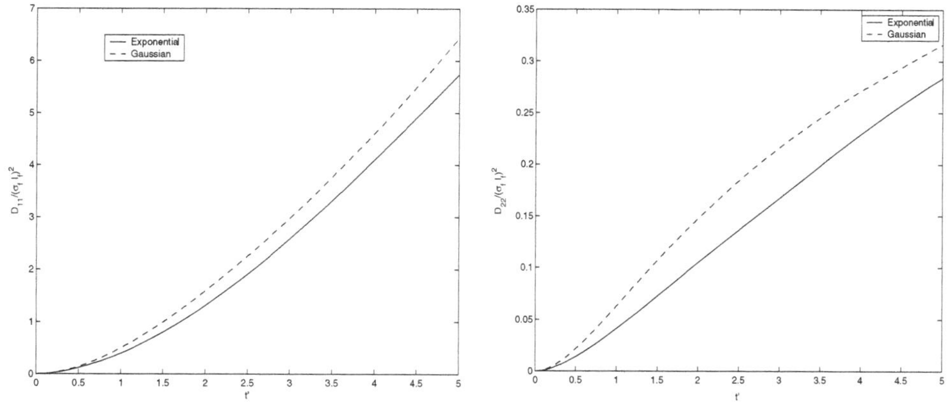

Figure 14.8. The dimensionless functions $D'_{11}(t')/\sigma_f^2$ and $D'_{22}(t')/\sigma_f^2$ (right panel). Small time.

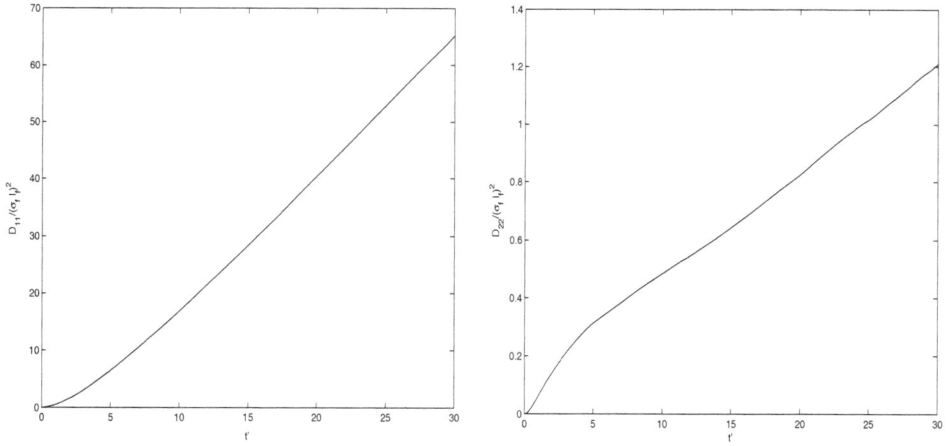

Figure 14.9. The dimensionless functions $D'_{11}(t')/\sigma_f^2$ and $D'_{22}(t')/\sigma_f^2$ (right panel). Large time. Gaussian spectrum S_{ff}, (14.14).

In Figure 14.10 we show the longitudinal (left panel) and transverse (right panel) Lagrangian correlation functions. Note that the transverse correlations have negative values after $t' \approx 1$, which may be connected with a trapping and which can lead to a deviation from the linear law of the transverse dispersion. To check it, we have calculated (see Figure 14.11) the integral

$$A_{jj}(\tau) = \int_0^\tau R_{jj}(\tau')d\tau'.$$

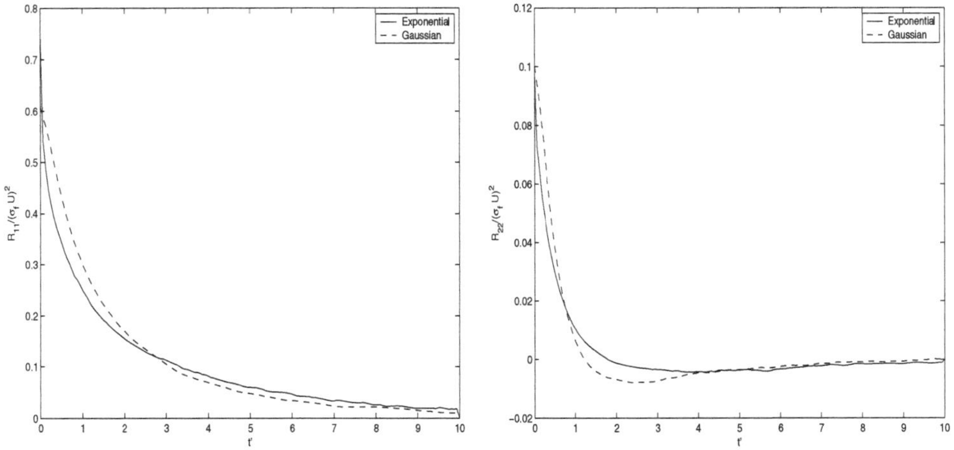

Figure 14.10. The dimensionless functions $R_{11}(t')/(\sigma_f U)^2$ (left panel) and $R_{22}(t')/(\sigma_f U)^2$ (right panel).

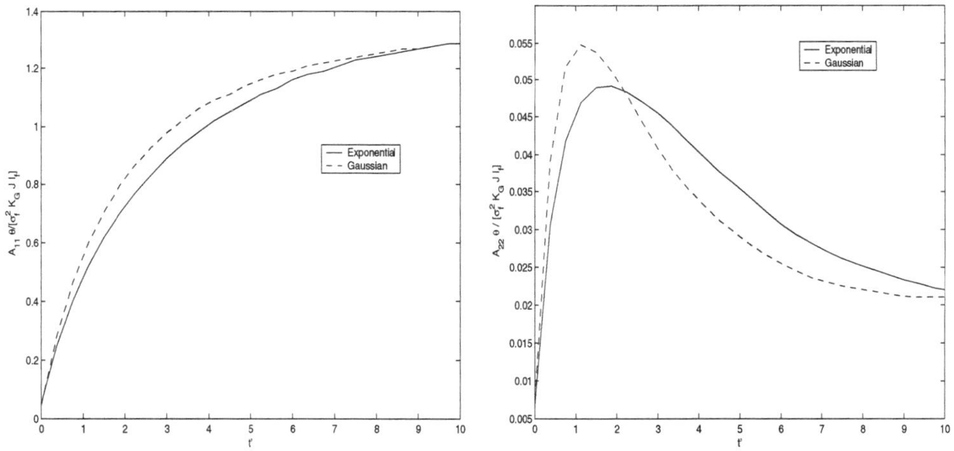

Figure 14.11. The dimensionless functions $A_{11}(t')\,\theta/(\sigma_f^2 K_G\; J\; I_f)$ (left panel) and $A_{22}(t')\,\theta/(\sigma_f^2 K_G\; J\; I_f)$ (right panel).

These results are in a good qualitative agreement with the results obtained by D. Chin and T. Wang in [23], although their calculations were limited by the grid size $h = I_f/2$, which is too large to provide high accuracy. Thus the D. Chin and T. Wang results underestimate the longitudinal dispersivity (left panel) to about 30 %, and the transverse dispersivity to about 15 % compared to our results.

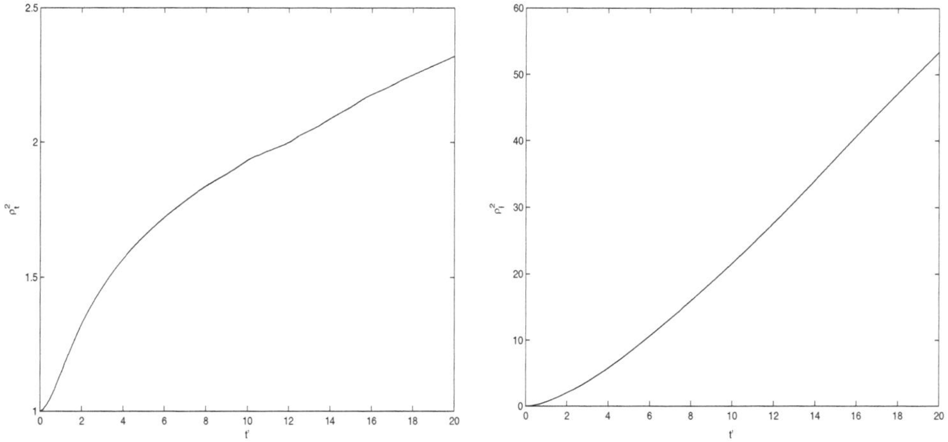

Figure 14.12. The mean square separation for two particles **X** and **Y** started at $(0, 0, 0)$ and $(0, 0, 1)$, respectively. In the left panel the transverse mean square separation is presented, and in the right panel the longitudinal mean square separation is shown.

Recall that Taylor's formula relating the dispersion tensor and the Lagrangian velocity covariance

$$D_{jl}(t) = 2\int_0^t (t-\tau)\, R_{jl}(\tau)\, d\tau$$

indicates that for large times the integral A_{jj} should not vanish (the linear dispersion behavior for large times). This is confirmed by the calculations presented in Figure 14.11.

Important Lagrangian statistical characteristic is the mean square separation for two particles. Let $\mathbf{X}(t') = (X_1(t'), X_2(t'), X_3(t'))$ and $\mathbf{Y}(t') = (Y_1(t'), Y_2(t'), Y_3(t'))$ be the Lagrangian trajectories of two particles initially separated by a vector r_0. The mean square separation $\rho^2(t')$ is defined as

$$\rho^2(t') = \langle (Y_1 - X_1)^2 + (Y_2 - X_2)^2 + (Y_3 - X_3)^2 \rangle.$$

Hence $\rho^2(t') = \rho_l^2(t') + \rho_t^2(t')$ where $\rho_l^2(t') = \langle (Y_1 - X_1)^2 \rangle$, and $\rho_t^2(t') = \langle (Y_2 - X_2)^2 + (Y_3 - X_3)^2 \rangle$ are the longitudinal and transverse mean square separations, respectively.

We show the functions $\rho_t^2(t')$ (left panel) and $\rho_l^2(t')$ (right panel) in Figure 14.12. Clearly, for sufficiently large time t' the two particles move independently, so we can expect that $\rho^2(t') = 2(D_l + D_t)\, t'$, according to classical dispersion law where D_l and D_t are the longitudinal and transverse diffusion coefficients, respectively. However for smaller times, the motion of two particles is correlated, and the dependence of $\rho_l^2(t')$ and $\rho_t^2(t')$ on t' is not linear. From the results presented in Figure 14.12 it can be estimated that the linear law is true approximately after $t' = 10$. For times $t' < 10$, the longitudinal dispersion is superdiffusional $\rho_l^2(t') = C_t\, t'^{1.4}$ with $C_t = 0.85$. More complicated is the structure of the transverse dispersion. There is a time interval $t' < 2$, where $\rho_t^2(t')$ seems to follow a linear behavior, and then it switches to the subdiffusional behavior in $2 \le t' \le 10$:

$$\rho_t^2(t') = 1.32 + C_l(t' - 2)^{0.7}, \qquad (14.16)$$

with $C_l = 0.145$. This agrees well with the behavior of the Lagrangian correlation functions presented in Figure 14.10, right panel. Indeed, it suggests that the first time interval of rapidly decaying correlation corresponds to the interval of linear behavior of $\rho_t^2(t')$, i. e., the diffusion regime is reached. Then, when the negative correlation reaches its maximum (about $t' = 2$), the trapping of particles results in the change of the function $\rho_t^2(t')$ from the linear behavior to the subdiffusional regime (14.16).

14.7 Conclusions and discussion

Stochastic numerical simulation technique for flow simulation through a 3D statistically isotropic porous media was developed without small perturbation assumptions. The hydraulic conductivity was modelled as an isotropic random field with a log-normal distribution and prescribed structure of the spectral functions. It was sampled by a Monte Carlo method based on a randomized spectral representation. The Darcy and continuity equations with the random hydraulic conductivity were solved numerically, using the successive over relaxation method in order to extract statistical characteristics of the flow. Hybrid averaging iwa used: we combined spatial and ensemble averaging to get an efficient numerical procedure. The method proposed enabled us to predict the applicability of the first-order approximation model derived under the assumption of small hydraulic conductivity fluctuations. Calculations of the longitudinal and transverse dispersions, the dispersivity, and the Lagrangian correlation functions have been carried out to extract the main statistical features of the flow. In particular, the calculations predict a subdiffusional behavior of the transverse dispersion and a superdiffusional behavior of the longitudinal dispersion.

It should be noted that the calculations take a lot of computer time, since the stochastic PDE is solved repetitively for many independent samples of the hydraulic conductivity field with the subsequent ensemble averaging to evaluate the desired statistical characteristics of the solution. A reasonable alternative would be a spatial averaging, but to make this approach realistic, the random field must have good ergodic properties. We applied a hybrid averaging by combining the ensemble and spatial averaging which has considerably decreased the cost of the algorithm. However we have not included here a detailed analysis of the ergodic properties of the randomized spectral models in 3D case; preliminary estimations show that the number of harmonics should be increased simultaneously with the refinement of the mesh used in the finite-difference method.

Chapter 15

A Lagrangian stochastic model for the transport in statistically homogeneous porous media

A new type of stochastic simulation models is developed for solving transport problems in saturated porous media, which is based on a generalized Langevin stochastic differential equation. A detailed derivation of the model is presented in the case when the hydraulic conductivity is assumed to be a random field with a log-normal distribution, being statistically isotropic in space. To construct a model consistent with this statistical information, we use the well-mixed condition which relates the structure of the Langevin equation and the probability density function of the Eulerian velocity field. Numerical simulations of various statistical characteristics such as the mean displacement, the displacement covariance tensor, and the Lagrangian correlation function are presented. These results are compared against the conventional direct simulation method.

15.1 Introduction

It is well known that stochastic models are well developed for solving transport problems in turbulent flows such as the transport in the atmospheric boundary layer (e. g., see [115, 181]). Stochastic models were constructed for a wide class of flows, in particular, to flows through porous media. (see [34,65]). To our knowledge, in the porous media transport only one type of stochastic models was used, namely, the random displacement method (RDM) for the hydrodynamic dispersion equation. It should be stressed that RDM can be applied only if the displacement covariance tensor is known (e. g., from measurements, or numerical simulation), and cannot be applied if the functionals of interest are evaluated at times comparable to the characteristic correlation scale of the flow. In contrast, the Lagrangian stochastic models based on the tracking particles in a random velocity field extracted from numerical solution of the flow equation (for brevity, we will call this model DSM, the direct simulation method) are free of these limitations, but the computational resources required are vast. Therefore, it is quite suggestive to construct a Langevin type stochastic model which is an approximation to DSM, and is written in the form of a stochastic differential equation for the position and velocity. It is worth mentioning that this approach is widely used in atmospheric transport problems. The basis for the Langevin type approach comes from the Kolmogorov similarity theory of fully developed turbulence [145], saying that in the inertial subrange the velocity structure tensor is a linear function in time. The linearity is the necessary condition to derive a Langevin type equation to mimic the behavior of the real Lagrangian trajectories. Therefore, the crucial point of the present study is to find out if in the porous media, this kind of linear law can be observed. This problem is studied by the DSM in Section 15.2. Detailed derivation of the Langevin type model is given in Section 15.3. The last section deals with the numerical simulations and comparisons with the direct simulation method.

15.2 Direct simulation method

In this section we derive Eulerian and Lagrangian statistical characteristics of the random velocity field in a porous medium. We first describe the model used to obtain the samples of the random velocity field and then analyze the properties of these random fields in the light of a Langevin type model.

15.2.1 Random flow model

Space scale

To define a random flow model in a porous medium, we first need to choose the space scale at which the velocity is to be studied. In hydrogeological literature a distinction is made between several space scales [30, 34]. We conventionally consider the following. (i) The microscopic scale or molecular scale where kinetics of molecules inside the pore volume plays a role. The solid walls of the pore space may be reflecting boundaries for the molecular velocities. (ii) The pore scale is the scale where the velocities are averaged over the pore volume. (iii) The macroscopic (or laboratory) scale is several orders of magnitude higher than the pore scale. It contains a sufficient number of pores to define the so-called "representative elementary volume" (REV) where macroscopic quantity like porosity, hydraulic conductivity, etc., can be defined. In a natural formation (field), the value of a parameter over the REV is assumed pointwise. (iv) The local (or formation or field) scale is the scale of a specific aquifer or field, containing one or several geological layers. (v) The regional scale may contain a system of aquifers or basins and may extend horizontally over several tens or hundreds of kilometers.

In this chapter we assume that the hydrogeological properties are defined over the REV at the macroscopic scale, and we derive a Langevin type model where these properties are assumed pointwise to predict transport over the field scale.

Darcy's law

In many general flow conditions, the phenomenological Darcy's law forms the basis of the theory of flow through porous media [34]. It is a consequence of the linearity of the equations of slow viscous flow which are obtained from the Navier–Stokes equations by neglecting the inertial terms. For time-independent flow conditions and saturated porous media, it is written as

$$\mathbf{q}(\mathbf{r}) = \theta(\mathbf{r})\,\mathbf{u}(\mathbf{r}) = -K(\mathbf{r})\,\nabla\phi(\mathbf{r}), \tag{15.1}$$

where $\mathbf{q}$, θ, $\mathbf{u}$, K, and ϕ are all macroscopic variables depending on space vector $\mathbf{r}$, and θ is the effective (or kinematic) porosity. This porosity takes into account the volumes of voids effectively concerned with groundwater flow (that is, for instance, without the dead-end pores and the adherence volume of the fluid to the grains). It is upper bounded by total porosity which is the volume occupied by the pores divided by the volume of the bulk medium. $\mathbf{q}$ is called Darcy's velocity and represents the groundwater flow rate, i. e., the volume of water crossing a unit area of porous medium per unit time; $\mathbf{q}$ is a measurable quantity, whereas $\mathbf{u}$ is the pore velocity, i. e., the flow rate per unit area of fluid (which is equivalent to consider that only fluid is present). ϕ is the hydraulic potential (or pressure head). It is defined by $\phi = p/\rho g + z$, where p is fluid pressure, ρ volumetric mass of the fluid, g the gravitational constant, and z the height. In ϕ, the kinematic term is always neglected due to small groundwater velocities in common applications. Finally, K, the proportionality coefficient between Darcy's velocity and the gradient of the hydraulic potential, is called hydraulic conductivity. This parameter (and also permeability) is recognized as a key parameter for groundwater flow. Several experimental techniques (mainly pumping tests, sedimentary analysis, but also seismic, geoelectrical or tracer methods) are used to intensively measure this parameter in the laboratory or in the field scale. These measurements have put in evidence the (highly) heterogeneous behavior of K in space and have suggested the use of stochastic models.

Random space function

Law [122] was presumably the first who used the stochastic approach in porous media and proposed, on the basis of core analysis data from a carbonate oil field reservoir a *log-normal* probability density function (p. d. f.) for K. Since this proposition, there is now a large body of direct evidence to support the statement that the p. d. f. for the hydraulic conductivity is log-normal [34, 58, 65, 121]. Hydraulic log conductivity $Y = \ln K$ is therefore commonly used and assumed to be distributed according to a Gaussian distribution $N(m_Y, \sigma_Y)$, where $m_Y = \langle Y \rangle$ and σ_Y is the standard deviation.

Another parameter appearing in (15.1) and considered in some models as a random field is the porosity θ. However its variability is recognized as being much smaller than hydraulic conductivity in common applications. However, a small amount of data is available for the stochastic properties of this parameter. Some linear laws, obtained on a speculative basis, have been proposed [58] to relate the porosity to log-hydraulic conductivity Y, suggesting then that porosity is normally distributed.

The flow equation

For a time-independent problem with no water source/sink, the continuity equation may be written as

$$\nabla \mathbf{q}(\mathbf{r}) = 0. \tag{15.2}$$

Combining this equation with Darcy's law (15.1), we obtain the flow equation inside the flow domain $\mathcal{D}$

$$\nabla \left[K(\mathbf{r}) \, \nabla \phi(\mathbf{r}) \right] = 0, \quad \mathbf{r} \in \mathcal{D} \tag{15.3}$$

with the following boundary conditions over the outer surface $\mathcal{S}$:

$$\phi(\mathbf{r}) = F_D(\mathbf{r}), \quad \frac{\partial \phi(\mathbf{r})}{\partial n} = F_N(\mathbf{r}), \quad \mathbf{r} \in \mathcal{S}_N. \tag{15.4}$$

Here $\mathcal{S}_D$ and $\mathcal{S}_N$ are parts of $\mathcal{S}$ where the Dirichlet and Neumann boundary conditions are used, respectively. F_D and F_N are given functions over $\mathcal{S}_D$ and $\mathcal{S}_N$. The solution $\phi(\mathbf{r})$ of the flow equation (15.3) with the boundary conditions (15.4) determines entirely the time-independent flow problem in a saturated porous medium, because the knowledge of the hydraulic potential $\phi(\mathbf{r})$ everywhere in $\mathcal{D}$ and over $\mathcal{S}$ yields the groundwater velocity by applying Darcy's law (15.1).

The hydraulic conductivity K in (15.3) is considered as a random field, the hydraulic potential $\phi(\mathbf{r})$ is therefore also a random field, and the velocity is a random vector field as well. In practice, F_D and F_N are often chosen as simple deterministic functions, and the boundaries are taken sufficiently far from the region of interest to avoid local effects due to the boundaries [184, 185]. Both the intensive practical measurements of the hydraulic conductivity in real applications and the central role played by this function in the flow equation (15.3) are favored for constructing a random flow model in porous media from a random conductivity model.

Generation of the random conductivity field is followed by the solution of the flow equation (15.3) and by the applcation of Darcy's law. This approach is commonly used by many authors in hydrogeology [211].

In this chapter, we will focus on the following model for $Y = \ln K$:

1. Y is Gaussian with constant mean $m_Y =< Y >$ and standard deviation σ_Y;
2. Y is statistically homogeneous and isotropic with an exponential auto-correlation function

$$C_{YY}(r) = \langle Y'(\mathbf{x}+\mathbf{r}) Y'(\mathbf{x}) \rangle = \sigma_Y^2 \exp\left(-\frac{r}{I_Y} \right), \tag{15.5}$$

where I_Y is a finite and given correlation length, and $r = |\mathbf{r}|$.

15.2.2 Numerical simulation

The numerical calculation of several realizations of random velocity fields and the simulation of particle trajectories by the DSM follow four principal steps : (i) generation of a hydraulic conductivity field with a prescribed statistical structure; (ii) evaluation of the Eulerian flow field by solving (15.3) with the boundary conditions (15.4); (iii) Identification of the particle's instantaneous position along the path line; (iv) evaluation of the statistical moments at fixed travel time or travel distance.

The accuracy of the solution depends on the numerical errors related to each step [211]. Step (i) is influenced by the choice of the generator procedure and by sampling frequency of the random field; step (ii) depends on the numerical method adopted in order to solve the flow equation (15.3) and also on the discretization; in step (iii), the error is related to the particle tracking procedure, while the convergence of step (iv) depends on the dimension of the sample used in the statistical computation.

In order to limit possible inaccuracy, we considered the following numerical procedure:

1. Generation of a homogeneous random hydraulic conductivity field $K(\mathbf{r})$ in a 3D domain with the characteristics

- $Y = \ln K$ is normal with mean $\langle Y \rangle = 3.4012$ and standard deviation $\sigma_Y = 1$;
- correlation function for log-hydraulic conductivity $C_{YY}(\mathbf{r}) = \langle Y'(\mathbf{x})Y'(\mathbf{x}+\mathbf{r})\rangle = \sigma_Y^2 \exp(-\frac{r}{I_Y})$ where $Y' = Y - \langle Y \rangle$ and $I_Y = 1$ is an isotropic correlation length;
- the porosity $\theta(x) = \theta = 0.5$ is constant;
- the K-field is generated by the randomized spectral formula (e. g., see [191]) with 8192 modes.

2. Solution of the groundwater flow equation for saturated conditions and stationary problem

$$\nabla\left[K(\mathbf{r})\nabla\phi(\mathbf{r})\right] = 0 \tag{15.6}$$

and boundary condition $\phi_{y=0} - \phi(y) = J\,y$, where $J = -\nabla\langle\phi\rangle\mathbf{1}_y$ is the mean hydraulic gradient. The pore velocity $\mathbf{u} = \frac{-K\,\nabla\phi}{\theta}$ is computed by the FORTRAN 90 code TRACE [251] which has been modified in order to write a finite element scheme for each component of the velocity vector so that there is no need to make additional finite difference approximations when evaluating derivatives. The hydraulic potential ϕ_0 is arbitrarily chosen ($\phi_0 = 100$ m) and the mean hydraulic gradient is fixed at $\mathbf{J} = 0.01\,\mathbf{1}_y$ implying a mean groundwater velocity along the y-axis and oriented towards positive y-values. The following normalized quantities are considered:

$$\widetilde{u} = \frac{u}{\frac{K_G J}{\theta}}, \quad \widetilde{r} = \frac{r}{I_Y}, \quad \widetilde{\phi} = \frac{\phi_0 - \phi(y)}{I_Y J}. \tag{15.7}$$

Equation (15.6) then becomes

$$\widetilde{\nabla}\left[e^{Y'}\,\widetilde{\nabla}\,\widetilde{\phi}\,\right] = 0, \tag{15.8}$$

with boundary conditions $\widetilde{\phi} = \widetilde{y}$. We have chosen the remaining parameters of the groundwater flow problem as follows:

$$\frac{K_G J}{\theta} = 0.6 \to K_G(= e^{\langle Y \rangle}) = 30, \quad \text{then} \quad \langle Y \rangle = \ln 30 = 3.4012.$$

The numerical values of the geometric mean $K_G = 30$ is a plausible value for hydraulic conductivity (expressed in m/day for instance) of an aquifer of moderate permeability.

3. Simulation of particle trajectories from the Eulerian velocity field. To construct a trajectory, we numerically solve the ordinary stochastic differential equation

$$\frac{d\mathbf{X}(t)}{dt} = \mathbf{u}(\mathbf{X}(t)) \tag{15.9}$$

with the initial conditions: $\mathbf{X}(t_0) = \mathbf{x}_0$ and $\mathbf{u}(\mathbf{X}(t_0)) = \mathbf{u}_0$. The velocity $\mathbf{u}$ is the Eulerian velocity obtained from the flow equation. For simplicity, we used the Euler scheme

$$\mathbf{X}(t_n) = \mathbf{X}(t_{n-1}) + \mathbf{u}(\mathbf{X}(t_{n-1}))\,\Delta t \tag{15.10}$$

where $\Delta t = t_n - t_{n-1}$.

We define three types of domains: (a) The groundwater flow domain $\mathcal{D}$ defined in the problem (15.6) with deterministic boundary conditions for $y = 0$ and $y = L$. Everywhere else, a deterministic "no flow boundary" is chosen, i. e., $\nabla\phi = 0$. In our test problem, the dimensions of this domain are $[0, 50] \times [0, 70] \times [0, 50]$ corresponding to 50 times the unit correlation length I_Y perpendicular to the mean velocity and 70 times this correlation length parallel to it. (b) The flow domain $\widetilde{\mathcal{D}}$ unperturbed by deterministic boundary conditions. In hydrogeological literature [184, 185], it is well known that beyond 3 to 4 correlation lengths (or even less for impervious boundary conditions in 3D) from the deterministic boundaries the effects of these deterministic boundary conditions on the stochastic behavior of pressure head (hydraulic potential) and therefore velocity may be neglected. We have considered here 5 correlation lengths leading to a domain $\widetilde{\mathcal{D}}$ of dimensions $[5, 45] \times [5, 65] \times [5, 45]$. (c) The flow domain Ω where the particle trajectories are initiated. We have empirically chosen a domain of dimensions $[15, 35] \times [15, 55] \times [15, 35]$. This is a domain Ω with isotropically minimum 20 correlation lengths over which a spatial mean will be sufficiently close to the ensemble mean [227].

To reproduce the stochastic behavior of the K-field, the grid size $h_x = h_y = h_z$ is chosen in the flow problem (15.6) to be $I_Y/4$. In practice, minimum $I_Y/3$ or $I_Y/4$ is recommended for h_x, h_y or h_z in the hydrological literature. We have then $201 \times 281 \times 201 = 11{,}352{,}681$ nodes in the flow problem for each realization of the random K-field. In Ω, a starting point x_0 is systematically selected at every node of the grid, that is, at 1,056,321 nodes. It is verified that all the trajectories remain inside $\widetilde{\mathcal{D}}$.

A maximum simulation time T is chosen. The time steps $\Delta t = t_n - t_{n-1}$ are variable and defined as

$$\Delta t = t_n - t_{n-1} = \mathrm{Min}\left[T - t_{n-1}; 0.5\,\frac{\mathrm{Min}(h_x, h_y, h_z)}{\|\mathbf{u}(t_{n-1})\|}\right]. \tag{15.11}$$

Simulation is stopped when $t_n > T$.

4. Evaluation of statistical moments at fixed travel time. We considered the following Lagrangian quantities: first, the Lagrangian correlation function for velocity. Two quantities are evaluated: the Lagrangian correlation function for the velocity component parallel to the mean flow $u_{\|}$ and the Lagrangian correlation function for the velocity component perpendicular to the mean flow $u_{\perp}$. They are defined by

$$C_{u_{\|}u_{\|}}(t) = \big\langle (u_{\|}(t) - \langle u_{\|}\rangle)\,(u_{\|}(0) - \langle u_{\|}\rangle)\big\rangle, \tag{15.12}$$

$$C_{u_{\perp}u_{\perp}}(t) = \big\langle (u_{\perp}(t) - \langle u_{\perp}\rangle)\,(u_{\perp}(0) - \langle u_{\perp}\rangle)\big\rangle. \tag{15.13}$$

Next we consider the Lagrangian velocity structure functions, which are defined as

$$\langle \Delta V_i(t)\,\Delta V_j(t) | X(t_0) = x_0; V(t_0) = u_0\rangle, \quad i, j = 1, 2, 3, \tag{15.14}$$

where $\Delta V_i(t) = V_i(X(t; t_0, x_0)) - V_i(t_0)$. The quantities

$$\Delta V_i(t_n)\,\Delta V_j(t_n) \quad i, j = 1, 2, 3 \quad \text{and for} \quad t_n \in [0, T] \tag{15.15}$$

are memorized in a table according to classes for the values of $u_{0\|}$ and $u_{0\perp}$. Taking into account the property of homogeneity of the random K-field and thus the random Eulerian velocity field, the arithmetic mean of the quantities (15.15) over the number of starting points multiplied by the number of realizations of the random K-field is an approximation of the conditional means

$$\big\langle \Delta V_i(t_n)\,\Delta V_j(t_n) | V(t_0) = u_0\big\rangle,$$

regrouped in classes for the values of $u_{0\|}$ and $u_{0\perp}$.

The quantities introduced above will be analyzed in Section 15.2.4 in more detail. We only mention here that in order to evaluate these quantities the time step is reduced in an appropriate way each time the simulation time overshoots the fixed times chosen at the beginning of the simulations.

15.2.3 Evaluation of Eulerian characteristics

Due to the symmetry of the flow problem, the Eulerian velocity (u, v, w) is conveniently decomposed into a longitudinal velocity component $u_{\|} \equiv v$ parallel to the mean flow and a perpendicular component $u_{\perp} = \sqrt{u^2 + w^2}$ perpendicular to the mean flow.

We are interested in the statistical properties of the Eulerian velocity field, i. e., mainly the probability density function $p_E(u, v, w)$. We decompose this p. d. f. as follows:

$$p_E(u, v, w) = \frac{1}{2\pi u_\perp} p_E^\parallel(u_\parallel)\, p_E^C(u_\perp|u_\parallel), \tag{15.16}$$

where $p_E^\parallel(u_\parallel)$ is the marginal p. d. f. of the Eulerian longitudinal velocity component $u_\parallel \equiv v$ and $p_E^C(u_\perp|u_\parallel)$ is the conditional p. d. f. of the Eulerian transverse velocity component $u_\perp$ under the condition that the longitudinal component is given.

By definition we have

$$\int_{-\infty}^{+\infty} du \int_{-\infty}^{+\infty} dw\, p_E(u, v, w) = 2\pi \int_0^{\infty} u_\perp\, p_E\, du_\perp = p_E^\parallel(v). \tag{15.17}$$

From the numerical procedure described in Section 15.2.2 above, we show in Figure 15.1 the histogram of the marginal p. d. f. $p_E^\parallel(u_\parallel)$ obtained with about 1 million samples of velocities. In Figure 15.2 we show the histogram of the conditional p. d. f. $p_E^C(u_\perp|u_\parallel)$ for different values of $u_\parallel$. Both figures are obtained for $\sigma_Y = 1$. Statistical calculations give $\langle u_\parallel \rangle = 0.73$, $\sigma_{u_\parallel}^2 = 0.33$; $\langle u_\perp \rangle = 0.246$ and $\sigma_{u_\perp}^2 = 0.059$.

We notice that the p. d. f.'s are asymmetric with quite heavy tails. For $p_E^\parallel(u_\parallel)$, negative $u_\parallel$ values are possible (with very small probability), meaning counter-current velocities. For $p_E^C(u_\perp|u_\parallel)$, the shape of the p. d. f.'s highly depends on the conditional value of $u_\parallel$: the higher the value of $u_\parallel$ (in absolute value) is, the larger is the interval of possible values for $u_\perp$.

It is well known in the hydrogeological literature that no exact analytical expressions can be found for the p. d. f. of the Eulerian velocity. Numerous theoretical studies [80, 183] suggest approximations for the first and second velocity moments from Darcy's law and the flow equation by applying perturbation methods of different order in σ_Y^2. However these studies cannot describe the type of velocity p. d. f. in general cases. These limitations can be avoided by numerical studies, like in [211], where the authors thoroughly analyzed the 1st, 2nd and 3rd moments of the velocity p. d. f.'s without giving hints about a possible p. d. f. family that could fit their numerical results.

We propose in this chapter to fit the numerical p. d. f.'s with generalized Weibull distributions. We assume that for a given σ_Y

$$p_E^C(u_\perp|u_\parallel) = \frac{p_2}{p_3^{p_1}\,\Gamma(p_1/p_2)} (u_\perp + p_4)^{p_1-1} \exp\left\{-\left[\frac{u_\perp + p_4}{p_3}\right]^{p_2}\right\}, \tag{15.18}$$

$$p_E^\parallel(u_\parallel) = \frac{q_2}{q_3^{q_1}\,\Gamma(q_1/q_2)} (u_\parallel + q_4)^{q_1-1} \exp\left\{-\left[\frac{u_\parallel + q_4}{q_3}\right]^{q_2}\right\}, \tag{15.19}$$

where p_3 and $q_3 > 0$; p_1, p_2, q_1 and $q_2 \geq 1$ and p_4 and $q_4 \geq 0$. p_E^C and $p_E^\parallel$ are generalized Weibull p. d. f.'s with shape parameters p_1 and q_1, exponents p_2 and q_2, scale parameters p_3 and q_3, and shift parameters p_4 and q_4.

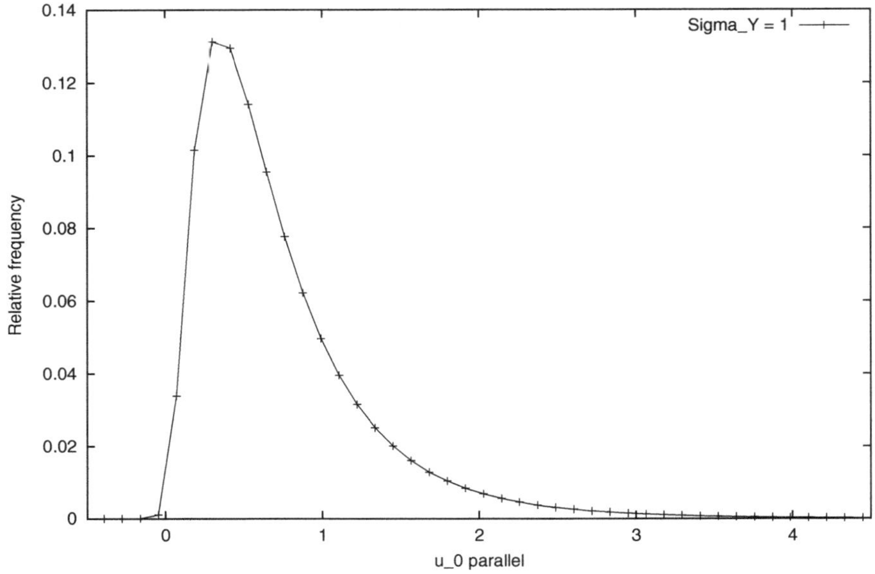

Figure 15.1. Histogram of the marginal p. d. f. $p_E(u_{0\|})$.

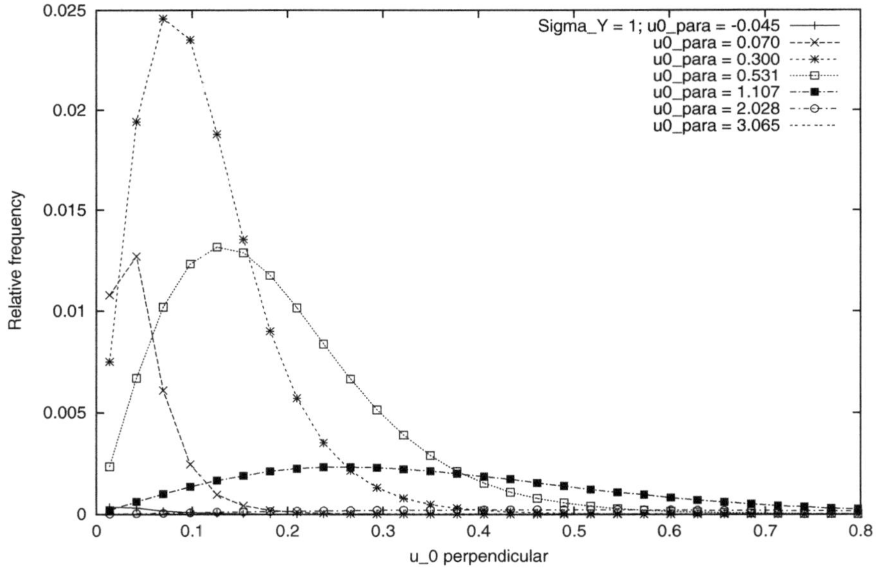

Figure 15.2. Histogram of the conditional p. d. f. $p_E(u_{0\perp}|u_{0\|})$.

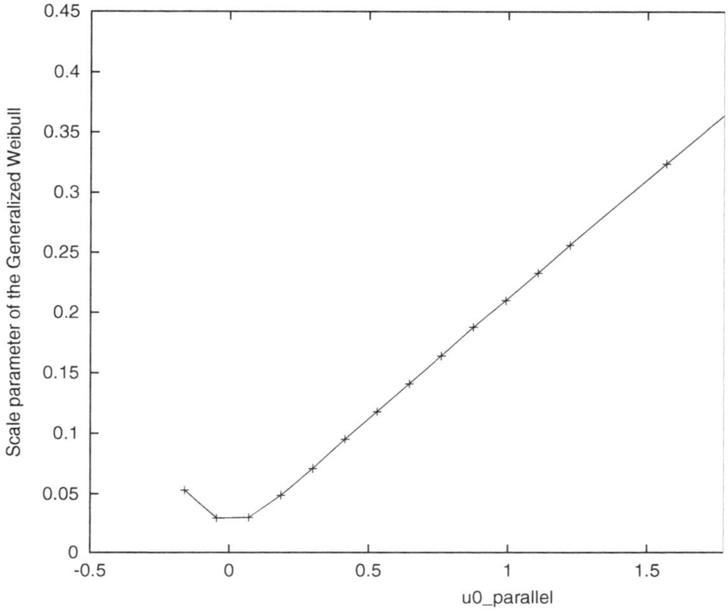

Figure 15.3. Dependence of the conditional p. d. f. $p_E(u_{0\perp}|u_{0\|})$ on $u_{0\|}$.

It should be noted that the choice of Weibull p. d. f.'s for p_E^C and $p_E^{\|}$ is not purely arbitrary; it is indeed a quite general p. d. f. family which includes the Gaussian and Gamma p. d. f.'s. In [211], it is shown that for $\sigma_Y^2 \ll 1$, both longitudinal and transversal velocities are distributed according to a Gaussian p. d. f.. Moreover, if the transversal components u and w are independently normally distributed with equal variances, we can show that $u_\perp = \sqrt{u^2 + w^2}$ is distributed according to the classical Weibull distribution with exponent $p_2 = 2$ and shape parameter $p_1 = p_2$.

In p_E^C, the parameters p_1 to p_4 depend on the value of the longitudinal velocity $u_\|$, the perpendicular velocity $u_\perp$ being nonnegative by definition, $p_4 = 0$ in (15.18).

From Figure 15.2, we can intuitively suggest that $u_\|$ is playing the role of a scale parameter for the p. d. f. p_E^C. A sensitivity analysis applied to the values of p_1, p_2 and p_3 after fitting of the numerical p. d. f.'s showed that p_3 is indeed the most sensitive parameter with respect to $u_\|$. In Figure 15.3 we show a law representing the dependency of p_3 with respect to $u_\|$ obtained after fitting the p. d. f.'s p_E^C for $u_\| = -0.2$ to 2. The relation is clearly linear in $u_\|$ for $u_\| > 0$. Negative values of $u_\|$ are very improbable, and we can assume the following law for the scale parameter p_3 to be a function of $u_\|$:

$$p_3(u_\|) = \lambda\, u_\| + \nu, \qquad (15.20)$$

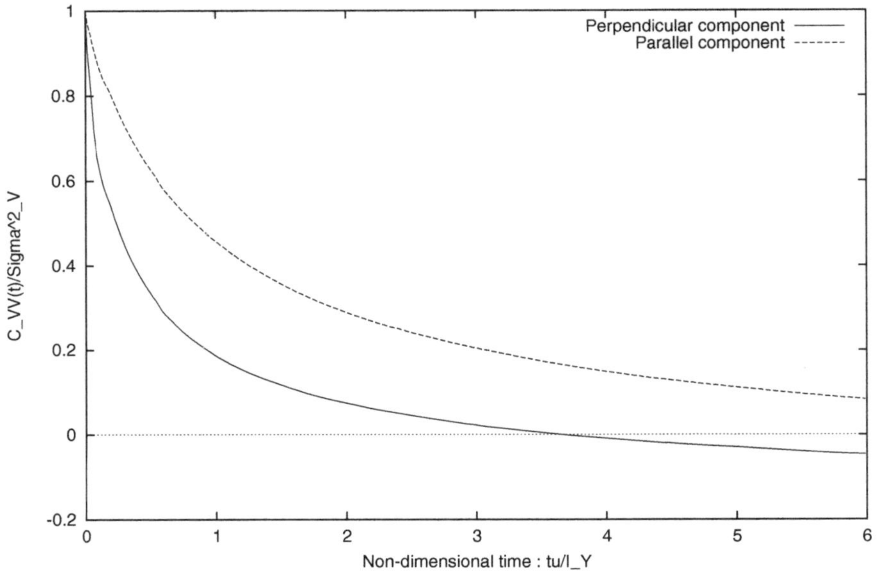

Figure 15.4. Lagrangian correlation functions for parallel and perpendicular velocity components.

where λ, $\nu > 0$ and $\nu > \lambda q_4$. In Figure 15.3, we find, for instance, $\lambda = 0.217$ and $\nu = 0.01$.

15.2.4 Evaluation of Lagrangian characteristics

By the direct simulation method described in Section 15.2.2, we have computed the Lagrangian correlation function $C_{u_\parallel u_\parallel}(t)$ for the component parallel to the mean flow $u_\parallel$ and the Lagrangian correlation function $C_{u_\perp u_\perp}(t)$ for the velocity component perpendicular to the mean flow $u_\perp$ defined in (15.12) and (15.13). In what follows we define $u = \langle u_\parallel \rangle$.

In Figure 15.4, the normalized correlation functions are shown versus nondimensional time; i.e., $C_{u_\parallel u_\parallel}/\sigma^2_{u_\parallel}$ and $C_{u_\perp u_\perp}/\sigma^2_{u_\perp}$ vs. ut/I_Y for $\sigma_Y = 1$. The results shown in this figure are similar to those obtained by Saladin and Fiorotto [211], who thoroughly studied the numerical accuracy of the DSM. From Figure 15.4, we can obtain an estimation of the Lagrangian correlation time for the longitudinal velocity $T_L \equiv 8.2$ in units I_Y/u. It is not very sensitive to the value of σ_Y, as also found in [211]. This important characteristic time determines the validity range of the approximations introduced in the random displacement model, as seen in Section 15.4.

The second quantity essential for the Langevin model developed in this chapter is the velocity structure function introduced in (15.14).

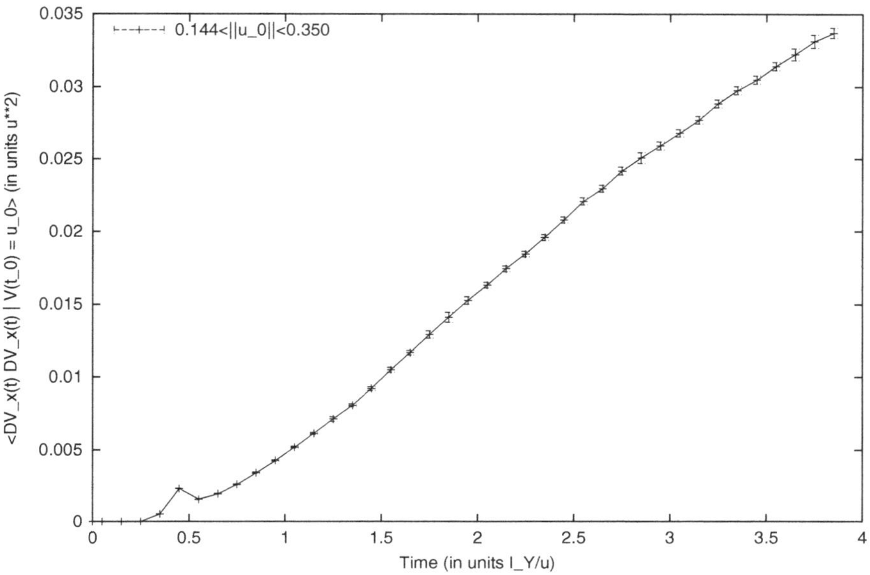

Figure 15.5. Velocity structure function vs. time (xx component); linear regime from $t \simeq 1.3$.

A series of calculations were realized by accumulating the results of $\Delta V_i(t_n)$ $\Delta V_j(t_n)$ for $T = 2$, 20 time intervals, 30 classes for $u_{0\parallel}$ and 35 classes for $u_{0\perp}$. Trajectories were initiated from each node of the grid in the domain Ω, i. e., at 1,056,321 points and a number of 40 realizations of the K and velocity fields were considered (generated from TRACE using parallel-processing over 64 CPUs).

In Figure 15.5 the velocity structure function $\langle \Delta V_x(t)\, \Delta V_x(t) | \mathbf{V}(t_0) = \mathbf{u}_0 \rangle$ is shown vs. time for $0.144 < u_0 < 0.350$, where u_0 is the norm of $\mathbf{u}_0$. The error bars appearing in this and the following figures represent the 90 % confidence interval of the results obtained after averaging over the trajectories. We recall here that the x-axis is perpendicular to the mean flow. Similar results are obtained for the function $\langle \Delta V_z(t)\, \Delta V_z(t) | \mathbf{V}(t_0) = \mathbf{u}_0 \rangle$, where z is the other axis perpendicular to the mean flow.

The calculations show that a linear behavior of this function vs. time can be identified. This linear behavior is appearing after a quadratic behavior vs. time. In Figure 15.6 the velocity structure function $\left\langle \Delta V_y(t)\, \Delta V_y(t) | \mathbf{V}(t_0) = \mathbf{u}_0 \right\rangle$ is also shown vs. time, for the same u_0 range. The y-axis is here parallel to the mean flow. Also in this figure a linear behavior of the structure function can be seen after a quadratic behavior and for the same time period.

The numerical results show that the values of the velocity structure functions $\langle \Delta V_x(t)\, \Delta V_y(t) | \mathbf{V}(t_0) = \mathbf{u}_0 \rangle$ and $\langle \Delta V_x(t)\, \Delta V_z(t) | \mathbf{V}(t_0) = \mathbf{u}_0 \rangle$ are much smaller and no clear linear behavior may be identified. By symmetry, similar results as$\langle \Delta V_x(t)\, \Delta V_y(t) | \mathbf{V}(t_0) = \mathbf{u}_0 \rangle$ are obtained for $\langle \Delta V_y(t)\, \Delta V_z(t) | \mathbf{V}(t_0) = \mathbf{u}_0 \rangle$.

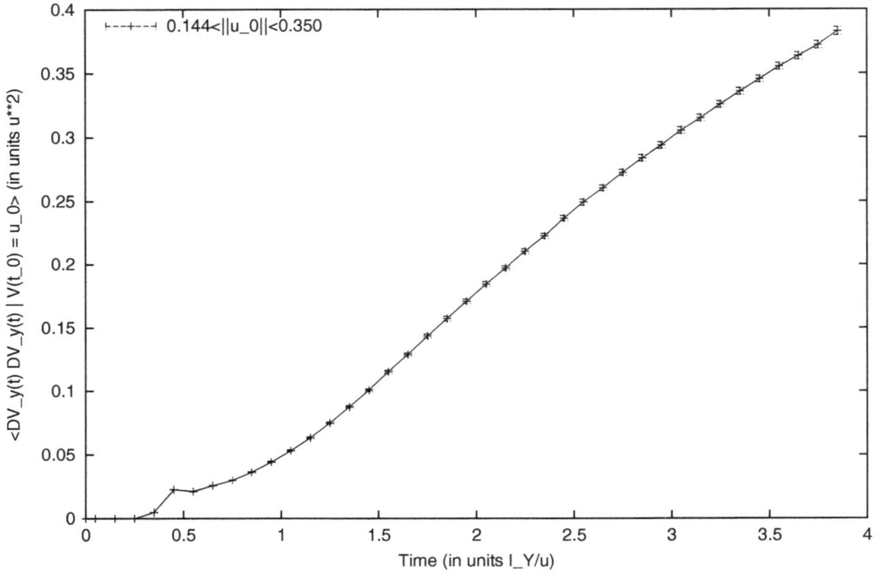

Figure 15.6. Velocity structure function versus time (yy component); linear regime from $t \simeq 1.3$.

In Figures 15.7 and 15.8, the velocity structure functions $\langle \Delta V_x(t)\, \Delta V_x(t) | \mathbf{V}(t_0) = \mathbf{u}_0 \rangle$, $\langle \Delta V_y(t)\, \Delta V_y(t) | \mathbf{V}(t_0) = \mathbf{u}_0 \rangle$ and $\langle \Delta V_x(t)\, \Delta V_y(t) | \mathbf{V}(t_0) = \mathbf{u}_0 \rangle$ are also shown vs. time for a higher intensity of $\mathbf{u}_0$; in this example $1.427 < u_0 < 1.629$.

From these results we see that a linear behavior of the velocity structure function may be identified. By considering that the groundwater velocity along the trajectory at early times is close to the value at the starting point u_0, we may deduce that between a distance along the path of about 40 to 70 % of the correlation length I_Y the linear regime is effective. This statement was verified for many different values of the starting velocity, values of σ_Y, and I_Y. For every class of velocity u_0, the tensor b with components

$$b_{ij} = \lim_{t \to \varepsilon} \frac{1}{t} \langle \Delta V_i(t)\, \Delta V_j(t) | \mathbf{X}(t_0) = \mathbf{x}_0; \mathbf{V}(t_0) = \mathbf{u}_0 \rangle \tag{15.21}$$

can be computed from the linear law observed in the Lagrangian velocity structure function. We computed the quantities b_{ij} from the different Lagrangian velocity structure functions versus time by applying the formula (15.21) for a time ε in the range $(0.4 \frac{I_Y}{||u_0||}, 0.7 \frac{I_Y}{||u_0||})$. The dependency of the components b_{ij} with respect to the normalized quantities defined in (15.7) can be written as

$$b_{ij} \equiv b_{ij}(\sigma_Y, \widetilde{u}_{0\parallel}, \widetilde{u}_{0\perp}). \tag{15.22}$$

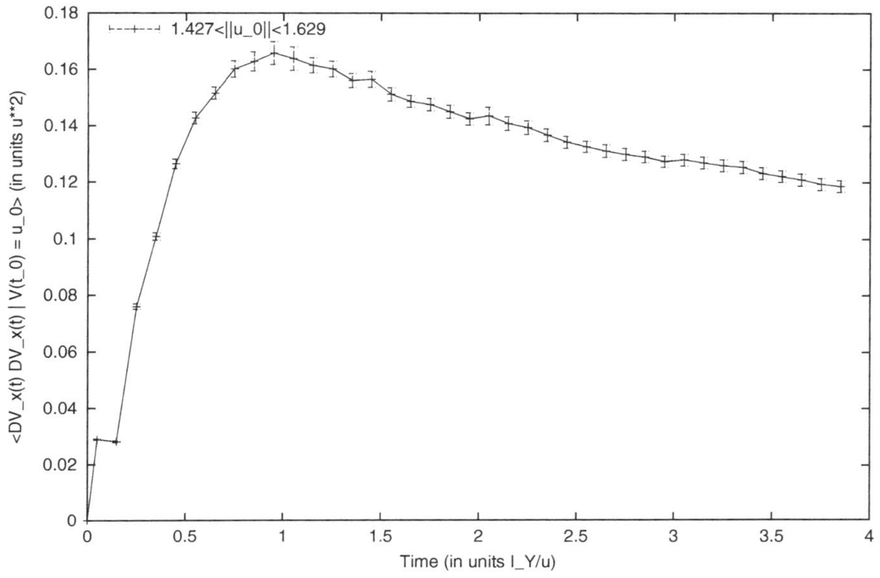

Figure 15.7. Velocity structure function versus time (xx component); linear regime for $t \in (0.4, 0.6)$.

We show as an example in Figure 15.8 to Figure 15.12 the behavior of b_{ij} (here b_{xx} and b_{yy}) with respect to $u_{0\parallel}$ and $u_{0\perp}$ for given values of $u_{0\perp}$ and $u_{0\parallel}$, respectively and for $\sigma_Y = 1$.

The curves have been obtained by a linear interpolation for the variables $u_{0\parallel}$ and $u_{0\perp}$. By comparing the Eulerian p. d. f.'s $p_E^{\parallel}(u_\parallel)$ and $p_E^C(u_\perp|u_\parallel)$ obtained in Section 15.2.3, we notice that in the zone of interest (where the density is nonzero, i. e., typically for $u_{0\parallel} \in [0.3, 1.2]$ and for $u_{0\perp} \in [0.05, 0.45]$ in this example) the values of b_{xx} and b_{yy} are smoothly varying. A first approximate model is to assume a constant value for these quantities. This is the model we consider in this chapter.

Finally, in Figure 15.13 we show as an example b_{xx} with regards to $u_{0\parallel}$ for $\sigma_Y = 0.5$. We see clearly that the dependency of b_{ij} with regards to σ_Y is much stronger than the dependency with regards to $u_{0\parallel}$.

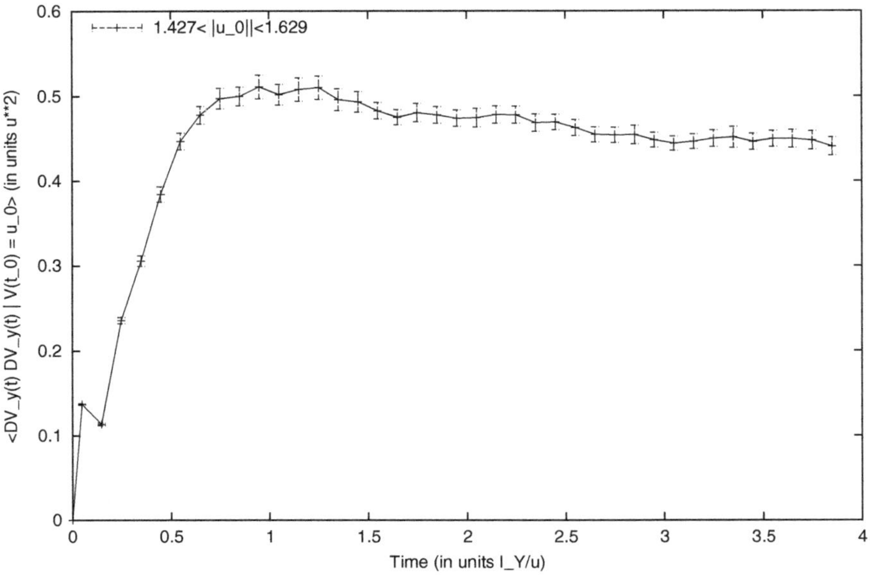

Figure 15.8. Velocity structure function vs. time (yy component); linear regime for $t \in (0.4, 0.6)$.

15.3 Construction of the Langevin-type model

15.3.1 Introduction

The Langevin-type models assume that the trajectory $(\mathbf{X}(t), \mathbf{V}(t))$ of a particle is governed by a stochastic differential equation of the Ito type

$$\begin{aligned} dX_i(t) &= V_i\, dt, \qquad (15.23)\\ dV_i(t) &= a_i(\mathbf{X}(t), \mathbf{V}(t), t)\, dt + \sigma_{ij}(\mathbf{X}(t), \mathbf{V}(t), t)\, dB_j(t), \qquad i = 1, 2, 3. \end{aligned}$$

The summation convention over the repeated indices is used. Here, a_i is the drift, σ_{ij} the diffusion terms, and B_j standard independent Wiener processes.

This kind of model is widely used in atmospheric turbulence simulations (e. g., see [115, 215, 237]). The motivation comes from the fact that the characteristic time of the acceleration correlations is much less than that for the velocity correlations, which is the case for the turbulent flow.

In porous medium, when dealing with laminar flows we cannot treat the velocity as turbulent. However, the acceleration direction is highly varying because of the pore structure inhomogeneity. Therefore, the acceleration and velocity fields can be considered as random flows, and the Langevin-type equations can be used to describe the

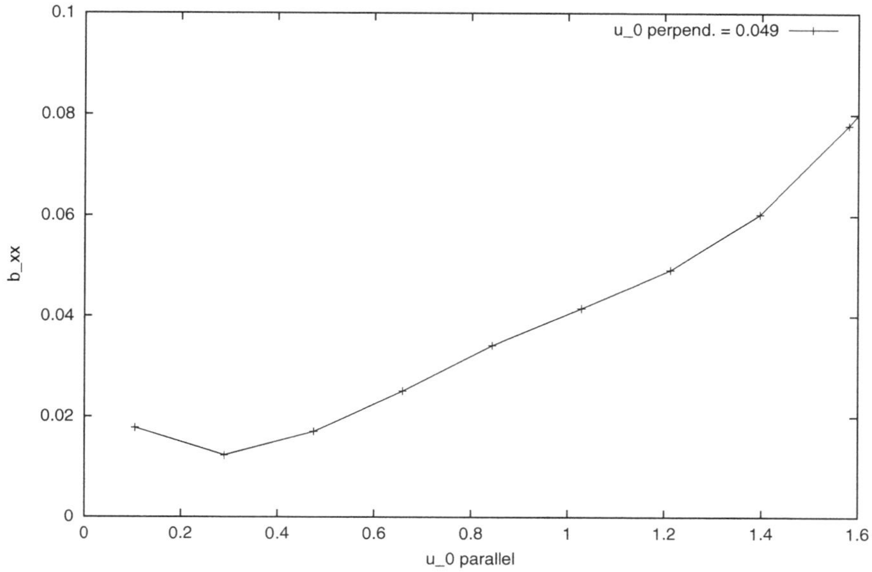

Figure 15.9. An interpolation for b_{xx}.

Lagrangian dynamics as in (15.24). The main difference, compared to the turbulence, is that the flow in porous media is extremely anisotropic, which results in a much more complicated form of the drift and diffusion terms. In particular, here we do not have this nice and simple diffusion term in the form of a constant C_0 coming from the Kolmogorov theory (e. g., see [181]).

Compared to RDM models, the Langevin type models involve more information about the statistics of velocity, e. g., it naturally uses the information about the p. d. f. of the Eulerian velocity field. Indeed, in case the velocity is incompressible, it is known that the Eulerian p. d. f. p_E is related to the coefficients of the Langevin equation (15.24) through the well-mixed condition [237]

$$\frac{\partial p_E}{\partial t} + u_i \frac{\partial p_E}{\partial x_i} + \frac{\partial}{\partial u_i}(a_i p_E) = \frac{1}{2} \frac{\partial^2 (b_{ij}\, p_E)}{\partial u_i \partial u_j}, \tag{15.24}$$

where $b_{ij} = \sigma_{ik}\sigma_{jk}$.

This complicates the model, but brings very important gains. The model is able to describe the transport in details for scales inside or compared to the Lagrangian time scale where the dispersion is not linear, so generally we are dealing here with the superdiffusion. For instance, it is possible to evaluate the concentration field not far from the source. Thus, these models are free of the Fickian hypothesis.

Derivation of the drift and diffusion terms is not simple. So far, even in the well-studied atmospheric turbulence, there is no theoretical approach which uniquely de-

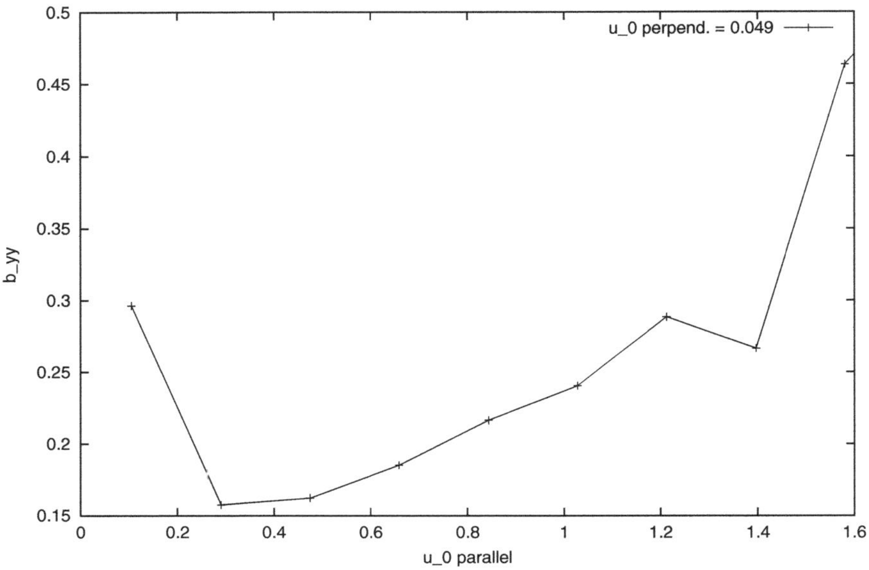

Figure 15.10. An interpolation for b_{yy}.

rives the expressions for these terms. Therefore, experimental and heuristical information is used to determine the model. In our case we evaluated the diffusion term from numerical solution of the flow equations reported in the previous section. In the next section we derive the drift term from the well-mixed condition (15.24).

15.3.2 Langevin model for an isotropic porous medium

We derive our Langevin model for a porous medium in which the flow conditions and properties of the random hydraulic conductivity field are similar to those described in Section 15.2, i. e., the mean horizontal flow is uniform over the domain $\mathcal{D}$, and it is assumed that the hydraulic conductivity field is log-normally distributed, with isotropic exponential correlation function.

To derive a unique Langevin model with these types of flow conditions we make the following general assumptions.

Let us construct the drift coefficients $a_1 = a_u, a_2 = a_v$, and $a_3 = a_w$ in the general form

$$a_u = \frac{1}{2p_E}\frac{\partial}{\partial u}(b_u^2 p_E) + \frac{u\phi}{p_E}, \quad a_v = \frac{1}{2p_E}\frac{\partial}{\partial v}(b_v^2 p_E) + \frac{\psi}{p_E},$$
$$a_w = \frac{1}{2p_E}\frac{\partial}{\partial w}(b_w^2 p_E) + \frac{w\phi}{p_E},$$

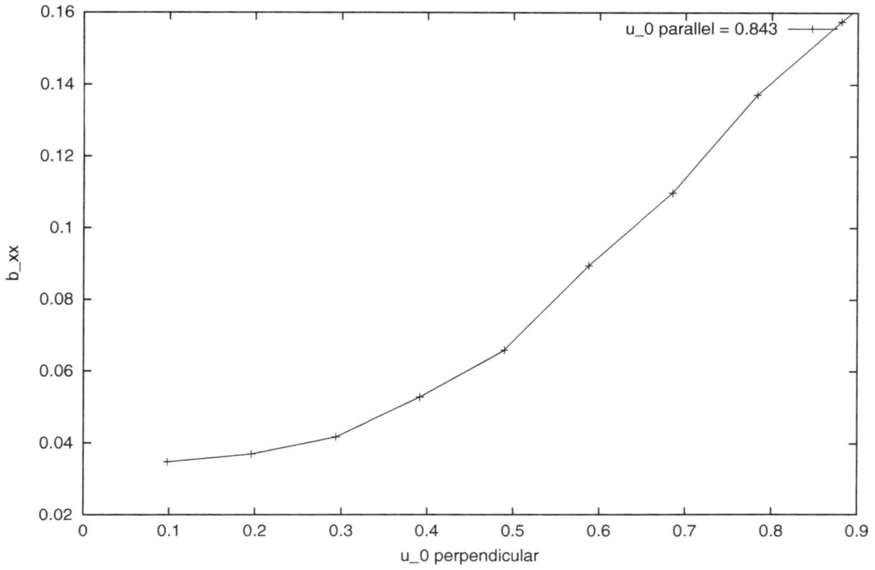

Figure 15.11. An interpolation for b_{xx}.

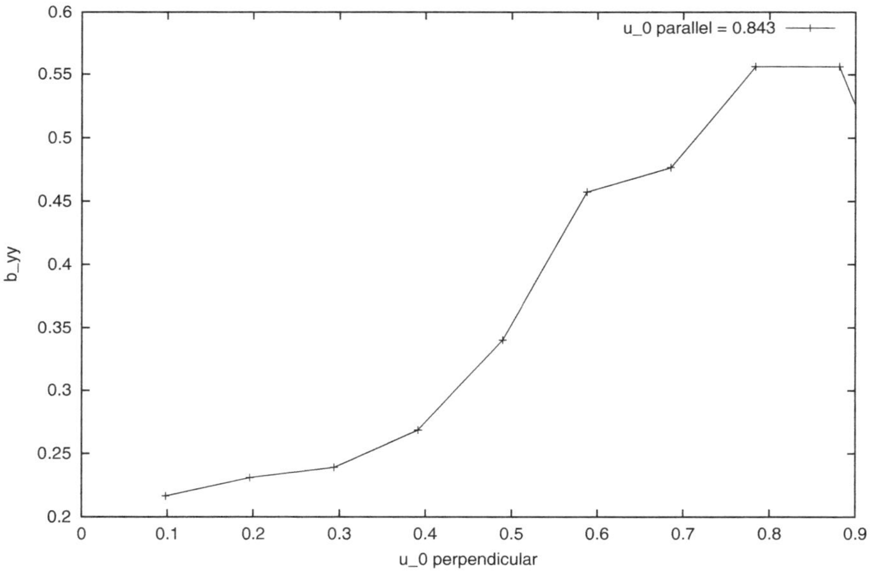

Figure 15.12. An interpolation for b_{yy}.

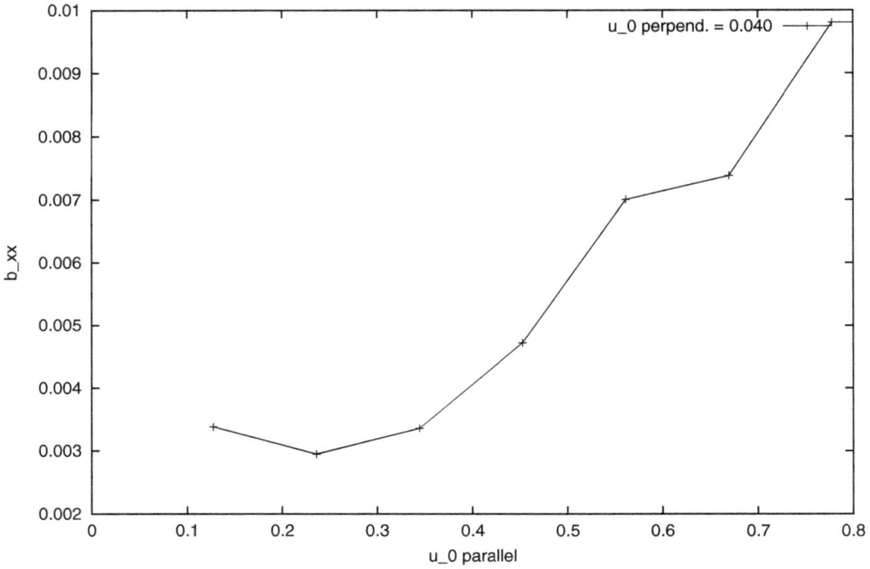

Figure 15.13. An interpolation for b_{xx}.

where the functions $\phi = \phi(u_\perp, v)$ and $\psi = \psi(u_\perp, v)$ should satisfy the equation (actually, the well-mixed condition)

$$\frac{\partial(u\phi)}{\partial u} + \frac{\partial\psi}{\partial v} + \frac{\partial(w\phi)}{\partial w} = 0.$$

The crucial point of the model is now as follows. We introduce the characteristic time scale of the transverse velocity τ_p as $\tau_p = \frac{\alpha I_Y}{\bar{U}}$, where $\bar{U}$ is the mean longitudinal velocity ($\bar{U} = \langle u_\parallel \rangle$), I_Y is the correlation length of the log hydraulic conductivity (in our case $I_Y = 1$, $\bar{U} = 0.73$), and α is a dimensionless constant.

This is the simplest choice motivated by dimensional arguments. Note that the dimension of ϕ/p_E is [time^{-1}], so we simply suggest that

$$\frac{\phi}{p_E} = -\frac{1}{\tau_p}. \tag{15.25}$$

Hence we have $\phi = -G\, p_E$ where we define the constant

$$G = \bar{U}/(\alpha I_Y). \tag{15.26}$$

We can then rewrite the well-mixed condition in the form

$$\frac{\partial\phi}{\partial u_\perp}\frac{u^2 + w^2}{u_\perp} + 2\phi + \frac{\partial\psi}{\partial v} = 0,$$

i. e.,

$$\frac{\partial \psi}{\partial v} = G p_E \left[2 + u_\perp \frac{\partial \ln p_E}{\partial u_\perp} \right].$$

From this, we find $\psi(u_\perp, v)$ by choosing $\psi(u_\perp, \infty) = 0$:

$$\psi(u_\perp, v) = -\int_v^\infty G p_E(u_\perp, v') \left[2 + u_\perp \frac{\partial \ln p_E}{\partial u_\perp} \right] dv'. \tag{15.27}$$

Thus the coefficients a_u, a_v, a_w are defined, and our model includes one free parameter α.

Let us now describe the numerical implementation. We have to solve the system

$$dX = u\,dt, \quad dY = v\,dt, \quad dZ = w\,dt, \tag{15.28}$$

$$du = a_u\,dt + b_u\,dB_u(t), \quad dv = a_v\,dt + b_v\,dB_v(t),$$
$$dw = a_w\,dt + b_w\,dB_w(t). \tag{15.29}$$

Note that we can write

$$a_u = u\,g(u_\perp, v), \quad a_v = h(u_\perp, v), \quad a_w = w\,g(u_\perp, v), \tag{15.30}$$

where

$$g(u_\perp, v) = -G + \frac{1}{2u_\perp}\frac{\partial b_u^2}{\partial u_\perp} + \frac{b_u^2}{2u_\perp}\frac{\partial \ln p_E}{\partial u_\perp}, \tag{15.31}$$

$$h(u_\perp, v) = \frac{1}{2}\frac{\partial b_v^2}{\partial v} + \frac{1}{2} b_v^2 \frac{\partial \ln p_E}{\partial v} - \frac{1}{p_E}\int_v^\infty G\,p_E(u_\perp, v') \left[2 + u_\perp \frac{\partial \ln p_E(u_\perp, v')}{\partial u_\perp} \right] dv', \tag{15.32}$$

and where the diffusion coefficients b_u^2, b_v^2 and b_w^2 are known from Section 15.3.

15.3.3 Expressions of the drift terms

We have seen in Section 15.2.3 that the p. d. f.'s of the Eulerian velocity may be fitted by the following Weibull densities:

$$p_E^C(u_\perp | u_\parallel) = \frac{p_2}{p_3(u_\parallel)^{p_1}\,\Gamma(p_1/p_2)} u_\perp^{p_1 - 1} \exp\left\{ -\left[\frac{u_\perp}{p_3(u_\parallel)} \right]^{p_2} \right\}, \tag{15.33}$$

$$p_E^\parallel(u_\parallel) = \frac{q_2}{q_3^{q_1}\,\Gamma(q_1/q_2)} (u_\parallel + q_4)^{q_1 - 1} \exp\left\{ -\left[\frac{u_\parallel + q_4}{q_3} \right]^{q_2} \right\}. \tag{15.34}$$

We have also seen that the dependence of p_E^C on $u_\parallel$ may be restricted to the scale parameter p_3 only, which depends linearly on $u_\parallel$ (see the relation (15.20)). Finally, from the Lagrangian characteristics studied in Section 15.2.4 we can model, as a first approximation, the diffusion coefficients by the following simple laws:

$$b_u^2 = b_1 + c_1\, u_\perp, \quad b_v^2 = b_2 + c_2\, u_\perp, \quad b_w^2 = b_u^2, \tag{15.35}$$

where b_1, b_2, c_1, and c_2 are constants.

From (15.30) and (15.32), and using the expressions (15.33) and (15.34) for p_E and (15.35) for b_v^2, we obtain for the longitudinal drift term $a_\parallel$:

$$\begin{aligned} a_\parallel = \frac{1}{2} b_\parallel^2 \Bigg\{ & \frac{1}{u_\parallel + q_4} \left[q_1 - 1 - q_2 \left(\frac{u_\parallel + q_4}{q_3} \right)^{q_2} \right] \\ & + \frac{1}{u_\parallel + \frac{\nu}{\lambda}} \left[p_2 \left(\frac{u_\perp}{\lambda\, u_\parallel + \nu} \right)^{p_2} - p_1 \right] \Bigg\} \\ & - \frac{G}{p_E(u_\perp, u_\parallel)} \int_{u_\parallel}^{\infty} p_E(u_\perp, u_\parallel')\, [\, p_1 - p_2 \left(\frac{u_\perp}{\lambda u_\parallel' + \nu} \right)^{p_2}]\, du_\parallel'. \end{aligned} \tag{15.36}$$

Similarly, again using the expressions (15.33) and (15.34) for p_E and (15.35) for b_u^2 and b_w^2, we obtain for the transverse drift components from (15.30) and (15.31): $a_u = u\, g(u_\perp, u_\parallel)$, and $a_w = w\, g(u_\perp, u_\parallel)$, where the friction function g is given by

$$g(u_\perp, u_\parallel) = -G + \frac{1}{2} \left\{ \frac{c_1}{u_\perp} + \frac{b_\perp^2}{u_\perp^2} \left[p_1 - 2 - p_2 \left(\frac{u_\perp}{\lambda\, u_\parallel + \nu} \right)^{p_2} \right] \right\}. \tag{15.37}$$

In the expressions (15.36) and (15.37) found above for the longitudinal drift term and the transverse friction function, we may find the following physical interpretations.

1. When $u_\parallel$ decreases to small values (close to $-q_4$), $a_\parallel$ becomes large and positive, whereas when $u_\parallel \gg q_4$, $a_\parallel$ becomes large in absolute value and negative. This behavior of the acceleration $a_\parallel$ with respect to $u_\parallel$ at the boundaries is expected to avoid divergence of the longitudinal velocity values towards large negative or positive values in the simulation of the Langevin model.

2. We see that g is large and positive for $u_\perp \to 0$, whereas g is large in absolute value and negative when $u_\perp \to \infty$ and $u_\parallel \to -q_4$. This behavior of g with respect to $u_\perp$ at the boundaries is expected to avoid divergence of the transverse velocity values towards zero or large positive values.

To make the simulations efficient we make the tabulation of the functions $g(u_\perp, u_\parallel)$ and $h(u_\perp, u_\parallel)$ in the domain K:

$$K = \big[\, (u_\perp, u_\parallel) : u_{\parallel\,\min} \le u_\parallel \le u_{\parallel\,\max},\ 0 \le u_\perp \le C_u\, u_\parallel \big],$$

where C_u is a constant, and $u_{\parallel\,\min}$ and $u_{\parallel\,\max}$ are the limits chosen so that the simulation gives stable results.

15.4 Numerical results and comparison against the DSM

It is important to verify that the new Langevin model discussed in this chapter is able to reproduce the behavior of some major characteristics of the transport of contaminants in porous media. We have therefore fitted the parameters of the Langevin model introduced in Section 15.3.2 to the dispersion curves vs. time. Figures 15.14 and 15.15 present the trajectories simulated by the DSM and the Langevin model, respectively. The Langevin model derived from equations (15.3.2), (15.30)–(15.32) mimic the real Lagrangian trajectories given by the DSM. Even if we see clearly that the structure of trajectories is quite different, the fact that the well-mixed condition is satisfied guarantees the correct consistency of the simulation results with the Eulerian p. d. f., which in turn assures that the mean concentrations should be close. By fixing the free parameter α in the definition of the constant G in (15.26)), it is possible to fit the Langevin model to some fundamental transport characteristics as transverse and longitudinal dispersion versus time. By using the results of Section 15.2.4, we identified the coefficients of the simple linear laws (15.35) modeling, as a first approximation, the diffusion coefficients $b_\parallel^2$ and $b_\perp^2$.

In the cases mentioned above we have found that the choice $C_u = 5.$, $u_{\parallel\,\min} = 0.05$, and $u_{\parallel\,\max} = 5.$ leads to stable calculations.

The model with tabulated drift coefficients (15.36) and (15.37) is quite fast, and with $\alpha = 4.$ it gives stable results which agreed well with the time behavior of the longitudinal and transversal dispersions. Figures 15.16 and 15.17 show the result of the comparison for small elapsed time for t up to 0.1 T_L, whereas Figure 15.18 and 15.19 show the results for a time t up to 1 T_L. Even if the agreement is not perfect (as for instance in the case of the transverse dispersion at early times), we see that for a first approximation made on the values of the diffusion coefficients $b_\parallel^2$ and $b_\perp^2$, the correct behavior at early and late times is obtained.

15.5 Conclusions

A stochastic simulation model based on Langevin type stochastic differential equations, widely used in applications such as contaminant transport of aerosols in the atmosphere, was generalized to the contaminant transport in porous media.

The crucial idea was in the assumption that even in the case of laminar flow in porous media, due to the extremely high heterogeneity, there exists an interval where the acceleration vector behaves like a random diffusion process governed by a stochastic differential equation. The main issue discussed in this chapter was the construction

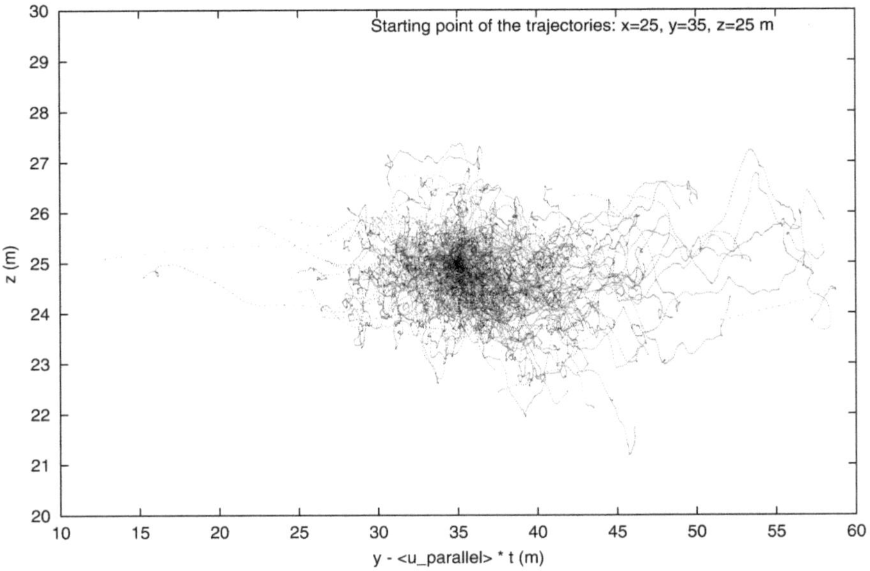

Figure 15.14. View of 500 DSM trajectories during 5 Lagrangian correlation time.

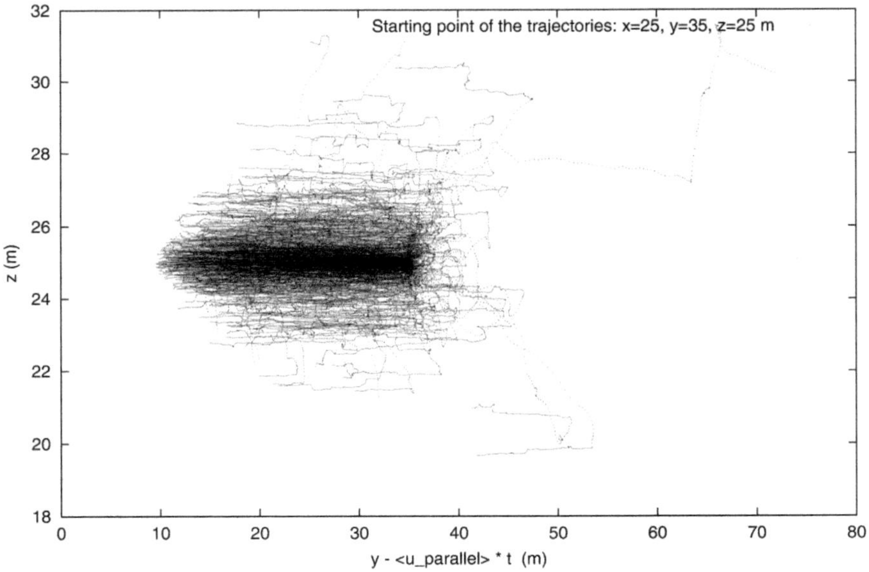

Figure 15.15. View of 500 Langevin trajectories during 5 Lagrangian correlation time.

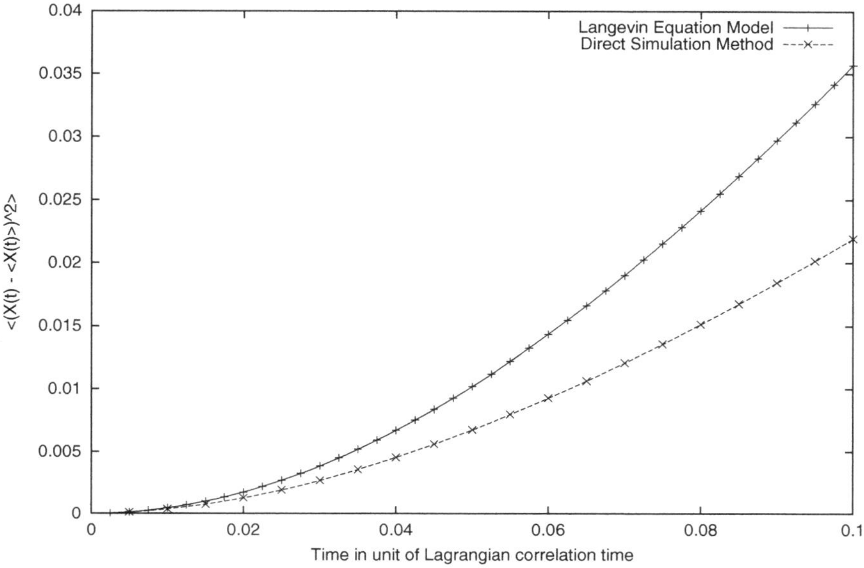

Figure 15.16. Comparison between Langevin model and DSM of the transverse dispersion component for small times.

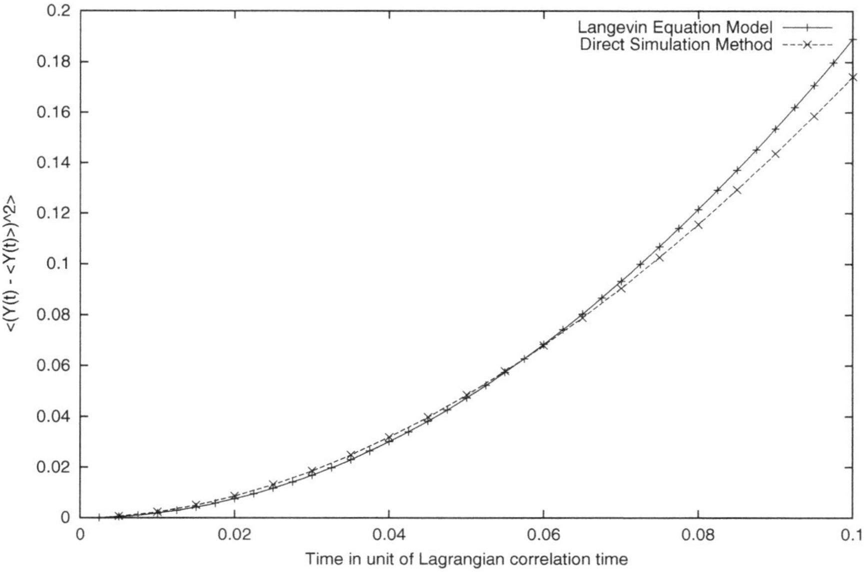

Figure 15.17. Comparison between Langevin model and DSM of the longitudinal dispersion component for small times.

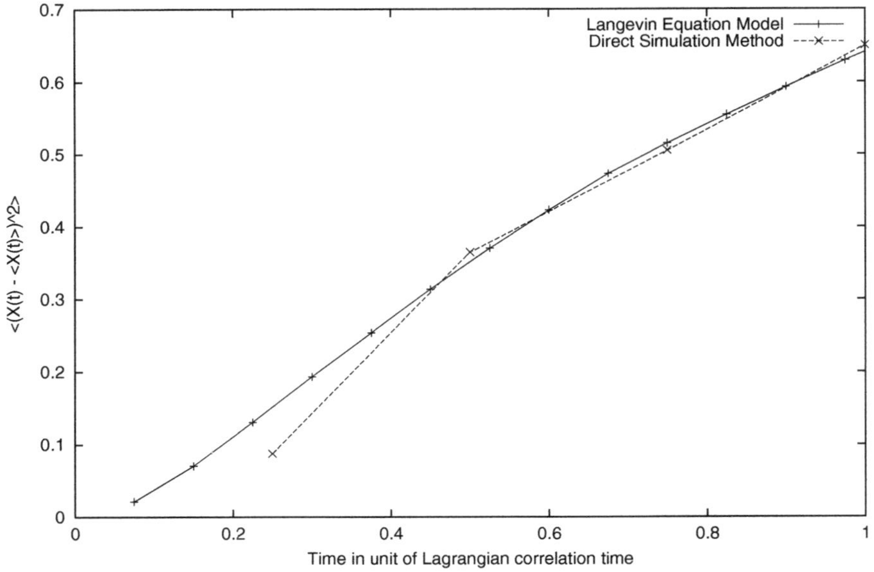

Figure 15.18. Comparison between Langevin model and DSM of the transverse dispersion component for large times.

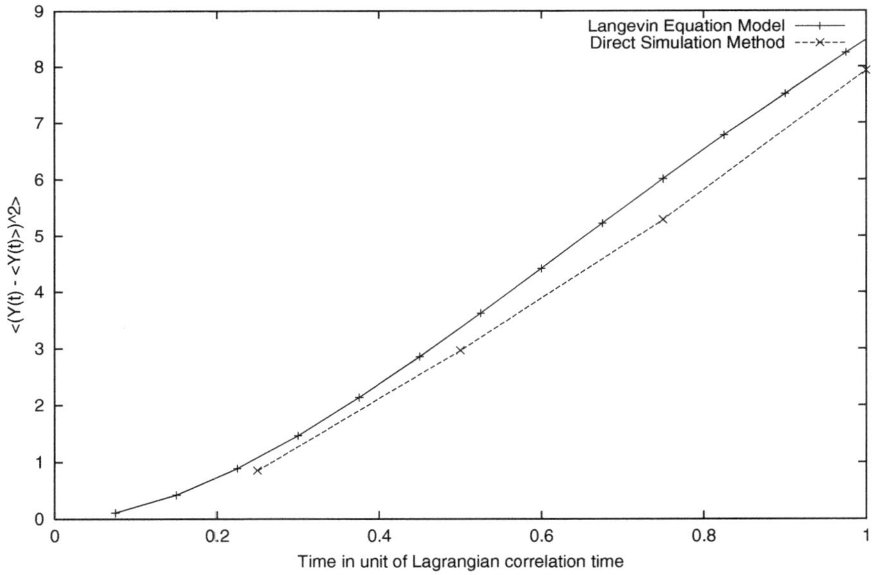

Figure 15.19. Comparison between Langevin model and DSM of the longitudinal dispersion component for large times.

of such a stochastic differential equation consistent with the statistical structure of the Eulerian velocity field. However the extraction of such a statistical information via DSM method applied to the Darcy equation which includes a random hydraulic conductivity was also a difficult task.

From the geostatistical models collected in the literature, and by applying a numerical groundwater flow model, it has been possible to determine the parameters of the Langevin model in porous medium. It was important to identify the Eulerian and Lagrangian characteristics of velocity and mainly to find out whether or not the Lagrangian structure function was presenting a linear behavior at early times in order to define the drift and diffusion terms of the Langevin model for velocity.

An important part of this work was to compare the new stochastic model to the direct simulation method, in order to see whether or not the Langevin model is able to mimic the transport characteristics of the contaminant transport. We have shown, from a very first approximation of the diffusion coefficients, that this was also possible.

Due to the computational efficiency of the Langevin model in comparison to the direct simulation method, and due to its potential generalization to many transport regimes (Fickian or not, for instance), it is believed that these first results are very promising.

Further work should include the generalization to several flow conditions (in an anisotropic log hydraulic conductivity structure, for instance) and should be compared to the common random displacement methods (based on the Fickian hypothesis) used in the hydrogeological literature.

It should be stressed that the advantage of the new stochastic model developed is not only in its numerical efficiency but also in the potential possibility to assimilate the detailed information about the velocity statistics.

Chapter 16

Coagulation of aerosol particles in intermittent turbulent flows

In this chapter we study the coagulation of particles in turbulent flows. The size distribution of particles is governed by the Smoluchowski equation with a random collision coefficient. The random coagulation coefficient is derived by a generalization of the approach suggested by Saffman and Turner [210]. The coagulation process is analysed in three main cases: (i) T_c, the characteristic coagulation time is much less than T_w, the characteristic Lagrangian time of the turbulent flow; (ii) conversely, $T_w \ll T_c$; and (iii) these times are of the same order $T_w \sim T_c$. A special stochastic time is introduced which drastically simplifies the analysis of the influence of the intermittency. A detailed numerical study is given for two cases with the known explicit solutions of the Smoluchowski equation. The numerical analysis in the turbulent collision regime is based on the stochastic algorithm presented in the book [191] and developed in [97, 116, 189, 195, 202].

16.1 Introduction

The coagulation processes of aerosol particles or clusters in spatially homogeneous case are governed by the Smoluchowski equation (e. g., see, [254]):

$$\frac{\partial n_l}{\partial t} = \frac{1}{2} \sum_{i+j=l} k_{ij} n_i n_j - n_l \sum_{i=1}^{\infty} k_{li} n_i, \tag{16.1}$$

with the initial conditions $n_l(0) = n_l^{(0)}$, $l = 1, 2, \ldots$.

Coagulation is a process by which two particles collide and adhere, or coagulate. We use the notation: $\{l\}$-cluster, or l-mer, for a cluster containing l monomers (or structural units); n_i, for the number density of the $\{i\}$-cluster; k_{ij}, for the coagulation coefficient characterizing the collision frequencies between an $\{i\}$- and a $\{j\}$-cluster. We will also use the notation δ_{ij} for the Kronecker function.

Under rather general assumptions about the coagulation coefficients k_{ij} there are known existence and uniqueness results for the solution to equation (16.1) (e. g., see [7]).

There are many different mechanisms that bring two particles to each other: Brownian diffusion, gravitational sedimentation, free molecule collisions, turbulent motion of the host gas, acoustic waves, the density, concentration and temperature gradients, particle electric charges, etc. We will deal here mainly with the case of coagulation of particles in a fully developed turbulence whose small-scale statistical structure is specified by $\bar{\varepsilon}$, the kinetic energy dissipation rate, and ν, the kinematic viscosity.

The structure of k_{ij} for different collision regimes is presented, e. g., in [254], and is well developed only in the case when there is no spatial dependence of the functions involved in the coagulation equation.

The Smoluchowski equation in the inhomogeneous case governing the coagulation of particles dispersed by a velocity field $\vec{v}(t,x)$ reads

$$\frac{\partial n_l(\mathbf{x},t)}{\partial t} + \vec{v}(\mathbf{x},t)\cdot\nabla n_l(\mathbf{x},t) = \frac{1}{2}\sum_{i+j=l} k_{ij}n_i n_j - n_l\sum_{i=1}^{\infty} k_{li}n_i, \tag{16.2}$$

where $n_l(\mathbf{x},t)$ is the concentration of clusters of size l, $l = 1,2,\dots$ at a point $\mathbf{x}$ at time t; $\vec{v}(\mathbf{x},t)$ is the velocity of the host gas, $k_{ij} = k_{ij}(\mathbf{x},t)$ is the coagulation coefficient. It is supposed, that the initial size distribution is given: $n_l(\mathbf{x},0) = n_l^0(\mathbf{x})$.

The coagulation coefficient is determined by the flow of the host gas in the neighborhood of the colliding particles. We assume that the flow is *incompressible* and it is *not disturbed* by the particles. Then N_{ij}, the number of collisions between the i- and j-clusters in a unit volume and unit time is defined by (e. g., see [254])

$$N_{ij}(\mathbf{x},t) = k_{ij}(\mathbf{x},t)n_i n_j = \frac{n_i n_j}{2}\int_{S_R(\mathbf{x})} |w_r(\mathbf{y})|\, dS_R(\mathbf{y}),$$

i. e.,

$$k_{ij}(\mathbf{x},t) = \frac{1}{2}\int_{S_R(\mathbf{x})} |w_r(\mathbf{y})|\, dS_R(\mathbf{y}), \tag{16.3}$$

where $R = r_i + r_j$ is the sum of the radii of colliding particles, $S_R(\mathbf{x})$ is the sphere of radius R centered at the point $\mathbf{x}$, $dS_R(\mathbf{y})$ is the surface measure at a point $\mathbf{y}$, and w_r is the radial component of the relative velocity:

$$w_r(\mathbf{y}) = \left(\vec{v}(\mathbf{x}+\mathbf{y},t) - \vec{v}(\mathbf{x},t), \frac{\mathbf{y}}{|\mathbf{y}|}\right). \tag{16.4}$$

Note that Saffman and Turner [210] described the turbulent coagulation regime in average (over the velocity fluctuations). They have defined the coagulation coefficient $\bar{k}_{ij}$ through $\langle N_{ij}\rangle$, the average of N_{ij} over the turbulent velocity fluctuations:

$$\bar{k}_{ij}(\mathbf{x},t) = \frac{1}{2}\left\langle \int_{S_R(\mathbf{x})} |w_r(\mathbf{y})|\, dS_R(\mathbf{y})\right\rangle. \tag{16.5}$$

Under the assumption that $\frac{\partial v_1}{\partial x_1}$ is normally distributed and that the velocity is isotropic in small scales, they obtained from (16.5)

$$\bar{k}_{ij}(\mathbf{x},t) \simeq 1.3(r_i + r_j)^3\sqrt{\frac{\langle\varepsilon\rangle}{\nu}}, \tag{16.6}$$

where $\langle\varepsilon\rangle$ is the average dissipation rate of the turbulent kinetic energy, and ν is the kinematic viscosity.

In our case, we deal with the random coagulation coefficient (16.3). Thus we need to evaluate the integral over the sphere. This evaluation is given in Appendix A of this chapter, which reads:

$$\int_{S_R(\mathbf{x})} |w_r(\mathbf{y})|\, dS_R(\mathbf{y}) = R^3 \sqrt{\frac{\varepsilon}{2\nu}}\, f(\xi), \tag{16.7}$$

where f is a dimensionless universal function of a dimensionless argument ξ (see the Appendix):

$$\xi = \frac{det(\hat{\tau})}{(\frac{\varepsilon}{2\nu})^{3/2}},$$

and $\hat{\tau} = \|\tau_{ij}\|_{i,j=1}^{3}$ is the deformation tensor of the velocity vector $\vec{v}(\mathbf{x},t) = (v_1(\mathbf{x},t), v_2(\mathbf{x},t), v_3(\mathbf{x},t))$:

$$\tau_{ij} = \frac{1}{2}\left(\frac{\partial v_i}{\partial x_j} + \frac{\partial v_j}{\partial x_i}\right),$$

ε is the dissipation rate of the kinetic energy: $\varepsilon = 2\nu \sum_{ij} \tau_{ij}^2$; ν is the kinematic viscosity.

It turns out (see the Appendix) that $f(\xi)$ can be approximated to within 2.3 % by a constant value. Thus we have

$$k_{ij}(\mathbf{x},t) = A\sqrt{\frac{\varepsilon(\mathbf{x},t)}{\nu}}\, R^3, \tag{16.8}$$

where $A = 2(\pi + 3)/9 \approx 1.3648$.

Remark 16.1. Note that from the expression (16.8) we find that

$$\bar{k}_{ij} = A\frac{\langle\sqrt{\varepsilon}\rangle}{\sqrt{\nu}} R^3. \tag{16.9}$$

Thus we see that this expression differs from (16.6). The expression (16.6) can be obtained (to within a constant factor) from (16.9) by additional assumptions on $\bar{\varepsilon}(\mathbf{x},t)$, assuming for instance that $\bar{\varepsilon}$ has a log-normal distribution.

Since we study the general case of coagulation in a fluctuating velocity field with k_{ij} given by (16.3), we will not only be able to find the true expectation of the solution, but also to find out when the *coagulation equation homogenization* happens. Such homogenization implies that the stochastic solution approaches the solution of the deterministic coagulation equation with $\bar{k}_{ij}$ defined in (16.9).

16.2 Analysis of the fluctuations in the size spectrum

Thus we deal with the Smoluchowski equation whose coagulation kernel is random:

$$\frac{\partial n_l(\mathbf{x},t)}{\partial t} + \vec{v}(\mathbf{x},t)\cdot\nabla n_l(\mathbf{x},t) = \frac{1}{2}\sum_{i+j=l} k_{ij} n_i n_j - n_l \sum_{i=1}^{\infty} k_{li} n_i, \tag{16.10}$$

where

$$k_{ij}(\mathbf{x},t) = A\sqrt{\frac{\varepsilon(\mathbf{x},t)}{\nu}} r_1^3 (i^{1/3} + j^{1/3})^3, \tag{16.11}$$

with $A = 1.3648$; r_1 is the radius of the monomer. Here $\vec{v}(\mathbf{x},t)$ is a turbulent velocity field which is assumed to be incompressible, statistically homogeneous, and stationary.

In Lagrangian coordinates induced by (e. g., see [146])

$$\frac{dX}{dt} = \vec{v}(X(t),t), \quad X(0) = \mathbf{x}_0 \tag{16.12}$$

the equation (16.10) reads

$$\frac{dN_l}{dt} = \frac{1}{2}\sum_{i+j=l} K_{ij} N_i N_j - N_l \sum_{i=1}^{\infty} K_{li} N_i, \quad N_l(0) = n_l^{(0)}(\mathbf{x}_0), \tag{16.13}$$

where $N_l(t) = N_l(t;\mathbf{x}_0) = n_l(X(t;\mathbf{x}_0),t)$ and

$$K_{ij}(t) = k_{ij}(X(t;\mathbf{x}_0),t) = A\sqrt{\frac{\mathcal{E}(t)}{\nu}} r_1^3 (i^{1/3} + j^{1/3})^3. \tag{16.14}$$

Here $\mathcal{E}(t) = \mathcal{E}(t,\mathbf{x}_0) = \varepsilon(X(t;\mathbf{x}_0),t)$ is the Lagrangian dissipation energy rate.

To avoid unessential complications, we consider here the case when $n_l^{(0)}(\mathbf{x}) = n_l^{(0)}$ does not depend on the spatial coordinates. Then the solution is also independent of the spatial coordinates and it is governed by the homogeneous Smoluchowski equation with the coagulation coefficient (16.14).

Let us consider the following deterministic equation in dimensionless form:

$$\frac{df_l}{d\tau} = \frac{1}{2}\sum_{i+j=l} \frac{\langle K_{ij}\rangle}{\langle K_{11}\rangle} f_i f_j - f_l \sum_{i=1}^{\infty} \frac{\langle K_{li}\rangle}{\langle K_{11}\rangle} f_i, \quad f_l(0) = \frac{n_l^{(0)}}{n_0}, \tag{16.15}$$

where $n_0 = \sum_{l=1}^{\infty} l n_l^{(0)}$. It is easy to find that by introducing a new time variable τ_t defined by

$$\frac{d\tau_t}{dt} = 8An_0 r_1^3 \sqrt{\frac{\mathcal{E}(t)}{\nu}} = B\sqrt{\frac{\mathcal{E}(t)}{\langle\varepsilon\rangle}}, \quad \tau(0) = 0, \tag{16.16}$$

we can express the solution to (16.13) as

$$N_l(t) = n_0 f_l(\tau_t), \quad \tau_t = B \int_0^t \sqrt{\frac{\mathcal{E}(s)}{\langle \varepsilon \rangle}}\, ds. \tag{16.17}$$

Here $B = 8An_0 r_1^3 \sqrt{\langle \varepsilon \rangle}/\sqrt{\nu}$.

Thus we came to a very important result: the random solution $N_l(t)$ is represented as a deterministic function of a random time τ_t. The function f_l can be obtained numerically, by the Monte Carlo [191–202] or a finite element method, (e. g., see [202]), hence the problem is reduced to the analysis of this random time.

In the case of incompressible homogeneous stationary velocity fields, the one-pint probability density function of the Lagrangian stochastic process $\mathcal{E}(t)$ and that of the Eulerian random field $\bar{\varepsilon}(\mathbf{x}, t)$ coincide [146], and therefore the function $\langle \sqrt{\mathcal{E}} \rangle$ is constant, and

$$\langle \tau_t \rangle = B \int_0^t \frac{\langle \sqrt{\mathcal{E}} \rangle}{\sqrt{\langle \varepsilon \rangle}}\, ds = B \frac{\langle \sqrt{\varepsilon} \rangle}{\sqrt{\langle \varepsilon \rangle}} t = B_1 t, \tag{16.18}$$

where $B_1 = B\langle \sqrt{\varepsilon} \rangle / \sqrt{\langle \varepsilon \rangle}$.

Let us formulate the *coagulation equation homogenization* problem. Generally, it can be formulated as follows: when and under what conditions the fluctuation $N_l(t) - \bar{N}_l(t)$ is negligibly small where $\bar{N}_l(t)$ is the solution to the following deterministic Smoluchowski equation

$$\frac{d\bar{N}_l}{dt} = \frac{1}{2} \sum_{i+j=l} \langle K_{ij} \rangle \bar{N}_i \bar{N}_j - \bar{N}_l \sum_{i=1}^{\infty} \langle K_{li} \rangle \bar{N}_i, \quad \bar{N}_l(0) = n_l^0. \tag{16.19}$$

Let us estimate the relative fluctuation through

$$\delta_l(t) = \frac{\left\{\langle (N_l(t) - \bar{N}_l(t))^2 \rangle\right\}^{1/2}}{\bar{N}_l(t)}.$$

Note that

$$\bar{N}_l(t) = n_0 f_l(\langle \tau_t \rangle) = n_0 f_l(B_1 t). \tag{16.20}$$

The first term of the Taylor expansion gives

$$\frac{f_l(\tau + \Delta\tau) - f_l(\tau)}{f_l(\tau)} = \frac{\Delta\tau}{\lambda_l(\tau)} + \cdots,$$

where

$$\lambda_l(\tau) = \frac{f_l(\tau)}{\frac{df_l(\tau)}{d\tau}}$$

is a scale characterizing the change in τ of the function $f_l(\tau)$. From this we find by (16.17), (16.20) (taking $\tau = \langle\tau_t\rangle$, $\Delta\tau = \tau_t' = \tau_t - \langle\tau_t\rangle$) that

$$\delta_l^2(t) = \frac{\langle(\tau_t')^2\rangle}{\lambda^2(\langle\tau_t\rangle)} + \cdots . \tag{16.21}$$

Now

$$\tau_t' = B\frac{\langle\sqrt{\varepsilon}\rangle}{\sqrt{\langle\varepsilon\rangle}}\int_0^t \xi(s)\,ds = B_1\int_0^t \xi(s)\,ds, \tag{16.22}$$

where

$$\xi(s) = \frac{\sqrt{\mathcal{E}(s)} - \langle\sqrt{\varepsilon}\rangle}{\langle\sqrt{\varepsilon}\rangle}. \tag{16.23}$$

By G. Taylor's formula [146] we can write

$$\langle(\tau_t')^2\rangle \simeq 2B_1^2\sigma_\xi^2 T_\xi\, t, \quad \text{if} \quad t \gg T_\xi,$$

where σ_ξ is the variance of ξ and

$$T_\xi = \int_0^\infty \frac{\langle\xi(s+s')\xi(s')\rangle}{\sigma_\xi^2}\,ds$$

is the integral time scale of the stochastic process $\xi(s)$. Thus we obtain in (16.21)

$$\delta_l(t) \simeq \sqrt{2}\sigma_\xi\frac{\sqrt{tT_\xi}}{T_l(t)}, \tag{16.24}$$

where

$$T_l(t) = \frac{|\lambda_l(\langle\tau_t\rangle)|}{B_1} = \frac{|\bar{N}_l(t)|}{|\frac{d\bar{N}_l(t)}{dt}|}$$

is the characteristic time scale of the function $\bar{N}_l(t)$. From this, it can be clearly seen that if $T_l(t)$ is much larger than T_ξ, the Lagrangian integral time scale of the stochastic process (16.23), then for the times $T_\xi \ll t \sim T_l(t)$ the quantity δ_l behaves like $\sqrt{tT_\xi}/T_l(t)$, which implies that the fluctuations of the random functions $N_l(t)$ are small and therefore $\langle N_l(t)\rangle \simeq \bar{N}_l(t)$. Note, however, that $T_\xi \ll t \sim T_l(t)$ is not considered to be a necessary condition for ensuring that $\delta_l(t)$ is small; it was required only to derive the exact asymptotics (16.24).

It is not difficult to find that

$$\delta_l(t) \le \sqrt{2}\sigma_\xi\frac{\sqrt{tT_\xi'}}{T_l(t)},$$

provided that $t \ll T_l^2(t)/T'_\xi$. Here

$$T'_\xi = \int_0^\infty \frac{|\langle \xi(s'+s)\xi(s')\rangle|}{\sigma_\xi^2}\, ds.$$

If $t \ll T_l^2(t)/T'_\xi$, then the last inequality shows that $\delta_l(t)$ is small.

Remark 16.2. Above we analysed the closeness between the expectation $\langle N_l(t)\rangle$ and the relevant deterministic function $\bar{N}_l(t)$. This closeness was controlled by the ratio $\sqrt{tT_\xi}/T_l(t)$. The same can be done for an arbitrary functional, say, $\Phi(N_l)$ (for instance, the total number of clusters, the mean cluster size, etc.). In this case the closeness between the expectation of this functional and $\Phi(\bar{N}_l)$ is controlled by $\sqrt{tT_\xi}/T_\Phi(t)$. Here

$$T_\Phi = \frac{|\bar{\Phi}(t)|}{|\frac{d\bar{\Phi}(t)}{dt}|}$$

is the characteristic time scale of the function $\bar{\Phi}(t) = \Phi(\bar{N}_l(t))$.

16.3 Models of the energy dissipation rate

Recall that here we deal with a spatially homogeneous, stationary and incompressible turbulent velocity field. As follows from the Section 16.2, the Lagrangian energy dissipation rate $\mathcal{E}(t)$ is entered into the relevant Smoluchowski equation. However, in the literature only the Eulerian energy dissipation rate has been treated in detail (e. g., see [74]). As for the Lagrangian models, to our knowledge there are two of them: (i) the log-normal model by Pope and Chen [171], and (ii) the multifractal model of Borgas and Sawford [16]. As another example we mention a Lagrangian stochastic model, which we developed for the boundary layer of atmosphere [115].

Let us describe the models we will use further.

16.3.1 The model by Pope and Chen (*P&Ch*)

The model to which we will refer as *P&Ch* assumes that

$$\mathcal{E}(t) = \langle \varepsilon \rangle \exp(\chi(t)), \tag{16.25}$$

where $\chi(t)$ is the solution to the following stochastic differential equation [171]

$$d\chi = -\frac{1}{T_\chi}(\chi - \langle \chi \rangle) + \sigma_\chi \sqrt{\frac{2}{T_\chi}}\, dB(t), \tag{16.26}$$

where T_χ is the integral time scale of the stochastic process $\chi(t)$, while σ_χ^2 and $\langle\chi\rangle$ are its variance and mean value, respectively. Since $\chi(t)$ is a stationary Gaussian process, we find from (16.25) that

$$\langle \mathcal{E}^p \rangle = \langle \varepsilon \rangle^p \exp\left\{\frac{1}{2}p^2\sigma_\chi^2 + p\langle\chi\rangle\right\}, \quad p > 0. \tag{16.27}$$

We used the fact that in the case considered here, the 1-point statistical characteristics of the Lagrangian energy dissipation rate coincide with that of the Eulerian energy dissipation rate [146]. In particular, $\langle \mathcal{E} \rangle = \langle \bar{\varepsilon} \rangle$.

Now,

$$\langle\chi\rangle = -\frac{1}{2}\sigma_\chi^2, \quad \sigma_\chi^2 = \ln\left(\frac{\langle\bar{\varepsilon}^2\rangle}{\langle\varepsilon\rangle^2}\right), \quad \langle\sqrt{\varepsilon}\rangle = \sqrt{\langle\varepsilon\rangle}\exp\left\{-\frac{\sigma_\chi^2}{8}\right\}. \tag{16.28}$$

Thus this model can be specified by the parameters: $\langle\varepsilon\rangle, \langle\bar{\varepsilon}^2\rangle$, and T_χ, or alternatively, by $\langle\varepsilon\rangle, \langle\sigma_\chi^2\rangle$, and T_χ. Note that in [171] the last variant was used, with $\sigma_\chi = 1$, and $T_\chi = 0.9T_w$, where T_w is the Lagrangian time scale of the vertical velocity fluctuations.

We will use both these variants. In the first variant, we will use the following relation (e. g., see [16, 74]):

$$\frac{\langle\bar{\varepsilon}^2\rangle}{\langle\varepsilon\rangle^2} = \left(\frac{T_w}{\tau_\eta}\right)^\mu, \tag{16.29}$$

where $\tau_\eta = \sqrt{\nu/\langle\varepsilon\rangle}$ is the Kolmogorov internal time scale and μ is the intermittency parameter ($\mu = 0.2 \div 0.3$).

Remark 16.3. By the definition of ξ (see (16.23)) we can find, using (16.27), that

$$\sigma_\xi^2 = \frac{\langle\varepsilon\rangle}{\langle\sqrt{\varepsilon}\rangle^2}\left(1 - \exp\left\{-\frac{\sigma_\chi^2}{4}\right\}\right).$$

The integral time scale T_ξ can be estimated from the following arguments.

Let $\zeta(t) = \exp\{\chi(t)\}$, where $\chi(t)$ is the Ornstein–Uhlenbeck process defined by (16.26). Then the following approximate relation holds [171]:

$$T_\zeta = T_\chi\left(1 - \frac{2}{9}\sigma_\chi^2\right),$$

where T_ζ is the integral time scale of the process $\zeta(t)$. This approximation is applicable if $\sigma_\chi^2 \le 2$. From this we find that

$$T_\xi = T_\chi\left(1 - \frac{1}{18}\sigma_\chi^2\right).$$

16.3.2 The model by Borgas and Sawford (*B&S*)

In this model, there are three input parameters: $\langle\varepsilon\rangle, T_w$, and $N_{\bar{\varepsilon}}$, a positive integer. The model can be described as follows. Let η, ρ, δ be independent random numbers constructed as

$$\eta = \begin{cases} a & \text{with probability} \quad \lambda \\ b & \text{with probability} \quad 1-\lambda, \end{cases}$$

$$\rho = \begin{cases} a' & \text{with probability} \quad \lambda' \\ b' & \text{with probability} \quad 1-\lambda', \end{cases}$$

$$\delta = \begin{cases} 0 & \text{with probability} \quad 0.5 \\ 1 & \text{with probability} \quad 0.5, \end{cases}$$

where $a = 0.7528$, $b = 0.3536$, $a' = 0.3951$, $b' = 0.1474$, $\lambda = \sqrt{a}$, $\lambda' = 1 - \sqrt{b}$.

Let us define a subdivision of the interval $\Delta = [0, T_w)$. First, let $\Delta_1^{(0)} = \Delta$. Then, for a positive integer i we construct a uniform subdivision of Δ into 2^i subintervals, $\Delta_1^{(i)}, \Delta_2^{(i)}, \dots, \Delta_{2^i}^{(i)}$. The interval $\Delta_j^{(i)}$ is divided in two equal subintervals: the left subinterval, $\Delta_j^{(i)'}$, and the right subinterval $\Delta_j^{(i)''}$.

The stochastic process $\mathcal{E}(t)$ on the interval $\Delta = [0, T_w)$ will be constructed recursively in $N_{\bar{\varepsilon}} + 1$ steps.

Step 0. Put $\mathcal{E}_0(t) \equiv \langle\varepsilon\rangle$, $t \in \Delta_1^{(0)}$.

Step 1. Simulate a triple $(\eta_1^{(0)}, \rho_1^{(0)}, \delta_1^{(0)})$ of independent random numbers as described above. Then define

$$\mathcal{E}_1(t) = \begin{cases} 2\mathcal{E}_0(t)[\delta_1^{(0)}\eta_1^{(0)} + (1-\delta_1^{(0)})\rho_1^{(0)}], & \text{if} \quad t \in \Delta_1^{(0)'} \\ 2\mathcal{E}_0(t)[(1-\delta_1^{(0)})\eta_1^{(0)} + \delta_1^{(0)}\rho_1^{(0)}], & \text{if} \quad t \in \Delta_1^{(0)''}. \end{cases}$$

Step 2. Let $(\eta_j^{(i-1)}, \rho_j^{(i-1)}, \delta_j^{(i-1)})$, $j = 1, \dots, 2^{i-1}$ be independent samples. In the interval $\Delta_j^{(i-1)}$ $(j = 1, \dots, 2^{i-1})$ we define

$$\mathcal{E}_i(t) = \begin{cases} 2\mathcal{E}_{i-1}(t)[\delta_j^{(i-1)}\eta_j^{(i-1)} + (1-\delta_j^{(i-1)})\rho_j^{(i-1)}], & \text{if} \quad t \in \Delta_j^{(i-1)'} \\ 2\mathcal{E}_{i-1}(t)[(1-\delta_j^{(i-1)})\eta_j^{(i-1)} + \delta_j^{(i-1)}\rho_j^{(i-1)}], & \text{if} \quad t \in \Delta_j^{(i-1)''}, \end{cases}$$

etc., until $i = N_{\bar{\varepsilon}}$. Finally, set $\mathcal{E}(t) = \mathcal{E}_{N_{\bar{\varepsilon}}}(t)$.

The following relations are known for this model [16]:

$$\langle \mathcal{E}^2(t) \rangle = \langle \varepsilon \rangle^2 \left(\frac{T_w}{\tau_\eta} \right)^{\mu},$$

where $\mu = 0.294$, and $\tau_\eta = 2^{-N_{\bar{\varepsilon}}} T_w$ is the inner time scale of the model. In addition, the integral time scale of $\mathcal{E}(t)$ satisfies

$$T_{\mathcal{E}} \approx C_{\bar{\varepsilon}} T_w \left(\frac{T_w}{\tau_\eta} \right)^{-\mu},$$

where $C_{\bar{\varepsilon}}$ is a constant which does not depend on the subdivision.

In conclusion let us make some remarks concerning the two described models.

1. The probability density function (pdf) of the model *P&Ch* is log-normal with a heavy tail, while for the p. d. f. of the model *B&S* this is not the case.
2. The correlation function of the model *P&Ch* is approximately exponentially decaying [171]. The correlation function of the *P&Ch* model is a power function.
3. The model *P&Ch* is much more convenient in numerical simulations. However it does not involve the dependence on the Reynolds number in contrast to the model *B&S*.

16.4 Monte Carlo simulation for the Smoluchowski equation in a stochastic coagulation regime

In Section 16.2 we have suggested theoretical analysis which provides qualitative estimations (see (16.24)) of the difference between the true expectations of the solutions to the stochastic Smoluchowski equation and the solutions of the relevant deterministic equations. These estimations show that two main cases may occur: (i) this difference is small which happens when $\sqrt{tT_\xi} \ll T_l(t)$, and (ii), this difference is not small if $\sqrt{tT_\xi} \simeq T_l(t)$. However, since the function $T_l(t)$ is generally very complicated, this criterion is not easy applied in practice. Therefore, we first investigate the case of a model Smoluchowski equation which has an explicit solution. In this case the criterion is also explicitly verified.

Let us consider the following Smoluchowski equation of the type (16.13) with the coagulation coefficient

$$K_{ij} = \frac{\beta}{n_0} \sqrt{\frac{\mathcal{E}(t)}{\nu}} \, (i + j), \tag{16.30}$$

with the initial condition $N_l(0) = n_0 \delta_{l1}$. Here $\beta = \frac{4}{3}\pi r_1^3 n_0$ is a dimensionless parameter, r_1 is the radius of the monomer, and n_0 is the initial number of monomers.

Note that β has a clear physical meaning; it is the total volume occupied by monomers in a unit volume.

As described in Section 16.1, our solution can be represented as

$$N_l(t) = n_0 f_l(\tau_t), \quad \tau_t = \frac{\beta}{\tau_\eta} \int_0^t \sqrt{\frac{\mathcal{E}(s)}{\langle \varepsilon \rangle}}\, ds, \tag{16.31}$$

where τ_t is the internal Kolmogorov time scale, and $f_l(\tau)$ is a dimensionless function of the dimensionless time τ satisfying the equation

$$\frac{df_l}{d\tau} = \frac{1}{2} \sum_{i+j=l} (i+j) f_i f_j - f_l \sum_{i=1}^{\infty} (l+i) f_i, \quad f_l(0) = \delta_{l1}. \tag{16.32}$$

Let us also consider the deterministic function $\bar{N}_l(t)$, which is the solution to (16.19) with the coagulation coefficients $\langle K_{ij} \rangle$, where K_{ij} is given by (16.30) and with the initial distribution $\bar{N}_l(0) = n_0 \delta_{l1}$. Here we have also

$$\bar{N}_l(t) = n_0 f_l(\langle \tau_t \rangle), \quad \langle \tau_t \rangle = \frac{\beta}{\tau_\eta} \frac{\langle \sqrt{\varepsilon} \rangle}{\sqrt{\langle \varepsilon \rangle}}\, t\,. \tag{16.33}$$

In this case, the exact solution to (16.32) is known [254]:

$$f_l(\tau) = n_0 \frac{l^{l-1}}{l!} [b(\tau)]^{l-1} \exp\{-\tau - l b(\tau)\}, \quad b(\tau) = 1 - \exp(-\tau). \tag{16.34}$$

Along the solution itself, we will be interested in the following functionals:

1. Total number of clusters:

$$\Sigma(t) = \sum_{l=1}^{\infty} N_l(t), \tag{16.35}$$

2. Mean cluster size:

$$\Lambda(t) = \frac{1}{\Sigma(t)} \sum_{l=1}^{\infty} l N_l(t). \tag{16.36}$$

In the case considered here these functions are explicitly known:

$$\Sigma(t) = n_0 \exp(-\tau_t), \quad \Lambda(t) = \exp(\tau_t). \tag{16.37}$$

In the deterministic case, these functionals of the solutions to the deterministic Smoluchowski equation are

$$\bar{\Sigma}(t) = \sum_{l=1}^{\infty} \bar{N}_l(t) = n_0 \exp(-\langle \tau_t \rangle) = n_0 \exp\left(-\frac{t}{T_c}\right), \tag{16.38}$$

$$\bar{\Lambda}(t) = \frac{1}{\bar{\Sigma}(t)} \sum_{l=1}^{\infty} l \bar{N}_l(t) = \exp(\langle \tau_t \rangle) = \exp\left(\frac{t}{T_c}\right), \tag{16.39}$$

where

$$T_c = \frac{\tau_\eta}{\beta} \frac{\sqrt{\langle\varepsilon\rangle}}{\langle\sqrt{\varepsilon}\rangle} \tag{16.40}$$

is an integral time scale of the coagulation process. Here $\tau_\eta = \sqrt{\frac{\nu}{\langle\varepsilon\rangle}}$ is the internal Kolmogorov time scale. In the calculations we take $\langle\varepsilon\rangle = 100$, $T_w = 10$, $\nu = 0.17$. *We use everywhere the* **CGS** *units.*

We will use the relation

$$\langle\bar{\varepsilon}^2\rangle = \langle\varepsilon\rangle^2 \left(\frac{T_w}{\tau_\eta}\right)^\mu, \tag{16.41}$$

where the intermittency parameter is chosen as $\mu = 0.25$.

Throughout this section (except for Section 16.4.4), the stochastic process $\mathcal{E}(t)$ will be simulated according to the model $P\&Ch$. The parameters were chosen as follows: $\langle\varepsilon\rangle = 100$, with the relevant value of $\langle\bar{\varepsilon}^2\rangle$ given by (16.41), and $T_\chi = 0.9T_w$.

In the process of coagulation governed by the random Smoluchowski equation (16.13) with the coagulation coefficient (16.30), two main time scales are involved: T_w, the Lagrangian time scale of the velocity fluctuations, and T_c, the integral time scale of the coagulation process. It is therefore very convenient to introduce a number γ characterising the rate of the coagulation: $\gamma = T_w/T_c$. We will consider four cases:

(i) *high rate of coagulation*, if $\gamma \sim 10$;

(ii) *moderate high rate*, if $\gamma \sim 1$;

(iii) *moderate low rate*, if $\gamma \sim 0.1$;

(iv) *low rate*, if $\gamma \sim 0.01$.

Here and in the following we use the notation $a \sim b$ to indicate that $a = C\,b$, where C is a constant of order 1. The radius of the monomer is taken as $r_1 = 10^{-4}$.

We have chosen the values $n_0 = 10^{10}$, $n_0 = 10^9$, $n_0 = 10^8$, and $n_0 = 10^7$, which correspond to the four coagulation rates mentioned above.

16.4.1 The total number of clusters and the mean cluster size

In this section we present the numerical analysis for $\Sigma(t)$ and $\Lambda(t)$.

Let us denote by $\delta_\Sigma(t)$ the relative difference between the average of the stochastic function $\Sigma(t)$ and the deterministic function $\bar{\Sigma}(t)$:

$$\delta_\Sigma(t) = \frac{\langle\Sigma(t)\rangle - \bar{\Sigma}(t)}{\bar{\Sigma}(t)}.$$

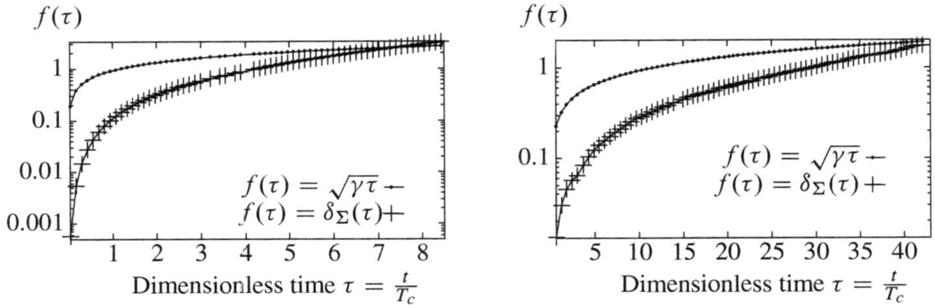

Figure 16.1. The relative difference $\delta_\Sigma(\tau)$ between the expectation of the total number of clusters in stochastic and deterministic cases. The coagulation rate $\gamma = 0.86$ (left panel), and $\gamma = 0.086$ (right panel).

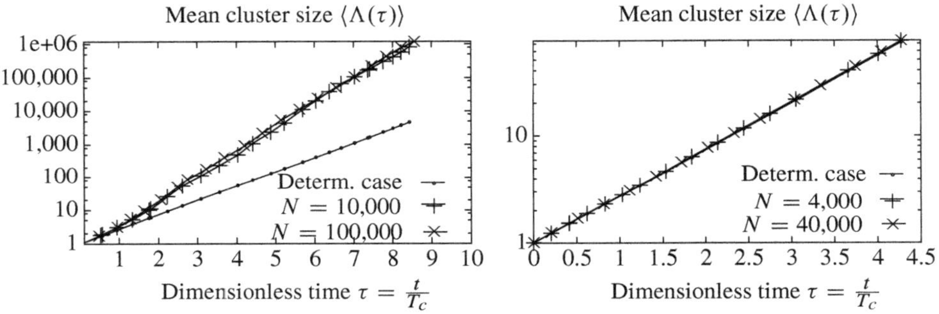

Figure 16.2. The expectation of the mean cluster size $\langle\Lambda(\tau)\rangle$ in stochastic ($N = 10^4$ and $N = 10^5$ samples) and deterministic cases; moderate high coagulation rate ($\gamma = 0.86$) (left panel) and $\gamma = 0.0086$ (right panel).

In Figure 16.1, left panel, this quantity (lower curve) is shown for the case of a moderate high rate of coagulation ($\gamma = 0.86$) as a function of the dimensionless time $\tau = t/T_c$. For convenience, we plot also the ratio $\sqrt{T_w t}/T_c = \sqrt{\gamma\tau}$ (upper curve).

It is seen from this curves that the deterministic and stochastic cases do not differ much if $\tau \leq 1$ (to within 10 %). After a certain time instant τ_0, the function $\sqrt{\gamma\tau}$ is close to 1, which indicates that after this time the difference becomes large. In Figure 16.1, right panel, the same curves are plotted in the case of a moderate low rate of coagulation ($\gamma = 0.086$). It is seen that the time instant $\tau_0 \sim 1/\gamma$ happens later, between 6 and 8.

Thus, even in the case of moderate low rate of coagulation the difference becomes large after the dimensional time $t_0 \sim \tau_0 T_c = T_c^2/T_w$.

The expectation of the mean cluster size $\langle\Lambda\rangle$ is shown in In Figure 16.2, left panel, in the case of a moderate high rate of coagulation ($\gamma = 0.86$). Two upper curves were

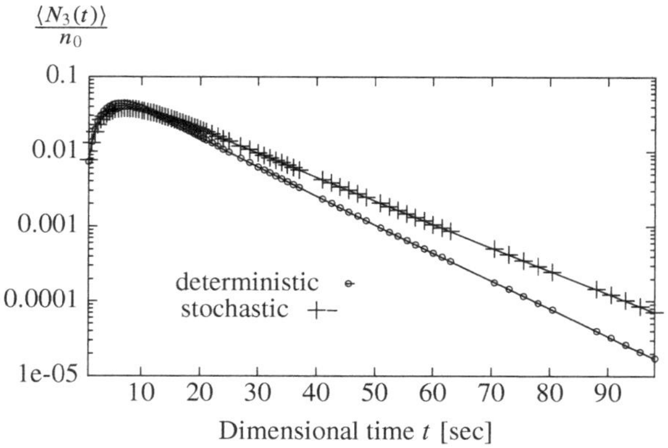

Figure 16.3. Concentration of 3-mers as a function of time; $\gamma = 0.86$.

obtained by Monte Carlo averaging over 10^4 and 10^5 samples, which shows that the statistical error is small enough. The lower curve is the deterministic function $\bar{\Lambda}(\tau)$. It is seen that after the time instant $\tau_0 \sim 1$ the difference begins to increase rapidly, and around $\tau = 8$ it reaches two orders of magnitude. Qualitatively the same picture was obtained for $\gamma = 0.085$.

For the low rate of coagulation ($\gamma = 0.0085$) the difference between the stochastic and deterministic functions is practically eliminated (see Figure 16.2, right panel).

Thus, in the case of these two functions the stochastic description can be well approximated by a deterministic equation for a low rate of coagulation at times up to several number of characteristic coagulation times T_c. The lower the coagulation rate is, the wider is the interval where the coagulation equation homogenization takes place.

It should be noted that this conclusion is made for these two specific functions, the total number of clusters and the mean cluster size. The situation can be drastically changed if the solution itself is considered. Let us consider some examples.

16.4.2 The functions $N_3(t)$ and $N_{10}(t)$

In Figure 16.3 we present $\langle N_3(t)\rangle/n_0$, the normalized concentration of 3-mers as a function of dimensional time for the moderate high rate of coagulation ($\gamma = 0.86$). The upper curve is obtained by Monte Carlo averaging over $N = 10^4$ samples, the lower curve is the function $\bar{N}_3(t)/n_0$. Note that the difference is seen for all times, although it is much larger for times $t \sim 5T_c$, $T_c = 11.68$ s. The difference is even more pronounced for the 10-mers.

The picture is different in the case of low rate of coagulation ($\gamma = 0.0086$). As shown in Figure 16.4, the difference is seen only for times up to $\sim T_c = 1168$ s. After this time instant the curves are very close. To explain such behavior we plot here in the

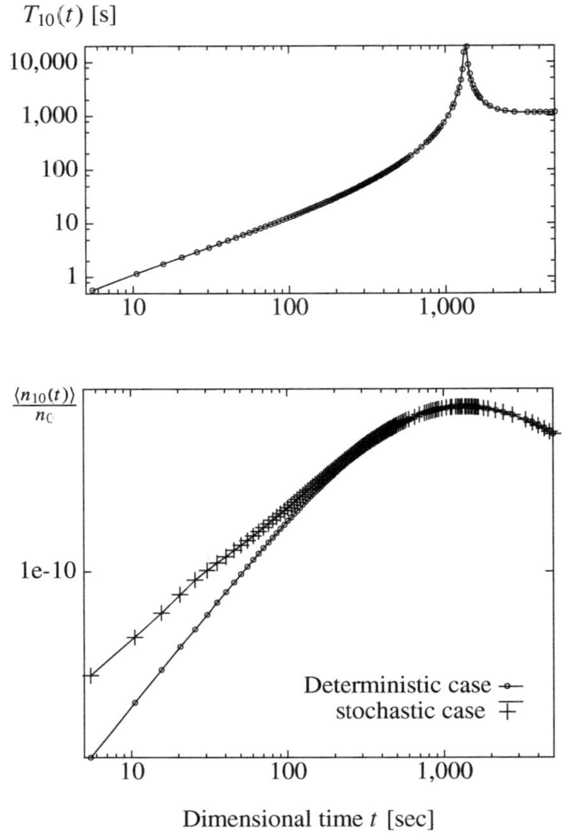

Figure 16.4. The concentration of 10-mers (lower picture) for stochastic and deterministic cases, $\gamma = 0.0086$ and the characteristic time scale $T_{10}(t)$ (upper picture).

upper picture also T_{10}, the characteristic time scale of coagulation for 10-mers defined in Section 16.4. From (16.24) it follows that the smaller $T_{10}(t) is$, the larger is the relative difference between the stochastic and deterministic cases.

To see a more detailed picture for clusters of different sizes, we turn to the size spectrum, i. e., the size distribution density $N_l(t)$.

16.4.3 The size spectrum N_l for different time instances

First we consider the case of low coagulation rate ($\gamma = 0.0086$), $T_c = 1168$.

In Figure 16.5 we plot the size spectrum for the time $t = T_w = 10$ s (left panel), and $t = 10T_w = 100$ s (right panel). The upper curve is obtained by the Monte Carlo calculations, the lower curve is the deterministic function $\bar{N}_l/n_0$. The difference

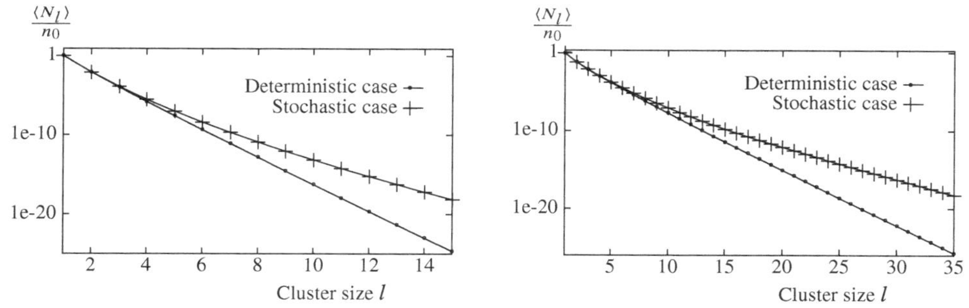

Figure 16.5. The size distribution for $\gamma = 0.0086$ at the time instant $t = T_w = 10$ s (left panel) and $t = 10T_w = 100$ s (right panel).

becomes rapidly larger with the growth of the cluster size. Note that the larger the time instances are, the wider is the size interval where the difference between the stochastic and deterministic cases is small. In calculations (which we do not show here) made for $t = 500T_w$ the curves are close up to the 500-mers.

Analogous results were obtained for the cases of moderate low coagulation rate ($\gamma = 0.086$, $T_c = 116.8$) and moderate high coagulation rate ($\gamma = 0.86$, $T_c = 11.68$). More details can be found in [116].

We only mention that in the case of moderate high coagulation rate the curves are different for all the considered sizes. The difference becomes even more pronounced in the case of high coagulation rate ($\gamma = 8.6$, $T_c = 1.17$).

16.4.4 Comparative analysis for two different models of the energy dissipation rate

As mentioned in Section 16.4.3, the models $B\&S$ and $P\&Ch$ differ in certain aspects, in particular, the tails of the distributions are quite different. Therefore, one may expect that this will lead to a change of the numerical results of the previous section. In this section we will compare the results for both models with common mean value. In the model $B\&S$ we take $T_w = 30$, $N_{\bar{\varepsilon}} = 8$, and the mean dissipation rate $\langle\varepsilon\rangle$ is determined from the relation $\tau_\eta = 2^{-N_{\bar{\varepsilon}}} T_w = (\nu/\langle\varepsilon\rangle)^{1/2}$. For $\nu = 0.17$ we have $\langle\varepsilon\rangle = 12.38$.

In the model $P\&Ch$ we choose the parameters from the following arguments. The mean is taken the same, $\langle\varepsilon\rangle = 12.38$; $T_\chi = 0.9T_w$, and the variance σ_χ^2 was taken equal to the variance of the random variable $\chi = \ln(\mathcal{E}/\langle\varepsilon\rangle)$ where $\mathcal{E}$ is taken from the model $B\&S$. This results in $\sigma_\chi = 1.7$.

First we show the results for the mean cluster size $\Lambda(t)$. The two models give similar results for low coagulation rates ($\gamma \sim 1$); see the relevant results in the previous section.

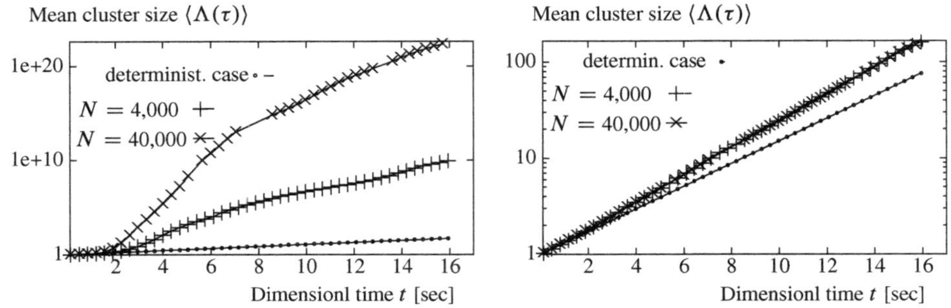

Figure 16.6. The expectation of the mean cluster size; the *P&Ch* model of energy dissipation rate. The case of high coagulation rate ($\gamma \sim 8$); $T_w = 30$ s, $T_c = 4.05$ s (left panel), *B&S* model; the high coagulation rate ($\gamma \sim 8$); $T_w = 30$ s, $T_c = 3.68$ s (right panel).

For high coagulation rates the models give essentially different results. In Figure 16.6 (left panel) the expectation of the mean cluster size was obtained for the model *P&Ch* with $N = 4{,}000$ and $N = 40{,}000$ samples. The lower curve is the deterministic function $\bar{\Lambda}(t)$.

The difference between the two upper curves should be additionally explained. Simple analysis of the explicit expression of the expectation $\langle\Lambda(\tau_t)\rangle$ shows that it behaves essentially differently at small and large times. For large (compared with T_χ) times the distribution of τ_t is approximately Gaussian, and the expectation $\langle\Lambda(\tau_t)\rangle = \langle\exp\{\tau_t\}\rangle$ is well defined. However for small times, the p. d. f. of τ_t has a heavy tale, since τ_t is proportional to $\sqrt{\mathcal{E}(0)}$, $\mathcal{E}(0)$ being log-normally distributed. Therefore, the expectation $\langle\Lambda(\tau_t)\rangle$ tends to infinity. This explains why the Monte Carlo calculations with $N = 40{,}000$ and 4,000 samples are so different, thus not providing stable results.

The model *B&S* gives essentially different results (see the right panel of Figure 16.6). Here we have very stable Monte Carlo results of $\langle\Lambda(\tau_t)\rangle$, and the deviation from the deterministic case is clearly seen, although not so large. This is a general situation: the log-normal model *P&Ch* leads to a larger difference between the stochastic and deterministic cases (see [116] for more details).

16.5 The case of a coagulation coefficient with no dependence on the cluster size

The case of a coagulation coefficient with no dependence of the cluster size may be useful in the study of the case of the turbulent coagulation coefficient generalized by [167]. Therefore, it is interesting to study this case in more detail, since it provides explicit expression of the solution.

Thus we assume that all the parameters remain the same as in the previous section, with only one difference, namely, the term $i + j$ is replaced with 1:

$$K_{ij} = \frac{\beta}{n_0}\sqrt{\frac{\mathcal{E}(t)}{\nu}}. \tag{16.42}$$

The exact solution to the Smoluchowski equation (16.15) in this case has the form

$$f_l(\tau) = \frac{b^{l-1}}{(1+b)^{l+1}}, \quad b = b(\tau) = \frac{\tau}{2}. \tag{16.43}$$

In calculations, we considered three cases of coagulation rate: $\gamma \sim 10$, 1, and 0.1, corresponding to the initial number density $n_0 = 5.10^9$, 5.10^8, and 5.10^7, respectively.

The results show that for high and moderate high coagulation rates at the times $t \sim T_c$ the difference between the stochastic and deterministic cases is large. In contrast to the case when $K_{ij} \sim (i + j)$, here even for the high coagulation rate at the times $t \gg T_c$ the difference between the stochastic and deterministic cases is very small. This can be explained by comparing the scales $T_l(t)$ of these two cases. From the exact solutions (16.34) and (16.43) we find that for large times $f_l(\tau) \sim \exp(-\tau)$ and $f_l(\tau) \sim 1/\tau^2$, respectively. Hence, if $t \gg T_c$, then $T_l(t) \simeq T_c$ and $T_l(t) \simeq t$, respectively. Therefore, from the relation (16.24) we conclude that in the case when the coagulation coefficient does not depend on i and j, the difference between the stochastic and deterministic cases is small for all cases of γ if the time is much larger than both the characteristic time scales T_w and T_c.

16.6 Simulation of coagulation processes in turbulent coagulation regime

Let us study the difference between the deterministic and stochastic cases in the case of turbulent coagulation regimes.

We will use the coefficient derived in the Appendix of this chaapter (see formula (16.14)):

$$K_{ij}(t) = A\sqrt{\frac{\mathcal{E}(t)}{\nu}}\, r_1^3 (i^{1/3} + j^{1/3})^3. \tag{16.44}$$

It should be noted that both in our and Turner and Saffman's considerations it was assumed that the statistical structure of the flow is considered undisturbed by the particles. This implies that the radii of the particles can not much differ, say the radii ratio is not larger than 3. In [167] a generalization is presented which is free of this restriction. The authors [167] derived a correction factor E to the coagulation coefficient of

Turner and Saffman, namely $E = 7.5(\rho_2/\rho_1)^2$, where ρ_1 and ρ_2 are the larger and smaller radius of the colliding particles. Thus if $i/3 > j$, then we put

$$K_{ij}(t) = E_{ij}\, A\sqrt{\frac{\mathcal{E}(t)}{\nu}}\, r_1^3 (i^{1/3} + j^{1/3})^3, \tag{16.45}$$

where

$$E_{ij} = 7.5\left(\frac{j}{i}\right)^2, \quad j = \min\{i,j\},\ i = \max\{i,j\}. \tag{16.46}$$

Otherwise, $E_{ij} = 1$.

Since the exact solutions to Smoluchowski equation with the kernels (16.44) and (16.45) are not known, we used the Monte Carlo stochastic particle method developed in [191–202] and recently extended to the inhomogeneous case [97].

It is interesting to note that the behavior of the turbulent coagulation coefficient (16.44) considered as a function of i and j is close to the case $K_{ij} \sim i + j$ treated in details in Section 16.4 while the corrected turbulent coagulation coefficient (16.45) is close to the case $K_{ij} \sim \text{const}$ treated in Section 16.5 in the sense that the relevant solutions behave qualitatively similar, as we will see later.

We present calculations both for the coefficient (16.44) and for (16.45). Here we used the model $P\&Ch$ with the the parameters chosen as in Section 16.4.4.

The calculations were carried out for three coagulation rates: (i) high coagulation rate, $\gamma = 11.15$, ($n_0 = 5.10^9, T_c = 0.89$), (ii) moderate high coagulation rate, $\gamma = 1.11$, ($n_0 = 5.10^8, T_c = 8.9$), (iii) moderate low coagulation rate, $\gamma = 0.11$, ($n_0 = 5.10^7, T_c = 89$), where

$$T_c = \frac{1}{B_1} = \frac{\tau_\eta}{8An_0r_1^3}\frac{\langle\varepsilon\rangle}{\langle\sqrt{\varepsilon}\rangle^2}.$$

The radius of the monomer was taken the same as in previous calculations: $r_1 = 10^{-4}$ cm.

Let us study the difference between the deterministic and stochastic cases for the coagulation coefficient (16.44). In Figure 16.7 the normalized concentration of 10-mers is shown as a function of time, for a high coagulation rate (left panel), and a moderate high coagulation rate (right panel). As in the case $K_{ij} \sim (i + j)$, the expectation differs from the deterministic solution both for small and large time instances. This picture remains true in the case of moderate high coagulation rate, but the difference becomes smaller for large times; see the right panel. Again, this situation is similar to that observed in the case of $K_{ij} \sim (i + j)$.

A qualitatively analogy remains true also for other functions like the expectation of the mean cluster size and the total number of clusters.

We turn to the case of the corrected coagulation coefficient (16.45).

In this case we calculated the normalized concentration of 10-mers as a function of time for the moderate low, moderate high, and high coagulation rates, respectively.

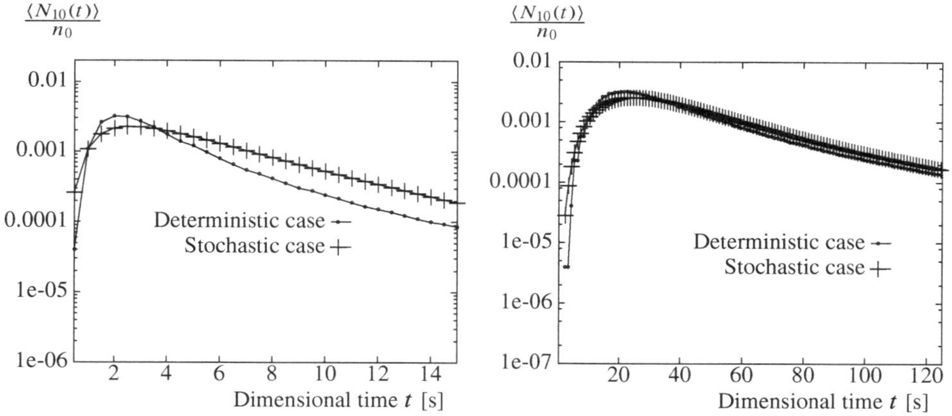

Figure 16.7. The concentration of 10-mers as a function of time, for the case of turbulent coagulation regime ((16.44)); high coagulation rate ($\gamma \sim 11$) (left panel), and moderate high coagulation rate $\gamma \sim 1$ (right panel).

At small times ($t \leq T_c$) the curves behave similar to the case with the coagulation coefficient (16.44). However, for the times $t \geq T_c$ this is not the case, and the curves look rather similar to the relevant curves for the case when the coagulation coefficient does not depend on i and j; see for details [116].

Thus we conclude that the case of turbulent coagulation coefficient (16.44) (without correction E_{ij}) is qualitatively described by the case $K_{ij} \sim (i + j)$. Therefore, the detailed analysis given in Section 16.4 may be useful in this case. The case of corrected coagulation coefficient (16.45) can be qualitatively analyzed on the basis of the calculation results presented in Section 16.5.

16.7 Conclusion

The coagulation processes in a turbulent regime were studied. The main purpose of the analysis was to distinguish between an averaged description and description, taking into account the fluctuating dissipation rate of the kinetic turbulent energy $\mathcal{E}(t)$. The averaged description was proposed by Saffman and Turner [210], and the coagulation coefficient for fluctuating $\mathcal{E}(t)$ was given in the present chapter by formula (16.11). It turns out that in many cases these two descriptions lead to essentially different average solutions. However, this is also often the case when both descriptions give close results. This last case, called here *coagulation equation homogenization*, was studied in detail, and conditions under which the homogenization takes place were derived.

We have described two cases of turbulent coagulation coefficient: (i) the coefficient (16.44), which is derived under assumption that the particles do not disturb the flow, and (ii) the coefficient (16.45), which involves a correction coefficient (given in [167]; the correction accounts for the structure of flow around the bigger particle).

In the case (i), for the high and moderate high coagulation rates, when $\gamma = T_w/T_c \sim 1$ (T_w is the characteristic Lagrangian time scale of the velocity fluctuations, and T_c is the characteristic time scale of the coagulation process), then the stochastic and deterministic descriptions lead to different behavior of the size spectrum. In the moderate low and low coagulation rates both descriptions lead to close results only for times well within the interval $(T_c, T_c/\gamma)$.

In case (ii), the difference is pronounced typically only in the initial time interval $(0, g\,T_c)$), where the dimensionless constant $g \sim 1$.

Two different models of the fluctuated dissipation rate of turbulent energy, the log-normal model $P\&Ch$ and the multifractal model $B\&S$, were used in our calculations. These models give qualitatively close results; however, the log-normal model leads as a rule to a larger difference between the stochastic and deterministic descriptions.

It should be noted that the turbulence was assumed to be spatially homogeneous and stationary in time. Besides, we studied only the coagulation mechanism due to relative motion of colliding particles in a turbulent velocity field. Thus we have not included the inertia of particles, the Brownian diffusion, gravitational sedimentation, etc. It is believed that the approach presented in this chapter can be extended to more general situations cited above.

16.8 Appendix. Derivation of the coagulation coefficient

Let us evaluate the coagulation coefficient

$$k_{ij}(\mathbf{x}, t) = \frac{1}{2} \int_{S_R(\mathbf{x})} |w_r(\mathbf{y})| \, dS_R(\mathbf{y}) \tag{16.47}$$

under the assumption that the velocity $\vec{v}(\mathbf{x}, t)$ is incompressible and the characteristic spatial scale of the velocity field is much larger than the size of the coagulating particles. First, expand the relative velocity:

$$\begin{aligned} &v_i(\mathbf{x}+\mathbf{y}, t) - v_i(\mathbf{x}, t) \\ &= \sum_{j=1}^{3} \frac{\partial v_i(\mathbf{x}, t)}{\partial x_j} y_j + \cdots \\ &= \frac{1}{2} \sum_{j=1}^{3} \left\{ \frac{\partial v_i(\mathbf{x}, t)}{\partial x_j} + \frac{\partial v_j(\mathbf{x}, t)}{\partial x_i} \right\} y_j + \frac{1}{2} \sum_{j=1}^{3} \left\{ \frac{\partial v_i(\mathbf{x}, t)}{\partial x_j} - \frac{\partial v_j(\mathbf{x}, t)}{\partial x_i} \right\} y_j + \cdots, \end{aligned} \tag{16.48}$$

where we omit the terms higher than the linear ones. Now, the radial component is expressed through the scalar product:

$$w_r(\mathbf{y}) = \left(\vec{v}(\mathbf{x}+\mathbf{y},t) - \vec{v}(\mathbf{x},t), \frac{\mathbf{y}}{|\mathbf{y}|}\right) = \sum_j \sum_i \frac{\partial v_i(\mathbf{x},t)}{\partial x_j} \frac{y_j y_i}{|\mathbf{y}|} + \cdots$$

$$= \sum_i \sum_j \tau_{ij} \frac{y_i y_j}{|y|} + \cdots. \tag{16.49}$$

Then we calculate the integral using the relevant coordinate system (by a rotation to the principal axes of $\hat{\tau}$):

$$\frac{1}{2}\int_{S_R(\mathbf{x})} |\sum_{ij} \tau_{ij} \frac{y_i y_j}{|y|}| \, dS_R(\mathbf{y})$$

$$= \frac{1}{2R}\int_{S_R(\mathbf{x})} |\lambda_1 y_1^2 + \lambda_2 y_2^2 + \lambda_3 y_3^2| \, dS_R(\mathbf{y})$$

$$= \frac{R^3}{2}\sqrt{\lambda_1^2+\lambda_1^2+\lambda_1^2}\int_{S_1(\mathbf{x})} |\mu_1 y_1^2 + \mu_2 y_2^2 + \mu_3 y_3^2| \, dS_1(y).$$

Here $\mu_i = \lambda_i / \sqrt{\lambda_1^2+\lambda_1^2+\lambda_1^2}$, where λ_i are the eigenvalues of the strain tensor $\hat{\tau}$, $i = 1, 2, 3$. Thus,

$$k_{ij} = \frac{R^3\sqrt{\lambda_1^2+\lambda_1^2+\lambda_1^2}}{2}\int_{S_1} |\mu_1 y_1^2 + \mu_2 y_2^2 + \mu_3 y_3^2| \, dS_1(y).$$

The last integral depends on one parameter, since $\mu_1^2 + \mu_2^2 + \mu_3^2 = 1$, and, by incopressibility, $\mu_1 + \mu_2 + \mu_3 = 0$.

Thus, we have to evaluate the integral over the sphere $S_1(\mathbf{x})$ for the whole region of this parameter. Let us treat this problem as follows.

The intersection of $\mu_1^2 + \mu_2^2 + \mu_3^2 = 1$ and $\mu_1 + \mu_2 + \mu_3 = 0$ is a circle which can be parametrized through

$$\mu_1 = \frac{1}{\sqrt{2}}\cos\theta + \frac{1}{\sqrt{6}}\sin\theta,\ \mu_2 = -\frac{1}{\sqrt{2}}\cos\theta + \frac{1}{\sqrt{6}}\sin\theta,\ \mu_3 = -\frac{2}{\sqrt{6}}\sin\theta,$$

where $\theta \in [0, 2\pi)$. Consequently,

$$\frac{1}{2}\int_{S_1} |\mu_1 y_1^2 + \mu_2 y_2^2 + \mu_3 y_3^2| dS_1 = 2\pi E_\omega |\mu_1(\theta)\omega_1^2 + \mu_2(\theta)\omega_2^2 + \mu_3(\theta)\omega_3^2| \equiv F(\theta),$$

where $\omega = (\omega_1, \omega_2, \omega_3)$ is a random unit isotropic vector, andE_ω means the expectation taken over the vector ω.

Monte Carlo calculations of the expectation shows that the function $F(\theta)$ has its maximum at $\theta = 0$ and its minimum at $\theta = \pi/2$. Explicit expression of the integral at these points gives $F(0) = 8/3\sqrt{2}$, $F(\pi/2) = 8\pi/9\sqrt{2}$. Thus,

$$\frac{8}{3\sqrt{2}} \leq \frac{1}{2}\int_{S_1} |\mu_1 y_1^2 + \mu_2 y_2^2 + \mu_3 y_3^2|\, dS_1(y) \leq \frac{8}{3\sqrt{2}}\frac{\pi}{3}.$$

Therefore we can approximate the integral by a constant which is the mean of $F(0)$ and $F(\pi/2)$ (to within 2.3%). This yields

$$K_{ij} = \frac{1}{2}\int_{S_R(\mathbf{x})} |w_r(\mathbf{y})|\, dS_R(\mathbf{y}) \simeq A\sqrt{\frac{\varepsilon}{\nu}}(r_i + r_j)^3 \tag{16.50}$$

with $A = 2(\pi + 3)/9 \approx 1.3648$.

Chapter 17

Stokes flows under random boundary velocity excitations

A viscous Stokes flow over a disc under random fluctuations of the velocity on the boundary is studied. We give exact Karhunen–Loève (K-L) expansions for the velocity components, pressure, stress, and vorticity, and the series representations for the corresponding correlation tensors. Both the white noise fluctuations and general homogeneous random excitations of the velocities prescribed on the boundary are studied. We analyze the decay of correlation functions in angular and radial directions, both for exterior and interior Stokes problems. Numerical experiments show the fast convergence of the K-L expansions. The results indicate that ignoring the stochastic fluctuations in boundary conditions dramatically underestimates the variance of the velocity and pressure in the interior/exterior of the domain. More details can be found in our papers [193, 194].

17.1 Introduction

It is well known that boundary value problems for PDEs with irregular coefficients, fluctuating source terms, or randomly excited boundary conditions are used in many fields of science and technology to describe uncertainty, probabilistic distribution of irregularities, or large ensembles of measurements under similar but randomly fluctuating conditions. (e. g., see [22, 53]). We mention also the analysis of synoptic meteorological data [131], simulation of spatially variable seismic ground motions experiencing loss of coherence [94], and statistical theory of fully developed turbulence [134, 138, 146]. The stochastic approach is intensively used in flows in porous media and soils governed by the Darcy equation with a random hydraulic conductivity coefficient [34, 99, 196], as well as biological tissues [258], and in geodesy [182, 212]. In electrical impedance tomography [89], some important problems are the evaluation of a global response to random boundary excitations and the estimation of local fluctuations of the solution fields. Similar analysis is made in the inverse problems of elastography [159, 192], recognition technology [72], acoustic scattering from rough surfaces [257], fluid dynamics [4], and reaction-diffusion equations with white noise boundary perturbations [229]. One interesting application is also the analysis of the dislocations in crystals by X-ray diffraction [87], The physical measurement method is based on the X-ray scattering from relaxed heteroepitaxial layers with the misfit dislocations randomly distributed at the interface between the layer and the substrate.

Flows over rough surfaces, e. g., over and in tubes with rough surfaces play an important role in a variety of applications [234]. Often the topography of such surfaces cannot be accurately described in all details because of either insufficient data or measurement errors or both. In such cases, this topography uncertainty can be efficiently handled by treating rough boundaries as random fields, so that an underlying physical phenomenon is described by deterministic or stochastic differential equations with random velocities near the boundary caused by the roughness. The results of the calculations which we present in this chapter indicate that ignoring the stochastic fluctuations in boundary conditions dramatically underestimates the variance of the velocity and pressure in the interior/exterior of the domain.

Many convenient and efficient methods based on the spectral decomposition can be applied in the case when the simulated random fields are homogeneous. Among those we mention both deterministic and randomized spectral methods (e. g., see [69, 104, 118, 191, 226]). Simulation of inhomogeneous random fields is less developed (see an overview in [193]). This much complicates the development of numerical solutions of PDEs involving inhomogeneous random fields. We deal in this chapter with 2D zero mean random fields $v(r,\theta)$ in a disc $S(0,R)$ which are homogeneous with respect to one coordinate, the angular coordinate θ, and inhomogeneous in the radial direction r. Random fields with this property are called partially homogeneous [191]. The correlation function $B_v = \langle v(r_1,\theta_1)v(r_2,\theta_2)\rangle$ is then depending on the difference $\theta = \theta_1 - \theta_2$. The partial spectral function is defined as the inverse Fourier transform with respect to θ:

$$S_v(\xi, r_1, r_2) = F^{-1}[B_v](\xi, r_1, r_2) = \frac{1}{2\pi}\int_0^{2\pi} B_v(\theta, r_1, r_2)\, e^{-\mathrm{i}\theta\xi}\, d\theta.$$

Assume we have no dependence on the variable r. The spectral representation of the random field is written in the form

$$v(\theta,\cdot) = \int_0^{2\pi} e^{\mathrm{i}\xi\theta}\, G(\xi,\cdot,\cdot)\, Z(d\xi), \tag{17.1}$$

where G is defined by $G(\xi,\cdot,\cdot)G^*(\xi,\cdot,\cdot) = S(\xi,\cdot,\cdot)$, the star sign stands for the complex conjugate, the sign $\cdot$ stands to recall the omission of the dependence on the variable r, and $Z(d\xi)$ is a white noise on $[0,2\pi]$.

To simulate homogeneous random field one commonly uses the randomized spectral methods (e. g., see [104, 191, 226]), which is based on the randomized calculation of the stochastic integral (17.1). Another method is based on the Riemann sums calculation with fixed cells. The integral is approximated by a finite sum

$$v(\theta,\cdot) \approx \sum_{i=1}^{N} \left[\zeta_i \sin(\xi_i\,\theta) + \eta_i \cos(\xi_i\,\theta)\right],$$

where ξ_i are deterministic nodes in the Fourier space and ζ_i and η_i are Gaussian variables with zero mean and relevant covariance.

In our case, however, we have the dependence on the variable r, and the random fields are partially inhomogeneous. In [191] we have extended the spectral randomization method to general partially homogeneous random fields, but the inhomogeneity in r was still a problem, which was assumed to be solved by other methods. In a sense it was a method which reduced the dimension of the problem (see for details [191]).

In the case we deal with, the partial spectral function has a special structure, namely, it depends on r_1 and r_2 as follows: $S(\xi, r_1, r_2) = G(\xi, r_1)G^*(\xi, r_2)$. Notice that if $\mathbf{v}$ is a vector random field, $S(\xi, r_1, r_2)$ is a partial spectral matrix, and this factorization means a product of two matrices where the star stands for the complex conjugate [191]. This enables constructtion of a simple extension of the randomization method, without solving the inhomogeneity problem in r, see [193,205]. This was first done in [193] by using the Karhunen–Loève expansion. It should be emphasized that in this case, when the partial spectral function factorizes, the spectral randomization method is more effective, since, unlike in the K-L expansion, there is no need to solve the eigenvalue problem for the correlation operator.

Assume now, without loss of generality, that a generally inhomogeneous random field $u(x)$ has a zero mean and a variance $E\, u^2(x)$ that is bounded. The Karhunen–Loève expansion has the form [259] $u(x) = \sum_{k=1}^{\infty} \sqrt{\lambda_k}\, \zeta_k\, h_k(x)$, where λ_k and $h_k(x)$ are the eigenvalues and eigenfunctions of the correlation function $B(x_1, x_2) = \langle u(x_1)\, u(x_2)\rangle$, and ζ_k is a family of random variables.

By definition, $B(x_1, x_2)$ is bounded, symmetric, and positive definite. For such kernels, the Hilbert–Schmidt theory says that the following spectral representation is valid:

$$B(x_1, x_2) = \sum_{k=1}^{\infty} \lambda_k\, h_k(x_1)\, h_k(x_2)$$

where the eigenvalues and eigenfunctions are the solutions of the following eigenvalue problem for the correlation operator:

$$\int B(x_1, x_2)\, h_k(x_1)\, dx_1 = \lambda_k\, h_k(x_2).$$

The eigenfunctions form a complete orthogonal set $\int h_i(x)\, h_j(x)\, dx = \delta_{ij}$ where δ_{ij} is the Kronecker delta-function. The family $\{\zeta_k\}$ is a set of uncorrelated random variables which are obviously related to h_k by

$$\zeta_k = \frac{1}{\sqrt{\lambda_k}} \int u(x)\, h_k(x)\, dx, \quad \langle \zeta_k \rangle = 0, \quad \langle \zeta_i\, \zeta_j \rangle = \delta_{ij}.$$

It is well known that the Karhunen–Loève expansion presents an optimal (in the mean square sense) convergence for any distribution of $u(x)$. If $u(x)$ is a zero mean Gaussian random field, then $\{\zeta_k\}$ is a family of standard Gaussian random variables. We assume in this study that the random fields are Gaussian. Some generalizations to non-Gaussian random fields are reported in [165].

In this chapter we deal with the Stokes incompressible flows under random excitations of the velocity prescribed on the boundary. Note that in the Stokes flow analysis there is a deep relation with the case of the Lamé equation described in [193], but there are also important differences, namely, (i) the problem is exterior, (ii) the velocity field is divergence-free, and (iii) the pressure in the Stokes flow is strongly correlated to the velocity fluctuations on the boundary which complicates the study of the cross-correlations. We consider also the interior problem where an additional boundary condition should be taken into account. Along the Karhunen–Loève expansions of the velocity and pressure, exact series representations are given for the vorticity and the stress tensor. Numerical experiments illustrate the rate of convergence of the K-L expansions and present the details of the correlation structure of different flow characteristics.

17.2 Exterior Stokes problem

Let us consider a linear Stokes problem in D_R^+, which is an exterior of D_R, a disc of radius R centered at the origin. On the boundary, which is the circle $S(0, R) = \partial D_R$, the velocity components g_1 , g_2 are prescribed, and the velocity vector $\mathbf{u}(\mathbf{x}) = (u_1(\mathbf{x}), u_2(\mathbf{x}))$ and the pressure $p(\mathbf{x})$ are to be found from the Stokes problem:

$$\begin{aligned} &\eta \Delta \mathbf{u}(\mathbf{x}) = \nabla p(\mathbf{x}), && \mathbf{x} \in D_R^+, \\ &\text{div } \mathbf{u}(\mathbf{x}) = 0, && \mathbf{x} \in D_R^+, \\ &\mathbf{u}(\mathbf{x}') = \mathbf{g}(\mathbf{x}'), && \mathbf{x}' \in S(0, R), \end{aligned} \tag{17.2}$$

where η is the dynamic viscosity.

Assume the prescribed boundary velocities g_i, $i = 1, 2$ are homogeneous random processes defined on $S(0, R)$, which implies that the correlation tensor $(B_g)_{ij} = \langle g_i(\theta_1) g_j(\theta_2) \rangle$ depends on the angular difference $\theta = \theta_1 - \theta_2$, and $\langle \mathbf{g}(\theta) \rangle = \text{const}$.

The solution of the problem (17.2), the velocity $\mathbf{u}$ and the pressure p are random fields, and our goal is to construct simulation formulas for these random fields and to find their main statistical characteristics, e. g., the stress and velocity correlation tensors, and the pressure correlation function.

Here we note that from the Poisson formula (17.3) which we will present in the next section it can be easily found that $\langle \mathbf{u} \rangle = \langle \mathbf{g} \rangle$, so without loss of generality we assume that $\langle \mathbf{g} \rangle = 0$. For simplicity, we deal here with Gaussian random fields, so we suppose that g_i are Gaussian random processes, which implies due to (17.3) that the solution $\mathbf{u}(\mathbf{x})$ and $p(\mathbf{x})$ are also Gaussian random fields. Then, these zero mean random fields are uniquely defined by their correlation functions. In the next section we obtain exact series representations for the correlation functions.

17.2.1 Poisson formula in polar coordinates

It is natural to use polar coordinates, $\mathbf{x} = re^{i\theta}$, where $r \geq R$ is the radial coordinate and θ is the angular coordinate, so that $u_i = u_i(r,\theta), i = 1,2$. Moreover, we turn also to polar coordinates for the velocity components, i. e., we introduce new velocities, u_r and u_θ by the transform $\mathbf{u} = u_r \mathbf{e}_r + u_\theta \mathbf{e}_\theta$, where $\mathbf{e}_r$, $\mathbf{e}_\theta$ are unit vectors in the directions r and θ, respectively. Then the vectors $(u_1, u_2)^T$ and $(u_r, u_\theta)^T$ are related through a rotation

$$\begin{pmatrix} u_1(r,\theta) \\ u_2(r,\theta) \end{pmatrix} = \begin{pmatrix} \cos\theta & -\sin\theta \\ \sin\theta & \cos\theta \end{pmatrix} \begin{pmatrix} u_r(r,\theta) \\ u_\theta(r,\theta) \end{pmatrix},$$

and conversely

$$\begin{pmatrix} u_r(r,\theta) \\ u_\theta(r,\theta) \end{pmatrix} = \mathcal{R}_\theta^T \begin{pmatrix} u_1(r,\theta) \\ u_2(r,\theta) \end{pmatrix},$$

where we use the notation for the rotation matrix

$$\mathcal{R}_\theta = \begin{pmatrix} \cos\theta & -\sin\theta \\ \sin\theta & \cos\theta \end{pmatrix}$$

and $\mathcal{R}_\theta^T$ means the transpose to $\mathcal{R}_\theta$.

We will use a spherical mean value relation for the velocity vector which is a generalization of the well-known Poisson integral formula

$$\begin{pmatrix} u_r(r,\theta) \\ u_\theta(r,\theta) \end{pmatrix} = \int_0^{2\pi} \begin{pmatrix} G_{11}(r;\theta-\varphi) & G_{12}(r;\theta-\varphi) \\ G_{21}(r;\theta-\varphi) & G_{22}(r;\theta-\varphi) \end{pmatrix} \begin{pmatrix} g_r(R\,e^{i\varphi}) \\ g_\theta(R\,e^{i\varphi}) \end{pmatrix} d\varphi. \tag{17.3}$$

The entries G_{ij} $(i,j = 1,2)$ can be derived explicitly [263]:

$$\begin{aligned} G_{11}(r;\theta) &= \cos\theta\, K(r,\theta) + \frac{r^2-R^2}{2r^2}\left\{\cos\theta\left(-r\frac{\partial}{\partial r}K(r,\theta)\right) + \sin\theta\,\frac{\partial}{\partial\theta}K(r,\theta)\right\}, \\ G_{12}(r;\theta) &= \sin\theta\, K(r,\theta) + \frac{r^2-R^2}{2r^2}\left\{\sin\theta\left(-r\frac{\partial}{\partial r}K(r,\theta)\right) - \cos\theta\,\frac{\partial}{\partial\theta}K(r,\theta)\right\}, \\ G_{21}(r;\theta) &= -\sin\theta\, K(r,\theta) + \frac{r^2-R^2}{2r^2}\left\{\sin\theta\left(-r\frac{\partial}{\partial r}K(r,\theta)\right) - \cos\theta\,\frac{\partial}{\partial\theta}K(r,\theta)\right\}, \\ G_{22}(r;\theta) &= \cos\theta\, K(r,\theta) - \frac{r^2-R^2}{2r^2}\left\{\cos\theta\left(-r\frac{\partial}{\partial r}K(r,\theta)\right) + \sin\theta\,\frac{\partial}{\partial\theta}K(r,\theta)\right\}, \end{aligned} \tag{17.4}$$

while the pressure $p(r,\theta)$ is related to the velocities $u_r(R,\theta)$ and $u_\theta(R,\theta)$ on the boundary by an analogous Poisson-type integral formula:

$$p(r,\theta) = \frac{2\eta}{r}\left\{\int_0^{2\pi} P_r(r;\theta-\varphi)u_r(Re^{i\varphi})\,d\varphi + \int_0^{2\pi} P_\theta(r;\theta-\varphi)u_\theta(Re^{i\varphi})d\varphi\right\} \tag{17.5}$$

where

$$P_r(r;\theta) = \cos\theta\left(-r\frac{\partial}{\partial r}K(r,\theta)\right) + \sin\theta\,\frac{\partial}{\partial\theta}K(r,\theta),$$

$$P_\theta(r;\theta) = \sin\theta\left(-r\frac{\partial}{\partial r}K(r,\theta)\right) - \cos\theta\,\frac{\partial}{\partial\theta}K(r,\theta).$$

In (17.4)-(17.5), $K(r,\theta)$ is the Poisson kernel for the harmonic equation in an exterior circular domain of radius R:

$$K(r,\theta) = \frac{r^2 - R^2}{2\pi(R^2 + r^2 - 2Rr\cos\theta}, \qquad r > R.$$

In the Karhunen–Loève expansion we will use series expansions of the kernels which can be readily derived starting from the well-known series [28]

$$K(r,\theta) \equiv \frac{1}{2\pi}\cdot\frac{1-\rho^2}{1-2\rho\cos(\theta)+\rho^2} = \frac{1}{2\pi} + \frac{1}{\pi}\sum_{k=1}^{\infty}\rho^k\cos(k\theta), \quad \rho < 1, \tag{17.6}$$

where we introduced the notation $\rho = R/r$.

To get a series expansions for the kernels, we use (17.6) and the following easily verified series representations:

$$-r\frac{\partial}{\partial r}K(r,\theta) = \frac{1}{\pi}\sum_{k=1}^{\infty}k\rho^k\cos(k\theta), \qquad \frac{\partial K(r,\theta)}{\partial\theta} = -\frac{1}{\pi}\sum_{k=1}^{\infty}k\rho^k\sin(k\theta).$$

Substitution of these series expansions in (17.3) and (17.5) leads to the desired expansions:

$$G_{11}(r;\theta) = \frac{\rho}{2\pi} + \frac{1}{2\pi}\sum_{k=1}^{\infty}\left[2\rho + \frac{k(1-\rho^2)}{\rho}\right]\rho^k\cos(k\theta),$$

$$G_{12}(r;\theta) = \frac{1}{2\pi}\sum_{k=1}^{\infty}\frac{(1-\rho^2)}{\rho}k\rho^k\sin(k\theta),$$

$$G_{21}(r;\theta) = \frac{1}{2\pi}\sum_{k=1}^{\infty}\frac{(1-\rho^2)}{\rho}(k-2)\rho^k\sin(k\theta),$$

$$G_{22}(r;\theta) = \frac{\rho}{2\pi} + \frac{1}{2\pi}\sum_{k=1}^{\infty}\left[\frac{2}{\rho} - \frac{k(1-\rho^2)}{\rho}\right]\rho^k\cos(k\theta), \tag{17.7}$$

and for the kernels P_r and P_θ:

$$P_r(r;\theta) = \frac{1}{\pi}\sum_{k=1}^{\infty}k\rho^k\cos[(k+1)\theta], \quad P_\theta(r;\theta) = \frac{1}{\pi}\sum_{k=1}^{\infty}k\rho^k\sin[(k+1)\theta].$$

Thus we can express the pressure through the boundary velocities g_r and g_θ by

$$p(r,\theta) = \int_0^{2\pi} P_1(r;\theta-\varphi) g_r(Re^{\mathrm{i}\varphi})\,d\varphi + \int_0^{2\pi} P_2(r;\theta-\varphi) g_\theta(Re^{\mathrm{i}\varphi})d\varphi \tag{17.8}$$

where

$$P_1(r;\theta) = \frac{2\eta}{\pi R}\sum_{k=1}^{\infty} k\rho^{k+1}\cos[(k+1)\theta],$$

$$P_2(r;\theta) = \frac{2\eta}{\pi R}\sum_{k=1}^{\infty} k\rho^{k+1}\sin[(k+1)\theta].$$

Let us introduce the notation

$$\lambda_{11}(\rho,k) = \rho + \frac{k(1-\rho^2)}{2\rho}, \qquad \lambda_{12}(\rho,k) = \frac{k(1-\rho^2)}{2\rho},$$

$$\lambda_{21}(\rho,k) = \frac{(k-2)(1-\rho^2)}{2\rho}, \quad \lambda_{22}(\rho,k) = \frac{1}{\rho} - \frac{k(1-\rho^2)}{2\rho}, \quad k = 1,2\ldots, \tag{17.9}$$

so that the series expansions (17.7) take the form

$$G_{11}(r;\theta) = \frac{\rho}{2\pi} + \frac{1}{\pi}\sum_{k=1}^{\infty}\lambda_{11}(\rho,k)\,\rho^k\cos(k\theta),$$

$$G_{12}(r;\theta) = \frac{1}{\pi}\sum_{k=1}^{\infty}\lambda_{12}(\rho,k)\,\rho^k\sin(k\theta),$$

$$G_{21}(r;\theta) = \frac{1}{\pi}\sum_{k=1}^{\infty}\lambda_{21}(\rho,k)\,\rho^k\sin(k\theta),$$

$$G_{22}(r;\theta) = \frac{\rho}{2\pi} + \frac{1}{\pi}\sum_{k=1}^{\infty}\lambda_{22}(\rho,k)\,\rho^k\cos(k\theta). \tag{17.10}$$

Let us verify that the velocity series representation is divergence-free:

$$\operatorname{div}\mathbf{u}(\mathbf{x}) = \frac{1}{r}\frac{\partial(ru_r)}{\partial r} + \frac{1}{r}\frac{u_\theta}{\partial\theta} = 0.$$

From (17.3) is follows that it is sufficient to show that

$$\frac{1}{r}\frac{\partial}{\partial r}\big[r\big(G_{11}(r;\theta) + G_{12}(r;\theta)\big)\big] + \frac{1}{r}\frac{\partial}{\partial\theta}\big[r\big(G_{21}(r;\theta) + G_{22}(r;\theta)\big)\big] = 0.$$

This in turn follows from the explicit series representations

$$-\frac{\partial}{\partial r}\big[r\big(G_{11}(r;\theta)\big] = \frac{\partial}{\partial \theta}\big[r\big(G_{21}(r;\theta)\big] = \frac{1}{2\pi}\sum_{k=1}^{\infty} k(k-2)(\rho^{k-1}-\rho^{k+1})\cos(k\theta)$$

and

$$-\frac{\partial}{\partial r}\big[r\big(G_{12}(r;\theta)\big] = \frac{\partial}{\partial \theta}\big[r\big(G_{22}(r;\theta)\big] = \frac{1}{2\pi}\sum_{k=1}^{\infty} \big(k(k-2)\rho^{k-1}-k^2\rho^{k+1}\big)\sin(k\theta).$$

Analogously, we can verify that the series representations for the velocity and pressure satisfy the Stokes equation (17.2).

17.3 K-L expansion of velocity

17.3.1 White noise excitations

In this section we will obtain the Karhunen–Loève expansion for the velocity vector over a set of eigenfunctions of the correlation operator. For this purpose we derive now some properties of the kernel functions G_{ij}.

First we note that from (17.9) we have the following property

$$\lambda_{11} - \lambda_{12} = \lambda_{21} + \lambda_{22} = \rho. \tag{17.11}$$

Let us consider the eigenvalue problem for the integral operator with the kernel G:

$$\int_0^{2\pi} \begin{pmatrix} G_{11}(\rho;\theta-\varphi) & G_{12}(\rho;\theta-\varphi) \\ G_{21}(\rho;\theta-\varphi) & G_{22}(\rho;\theta-\varphi) \end{pmatrix} \begin{pmatrix} h_1(\varphi) \\ h_2(\varphi) \end{pmatrix} d\varphi = \lambda \begin{pmatrix} h_1(\theta) \\ h_2(\theta) \end{pmatrix}. \tag{17.12}$$

Theorem 17.1. *The eigenvalue problem (17.12) has the following system of eigenvalues and corresponding eigenfunctions* ($k = 1, 2, \dots$)*:*

$$\lambda_{2k-1} = \lambda_{2k} = \rho^{k+1}, \quad \begin{pmatrix} h_{1,2k-1} \\ h_{2,2k-1} \end{pmatrix} = \begin{pmatrix} \cos k\theta \\ \sin k\theta \end{pmatrix}, \quad \begin{pmatrix} h_{1,2k} \\ h_{2,2k} \end{pmatrix} = \begin{pmatrix} \sin k\theta \\ -\cos k\theta \end{pmatrix},$$

and for $\lambda = \rho$, *the column-vector* $(1,1)^T$ *is the corresponding eigenfunction.*

Proof. From the expansions (17.7) we find that

$$\begin{aligned}
\int_0^{2\pi} G_{11}(\rho;\theta-\varphi) \begin{pmatrix} \sin k\varphi \\ \cos k\varphi \end{pmatrix} d\varphi &= \lambda_{11}(\rho,k) \begin{pmatrix} \sin k\theta \\ \cos k\theta \end{pmatrix}, \\
\int_0^{2\pi} G_{12}(\rho;\theta-\varphi) \begin{pmatrix} \cos k\varphi \\ \sin k\varphi \end{pmatrix} d\varphi &= \lambda_{12}(\rho,k) \begin{pmatrix} \sin k\theta \\ -\cos k\theta \end{pmatrix}, \\
\int_0^{2\pi} G_{21}(\rho;\theta-\varphi) \begin{pmatrix} \sin k\varphi \\ \cos k\varphi \end{pmatrix} d\varphi &= \lambda_{21}(\rho,k) \begin{pmatrix} -\cos k\theta \\ \sin k\theta \end{pmatrix}, \\
\int_0^{2\pi} G_{22}(\rho;\theta-\varphi) \begin{pmatrix} \cos k\varphi \\ \sin k\varphi \end{pmatrix} d\varphi &= \lambda_{22}(\rho,k) \begin{pmatrix} \cos k\theta \\ \sin k\theta \end{pmatrix}.
\end{aligned} \tag{17.13}$$

Now, by substituting these equalities in the eigenvalue problem (17.12) and using the property (17.11), we find the solution of the eigenvalue problem for $k = 1, 2 \ldots$. The existence of the eigenfunction $(1,1)^T$ for $\lambda = \rho$ follows from the properties

$$\int_0^{2\pi} G_{11}(\rho; \theta - \varphi) \cdot 1 \, d\varphi = \rho, \quad \int_0^{2\pi} G_{22}(\rho; \theta - \varphi) \cdot 1 \, d\varphi = \rho,$$

$$\int_0^{2\pi} G_{12}(\rho; \theta - \varphi) \cdot 1 \, d\varphi = 0, \quad \int_0^{2\pi} G_{21}(\rho; \theta - \varphi) \cdot 1 \, d\varphi = 0.$$

The proof is complete. □

We now turn to the derivation of the correlation tensor of the solution:

$$\begin{aligned} B_u(\rho_1, \theta_1; \rho_2, \theta_2) &= \langle \mathbf{u}(r_1, \theta_1) \otimes \mathbf{u}(r_2, \theta_2) \rangle \\ &\equiv \left\langle \begin{pmatrix} u_r(r_1, \theta_1) \\ u_\theta(r_1, \theta_1) \end{pmatrix} (u_r(r_2, \theta_2) \,, \, u_\theta(r_2, \theta_2)) \right\rangle \end{aligned} \tag{17.14}$$

assuming the boundary random vector function $\mathbf{g}$ has a Gaussian distribution specified by the zero mean and covariance tensor

$$B_g(\varphi_1, \varphi_2) = \langle \mathbf{g}(\varphi_1) \otimes \mathbf{g}(\varphi_2) \rangle = \left\langle \begin{pmatrix} g_r(\varphi_1) \\ g_\theta(\varphi_1) \end{pmatrix} \; (g_r(\varphi_2) \,, \, g_\theta(\varphi_2)) \right\rangle.$$

We use here and in the following the notation $\mathbf{v} \otimes \mathbf{u}$ for a tensor product of two vectors: $\mathbf{v} \otimes \mathbf{u} = \mathbf{v}\mathbf{u}^T$.

Substituting the representation (17.3) in (17.14) and changing the relevant product of integral expressions by double integrals, we get

$$B_u(\rho_1, \theta_1; \rho_2, \theta_2) = \int_0^{2\pi} \int_0^{2\pi} G(\rho_1; \theta_1 - \varphi') \, B_g(\varphi', \varphi'') \, G^T(\rho_2; \theta_2 - \varphi'') \, d\varphi' \, d\varphi''. \tag{17.15}$$

Now let us consider the case of white noise excitations. In this case, the boundary vector function $\mathbf{g}$ is a white noise, hence,

$$B_g(\varphi_1, \varphi_2) = \begin{pmatrix} \delta(\varphi_1 - \varphi_2) & 0 \\ 0 & \delta(\varphi_1 - \varphi_2) \end{pmatrix}. \tag{17.16}$$

Note that this property follows from the assumption that in the Cartesian coordinates. g_1 and g_2 are independent white noise processes.

Then, from (17.15) we obtain

$$B_u(\rho_1, \theta_1; \rho_2, \theta_2) = \int_0^{2\pi} G(\rho_1; \theta_1 - \varphi) \, G^T(\rho_2; \theta_2 - \varphi) \, d\varphi. \tag{17.17}$$

Theorem 17.2. *The exact Karhunen-.Loève expansions for the entries of the correlation tensor* (17.14), *and the random field* $(u_r, u_\theta)^T$ *which solves the Stokes equation under the boundary white noise excitations with the covariance tensor (17.16) are given by*

$$\big(B_u(\rho_1,\theta_1;\rho_2,\theta_2)\big)_{ij} = \frac{\rho_1\rho_2}{2\pi}\,\delta_{ij} + \frac{1}{\pi}\sum_{k=1}^{\infty}\Lambda_{ij}(\rho_1,\rho_2;k)\,\rho_1^k\rho_2^k \sin\left[\frac{\pi}{2}\,\delta_{ij} + k(\theta_2-\theta_1)\right], \quad (17.18)$$

for $i, j = 1, 2$, *where*

$$\begin{aligned}
\Lambda_{11}(\rho_1,\rho_2;k) &= \lambda_{11}(\rho_1,k)\,\lambda_{11}(\rho_2,k) + \lambda_{12}(\rho_1,k)\,\lambda_{12}(\rho_2,k),\\
\Lambda_{12}(\rho_1,\rho_2;k) &= \lambda_{11}(\rho_1,k)\,\lambda_{21}(\rho_2,k) - \lambda_{12}(\rho_1,k)\,\lambda_{22}(\rho_2,k),\\
\Lambda_{21}(\rho_1,\rho_2;k) &= \lambda_{22}(\rho_1,k)\,\lambda_{12}(\rho_2,k) - \lambda_{21}(\rho_1,k)\,\lambda_{11}(\rho_2,k),\\
\Lambda_{22}(\rho_1,\rho_2;k) &= \lambda_{22}(\rho_1,k)\,\lambda_{22}(\rho_2,k) + \lambda_{21}(\rho_1,k)\,\lambda_{21}(\rho_2,k), \quad (17.19)
\end{aligned}$$

and

$$\begin{aligned}
u_r(r,\theta) &= \frac{\xi_0\,\rho}{\sqrt{2\pi}} + \frac{1}{\sqrt{\pi}}\sum_{k=1}^{\infty}\lambda_{11}(\rho,k)\,\rho^k[\xi_k\cos k\theta + \eta_k\sin k\theta]\\
&\quad + \frac{1}{\sqrt{\pi}}\sum_{k=1}^{\infty}\lambda_{12}(\rho,k)\,\rho^k[-\eta'_k\cos k\theta + \xi'_k\sin k\theta], \quad (17.20)
\end{aligned}$$

$$\begin{aligned}
u_\theta(r,\theta) &= \frac{\xi'_0\,\rho}{\sqrt{2\pi}} + \frac{1}{\sqrt{\pi}}\sum_{k=1}^{\infty}\lambda_{21}(\rho,k)\,\rho^k\big[-\eta_k\cos k\theta + \xi_k\sin k\theta\big]\\
&\quad + \frac{1}{\sqrt{\pi}}\sum_{k=1}^{\infty}\lambda_{22}(\rho,k)\,\rho^k\big[\xi'_k\cos k\theta + \eta'_k\sin k\theta\big]. \quad (17.21)
\end{aligned}$$

Here $\{\xi_k,\eta_k\}$ *and* $\{\xi'_k,\eta'_k\}$, $k = 0,1,2,\ldots$ *are two independent families of standard independent Gaussian random variables. Thus the random field is homogeneous with respect to the angular variable, and the respective partial spectral matrix* $S(\rho_1,\rho_2;k)$ *is defined by its entries:* $S_{mm}(\rho_1,\rho_2;k) = \frac{1}{\pi}\Lambda_{mm}\rho_1^k\rho_2^k$, $S_{mm}(\rho_1,\rho_2;0) = \rho_1\rho_2/2\pi$, *and for* $n \neq m$ *the spectrum is pure imaginary:* $S_{mn}(\rho_1,\rho_2;k) = \mathrm{i}\frac{1}{\pi}\Lambda_{mn}\rho_1^k\rho_2^k$, $m, n = 1, 2$.

Proof. The expansion of the correlation tensor is obtained as follows. We substitute the representations (17.10) in (17.17) and use the eigenfunction properties (17.13). This immediately yields the exact series representation given in the theorem.

To construct the explicit simulation formula (17.20), (17.21) for our random field, we first split it into two independent Gaussian random fields:

$$\mathbf{u}(r,\theta) = \mathbf{V}_1(r,\theta) + \mathbf{V}_2(r,\theta).$$

We will now show that for each of these random fields we can obtain a Karhunen–Loève expansion.

We introduce two pairs of families of single mode vector functions: the first one,

$$\mathbf{h}_{1k}(\rho,\theta) = \begin{pmatrix}\lambda_{11}(\rho,k)\,\cos k\theta\\ \lambda_{21}(\rho,k)\,\sin k\theta\end{pmatrix}, \quad \tilde{\mathbf{h}}_{1k}(\rho,\theta) = \begin{pmatrix}\lambda_{11}(\rho,k)\,\sin k\theta\\ -\lambda_{21}(\rho,k)\,\cos k\theta\end{pmatrix}, \tag{17.22}$$

and the second one,

$$\mathbf{h}_{2k}(\rho,\theta) = \begin{pmatrix}-\lambda_{12}(\rho,k)\,\cos k\theta\\ \lambda_{22}(\rho,k)\,\sin k\theta\end{pmatrix}, \quad \tilde{\mathbf{h}}_{2k}(\rho,\theta) = \begin{pmatrix}\lambda_{12}(\rho,k)\,\sin k\theta\\ \lambda_{22}(\rho,k)\,\cos k\theta\end{pmatrix}. \tag{17.23}$$

Here the modes are indexed by $k = 1,2\ldots$, while the subindexes $_1$ and $_2$ stand for the first and second series of eigenfunctions. Each of the four systems of functions (17.22), (17.23) is orthogonal, but all the functions do not form one system of orthogonal functions. We can, however, construct from these basis functions two orthogonal systems, since these vectors are pairwise orthogonal:

$$\int_0^1 d\rho \int_0^{2\pi} d\theta\ \mathbf{h}_{1k}\cdot\tilde{\mathbf{h}}_{1j} = 0, \quad \int_0^1 d\rho \int_0^{2\pi} d\theta\ \mathbf{h}_{2k}\cdot\tilde{\mathbf{h}}_{2j} = 0 \quad \forall j,k = 1,2\ldots.$$

To complete the systems we add two orthogonal vectors

$$\mathbf{h}_0 = \begin{pmatrix}\frac{\rho}{\sqrt{2\pi}}\\ 0\end{pmatrix}, \quad \tilde{\mathbf{h}}_0 = \begin{pmatrix}0\\ \frac{\rho}{\sqrt{2\pi}}\end{pmatrix}.$$

It is now a matter of technical evaluations to find that the correlation tensor can be represented in the form

$$B_u(\rho_1,\theta_1;\rho_2,\theta_2) = \mathbf{h}_0(\rho_1)\,\mathbf{h}_0^T(\rho_2) \tag{17.24}$$

$$+\frac{1}{\pi}\sum_{k=1}^{\infty}\{\mathbf{h}_{1k}(\rho_1,\theta_1)\mathbf{h}_{1k}^T(\rho_2,\theta_2) + \tilde{\mathbf{h}}_{1k}(\rho_1,\theta_1)\,\tilde{\mathbf{h}}_{1k}^T(\rho_2,\theta_2)\}\rho_1^k\rho_2^k$$

$$+\ \tilde{\mathbf{h}}_0(\rho_1)\,\tilde{\mathbf{h}}_0^T(\rho_2) \tag{17.25}$$

$$+\frac{1}{\pi}\sum_{k=1}^{\infty}\{\mathbf{h}_{2k}(\rho_1,\theta_1)\mathbf{h}_{2k}^T(\rho_2,\theta_2) + \tilde{\mathbf{h}}_{2k}(\rho_1,\theta_1)\,\tilde{\mathbf{h}}_{2k}^T(\rho_2,\theta_2)\}\rho_1^k\rho_2^k.$$

This follows from the easily verified representations

$$\mathbf{h}_{1k}(\rho_1,\theta_1)\,\mathbf{h}_{1k}^T(\rho_2,\theta_2) + \tilde{\mathbf{h}}_{1k}(\rho_1,\theta_1)\,\tilde{\mathbf{h}}_{1k}^T(\rho_2,\theta_2) =$$

$$\begin{pmatrix}\lambda_{11}(\rho_1,k)\,\lambda_{11}(\rho_2,k)\,\cos[k(\theta_2-\theta_1)] & \lambda_{11}(\rho_1,k)\,\lambda_{21}(\rho_2,k)\,\sin[k(\theta_2-\theta_1)]\\ -\lambda_{21}(\rho_1,k)\,\lambda_{11}(\rho_2,k)\,\sin[k(\theta_2-\theta_1)] & \lambda_{21}(\rho_1,k)\,\lambda_{22}(\rho_2,k)\,\cos[k(\theta_2-\theta_1)]\end{pmatrix}$$

and

$$\mathbf{h}_{2k}(\rho_1,\theta_1)\,\mathbf{h}_{2k}^T(\rho_2,\theta_2) + \tilde{\mathbf{h}}_{2k}(\rho_1,\theta_1)\,\tilde{\mathbf{h}}_{2k}^T(\rho_2,\theta_2)$$
$$= \begin{pmatrix} \lambda_{12}(\rho_1,k)\,\lambda_{12}(\rho_2,k)\,\cos[k(\theta_2-\theta_1)] & -\lambda_{12}(\rho_1,k)\,\lambda_{22}(\rho_2,k)\,\sin[k(\theta_2-\theta_1)] \\ \lambda_{22}(\rho_1,k)\,\lambda_{12}(\rho_2,k)\,\sin[k(\theta_2-\theta_1)] & \lambda_{22}(\rho_1,k)\,\lambda_{22}(\rho_2,k)\,\cos[k(\theta_2-\theta_1)] \end{pmatrix}.$$

So we can see that the first and the second pairs of lines ((17.24) and (17.25), respectively) present the covariances of the first and second vectors in our splitting, respectively:

$$B_u = \langle \mathbf{u}(r_1,\theta_1)\;\mathbf{u}^T(r_2,\theta_2)\rangle = \langle \mathbf{V}_1(r_1,\theta_1)\,\mathbf{V}_1^T(r_2,\theta_2)\rangle + \langle \mathbf{V}_2(r_1,\theta_1)\mathbf{V}_2^T(r_2,\theta_2)\rangle,$$

and thus

$$\begin{aligned} B_{V_1} &= \mathbf{h}_0(\rho_1)\,\mathbf{h}_0^T(\rho_2) \\ &\quad + \frac{1}{\pi}\sum_{k=1}^{\infty}\{\mathbf{h}_{1k}(\rho_1,\theta_1)\,\mathbf{h}_{1k}^T(\rho_2,\theta_2) + \tilde{\mathbf{h}}_{1k}(\rho_1,\theta_1)\,\tilde{\mathbf{h}}_{1k}^T(\rho_2,\theta_2)\}\rho_1^k\rho_2^k, \\ B_{V_2} &= \tilde{\mathbf{h}}_0(\rho_1)\,\tilde{\mathbf{h}}_0^T(\rho_2) \\ &\quad + \frac{1}{\pi}\sum_{k=1}^{\infty}\{\mathbf{h}_{2k}(\rho_1,\theta_1)\,\mathbf{h}_{2k}^T(\rho_2,\theta_2) + \tilde{\mathbf{h}}_{2k}(\rho_1,\theta_1)\,\tilde{\mathbf{h}}_{2k}^T(\rho_2,\theta_2)\}\rho_1^k\rho_2^k, \end{aligned}$$

where $B_{V_1} = \langle \mathbf{V}_1(r_1,\theta_1)\,\mathbf{V}_1^T(r_2,\theta_2)\rangle \quad B_{V_2} = \langle \mathbf{V}_2(r_1,\theta_1)\mathbf{V}_2^T(r_2,\theta_2)\rangle$.

Note that each part, i. e., B_{V_1} and B_{V_2}, is represented as an orthogonal-mode expansion. Therefore, we can construct a K-L-expansion for our random fields $\mathbf{V}_1$ and $\mathbf{V}_2$.

We have not yet normalized the eigenfunctions. We can do thi by dividing the angular modes by $\sqrt{\pi}$, and the radial modes by $\Delta_1(k) = \int_0^1(\lambda_{11}^2 + \lambda_{21}^2)\,d\rho$ for the first family of eigenfunctions (17.22), and by $\Delta_2(k) = \int_0^1(\lambda_{12}^2 + \lambda_{22}^2)\,d\rho$ for the second family of eigenfunctions (17.23). We then collect the orthonormal eigenmodes in two families:

$$\mathcal{H}_{2k-1}^{(1)} = \frac{1}{\sqrt{\Delta_1(k)\,\pi}}\;\mathbf{h}_{1k}(\rho,\theta),\quad \mathcal{H}_{2k}^{(1)} = \frac{1}{\sqrt{\Delta_1(k)\,\pi}}\;\tilde{\mathbf{h}}_{1k}(\rho,\theta),\quad k = 1,2,\ldots,$$

and

$$\mathcal{H}_{2k-1}^{(2)} = \frac{1}{\sqrt{\Delta_2(k)\,\pi}}\;\mathbf{h}_{2k}(\rho,\theta),\quad \mathcal{H}_{2k}^{(2)} = \frac{1}{\sqrt{\Delta_2(k)\,\pi}}\;\tilde{\mathbf{h}}_{2k}(\rho,\theta),\quad k = 1,2,\ldots.$$

Then, the orthonormal functions $\mathcal{H}_k^{(1)}$ and $\mathcal{H}_k^{(2)}$ are eigenfunctions of the covariance tensors B_{V_1} and B_{V_2}, respectively, with the corresponding eigenvalues $\Delta_1(k)$ and $\Delta_2(k)$:

$$\int_0^1\int_0^{2\pi} B_{V_m}\cdot\mathcal{H}_k^{(m)}(\rho_2,\theta_2)\,d\rho_2\,d\theta_2 = \Delta_m(k)\,\mathcal{H}_k^{(m)}(\rho_1,\theta_1),\quad m = 1,2.$$

We thus end up with the K-L-expansion for the random field $\mathbf{V}_1(r,\theta)$ in the form

$$\mathbf{V}_1(r,\theta) = \sum_{k=1}^{\infty} \zeta_k \mathcal{H}_k^{(1)}(\rho,\theta),$$

where ζ_k are Gaussian random variables such that

$$\langle \zeta_k \zeta_j \rangle = \Delta_1(k)\, \delta_{jk},$$

and the same for $\mathbf{V}_2(r,\theta)$.

Putting these expansions together we conclude with the desired representation (17.20), (17.21).

Finally note that the spectra given in the theorem are obtained immediately from the expansion of the correlation function. This completes the proof of Theorem 17.2. □

17.3.2 General case of homogeneous excitations

So far we have studied the case where g_r and g_θ were two independent white noise processes. We will now show that the general situation when g_r and g_θ are some arbitrary possibly dependent homogeneous processes is basically derived from the white noise case.

Assume we are given two homogeneous zero mean processes g_r and g_θ with the correlation tensor $B_g(\varphi_2 - \varphi_2)$ (with the entries $B_{g,ij}$, $i,j = 1,2$) defined on the boundary. As shown above, the correlation tensor B_u is related to B_g by the double integral representation (17.15). Changing the integration variable φ'' to a new one, ψ by $\varphi'' - \varphi' = \psi$, for $\vec{u} = (u_\rho, u_\theta)^T$ we obtain from (17.15)

$$B_u(\rho_1,\theta_1;\rho_2,\theta_2) = \int_0^{2\pi}\int_0^{2\pi} G(\rho_1;\theta_1-\varphi')\, B_g(\psi)\, G^T(\rho_2;\theta_2-\psi-\varphi')\, d\varphi'\, d\psi. \tag{17.26}$$

The inner integral with respect to φ' can be evaluated explicitly using the series expansions for the kernel $G(\rho,\theta)$ given above in (17.7).

It is convenient to rewrite the relation (17.26) in a different form. We now take the entries of the correlation tensor B_u, and define a column vector function $\hat{\mathbf{B}}_u$ as follows $\hat{\mathbf{B}}_u = (B_{u,11}, B_{u,12}, B_{u,21}, B_{u,22})^T$. Analogously, we use the notation $\hat{\mathbf{B}}_g$ for the column vector $\hat{\mathbf{B}}_g = (B_{g,11}, B_{g,12}, B_{g,21}, B_{g,22})^T$.

Theorem 17.3. *Under the assumptions that the boundary random functions g_r and g_θ are statistically homogeneous, with the given correlation tensor $B_g(\varphi)$, the solution of the Stokes problem is a random field partially homogeneous with respect to the* *angular coordinate, and the correlation functions $\hat{\boldsymbol{B}}_u$ and $\hat{\boldsymbol{B}}_g(\varphi)$ are related by the convolution*

$$\hat{\boldsymbol{B}}_u(\rho_1,\rho_2;\theta) = \int_0^{2\pi} A(\rho_1,\rho_2;\theta-\psi)\,\hat{\boldsymbol{B}}_g(\psi)\,d\psi, \tag{17.27}$$

where the matrix A is explicitly given by its entries (17.30)–(17.33) given below.

Proof. In the introduced notation, the equality (17.26) takes the form

$$\hat{\mathbf{B}}_u(\rho_1,\theta_1;\rho_2,\theta_2) = \int_0^{2\pi}\int_0^{2\pi} G(\rho_1;\theta_1-\varphi')\otimes G(\rho_2;\theta_2-\psi-\varphi')\,\hat{\mathbf{B}}_g(\psi)\,d\varphi'\,d\psi. \tag{17.28}$$

Here we denote by $\otimes$ a tensor product of two matrices, which is defined as a 4×4 matrix, represented as a 2×2-block matrix each block being a 2×2 matrix of the form $G_{ij}(\rho_1;\theta_1-\varphi')G(\rho_2;\theta_2-\psi-\varphi')$, $i,j=1,2$.

The entries a_{ij}, $i,j=1,2,3,4$ of the matrix

$$A = \int_0^{2\pi} G(\rho_1;\theta_1-\varphi')\otimes G(\rho_2;\theta_2-\psi-\varphi')\,d\varphi' \tag{17.29}$$

can be found explicitly. Indeed, substituting the series representation of the matrix G given by (17.7) in (17.29) we obtain

$$\begin{aligned}
a_{11} &= \frac{\rho_1\rho_2}{2\pi} + \frac{1}{\pi}\sum_{k=1}^{\infty}\lambda_{11}(\rho_1,k)\lambda_{11}(\rho_2,k)\,\rho_1^k\rho_2^k\cos[k(\theta_2-\theta_1-\psi)],\\
a_{12} &= \frac{1}{\pi}\sum_{k=1}^{\infty}\lambda_{11}(\rho_1,k)\lambda_{12}(\rho_2,k)\,\rho_1^k\rho_2^k\sin[k(\theta_2-\theta_1-\psi)],\\
a_{13} &= -\frac{1}{\pi}\sum_{k=1}^{\infty}\lambda_{12}(\rho_1,k)\lambda_{11}(\rho_2,k)\,\rho_1^k\rho_2^k\sin[k(\theta_2-\theta_1-\psi)],\\
a_{14} &= \frac{1}{\pi}\sum_{k=1}^{\infty}\lambda_{12}(\rho_1,k)\lambda_{12}(\rho_2,k)\,\rho_1^k\rho_2^k\cos[k(\theta_2-\theta_1-\psi)],
\end{aligned} \tag{17.30}$$

$$\begin{aligned}
a_{21} &= \frac{1}{\pi}\sum_{k=1}^{\infty}\lambda_{11}(\rho_1,k)\lambda_{21}(\rho_2,k)\,\rho_1^k\rho_2^k\sin[k(\theta_2-\theta_1-\psi)],\\
a_{22} &= \frac{\rho_1\rho_2}{2\pi} + \frac{1}{\pi}\sum_{k=1}^{\infty}\lambda_{11}(\rho_1,k)\lambda_{22}(\rho_2,k)\,\rho_1^k\rho_2^k\cos[k(\theta_2-\theta_1-\psi)],\\
a_{23} &= \frac{1}{\pi}\sum_{k=1}^{\infty}\lambda_{12}(\rho_1,k)\lambda_{21}(\rho_2,k)\,\rho_1^k\rho_2^k\cos[k(\theta_2-\theta_1-\psi)],
\end{aligned}$$

$$a_{24} = -\frac{1}{\pi}\sum_{k=1}^{\infty} \lambda_{12}(\rho_1, k)\lambda_{22}(\rho_2, k)\, \rho_1^k \rho_2^k \sin\,[k(\theta_2 - \theta_1 - \psi)], \tag{17.31}$$

$$a_{31} = -\frac{1}{\pi}\sum_{k=1}^{\infty} \lambda_{21}(\rho_1, k)\lambda_{11}(\rho_2, k)\, \rho_1^k \rho_2^k \sin\,[k(\theta_2 - \theta_1 - \psi)],$$

$$a_{32} = \frac{1}{\pi}\sum_{k=1}^{\infty} \lambda_{21}(\rho_1, k)\lambda_{12}(\rho_2, k)\, \rho_1^k \rho_2^k \cos\,[k(\theta_2 - \theta_1 - \psi)],$$

$$a_{33} = \frac{\rho_1\rho_2}{2\pi} + \frac{1}{\pi}\sum_{k=1}^{\infty} \lambda_{22}(\rho_1, k)\lambda_{11}(\rho_2, k)\, \rho_1^k \rho_2^k \cos\,[k(\theta_2 - \theta_1 - \psi)],$$

$$a_{34} = \frac{1}{\pi}\sum_{k=1}^{\infty} \lambda_{22}(\rho_1, k)\lambda_{12}(\rho_2, k)\, \rho_1^k \rho_2^k \sin\,[k(\theta_2 - \theta_1 - \psi)], \tag{17.32}$$

$$a_{41} = \frac{1}{\pi}\sum_{k=1}^{\infty} \lambda_{21}(\rho_1, k)\lambda_{21}(\rho_2, k)\, \rho_1^k \rho_2^k \cos\,[k(\theta_2 - \theta_1 - \psi)],$$

$$a_{42} = -\frac{1}{\pi}\sum_{k=1}^{\infty} \lambda_{21}(\rho_1, k)\lambda_{22}(\rho_2, k)\, \rho_1^k \rho_2^k \sin\,[k(\theta_2 - \theta_1 - \psi)],$$

$$a_{43} = \frac{1}{\pi}\sum_{k=1}^{\infty} \lambda_{22}(\rho_1, k)\lambda_{21}(\rho_2, k)\, \rho_1^k \rho_2^k \sin\,[k(\theta_2 - \theta_1 - \psi)],$$

$$a_{44} = \frac{\rho_1\rho_2}{2\pi} + \frac{1}{\pi}\sum_{k=1}^{\infty} \lambda_{22}(\rho_1, k)\lambda_{22}(\rho_2, k)\, \rho_1^k \rho_2^k \cos\,[k(\theta_2 - \theta_1 - \psi)]. \tag{17.33}$$

Thus we see from these formulas that the entries of the matrix A depend on the difference $\theta = \theta_2 - \theta_1$, hence the correlation tensor B_u also depends on $\theta = \theta_2 - \theta_1$, and from (17.28), (17.29) we obtain the desired convolution representation (17.27). The proof of Theorem 17.3 is complete. □

Remark 17.1. Note that if the boundary correlation tensor B_g is given by its spectral expansion, we can express the correlation tensor of the solution through the spectra. For instance, assuming the spectral tensor is real-valued, so that

$$B_{g,ij}(\varphi'' - \varphi) = \frac{f_{ij}(0)}{2\pi} + \frac{1}{\pi}\sum_{k=1}^{\infty} f_{ij}(k)\, \cos\, k(\varphi'' - \varphi'), \quad i, j = 1, 2, \tag{17.34}$$

we can derive a general formula for the covariance tensor by substituting this expansion in (17.26). After routine evaluations we obtain the general formulas

$$B_{11} = \frac{f_{11}(0)\,\rho_1\,\rho_2}{2\pi} + \frac{1}{\pi}\sum_{k=1}^{\infty} \rho_1^k\,\rho_2^k\left(\Lambda_{11}^c\ \cos[k(\theta_2-\theta_1)] + \Lambda_{11}^s\ \sin[k(\theta_2-\theta_1)]\right),$$

$$B_{12} = \frac{1}{\pi}\sum_{k=1}^{\infty} \rho_1^k\,\rho_2^k\left(\Lambda_{12}^c\ \cos[k(\theta_2-\theta_1)] + \Lambda_{12}^s\ \sin[k(\theta_2-\theta_1)]\right),$$

$$B_{21} = \frac{1}{\pi}\sum_{k=1}^{\infty} \rho_1^k\,\rho_2^k\left(\Lambda_{21}^c\ \cos[k(\theta_2-\theta_1)] + \Lambda_{21}^s\ \sin[k(\theta_2-\theta_1)]\right),$$

$$B_{22} = \frac{f_{22}(0)\,\rho_1\,\rho_2}{2\pi} + \frac{1}{\pi}\sum_{k=1}^{\infty} \rho_1^k\,\rho_2^k\left(\Lambda_{22}^c\ \cos[k(\theta_2-\theta_1)] + \Lambda_{22}^s\ \sin[k(\theta_2-\theta_1)]\right),$$

where

$$\begin{aligned}
\Lambda_{11}^c &= f_{11}\lambda_{11}^1\lambda_{11}^2 + f_{22}\lambda_{12}^1\lambda_{12}^2, & \Lambda_{11}^s &= f_{21}\lambda_{11}^1\lambda_{12}^2 - f_{12}\lambda_{12}^1\lambda_{11}^2,\\
\Lambda_{12}^c &= f_{21}\lambda_{12}^1\lambda_{21}^2 + f_{12}\lambda_{11}^1\lambda_{22}^2, & \Lambda_{12}^s &= f_{11}\lambda_{11}^1\lambda_{21}^2 - f_{22}\lambda_{12}^1\lambda_{22}^2,\\
\Lambda_{21}^c &= f_{21}\lambda_{22}^1\lambda_{11}^2 + f_{12}\lambda_{21}^1\lambda_{12}^2, & \Lambda_{21}^s &= f_{22}\lambda_{22}^1\lambda_{12}^2 - f_{11}\lambda_{21}^1\lambda_{11}^2,\\
\Lambda_{22}^c &= f_{22}\lambda_{22}^1\lambda_{22}^2 + f_{11}\lambda_{21}^1\lambda_{21}^2, & \Lambda_{22}^s &= f_{21}\lambda_{22}^1\lambda_{21}^2 - f_{12}\lambda_{21}^1\lambda_{22}^2.
\end{aligned}$$

Here we use the notations $\lambda_{ij}^m = \lambda_{ij}(\rho_m, k)$, $m = 1, 2$.

Analogous arguments can be used to derive the K-L expansion for the solution. Let us do thi in matrix form, to save a space. Let Φ be the spectral matrix with the entries f_{ij} defined in (17.34), and let $\Phi^{1/2}$ be the "square root" of Φ, i. e., $\Phi = \Phi^{1/2}(\Phi^{1/2})^*$, where * stands for the conjugate transpose. The same notation is used for the spectral matrix $S(\rho_1, \rho_2; k)$ defined at the end of Theorem 17.2 by its entries $S_{ij}(\rho_1, \rho_2; k)$, $S(\rho_1, \rho_2; k) = S^{1/2}(\rho_1, k)(S^{1/2}(\rho_2, k))^*$ where the matrix $S^{1/2}(\rho, k)$ has the following simple form:

$$S^{1/2}(\rho, k) = \frac{\rho^k}{\sqrt{\pi}}\begin{pmatrix} \lambda_{11}(\rho, k) & -\mathrm{i}\lambda_{12}(\rho, k)\\ -\mathrm{i}\lambda_{21}(\rho, k) & \lambda_{22}(\rho, k)\end{pmatrix}.$$

Using this factorization we can find by direct evaluations that the random field

$$\begin{pmatrix} u_r(\rho,\theta)\\ u_\theta(\rho,\theta)\end{pmatrix} = \frac{\xi_0\rho}{\sqrt{2\pi}}\mathbf{I} + \sum_{k=1}^{\infty}\Phi^{1/2}(k)\Big\{\big[\Re S^{1/2}(\rho,k)\,\boldsymbol{\xi}_k + \Im S^{1/2}(k)\,\boldsymbol{\eta}_k\big]\cos k\theta + [\Re S^{1/2}(\rho,k)\boldsymbol{\eta}_k - \Im S^{1/2}(k)\,\boldsymbol{\xi}_k]\sin k\theta\Big\} \quad (17.35)$$

has the desired correlation tensor.

Here we use the notation $\Re S^{1/2}$ and $\Im S^{1/2}$ for the real and imaginary part of $S^{1/2}$, respectively, $\mathbf{I}$ is the identity matrix, and $\boldsymbol{\xi}_k$, $\boldsymbol{\eta}_k$ are two independent families of independent standard Gaussian 2D vectors.

Note that in a particular case when the boundary excitations g_r and g_θ are independent, the spectral matrix $\Phi(k)$ is diagonal, and the simulation formula (17.35) takes the form of (17.20), (17.21) where in (17.20), λ_{11} should be changed with $\lambda_{11}\Phi_{11}^{1/2}$, and λ_{12}, with $\lambda_{12}\Phi_{22}^{1/2}$; in (17.21), λ_{22} should be changed with $\lambda_{22}\Phi_{22}^{1/2}$, and λ_{21}, with $\lambda_{21}\Phi_{11}^{1/2}$.

Remark 17.2. Note that using the relation between the vectors in polar and rectangular coordinates,

$$\begin{pmatrix} u_1(r,\theta) \\ u_2(r,\theta) \end{pmatrix} = \mathcal{R}_\theta \begin{pmatrix} u_\rho(r,\theta) \\ u_\theta(r,\theta \end{pmatrix}$$

we can easily relate the desired statistical characteristics in these two coordinate systems. For example, the covariance tensors are related as follows:

$$B_{(u_1,u_2)}(\rho_1,\rho_2;\theta_1,\theta_2) = \mathcal{R}_{\theta_1}\, B_{(u_r,u_\theta)}(\rho_1,\rho_2;\theta_1,\theta_2)\, \mathcal{R}_{\theta_2}^T. \tag{17.36}$$

The K-L-expansion in the rectangular coordinates is also obtained directly from the K-L-expansion of the random field in the polar coordinates on the basis that the eigenfunctions are related by $h_{rectangular} = \mathcal{R}_\theta\, h_{polar}$ and $\tilde{h}_{rectangular} = \mathcal{R}_\theta\, \tilde{h}_{polar}$.

Let us write down here the relation (17.36) in detail. We denote the entries of the covariance matrix $B_{(u_1,u_2)}$ by B_{ij}^{rec}, and the entries of the covariance matrix $B_{(u_r,u_\theta)}$ by B_{ij}^{pol}. From (17.36) we obtain

$$\begin{aligned} B_{11}^{rec} &= \cos\theta_1\cos\theta_2\, B_{11}^{pol} - \cos\theta_1\sin\theta_2\, B_{12}^{pol} \\ &\quad - \sin\theta_1\cos\theta_2\, B_{21}^{pol} + \sin\theta_1\sin\theta_2\, B_{22}^{pol}, \\ B_{12}^{rec} &= \cos\theta_1\sin\theta_2\, B_{11}^{pol} + \cos\theta_1\cos\theta_2\, B_{12}^{pol} \\ &\quad - \sin\theta_1\sin\theta_2\, B_{21}^{pol} - 0.05cm - \sin\theta_1\cos\theta_2\, B_{22}^{pol}, \\ B_{21}^{rec} &= \sin\theta_1\cos\theta_2\, B_{11}^{pol} - \sin\theta_1\sin\theta_2\, B_{12}^{pol} \\ &\quad + \cos\theta_1\cos\theta_2\, B_{21}^{pol} - \cos\theta_1\sin\theta_2\, B_{22}^{pol}, \\ B_{22}^{rec} &= \sin\theta_1\sin\theta_2\, B_{11}^{pol} + \sin\theta_1\cos\theta_2\, B_{12}^{pol} \\ &\quad + \sin\theta_2\cos\theta_1\, B_{21}^{pol} + \cos\theta_1\cos\theta_2\, B_{22}^{pol}. \end{aligned}$$

17.4 Correlation function of the pressure

17.4.1 White noise excitations

From (17.8), it is easy to derive an explicit representation for the correlation function of the pressure $p(r,\theta)$. Indeed, let us denote by B_p the correlation function of the pressure: $B_p(r_1,\theta_1;r_2,\theta_2) = \langle p(r_1,\theta_1)p(r_2,\theta_2)\rangle$.

Theorem 17.4. *Assume that the functions g_r and g_θ prescribed on the boundary are independent Gaussian white noise random processes defined on the circle $S(0,R)$: $B_\varphi=\langle g_i(\varphi_1)\,g_j(\varphi_2)\rangle = \delta_{ij}\,\delta(\varphi_2-\varphi_1)$, $i,j = 1,2$, where we mean that $g_1 = g_r$, $g_2 = g_\theta$. Then $B_p(r_1,\theta_1;r_2,\theta_2)$, the correlation function of the pressure $p(r,\theta)$ has the following series representation:*

$$B_p = B_p(r_1,\theta_1;r_2,\theta_2) = \frac{8\eta^2}{\pi R^2}\sum_{k=1}^{\infty} k^2\rho_1^{k+1}\,\rho_2^{k+1}\,\cos[(k+1)(\theta_2-\theta_1)]. \quad (17.37)$$

Thus the correlation function (17.37) depends on the difference $\theta = \theta_2-\theta_1$ and the product $\rho = \rho_1\rho_2$: $B_p = B_p(\rho,\theta)$. Moreover, it can be explicitly expressed as

$$B_p(\rho,\theta) = \frac{8\eta^2}{\pi R^2}\left[\rho\cos(\theta)\frac{\partial^2}{\partial\theta^2}B_\Delta(\rho,\theta) + \rho^2\sin(\theta)\frac{\partial^2}{\partial\theta\,\partial\rho}B_\Delta(\rho,\theta)\right], \quad (17.38)$$

where $B_\Delta(\rho,\theta)$ is the correlation function for the Laplace equation under white noise boundary excitations derived in [193]*:*

$$B_\Delta(\rho,\theta) = \frac{1}{2\pi}\frac{1-\rho^2}{1-2\rho\cos\theta+\rho^2}.$$

Proof. Changing the product of integrals by double integrals and using the definition of the white noise we find

$$\begin{aligned} B_p(r_1,\theta_1;r_2,\theta_2) &= \langle p(r_1,\theta_1)p(r_2,\theta_2)\rangle \quad (17.39)\\ &= \left[\left\{\int_0^{2\pi} P_1(r_1;\theta_1-\varphi')g_r(\varphi')d\varphi' + \int_0^{2\pi} P_2(r_1;\theta_1-\varphi')g_\theta(\varphi')d\varphi'\right\}\right.\\ &\quad\left.\times\left\{\int_0^{2\pi} P_1(r_2;\theta_2-\varphi'')g_r(\varphi'')d\varphi'' + \int_0^{2\pi} P_2(r_2;\theta_2-\varphi'')g_\theta(\varphi'')d\varphi''\right\}\right] \end{aligned}$$

$$= \int_0^{2\pi}\int_0^{2\pi} \Big\{ P_1(r_1;\theta_1-\varphi')P_1(r_2;\theta_2-\varphi'')\langle g_r(\varphi')g_r(\varphi'')\rangle$$
$$+ P_1(r_1;\theta_1-\varphi')P_2(r_2;\theta_2-\varphi'')\langle g_r(\varphi')g_\theta(\varphi'')\rangle$$
$$+ P_2(r_1;\theta_1-\varphi')P_1(r_2;\theta_2-\varphi'')\langle g_\theta(\varphi')g_r(\varphi'')\rangle$$
$$+ P_2(r_1;\theta_1-\varphi')P_2(r_2;\theta_2-\varphi'')\langle g_\theta(\varphi')g_\theta(\varphi'')\rangle \Big\} d\varphi' d\varphi''$$
$$= \int_0^{2\pi} P_1(r_1;\theta_1-\varphi)P_1(r_2;\theta_2-\varphi)d\varphi + \int_0^{2\pi} P_2(r_1;\theta_1-\varphi)P_2(r_2;\theta_2-\varphi)d\varphi .$$

Now we substitute the expansions for P_1 and P_2 given by (17.8). This yields

$$B_p(r_1,\theta_1;r_2,\theta_2) = \frac{8\eta^2}{\pi R^2}\sum_{k=1}^{\infty} k^2\rho_1^{k+1}\rho_2^{k+1}\cos[(k+1)(\theta_2-\theta_1)].$$

The exact expression (17.38) follows from (17.37) and the series representation for the correlation function of the solution to the Laplace equation under white noise excitations derived in my recent paper [193]:

$$B_\Delta = \frac{1}{2\pi} + \frac{1}{\pi}\sum_{k=1}^{\infty}\rho^k \cos[k(\theta_2-\theta_1)] = \frac{1}{2\pi}\cdot\frac{1-\rho^2}{1-2\rho\cos\theta+\rho^2},$$

where we recall that $\rho = \rho_1\rho_2$, and $\theta = \theta_2 - \theta_1$.

The proof is complete. □

It is interesting to note that we could obtain expressions for the pressure by substituting formally a generalized representation of the boundary white noises on the circle

$$g_r(\varphi) = \frac{\xi_0}{\sqrt{2\pi}} + \frac{1}{\sqrt{\pi}}\sum_{k=1}^{\infty}[\xi_k \cos k\varphi + \eta_k \sin k\varphi]$$

$$g_\theta(\varphi) = \frac{\xi_0'}{\sqrt{2\pi}} + \frac{1}{\sqrt{\pi}}\sum_{k=1}^{\infty}[\xi_k' \cos k\varphi + \eta_k' \sin k\varphi],$$

where ξ_l and η_k are families of independent standard Gaussian random variables, into the Poisson formula (17.8) with the kernels $P_1(r;\theta)$ and $P_2(r;\theta)$. Indeed, a little algebra results in the representation for the random pressure:

$$p(r,\theta) = \frac{4\eta}{\sqrt{2\pi}R}\sum_{k=1}^{\infty} k\rho^{k+1}\{\xi_k\cos[(k+1)\theta] + \eta_k\sin[(k+1)\theta]\}.$$

Note that from this representation the series expansion (17.37) is easily obtained by direct evaluation.

17.4.2 Homogeneous random boundary excitations

We extend the result of the Theorem 17.4 to the case of general homogeneous random boundary excitations. So we suppose that the velocity $\mathbf{g}(\psi)$ prescribed on the circle $S(0, R)$ is a real-valued vector zero mean homogeneous random process. Let us denote the correlation tensor of $\mathbf{g}$ by $B^{(g)}(\psi)$ with the entries $B_{ij}^{(g)}(\psi) = \langle g_i(\psi_1)g_j(\psi_1 + \psi))\rangle$, $i, j = 1, 2$, where we mean that $g_1 = g_r$, and $g_2 = g_\theta$.

Our goal is now to relate the correlation function of the pressure and the input correlation tensor $B^{(g)}$.

Theorem 17.5. *The correlation function of the pressure is represented through the correlation tensor of the homogeneous boundary velocity excitations by the following convolution formula:*

$$\begin{aligned} B_p(r_1, \theta_1; r_2, \theta_2) &= B_p(\rho_1\rho_2, \theta_1 - \theta_2) \qquad (17.40)\\ &= \frac{4\eta^2}{\pi R^2}\Bigg[\int_0^{2\pi}\sum_{k=1}^{\infty} k^2\rho_1^{k+1}\rho_2^{k+1}\cos[(k+1)(\theta_1-\theta_2-\psi)]\{B_{11}^{(g)}(\psi)+B_{22}^{(g)}(\psi)\}\,d\psi \\ &\quad + \int_0^{2\pi}\sum_{k=1}^{\infty} k^2\rho_1^{k+1}\rho_2^{k+1}\sin[(k+1)(\theta_1-\theta_2-\psi)]\{B_{21}^{(g)}(\psi) - B_{12}^{(g)}(\psi)\}\,d\psi\Bigg], \end{aligned}$$

hence the pressure is a random field homogeneous with respect to the angular variable.

Proof. The proof follows the same way presented above by starting with the expression (17.39), and making the change of variable $\psi = \varphi' - \varphi''$, where φ' is fixed (i. e., ψ is fixed), and the inner integrals are taken explicitly with respect to the variable φ''. Using the orthogonality of the functions $\sin k\varphi''$ and $\cos k\varphi''$ we get (17.40). □

17.4.3 Vorticity and stress tensor

The stress tensor for the Stokes flow has the form [124]

$$\sigma_{ik} = -p\delta_{ik} + \eta\left(\frac{\partial u_i}{\partial x_k} + \frac{\partial u_k}{\partial x_i}\right), \quad i, k = 1, 2,$$

where p is the pressure, and δ_{ik} is the Kronecker symbol.

Recall that in polar coordinates we use

$$u_1 = u_r \cos\theta - u_\theta \sin\theta, \qquad u_2 = u_r \sin\theta + u_\theta \cos\theta,$$

and

$$\frac{\partial}{\partial x} = \cos\theta\,\frac{\partial}{\partial r} - \sin\theta\,\frac{\partial}{r\partial\theta}, \quad \frac{\partial}{\partial y} = \sin\theta\,\frac{\partial}{\partial r} + \cos\theta\,\frac{\partial}{r\partial\theta}.$$

Now choosing the axis x along the vector direction defined by θ we find

$$\frac{\partial u_1}{\partial x} = \frac{\partial u_r}{\partial r}, \quad \frac{\partial u_2}{\partial y} = \frac{1}{r}\frac{\partial u_\theta}{\partial \theta} + \frac{u_r}{r},$$
$$\frac{\partial u_2}{\partial x} + \frac{\partial u_1}{\partial y} = \frac{\partial u_\theta}{\partial r} + \frac{1}{r}\frac{\partial u_r}{\partial \theta} - \frac{u_\theta}{r}, \tag{17.41}$$

and hence the entries of the stress tensor become

$$\sigma_{rr} = -p + 2\eta\frac{\partial u_r}{\partial r}, \qquad \sigma_{\theta\theta} = -p + 2\eta\Big(\frac{1}{r}\frac{\partial u_\theta}{\partial \theta} + \frac{u_r}{r}\Big),$$
$$\sigma_{r\theta} = \eta\Big(\frac{1}{r}\frac{\partial u_r}{\partial \theta} + \frac{\partial u_\theta}{\partial r} - \frac{u_\theta}{r}\Big). \tag{17.42}$$

Using the series expansions for the kernels G_{ij} and the pressure obtained above, it is easy to derive series expansion representations for the correlation tensor for all the strain entries. Let us do this for the vorticity.

From (17.41) we get for the vorticity

$$\omega_{r\theta} = \frac{\partial u_2}{\partial x} - \frac{\partial u_1}{\partial y} = \frac{1}{r}\Big[\frac{\partial}{\partial r}(r u_\theta) - \frac{\partial u_r}{\partial \theta}\Big].$$

Our goal is to find the correlation tensor of vorticity.

Theorem 17.6. *The correlation function of the pressure, B_p and the correlation function of the vorticity, $B_{\omega_{r\theta}}$ are related by*

$$B_{\omega_{r\theta}} = \frac{1}{2\eta^2}B_p = \frac{4}{\pi R^2}\sum_{k=1}^{\infty} k^2\rho_1^{k+1}\rho_2^{k+1}\cos[(k+1)(\theta_2-\theta_1)]. \tag{17.43}$$

Proof. We start by substituting the series expansions (17.10) for the kernels G_{ij} to get the expressions

$$\frac{1}{r}\frac{\partial}{\partial r}(r u_\theta) = \frac{1}{2\pi R}\int_0^{2\pi}\sum_{k=1}^{\infty}\big[-(k-2)^2\rho^k + k(k-2)\rho^{k+2}\big]\sin[(k(\theta-\varphi)]\, g_r(\varphi)\, d\varphi$$
$$+ \frac{1}{2\pi R}\int_0^{2\pi}\sum_{k=1}^{\infty}\big[(k-2)^2\rho^k - k^2\rho^{k+2}\big]\cos[(k(\theta-\varphi)]\, g_\theta(\varphi)\, d\varphi$$

and

$$\frac{1}{r}\frac{\partial u_r}{\partial \theta} = \frac{1}{2\pi R}\int_0^{2\pi}\sum_{k=1}^{\infty}\big[-k^2\rho^k + k(k-2)\rho^{k+2}\big]\sin[(k(\theta-\varphi)]\, g_r(\varphi)\, d\varphi$$
$$+ \frac{1}{2\pi R}\int_0^{2\pi}\sum_{k=1}^{\infty}\big[k^2\rho^k - k^2\rho^{k+2}\big]\cos[(k(\theta-\varphi)]\, g_\theta(\varphi)\, d\varphi.$$

This yields

$$\omega_{r\theta} = \frac{2}{\pi R}\int_0^{2\pi}\sum_{k=1}^{\infty} k\rho^{k+1}\sin[(k+1)(\theta-\varphi)]\,g_r(\varphi)\,d\varphi - \frac{2}{\pi R}\int_0^{2\pi}\sum_{k=1}^{\infty} k\rho^{k+1}\cos[(k+1)(\theta-\varphi)]\,g_\theta(\varphi)\,d\varphi.$$

From this we readily find the correlation function of the vorticity:

$$B_{\omega_{r\theta}}(r_1,\theta_1;r_2,\theta_2) = \frac{4}{\pi R^2}\sum_{k=1}^{\infty} k^2\rho_1^{k+1}\rho_2^{k+1}\cos[(k+1)(\theta_2-\theta_1)]$$

which by comparing with the expansion (17.37) gives the desired result (17.43). □

Radial stress

Analogously, the expansions for the correlations of the stress tensor can be obtained. For instance, let us consider the radial stress $\sigma_{rr} = -p + 2\eta\,\partial u_r/\partial r$ given by the relation (17.42). We have obviously

$$B_{\sigma_{rr}} = \langle\sigma_{rr}(r_1,\theta_1)\sigma_{rr}(r_2,\theta_2)\rangle = B_p + 4\eta^2\left\langle\frac{\partial u_r(r_1,\theta_1)}{\partial r}\,\frac{\partial u_r(r_2,\theta_2)}{\partial r}\right\rangle - 2\eta\left\langle p(r_1,\theta_1)\frac{\partial u_r(r_2,\theta_2)}{\partial r}\right\rangle - 2\eta\left\langle\frac{\partial u_r(r_1,\theta_1)}{\partial r}p(r_2,\theta_2)\right\rangle \quad (17.44)$$

where B_p is the correlation function of the pressure derived in Theorem 17.4.

Taking the radial derivative we get

$$\frac{\partial u_r}{\partial r}(r,\theta) = \frac{1}{2\pi R}\int_0^{2\pi} \times\left\{-\rho^2 + \sum_{k=1}^{\infty}\left[(k+1)(k-2)\rho^{k+2} - k(k-1)\rho^k\right]\cos[k(\theta-\varphi)]\right\} \times g_r(\varphi)d\varphi + \frac{1}{2\pi\,R}\int_0^{2\pi}\sum_{k=1}^{\infty}\left[k(k+1)\rho^{k+2} - k(k-1)\rho^k\right] \times \sin[k(\theta-\varphi)]\,g_\theta(\varphi)d\varphi. \quad (17.45)$$

For the pressure we have

$$p(r,\theta) = \frac{2\eta}{\pi R}\int_0^{2\pi} \times \left\{ \sum_{k=1}^{\infty} k\rho^{k+1} \cos[(k+1)(\theta-\varphi)]\, g_r(Re^{i\varphi}) + \sum_{k=1}^{\infty} k\rho^{k+1} \sin[(k+1)(\theta-\varphi)]\, g_\theta(Re^{i\varphi}) \right\} d\varphi. \tag{17.46}$$

From (17.45) we find

$$4\eta^2 \left\langle \frac{\partial u_r(r_1,\theta_1)}{\partial r} \frac{\partial u_r(r_2,\theta_2)}{\partial r} \right\rangle = \frac{2\eta^2\rho_1^2\rho_2^2}{\pi R^2} + \frac{4\eta^2\rho_1^2\rho_2^2}{\pi R^2}\cos(\theta_2-\theta_1) + \frac{\eta^2}{\pi R^2}\sum_{k=1}^{\infty} \rho_1^{k+1}\rho_2^{k+1}\Lambda_{k+1}(\rho_1,\rho_2)\cos[(k+1)(\theta_2-\theta_1)]. \tag{17.47}$$

Here

$$\Lambda_k(\rho_1,\rho_2) = \lambda_1(k,\rho_1)\lambda_1(k,\rho_2) + \lambda_2(k,\rho_1)\lambda_2(k,\rho_2),$$

where

$$\lambda_1(k,\rho) = (k+1)(k-2)\rho^2 - k(k-1), \quad \lambda_2(k,\rho) = k(k+1)\rho^2 - k(k-1).$$

Finally, from (17.45) and (17.46) we find that the sum of two last terms in (17.44) has the form

$$-2\eta \left\langle p(r_1,\theta_1)\frac{\partial u_r(r_2,\theta_2)}{\partial r} \right\rangle - 2\eta \left\langle \frac{\partial u_r(r_1,\theta_1)}{\partial r} p(r_2,\theta_2) \right\rangle = -\frac{2\eta^2}{\pi R^2}\sum_{k=1}^{\infty} k\, \rho_1^{k+1}\rho_2^{k+1}\hat{\Lambda}_{k+1}(\rho_1,\rho_2)\cos[(k+1)(\theta_2-\theta_1), \tag{17.48}$$

where

$$\hat{\Lambda}_k(\rho_1,\rho_2) = \lambda_1(k,\rho_1) + \lambda_2(k,\rho_1) + \lambda_1(k,\rho_2) + \lambda_2(k,\rho_2).$$

From the series representations (17.37), (17.47), and (17.48) we finally obtain

$$B_{\sigma_{rr}} = \frac{2\eta^2}{\pi R^2}\{\rho_1^2\rho_2^2 + 2\rho_1^3\rho_2^3\cos(\theta_2-\theta_1)\} + \frac{\eta^2}{\pi R^2} \times \sum_{k=1}^{\infty} \left[k^2 + \Lambda_{k+1}(\rho_1,\rho_2) - 2k\hat{\Lambda}_{k+1}(\rho_1,\rho_2)\right]\rho_1^{k+1}\rho_2^{k+1} \cos[(k+1)(\theta_2-\theta_1)]. \tag{17.49}$$

17.5 Interior Stokes problem

Let us now consider the interior Stokes problem in the disc D_R^- of radius R centered at the origin. On the boundary, which is the circle $S(0, R) = \partial D_R$, the velocity components g_1 , g_2 are prescribed, and the velocity vector $\mathbf{u}(\mathbf{x}) = (u_1(\mathbf{x}), u_2(\mathbf{x}))$ and the pressure $p(\mathbf{x})$ are to be found from the Stokes problem:

$$\begin{aligned} &\eta\Delta\mathbf{u}(\mathbf{x}) = \nabla p(\mathbf{x}), &&\mathbf{x} \in D_R^-, \\ &\mathrm{div}\ \mathbf{u}(\mathbf{x}) = 0, &&\mathbf{x} \in D_R^-, \\ &\mathbf{u}(\mathbf{x}') = \mathbf{g}(\mathbf{x}'), &&\mathbf{x}' \in S(0, R), \end{aligned} \tag{17.50}$$

where η is the dynamic viscosity, and the velocity components g_1 and g_2 are supposed to be zero mean random processes on the circle $S(0, R)$.

We use again polar coordinates, $r \le R$ is the radial coordinate, and θ is the angular coordinate. The Poisson integral formula for the interior problem has the form analogous to (17.3) (e. g., see [263])

$$\begin{pmatrix} u_r(r,\theta) \\ u_\theta(r,\theta) \end{pmatrix} = \int_0^{2\pi} \begin{pmatrix} G_{11}(r;\theta-\varphi) & G_{12}(r;\theta-\varphi) \\ G_{21}(r;\theta-\varphi) & G_{22}(r;\theta-\varphi) \end{pmatrix} \begin{pmatrix} g_r(R\,e^{\mathrm{i}\varphi}) \\ g_\theta(R\,e^{\mathrm{i}\varphi}) \end{pmatrix} d\varphi,$$

where the entries G_{ij} $(i, j = 1, 2)$ have the form

$$G_{11}(r;\theta) =$$
$$\cos\theta\,K(r,\theta) + \frac{R^2-r^2}{2r^2}\left\{\cos\theta\left(r\frac{\partial}{\partial r}\,K(r,\theta)\right) - \sin\theta\,\frac{\partial}{\partial\theta}K(r,\theta)\right\} - \frac{R^2-r^2}{2\pi Rr},$$

$$G_{12}(r;\theta) =$$
$$\sin\theta\,K(r,\theta) + \frac{R^2-r^2}{2r^2}\left\{\sin\theta\left(r\frac{\partial}{\partial r}\,K(r,\theta)\right) + \cos\theta\,\frac{\partial}{\partial\theta}K(r,\theta)\right\},$$

$$G_{21}(r;\theta) =$$
$$-\sin\theta\,K(r,\theta) + \frac{R^2-r^2}{2r^2}\left\{\sin\theta\left(r\frac{\partial}{\partial r}\,K(r,\theta)\right) + \cos\theta\,\frac{\partial}{\partial\theta}K(r,\theta)\right\},$$

$$G_{22}(r;\theta) =$$
$$\cos\theta\,K(r,\theta) - \frac{R^2-r^2}{2r^2}\left\{\cos\theta\left(r\frac{\partial}{\partial r}\,K(r,\theta)\right) - \sin\theta\,\frac{\partial}{\partial\theta}K(r,\theta)\right\} + \frac{R^2-r^2}{2\pi Rr}.$$

The pressure $p(r,\theta)$ is related to the velocities $u_r(R,\theta)$ and $u_\theta(R,\theta)$ on the boundary by analogous Poisson-type integral formula:

$$\begin{aligned} p(r,\theta) = &-\frac{2\eta}{r}\left\{\int_0^{2\pi} P_r(r;\theta-\varphi)u_r(Re^{\mathrm{i}\varphi})\,d\varphi + \int_0^{2\pi} P_\theta(r;\theta-\varphi)u_\theta(Re^{\mathrm{i}\varphi})d\varphi\right\} \\ &+ \frac{1}{2\pi}\int_0^{2\pi} p(R,\varphi)\,d\varphi \end{aligned} \tag{17.51}$$

where

$$P_r(r;\theta) = \cos\theta \left(r \frac{\partial}{\partial r} K(r,\theta) \right) - \sin\theta \frac{\partial}{\partial \theta} K(r,\theta),$$

$$P_\theta(r;\theta) = \sin\theta \left(r \frac{\partial}{\partial r} K(r,\theta) \right) + \cos\theta \frac{\partial}{\partial \theta} K(r,\theta).$$

The function $K(r,\theta)$ is the Poisson kernel for the harmonic equation in the circular domain of radius R:

$$K(r,\theta) = \frac{R^2 - r^2}{2\pi(R^2 + r^2 - 2Rr\cos\theta}, \quad 0 \le r \le R.$$

We define here $\rho = r/R$, so that $\rho < 1$, and in the notation

$$\lambda_{11}(\rho,k) = \frac{1}{\rho} + \frac{k(1-\rho^2)}{2\rho}, \quad \lambda_{12}(\rho,k) = -\frac{k(1-\rho^2)}{2\rho},$$
$$\lambda_{21}(\rho,k) = -\frac{(k+2)(1-\rho^2)}{2\rho}, \quad \lambda_{22}(\rho,k) = \rho - \frac{k(1-\rho^2)}{2\rho},$$
$$k = 1, 2 \ldots, \tag{17.52}$$

we obtain the series expansions in the form

$$G_{11}(r;\theta) = \frac{\rho}{2\pi} + \frac{1}{\pi}\sum_{k=1}^{\infty} \lambda_{11}(\rho,k)\rho^k \cos(k\theta),$$

$$G_{12}(r;\theta) = \frac{1}{\pi}\sum_{k=1}^{\infty} \lambda_{12}(\rho,k)\,\rho^k \sin(k\theta),$$

$$G_{21}(r;\theta) = \frac{1}{\pi}\sum_{k=1}^{\infty} \lambda_{21}(\rho,k)\,\rho^k \sin(k\theta),$$

$$G_{22}(r;\theta) = \frac{\rho}{2\pi} + \frac{1}{\pi}\sum_{k=1}^{\infty} \lambda_{22}(\rho,k)\,\rho^k \cos(k\theta).$$

Now we note that all the statements presented for the exterior problem hold true for the interior problem up to the Section 17.4; we just have to replace the coefficients λ_{ij} in the functions G_{ij} by the expressions (17.5). The pressure however should be analyzed separately.

Indeed, the Poisson formula (17.51) for the interior problem has a structure different from that of the exterior problem; compare with (17.5).

After substituting the series expansions for the kernels we find that

$$p(r,\theta) = -\frac{\eta}{\pi R}\left\{\int_0^{2\pi}\left[1+\sum_{k=1}^{\infty}(k+1)\rho^k\cos[(k(\theta-\varphi)]\right]u_r(\varphi)d\varphi + \int_0^{2\pi}\left[\sum_{k=1}^{\infty}(k+1)\rho^k\sin[(k(\theta-\varphi)]\right]u_\theta(\varphi)d\varphi\right\} + \frac{1}{2\pi}\int_0^{2\pi}p(R,\varphi)d\varphi. \quad (17.53)$$

Integrating both parts of this equality from $\theta = 0$ to 2π we conclude that

$$\int_0^{2\pi} u_r(R,\theta)d\theta = 0, \quad (17.54)$$

which is just the consistency condition for the interior Dirichlet problem of the Stokes equation

$$\int_{S(0,R)} \vec{u}\cdot\mathbf{n}\,ds = 0,$$

which does not appear in the case of the exterior Dirichlet problem.

From this we see that the random process on the boundary $p(R,\theta)$ is homogeneous with respect to the angular coordinate, with a constant mean $p_0 = \bar{p}(R,\cdot)$, and moreover, by (17.53), $\bar{p}(r,\theta) = p_0$ inside the disc.

From the representation (17.53) we can find the correlation function of the pressure. The last term $\bar{P} = \frac{1}{2\pi}\int_0^{2\pi} p(R,\varphi)d\varphi$ in this expression is just a random variable which is normally distributed, so it does not affect the angular behavior of the correlation function. Note also that due to the property (17.54) we can assume without loss of generality that the mean of the pressure is zero. Thus it is sufficient to find the correlation function

$$B_p = \langle (p(r_1,\theta_1)-\bar{P})(p(r_2,\theta_2)-\bar{P})\rangle.$$

Direct evaluation yields by (17.53)

$$B_p = \frac{2\eta^2}{\pi R^2}\sum_{k=1}^{\infty}(k+1)^2\rho_1^k\rho_2^k\cos[k(\theta_2-\theta_1)].$$

17.6 Numerical results

In this section we give the results of numerical calculations for the Stokes problem with white noise boundary excitations. The angular and radial dependencies of the correlations for the velocity, pressure, and radial stress are presented. We also compare these functions for the interior and exterior problems.

Let us begin with the velocity correlation tensor for the exterior Stokes problem. In Figure 17.1 we present the angular behavior of the correlation functions B_{11} (left panel) and B_{22} (right panel), for different values of ρ_1 and ρ_2, calculated according to the series representation (17.18), with the number of retained terms $n = 40$. It should be noted that this number of terms ensures that the error is less than 0.1 %. We have also calculated these functions by a direct Monte Carlo algorithm based on the K-L representation (17.20), (17.21), but we do not plot these functions since the curves are practically coincident with the results obtained by the series representation (17.18). To keep the error of about 0.1 %, we had to increase the number of the retained terms n as ρ_1 and ρ_2 go closer to 1, i. e., when the correlations are calculated closer to the boundary. However in all our calculations, n was not larger than 200.

An interesting feature of the correlation function B_{22} is that, unlike B_{11}, there is a peak around the angle $\theta = 15°$ for a certain range of the radial values of ρ_1 and ρ_2. In Figure 17.2, we present these correlations vs. the angle θ, for fixed and equal values of ρ_1 and ρ_2. It is clearly seen that as the distance to the boundary increases (ρ decreases), the fluctuation intensity decreases while the correlation length increases. In Figure 17.3 the cross-correlations B_{12} and B_{21} are shown. Here we can see that the property $B_{ij}(\theta, \rho_1, \rho_2) = B_{ji}(\theta, \rho_2, \rho_1)$ holds true (left panel). In the right panel of Figure 17.3 we show the function B_{21} for different values of ρ_1, ρ_2.

Let us turn to the pressure correlation function. To illustrate the rate of convergence of the K-L expansion, we compare in Figure 17.4 the pressure correlation $B_p(\rho_1, \rho_2; \theta)$ for different values of the retained terms n in the K-L series representation (17.37), with the exact result presented by (17.38). It is seen that a small number of terms is sufficient for a good approximation even near the boundary where the decrease of correlations is very fast (see the right panel). We can analyze these fast decorrelations by comparing them with the results presented in the left and right panels of Figure 17.5. Here it is also seen that the larger the distance to the boundary is, the smaller the intensity fluctuations and the larger the correlation length.

The rate of convergence of the K-L expansion is closely related to the smoothness of the correlation kernel and L, the correlation length of the process. For example, in [132] it was reported that for the particular case $B(x_1, x_2) = \sigma\, e^{-|x_2 - x_1|/L}$ an upper bound for the relative error in variance ε of the process represented by its K-L expansion is given by $\varepsilon \le \frac{4}{\pi^2} \frac{1}{n} \frac{1}{L}$, where n is the number of retained terms.

The correlation function of the radial stress, $B_{\sigma_{rr}}$, calculated according to the K-L representation (17.49), is presented in Figure 17.6 for different values of ρ_1 and ρ_2: we fix the value $\rho_1 = 0.99$, and vary the value of ρ_2 from $\rho_2 = 0.1$ (small values in the left panel), to $\rho_2 = 0.7$ (large values in the right panel). Here we can see how strong the correlation function is changed when varying the radial coordinate.

Let us consider the interior Stokes problem. The correlation function of the velocity is presented in Figures 17.7 and 17.8. In Figure 17.7 we present the velocity correlation functions B_{11} (left panel) and B_{22} (right panel) for equal values of the radial coordinate, where $\rho = \rho_1 = \rho_2$ is varied from 0.4 to 0.85. The same curves are shown

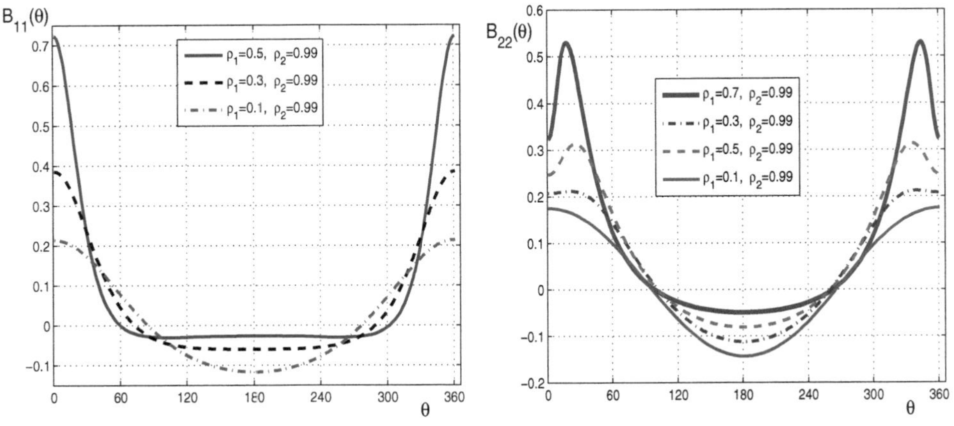

Figure 17.1. Angular correlation function for B_{11} (left panel) and B_{22} (right panel), for different values of ρ_1 and ρ_2.

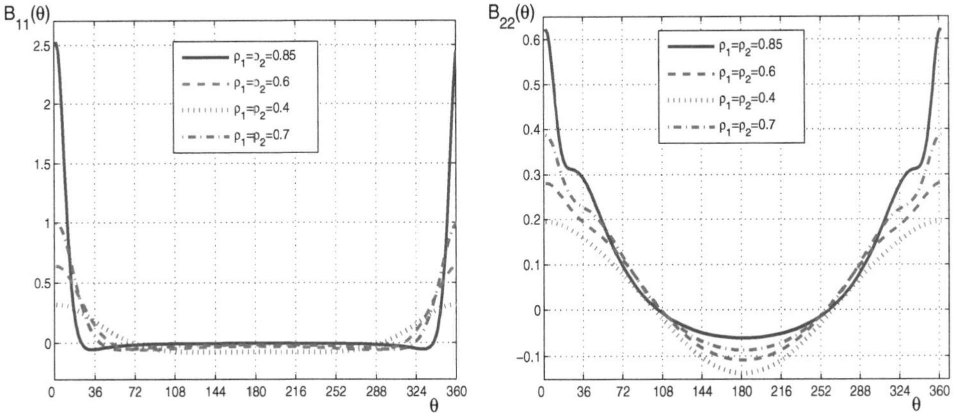

Figure 17.2. Angular correlation function for B_{11} (left panel) and B_{22} (right panel), for equal values of ρ_1 and ρ_2.

in Figure 17.8, but for different values of the radial coordinates: we fix $\rho_2 = 0.99$, and vary ρ_1 from 0.1 to 0.7. Comparing the results for these two cases, we note that there is no qualitative difference between the correlations B_{11} in Figure 17.7 and 17.8, while the functions B_{22} do differ dramatically. This illustrates how complex the behavior of inhomogeneous random fields can be.

This is also illustrated by the results presented in Figure 17.9 where we show the cross-correlations B_{12} and B_{21} of equal (left panel) and different (right panel) values of ρ_1, ρ_2.

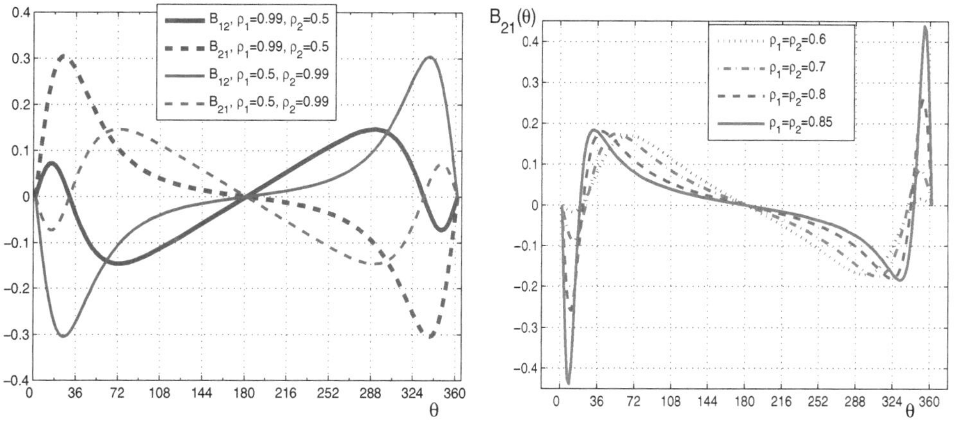

Figure 17.3. Angular correlation functions B_{12} and B_{21}, for different values of ρ_1 and ρ_2 (left panel), and B_{21}, for equal values of ρ_1 and ρ_2 (right panel).

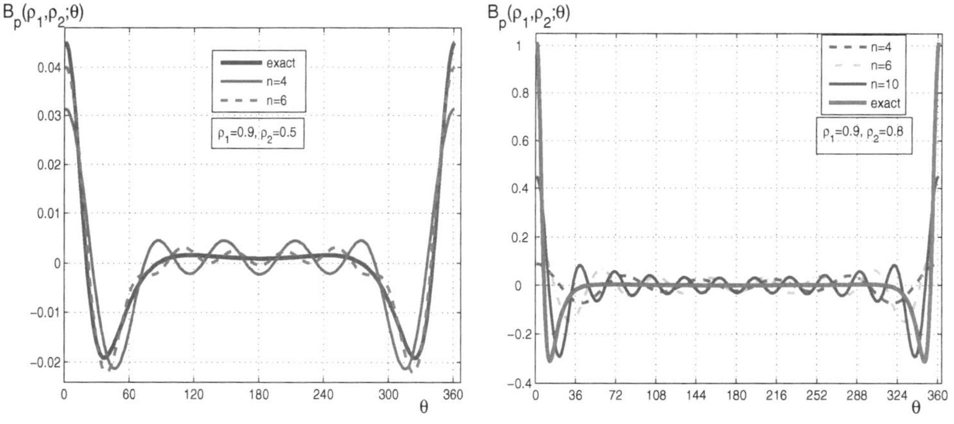

Figure 17.4. Comparison of the K-L-approximation with the exact result: angular correlation function of the pressure $B_p(\rho_1, \rho_2; \theta)$, for different values of ρ_1 and ρ_2, and for different number of terms in the K-L expansion.

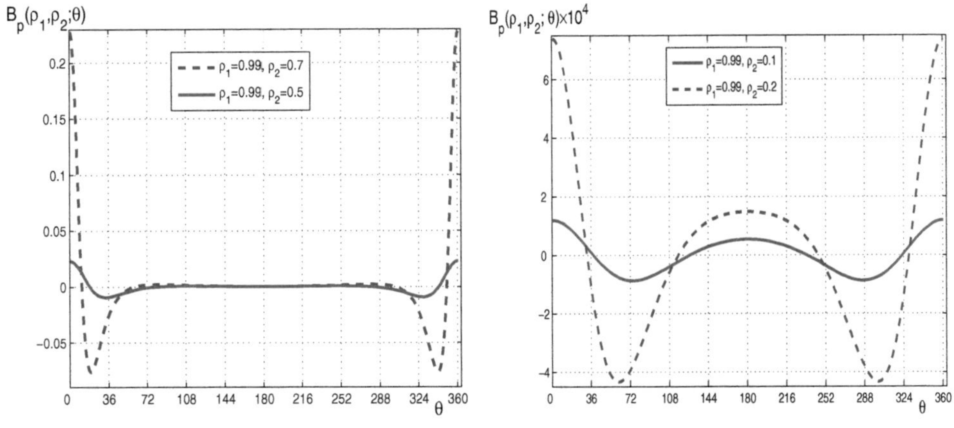

Figure 17.5. Angular correlation function of the pressure $B_p(\rho_1, \rho_2; \theta)$, for different values of ρ_1 and ρ_2.

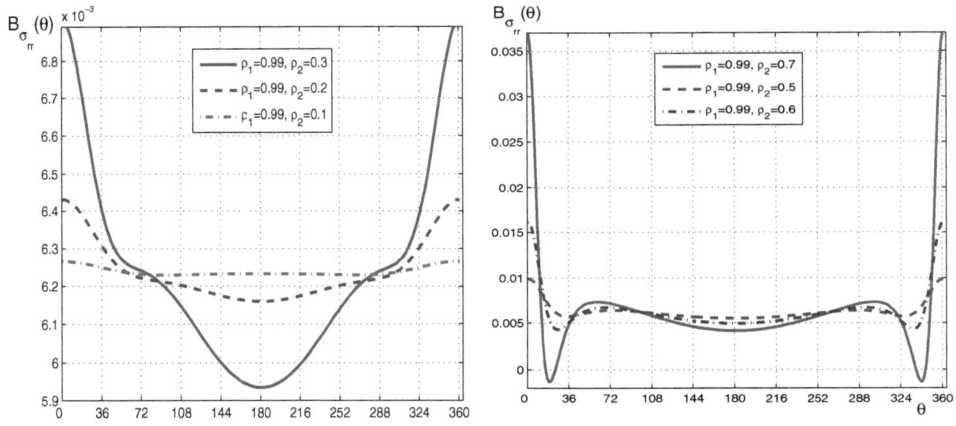

Figure 17.6. Angular behavior of the correlation function of the radial stress $B_{\sigma_{rr}}$, for different values of ρ_1 and ρ_2 with ρ_1 fixed near the boundary, and varying second point: far from the boundary (small values of ρ_2, left panel), and closer to the boundary (larger values of ρ_2, right panel).

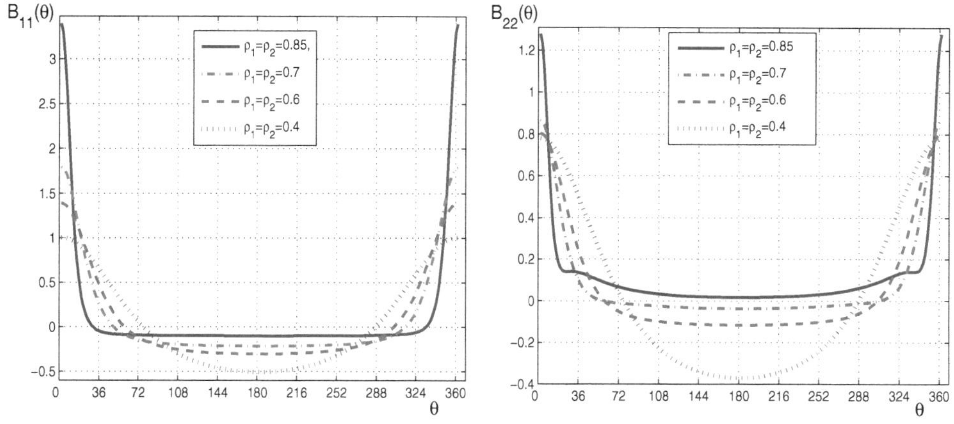

Figure 17.7. The velocity correlation functions B_{11} (left panel), and B_{22} (right panel), for equal values of ρ_1 and ρ_2. Interior Stokes problem.

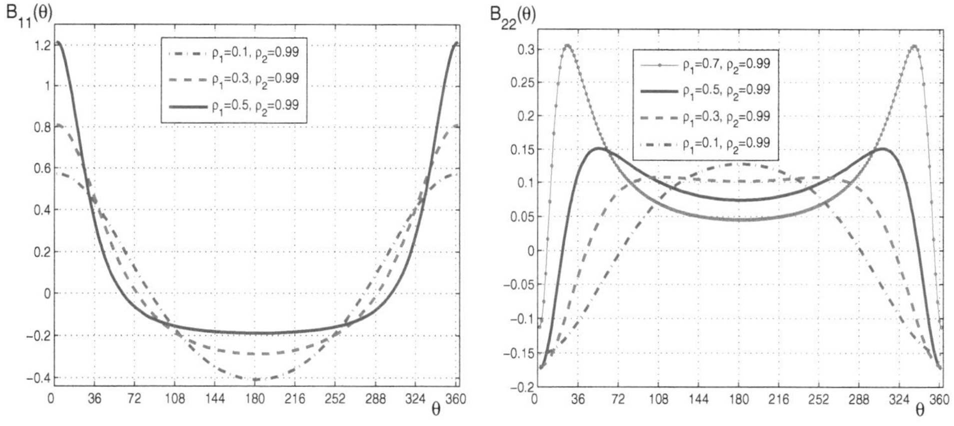

Figure 17.8. The velocity correlation functions B_{11} (left panel), and B_{22} (right panel), for different values of ρ_1 and ρ_2. Interior Stokes problem.

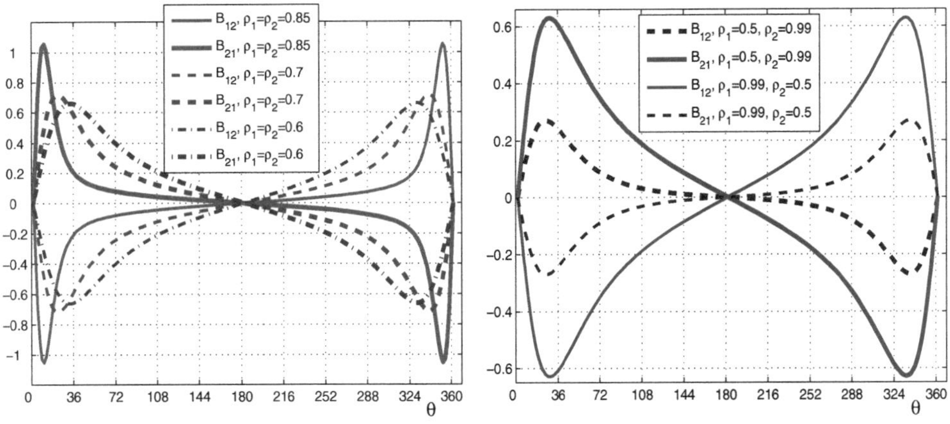

Figure 17.9. The velocity cross-correlation functions B_{12} and B_{21}, for equal values of ρ_1 and ρ_2 (left panel), and different values of ρ_1 and ρ_2 (right panel). Interior Stokes problem.

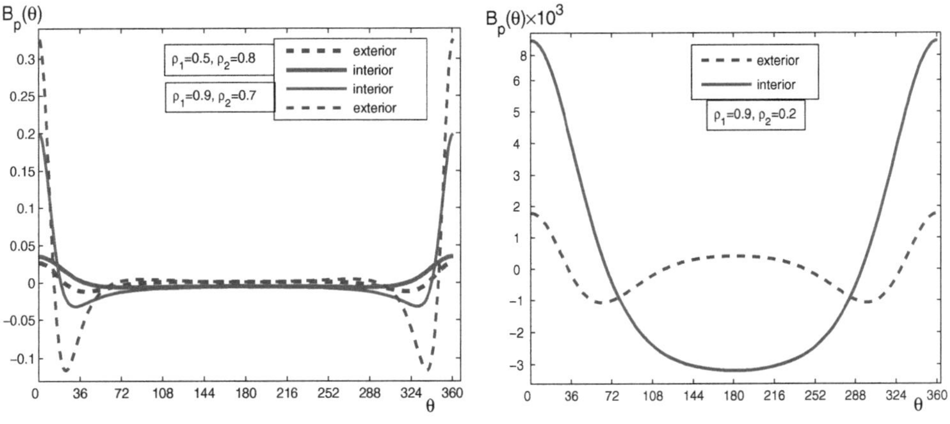

Figure 17.10. Comparison of the pressure correlation functions for the exterior and interior Stokes problems, for different values of ρ_1 and ρ_2.

It is also interesting to compare the correlation functions for the interior and exterior problems, which can be done by a comparative analysis of results presented in Figures 17.1–17.6 and those shown in Figures 17.7–17.9. We present here only the pressure correlation functions, see Figure 17.10, for different values of ρ_1, ρ_2.

Bibliography

[1] R. Ababou, D. McLaughlin, L. W. Gelhar, and A. F. B. Tompson, Numerical Simulation of Three-Dimensional Saturated Flow in Randomly Heterogeneous Porous Media, *Transport in Porous Media* **4** (1989), 549–565.

[2] M. Anders and M. Hori, Stochastic finite element method for elasto-plastic body, *International Journal for Numerical Methods in Engineering* **46**(11) (1999), 1897–1916.

[3] F. Anselmet, Y. Gange, E. Hopfinger, and R. A. Antonia, High-order velocity structure functions in turbulent shear flows, *J. Fluid Mech.* **140**(1) (1984), 63–95.

[4] N. Aubry, On the hidden beauty of the proper orthogonal decomposition. *Theoretical and Computational Fluid Dynamics* **2** (1991), 339–352.

[5] L. B. Axell and O. Liungman, A one-equation turbulence model for geophysical applications: comparison with data and the $k - \varepsilon$ model, *Environmental Fluid Mechanics*, **1** (2001), 71–106.

[6] D. Baldocchi, Flux footprints within and over forest canopies, *Boundary-Layer Meteorol.* **85** (1997), 273–292.

[7] J. M. Ball and J. Carr, The discrete coagulation-fragmentation equations: existence, uniqueness and density concentration, *J. Stat. Phys.* **61** (1990), 203–234.

[8] A. A. Bark, L. W. Gelhar, A. L. Gutjahr, and J. R. MacMillan, Stochastic analysis of spatial variability in subsurface flows, 1, Comparison of one- and three-dimensional flows, *Water Resour. Res.* **14**(2) (1978), 263–271.

[9] A. T. Barucha-Reid, *Random Integral Equations*, Academic Press, New York, 1972.

[10] G. K. Batchelor, Diffusion in a field of homogeneous turbulence. II. The relative motion of particles, *Proc. Camb. Phil. Soc.* **48** (1952), 345–362.

[11] J. Bear, *Dynamics of Fluids in Porous Media*, American Elsevier, New York, 1972.

[12] A. Bellin, P. Saladin, and A. Rinaldo, Simulation of dispersion in heterogeneous porous formation: Statistics, first-order theories, convergence of computations. *Water Resources Resources* **28**(9) (1992), 2211–2227.

[13] G. Berkooz, P. Holmes, and J. L. Lumley, The proper orthogonal decomposition in the analysis of turbulent flows. *Annu. Rev. Fluid Mech.* **25** (1993), 539–575.

[14] M. S. Borgas, Peak-to-Mean Fluctuations of Chemical Concentrations in Plumes in Turbulent Flows, Abstracts of the Workshop Stochastic models for turbulent transport processes, WIAS, Berlin, 2000, *Monte Carlo Methods and Applications*, **6**(4) (2000) 349–358.

[15] M. S. Borgas and B. L. Sawford, A family of stochastic models for two-particle dispersion in isotropic homogeneous stationary turbulence, *J. Fluid. Mech.* **279** (1994a), 69–99.

[16] M. S. Borgas and B. L. Sawford, Stochastic equations with miltifractal random increments for modeling turbulent dispersion. *Phys. Fluids* **6**(N2) (1994b), 618–633.

[17] M. S. Borgas and P. K. Yeung, Conditional fluid-particle accelerations in turbulence, *Theoret. Comput. Fluid Dynamics* **N11** (1998), 69–93.

[18] N. A. Buglanova and O. Kurbanmuradov, Convergence of the Randomized Spectral Models of Homogeneous Gaussian Random Fields, *Monte Carlo Methods and Appl.*, **1**(3) (1995), 173–201.

[19] N. L. Bysova, E. K. Garger, and V. N. Ivanov, Experimental studies of atmospheric diffusion and calculation of pollutant dispersion, *Gidrometeoizdat.* **L** (1991), 250 (in Russian).

[20] P. C. Chatwin and P. J. Sillivan, The relative diffusion of a cloud of passive contaminant in incompressible turbulent flow, *J. Fluid Mech.* **91** (1979), 337–355.

[21] F. H. Chaudhry and R. N. Moroney, A laboratory study of diffusion in stably stratified flow. *Atmos. Environ.* **7** (1973) 441–454.

[22] A. H. D. Cheng and O. E. Lafe, Boundary Element Solution for Stochastic Groundwater Flow: Random Boundary Condition and Recharge, *Water Resources Research* **27**(2) (1991), 231–242.

[23] D. A. Chin and T. Wang, An investigation of the validity of the first-order stochastic dispersion theories in isotropic porous media, *Water Resour. Res.* **28**(6) (1972), 1531–1542.

[24] A. J. Chorin, *Computational Fluid mechanics, Selected Papers*, Academic Press, Boston San Diego New York, 1989.

[25] C. Cameron, Relative efficiency of Gaussian stochastic process sampling procedures, *J. Comput. Phys.* **192**(2) (2003), 546–569.

[26] R. A. Carmona, S. A. Grishin, and S. A. Molchanov. Massively parallel simulations of motions in a Gaussian velocity field, in: R. J. Adler, P. Mueller, and B. L. Rozovskii (eds.) *Stochastic modelling in physical oceanography*, Progr. Prob. 39, pp. 47–68, Birkhäuser, Boston, 1996.

[27] C. K. Chui, *An Introduction to Wavelets*, Academic Press, San Diego, 1992.

[28] R. Courant and D. Hilbert, *Methods of Mathematical Physics*, vol. 2, Wiley, New York, 1989.

[29] H. Cramer and M. H. Leadbetter, *Stationary and related stochastic processes,* John Wiley, New York, 1967.

[30] J. H. Cushman, *Dynamics of Fluids in Hierarchical Porous Media*, Academic Press, London, 1990.

[31] R. C. Dalang and O. Levesque, Second-order linear hyperbolic SPDEs driven by isotropic Gaussian noise on a sphere, *The Annals of Probability* **32**(1B) (2004), 1068–1099.

[32] G. Dagan, Solute transport in heterogeneous porous formations, *J. Fluid Mech.* **145** (1984), 151–177.

[33] G. Dagan, A note of the higher-order corrections of the head covariances in steady aquifer flow, *Water Resour. Res.* **21** (1985), 573–578.

[34] G. Dagan, *Flow and Transport in Porous Formations*, Springer-Verlag, Berlin Heidelberg, 1989.

[35] G. Dagan, Spatial moments, Ergodicity, and Effective Dispersion, *Water Resour. Res.* **26**(6) (1990), 1281–1290.

[36] I. Daubechies, *Ten Lectures On Wavelets*, CBS-NSF Regional Conferences in Applied Mathematics 61, SIAM, 1992.

[37] K. Deimling, Sample solutions of stochastic ordinary differential equations, *Stoshastic Analysis and Appl.* **3** (1985), 15–21.

[38] F.-W. Deng and J. H. Cushman, On higher-order corrections to the flow velocity covariance tensor, *Water Resour. Res.*, **31** (1995), 1659–1672.

[39] G. Deodatis and M. Shinozuka, Simulation of stochastic processes by spectral representation, *Applied Mechanics Reviews* **44**(4) (1991), 191–204.

[40] M. W. Davis, Production of conditional simulations via the LU trangular decomposition of the covariance matrix, *Math. Geol.* **19** (1987), 91–98.

[41] C. R. Dietrich and G. N. Newsam, Fast and Exact Simulation of Stationary Gaussian Processes through Circulant Embedding of the Covariance Matrix, *Siam J. Sci. Comput.* **18**(4) (1997), 1088–1107.

[42] I. T. Drummond, S. Duane, and R. R. Horgan, Scalar diffusion in simulated helical turbulence with molecular diffusivity, *J. Fluid Mech.* **138** (1984), 75–91.

[43] S. Du, B. L. Sawford, J. D. Wilson, and D. J. Wilson, A determination of the Kolmogorov constant (C_0) for the Lagrangian velocity structure function, using a second-order Lagrangian stochastic model for decaying homogeneous, isotropic turbulence, *Phys. Fluids* **7** (1985) 3083–3090.

[44] P. A. Durbin, A stochastic model of two-particle dispersion and concentration fluctuations in homogeneous turbulence, *J. Fluid Mech.* **100** (1980), 279–302.

[45] P. A. Durbin, Analysis of the decay of temperature fluctuations in isotropic turbulence, *Phys. Fluids* **25**(8) (1982), 1328–1332.

[46] G. D. Egbert and M. B. Baker, Comments on paper "The effect of Gaussian particle-pair distributions functions in the statistical theory of concentration fluctuations in homogeneous turbulence" by B. L. Sawford (*Q. J. R. Met. Soc.* **109** (1983), 339–353), *Q. J. R. Met. Soc.* **110** (1984), 1195–1199.

[47] F. W. Elliott Jr, D. J. Horntrop, and A. J. Majda, A Fourier-wavelet Monte Carlo method for fractal random fields, *J. Comp. Phys.* **132**(2) (1997a), 384–408.

[48] F. W. Elliott Jr, D. J. Horntrop, and A- J. Majda, Monte Carlo methods for turbulent tracers with long range and fractal random velocity fields, *Chaos* **7**(1) (1997b), 39–48.

[49] F. W. Elliott Jr and A. J. Majda, A wavelet Monte Carlo method for turbulent diffusion with many spatial scales, *J. Comp. Phys.* **113**(1) (1994), 82–111.

[50] F. W. Elliott Jr and A. J. Majda, A new algorithm with plane waves and wavelets for random velocity fields with many spatial scales, *J. Comp. Phys.* **117** (1995), 146–162.

[51] F. W. Elliott Jr and A. J. Majda, Pair dispersion over an inertial range spanning many decades, *Phys. Fluids* **8**(4) (1996), 1052–1060.

[52] F. Farkhooi, M. F. Strube-Bloss, and M. P. Nawrot, Serial correlation in neural spike trains: Experimental evidence, stochastic modeling, and single neuron variability, *Phys. Review E* **79** (2009), 021905.

[53] B. F. Farell and P. J. Ioannou, Stochastic forcing of the linearized Navier–Stokes equations, *Phys. Fluids A* **5**(11) (1993), 2600–2609.

[54] G. A. Fenton,, Error evaluation of three random field generators, *ASCE J. Engineering Mechanics* **120**(12) (1994), 2487–2497.

[55] T. K. Flesch, The footprint for flux measurements, from backward Lagrangian stochastic models, *Boundary-Layer Meteorology*, **78** (1996), 399–404.

[56] T. K. Flesch and G. D. Wilson, A two-dimensional trajectory-simulation model for non-Gaussian, inhomogeneous turbulence within plant canopies, *Boundary-Layer Meteorology* **61** (1992), 349–374.

[57] T. K. Flesch, J. D. Wilson, and E. Yee, Backward-time Lagrangian stochastic dispersion models and their application to estimate gaseous emissions, *J. Applied Meteorology* **34** (1995), 1320–1332.

[58] R. A. Freeze, A stochastic-conceptual analysis of one-dimensional groundwater flow in nonuniform homogeneous media, *Water Resources Research* **11**(5) (1975), 725–741.

[59] M. I. Freidlin and A. D. Wentzell, *Random Perturbations of Dynamical Systems*, Springer-Verlag, New York Berlin, 1983.

[60] A. Friedman, *Stochastic Differential Equations and Applications*, Academic Press, New York, 1975.

[61] U. Frish, *Turbulence,* Cambridge University Press, 1996.

[62] J. C. H. Fung, Shear flow turbulence structure and its Lagrangian statistics, *Fluid Dynamics Research* **17** (1996), 147–180.

[63] J. C. H. Fung, J. C. R. Hunt, N. A. Malik, and R. J. Perkins.,Kinematic simulation of homogenous turbulence by unsteady random Fourier modes, *J. Fluid Mech.* **236** (1992), 281–318.

[64] E. K. Garger and G. P. Zhukov, On vertical pollution dispersion from a local source in the atmospheric surface layer, *Izv. Acad. Nauk SSSR, ser. FAO* **22**(2) (1986), 115–123 (in Russian).

[65] L. W. Gelhar, *Stochastic Subsurface Hydrology*, Prentice-Hall, Englewood Cliffs, NJ, 1993.

[66] L. J. Gelhar and C. L. Axness, Three-dimensional stochastic analysis of macrodispersion in aquifers, *Water Resour. Res.* **19** (1983), 161–180.

[67] L. W. Gelhar, A. L. Gutjahr, and R. L. Naff, Stochastic analysis of macro-dispersion in a stratified aquifer, *Water Resour. Res.* **19** (1979), 161–180.

[68] W. Gerstner and W. M. Kistler, *Spiking Neuron Models. Single neurons, populations, placticity*, Cambridge University Press, 2002.

[69] R. G. Ghanem and P. D. Spanos, *Stochastic Finite Elements: A Spectral Approach*, Springer Verlag, New York, 1991.

[70] I. I. Gikhman and A. V. Skorohod, *Introduction to the Theory of Random Processes*, W. B., Sounders Company, Philadelphia, 1969.

[71] I. I. Gikhman and A. V. Skorohod, *Stochastic Differential Equations*, 4th ed., Springer-Verlag, Heidelberg New York Berlin, 1995.

[72] A. Giordano and M. Uhrig, Human face recognition technology using the Karhunen–Loève expansion technique, Regis University. Denver, Colorado, available at http://www.rose-hulman.edu/mathjournal/archives/2006/vol7-n1/paper11/v7n1-11pd.pdf (last accessed Sept. 22, 2012).

[73] A. S. Gurvich, profiles of the wind velocity and temperature in the atmospheric surface layer, *Izv. Acad. Nauk SSSR, ser. FAO* **1**(1), 1965 (in Russian).

[74] U. Frisch, *Turbulence*, Cambridge University Press, 1996.

[75] P. Hartman, *Ordinary Differential Equations*, Wiley, New York, 1964.

[76] A. E. Hassan, J. H. Cushman, and J. W. Delleur, A Monte Carlo assessment of Eulerian flow and transport perturbation models, *Water Resour. Res.* **34** (1998), 1143–1163.

[77] B. M. O. Heppe, Generalized Langevin equation for relative turbulent dispersion, *J. Fluid Mech.* **357** (1998), 167–198.

[78] E. Hernandez and G. Weiss, *A First Course on Wavelets*, CRC Press, Inc., London New York Washington, 1996.

[79] T. W. Horst and J. C. Wei,, Footprint estimation for scalar flux measurements in the atmospheric surface layer, *Boundary-Layer Meteorology* **59** (1992), 279–296.

[80] K.-C. Hsu, A general method for obtaining analytical expressions for the first-order velocity covariance in heterogeneous porous media, *Water Resources Research* **35**(7) (1999), 2273–2277.

[81] K.-C. Hsu and S. P. Neuman, Second-order expressions for velocity moments in two and three-dimensional statistical anisotropic media, *Water Resources Research* **33**(4) (1997), 625–637.

[82] K.-C. Hsu, D. Zhang, and S. P. Neuman, Higher-order effects on flow and transport in randomly heterogeneous porous media, *Water Resources Research* **32**(3) (1996), 571–582.

[83] S. P. Huang, S. T. Quek, and K. K. Phoon, Convergence study of the truncated Karhunen–Loève expansion for simulation of stochastic processes. International Journal for numerical methods in engineering, **52** (2001), 1029–1043.

[84] I. A. Ibragimov and Y. V. Linnik, *Independent and stationary dependent variables*, Nauka, Moscow, 1965 (in Russian).

[85] I. A. Ibragimov and Y. A. Rozanov, *Gaussian stochastic processes,* Nauka, Moscow, 1970 (in Russian).

[86] V. P. Iljin, *Finite difference and finite volume element methods for elliptic equations*, Institute of Comp. Math. amd Math. Geophysics, Russian Academy of Sciences, Siberian Branch, Novosibirsk, 2001 (in Russian).

[87] V. M. Kaganer, O. Brandt, and K K. Sabelfeld, X-ray diffraction peaks from partially ordered misfit dislocations, *Phys. Rev. B* **80** (2009), 033306.

[88] J. C. Kaimal and J. J. Finnigan, *Atmospheric Boundary Layer Flows. Their Structure and Measurement*, Oxford University Press, New York, 1994.

[89] J. Kaipio, V. Kolehmainen, E. Somersalo, and M. Vauhkonen, Statistical inversion and Monte Carlo sampling methods in electrical impedance tomography. Inverse problems, **16** (2000), 1487–1522.

[90] J. Kaiser, New network aims to take the World's CO2 pulse, *Science* **281** (1998), 506–507.

[91] H. Kaplan and N. Dinar, A three-dimensional stochastic model for concentration fluctuations statistics in isotropic homogeneous turbulence, *J. Comp. Physics* **79**(2) (1988), 317–335.

[92] K. Karhunen, Zur Spektraltheorie stochastischer prozesse, *Ann. Acad. Sci. Fennicae* **A1** (1946), 34.

[93] R. Z. Khas'minsky, *Stochastic stability of Differential equations,* Sijthoff and Noordhoff, Alpen aan den Rijn, 1980.

[94] L. S. Katafygiotis, A. Zerva, and A. A. Malyarenko, Simulation of homogeneous and partially isotropic random fields, *J. Engineering Mechanics* **125**(10) (1999), 1180–1189.

[95] G. K. Katul and J. D. Albertson,, An investigation of higher-order closure models for a forested canopy, *Boundary-Layer Meteorology* **89** (1998), 47–74.

[96] V. I. Klyatskin, *Stochastic equations and waves in randomly inhomogeneous media*, Nauka, Moscow, 1980 (in Russian).

[97] A. A. Kolodko, K. K. Sabelfeld, and W. Wagner, A stochastic method for solving Smoluchowski coagulation equation, *Mathematics and Computers in Simulation* **49** (1999), 57–79.

[98] D. Kolyukhin and K. Sabelfeld, Stochastic Eulerian model for the flow simulation in porous media. Unconfined aquifers, *Monte Carlo Mand Applications* **10**(3–4) (2004), 345–357.

[99] D. Kolyukhin and K. Sabelfeld, Stochastic flow simulation in 3D Porous media, *Monte Carlo Methods and Applications* **11**(1) (2005), 15–38.

[100] D. D. Kosambi, Statistics in function space, *J. Indian Math. Soc.* **7** (1943), 76–88.

[101] R. H. Kraichnan, Dispersion of particle pairs in homogeneous turbulence, *Phys. Fluids* **9** (1966), 1728–1752.

[102] R. H. Kraichnan, Diffusion by a random velocity field, *Phys. Fluids* **13**(1) (1970), 22–31.

[103] P. R. Kramer, A Review of Some Monte Carlo Simulation Methods for Turbulent Systems, *Monte Carlo Methods and Appl.* **7**(3–4) (2001), 229–244.

[104] P. Kramer, O. Kurbanmuradov, and K. Sabelfeld, Comparative Analysis of Multiscale Gaussian Random Field Simulation Algorithms, *J. Computational Physics* **226** (2007), 897–924.

[105] N.V. Krylov and B. L. Rozovskii, Stochastic partial differential equations and diffusion processes, *Uspekhi Mat. Nauk.* **37** (1982), 75–95 (in Russian).

[106] O. A. Kurbanmuradov, Weak Convergence of Randomized Spectral Models of Gaussian Random Vector Fields, *Bull. Novosibirsk Computing Center, Numerical Analysis* **4**, (1993) 19–25.

[107] O. A. Kurbanmuradov, Weak Convergens of Approximate Models of Random Fields. *Russian Journal of Numerical Analysis and Mathematical Modelling* **10**(6) (1995a), 500–517.

[108] O. A. Kurbanmuradov, A new Lagrangian model of two-particle relative turbulent dispersion, *Monte Carlo Methods and Applications* **1**, (1995b), 101–136.

[109] O. A. Kurbanmuradov, A 3D stochastic model of relative dispersion of particle pairs in local-isotropic turbulence, *Monte Carlo Methods and Applications* **1**(4) (1995c), 301–329.

[110] O. A. Kurbanmuradov, A Lagrangian stochastic model of two-particle relative dispersion in high-Reynolds number turbulence, *Monte Carlo Methods and Applications* **3**(1) (1997), 37–52.

[111] O. A. Kurbanmuradov, U. Rannik, K. K. Sabelfeld, and T. Vesala, Direct and adjoint Monte Carlo for the footprint problemm *Monte Carlo Methods and Applications* **5**(2) (1999), 85–112.

[112] O.A. Kurbanmuradov and K. K. Sabelfeld, Direct and backward random walk algorithms for solving of the problem of admixture spread in random velocity field, preprint no. 506, Computing Center, Novosibirsk, 1984 (in Russian).

[113] O. A. Kurbanmuradov and K. K. Sabelfeld, Statistical modelling of turbulent motion of particles in random velocity fields, *Soviet J. Numerical Analysis and Math. Modelling* **4**(1) (1989), 53–68.

[114] O. A. Kurbanmuradov and K. K. Sabelfeld, Stochastic Lagrangian models of relative dispersion of a pair of fluid particles in turbulent flows, *Monte Carlo Methods and Applications* **1**(2) (1995), 101–136.

[115] O. A. Kurbanmuradov and K. K. Sabelfeld, Lagrangian stochastic models for turbulent dispersion in the atmospheric boundary layer, *Boundary-Layer Meteorolology* **97**(2) (2000a), 191–218.

[116] O. A. Kurbanmuradov and K. K. Sabelfeld, Coagulation of aerosol particles in intermittent turbulent flows, *Monte Carlo Methods and Applications* **6**(3) (2000b), 211–253.

[117] O. A. Kurbanmuradov and K. K. Sabelfeld, Convergence of Fourier-Wavelet models for Gaussian random processes, *SIAM J. Numerical Analysis* **46**(6) (2008), 3084–3112.

[118] O. A. Kurbanmuradov and K. K. Sabelfeld, Stochastic spectral and Fourier-wavelet methods for vector Gaussian random field, *Monte Carlo Methods and Applications* **12**(5–6) (2006) , 395–446.

[119] O. A. Kurbanmuradov, K. Sabelfeld, and D. Koluhin, Stochastic Lagrangian models for two-particle motion in turbulent flows. Numerical results, *Monte Carlo Methods Appl.* **3**(3) (1997), 199–223.

[120] O. A. Kurbanmuradov, K. Sabelfeld, O. Smidts, and H. A. Vereecken, Lagrangian stochastic model for transport in statistically homogeneous porous media, *Monte Carlo Methods and Applications* **9**(4) (2003), 341–366.

[121] A. Lallemand-Barrès and P. Peaudecerf, Recherche des relations entre la valeur de la dispersivité macroscopique d'un milieu aquifère, ses autres caractéristiques et les conditions de mesure, *Bulletin du B.R. G. M.* **3**(4) (1978), 277–284.

[122] J. Law, A statistical approach to the interstitial heterogeneity of sand reservoirs, *Trans. AIME* **155** (1944), 202–222.

[123] R. G. Lamb, A scheme for simulating particle pair motions in turbulent fluid, *J. Comput. Physics* **39** (1980), 329–346.

[124] L. D. Landau and E. M. Lifshitz, *Theoretical Physics. Hydrodynamics*. Nauka, Moscow, 1988 (in Russian).

[125] M. Larcheveque and M. Lesieur, The application of eddy-damped markovian closures to the problem of dispersion of particle pairs, *J. Mec.* **20** (1981), 113–134.

[126] M. Y. Leclerc and G. W. Thurtell,, Footprint prediction of scalar fluxes using a Markovian analysis, *Boundary-Layer Meteorol.* **52** (1990), 247–258.

[127] B. J. Legg, Turbulent dispersion from an elevated line source: Markov chain simulations of concentrtaion and flux profiles, *Quart. J. Roy. Meteorol. Soc.* **109** (1983) 645–660.

[128] A. Li and G. Ahmadi, Dispersion and deposition of spherical particles from point sources in a turbulent channel flow, *Aerosol Science and Technology* **16** (1992), 209–236.

[129] C. C. Lin and W. H. Reid, Turbulent flow, theoretical aspects. in: S. Flügge (ed.), *Handbuch der Physik, VIII/2* , Springer-Verlag, Berlin, 1963.

[130] M. Loève, Functions aleatorie de second ordre, *Comptes rendus Acad. Sci., Paris* (1945), pp. 220, in: Lecture Notes in Computer Science 3038, pp. 827–834, Springer-Verlag, 2004.

[131] E. N. Lorenz, Empirical orthogonal functions and statistical weather prediction, Report 1, Statistical Forecasting Project, MIT, 1956.

[132] D. Lucor, C. H. Su, and G. E. Karniadakis, Karhunen–Loève representation of periodic second-order autoregressive processes, in: *Proc. International Conference on Computational Science*, pp. 827–834, Springer-Verlag, 2004.

[133] A. K. Luhar and R. E. Britter, A random walk model for dispersion in inhomogeneous turbulence in a convective boundary layer, *Atmospheric Environment* **23**(9) (1981), 1911–1924.

[134] J. L. Lumley, The structure of homogeneous turbulent flows, in: A. M. Yaglom and V. I. Tatarsky (eds.). *Atmospheric Turbulence and Radio Wave Propagation*, p. 166, Nauka, Moscow, 1967.

[135] T. S. Lundgren, Turbulent pair dispersion and scalar diffusion, *J. Fluid. Mech.* **111** (1981), 27–57.

[136] A. Majda, Random shearing direction models for isotropic turbulent diffusion. *J. Stat. Phys.* **75** (1994), 1153.

[137] G. Matheron and G. de Marsily. Is transport in porous media always diffusive? A counter example.,*Water Resour. Res.*, **16**(5) (1980), 901–917.

[138] C. Maury, P. Gardonio, and S. J. Elliot, A wavenumber approach to modelling the response of a randomly excited panel. II. Application to aircraft panels excited by a turbulent boundary layer. *J. Sound and Vibration* **252**(1) (2002), 115–139.

[139] M. R. Maxey, The motion of small spherical particles in a cellular flow field, *Phys. Fluids* **30** (1987), 1915–1928.

[140] A. McCoy, A numerical study of turbulent diffusion, PhD thesis, University of California at Berkley, 1975.

[141] Y. Meyer. *Ondelettes et opérateurs. I: Ondelettes*, Hermann, Paris, 1990. (English translation: *Wavelets and operators*, Cambridge University Press, 1992).

[142] G. A. Mikhailov, Approximate models of random processes and fields, *Russian J. Comp. Mathem. and mathem. Physics* **23**(3) (1983), 558–566 (in Russian).

[143] G. A. Mikhailov, *Optimization of Weighted Monte Carlo Methods,* Springer-Verlag, Heidelberg New York Berlin, 1992.

[144] G. A. Mikhailov and K. K. Sabelfeld, On a numerical modelling of the pollution in stochastic velocity fields, *Izv. Acad. Nauk SSSR, Fizika atmos. i okeana* **16**(3) (1980), 229–235 (in Russian).

[145] A. S. Monin and A. M. Yaglom, *Statistical Fluid Mechanics,* vol. 1, MIT Press, Cambridge, MA, 1971.

[146] A. S. Monin and A. M. Yaglom, *Statistical Fluid Mechanics,* vol. 2, MIT Press, Cambridge, MA, 1975.

[147] P. Monti and G. Luezzi, A closure to derive a three-dimensional well-mixed trajectory model for non-Gaussian, inhomogeneous turbulence, *Boundary-Layer Meteorol.* **80** (1995), 311–331.

[148] R. L. Naff and A. V. Vecchia, Stochastic analysis of three-dimensional flow in a bounded domain, *Water Resour. Res.* **22** (1986), 695.

[149] H. Nakao, Relative diffusion in turbulent flow by using the effective Hamiltonian method, *J. Physical Society of Japan* **60** (1991), 2942–2950.

[150] L. Nielsen and R. F. Landel, *Mechanical properties of polymers and composites*. 2nd edn., Marcel Dekker, New York. 1994.

[151] S. M. Nikolskiĭ, *Approximation of function of several variables and imbedding theorems*, 2nd edn., Nauka, Moscow, 1977.

[152] E. A. Novikov, Random force method in turbulence theory, *Zh. Eksp. Teor. Fiz.* **44** (1963), 2159.

[153] E. A. Novikov, A relation between the Lagrangian and Eulerian description of turbulence, *Appl. Math. Mech.* **33**(5) (1969), 887–888 (in Russian).

[154] E. A. Novikov, The Lagrangian–Eulerian probability relations and the random force method for nonhomogeneous turbulence, *Phys. Fluids* **29**(12) (1986), 3907–3909.

[155] E. A. Novikov, Two-particle description of turbulence, Markov property, and intermittency, *Phys. Fluids A* **1**(2) (1989), 326–330.

[156] E. A. Novikov, The effects of intermittency on statistical characteristics of turbulence and scale similarity of breakdown coefficients, *Phys. Fluids A* **2**(2) (1990), 814.

[157] A. M. Obukhov, Description of turbulence in terms of Lagrangian variables, *Adv. Geophys.* **6** (1959), 113–115.

[158] B. Oksendal, *Stochastic differential equations*, 5th edn., Springer-Verlag, Berlin Heidelberg New York, 1998.

[159] J. Ophir, S. Alam, B. Garra, et al., Elastography: Imaging the elastic properties of soft tissues with ultrasound, *J. Med. Ultrasonics* **29** (2002), 155–171.

[160] S. A. Orszag and G. S. Patterson Jr.. Numerical Simulation of Three-Dimensional Homogeneous Isotropic Turbulence., *Phys. Rev. Lett.* **28**(2) (1972) 76–79.

[161] H. A. Panofsky and J. A. Dutton, *Atmospheric Turbulence*, Wiley, New York, 1984.

[162] F. Pasquill and F. B. Smith, *Atmospheric Diffusion,* 3rd edn., John Wiley & Sons, Chichester, 1983

[163] G. Pedrizzetti, Quadratic Markov modeling for intermittent turbulence, *Phys. Fluids A* **11**(6) (1999), 1694–1696.

[164] G. Pedrizzetti and E. A. Novikov, On Markov modelling of turbulence, *J. Fluid Mech.* **280** (1989), 69–93.

[165] K. K. Phoon, H. W. Huang, and S. T. Quek, Comparison betveen Karhunen–Loève and wavelet expansions for simulation of Gaussian processes, *Computers and Structures* **82** (2004), 985–991.

[166] K. K. Phoon, H. W. Huang, and S. T. Quek. Simulation of strongly non-Gaussian processes using Karhunen–Loève expansion, *Probabilistic Engineering Mechanics* **20** (2005), 188–198.

[167] D. Pnueli, C. Gutfinger, and M. Fichman, A Turbulent-Brownian model for aerosol coagulation, *Aerosol Science and Technology* **14** (1991), 201–209.

[168] F. Poirion and C. Soize, Numerical methods and mathematical aspects for simulation of homogenous and non homogenous Gaussian vector fields, in: P. Kree and W. Wedig

(eds.), *Probabilistic methods in applied physics*, Lecture Notes in Physics 451, pp. 17–53. Springer-Verlag, Berlin, 1995.

[169] S. B. Pope, Consistency conditions for random-walk models of turbulent dispersion, *Phys. Fluids* **30** (1987), 2374–2379.

[170] S. B. Pope, Lagrangian PDF Methods for Turbulent Flows, *Annu. Rev. Fluid Mech.* **26** (1994), 23–63.

[171] S. B. Pope and Y. L. Chen, The velocity-dispersion probability density function model for turbulent flows, *Phys. Fluids A* **2**(8) (1990), 1437–1449.

[172] V. S. Pougachev, General theory of the correlations of random functions, *Izv. Akad. Nauk SSSR, Ser. Mat.* **17** (1953), 1401–1402.

[173] Ü. Rannik, M. Aubinet, O. A. Kurbanmuradov, K. Sabelfeld, T. Markkanen, and T. Vesala, Footprint analysis for measurements over a heterogeneous forest, *Boundary-Layer Meteorology* **97**(1) (2000), 137–166.

[174] Ü. Rannik, T. Markkanen, J. Raittila, P. Hari, and T. Vesala, Turbulence statistics inside and over forest: influence on footprint prediction, *Boundary-Layer Meteorol.* (2003), accepted.

[175] M. R. Raupach, Canopy transport processes, in: W. L. Steffen and O. T. Denmead (eds.), *Flow and Transport in the Natural Environment*, pp. 95–127, Springer-Verlag, New York, 1988.

[176] M. R. Raupach and B. J. Legg, Turbulent diffusion from an elevated line source: measurements of wind-concentration moments and budgets, *J. Fluid Mech.* **136** (1983), 111–137.

[177] A. M. Reynolds, On the formulation of Lagrangian stochastic models of scalar dispersion within plant canopies, *Boundary-Layer Meteorol.* **86** (1997), 333–344.

[178] A. M. Reynolds, On trajectory curvature as a selection criterion for valid Lagrangian stochastic dispersion models, *Boundary-Layer Meteorol.* **88** (1998), 77–86.

[179] N. E. Rider, Eddy diffusion of momentum, water vapour and heat near the ground, *Phil. Trans. Roy. Soc. A* **246** (1954), 481–501.

[180] J. de la Rocha, B. Doiron, E. S. .Brown, K. Josi, and A. Reyes, Correlation between neural spike trains increases with firing rate, *Nature* **448** (2007), 802–806.

[181] H. Rodean, Stochastic lagrangian models of turbulent diffusion, *Meteorological Monographs* **26**(48) (1996), 84 pp.

[182] Y. A. Rozanov and F. Sanso, The analysis of the Neumann and the oblique derivative problem: the theory of regularization and its stochastic version, *J. Geodesy* **75** (2001), 391–398.

[183] Y. Rubin, Stochastic modeling of macrodispersion in heterogeneous porous media, *Water Resources Research* **26**(1) (1990), 133–141.

[184] Y. Rubin and G. Dagan, Stochastic analysis of boundaries effects on head spatial variability in heterogeneous aquifers: 1. Constant head boundary, *Water Resources Research* **24**(10) (1988), 1689–1697.

[185] Y. Rubin and G. Dagan, Stochastic analysis of boundaries effects on head spatial variability in heterogeneous aquifers: 2. Impervious boundary, *Water Resources Research* **25**(4) (1989), 707–712.

[186] Y. Rubin and G. Dagan, A note on head and velocity covariances in three-dimensional flow through heterogeneous anisotropic porous media, *Water Resources Research* **28**(5) (1992), 1463–1470.

[187] S. Running, A blueprint for improved global change monitoring of the terrestrial biosphere, *Earth. Obs.* **10** (1998), 8–12.

[188] D. Russo, On the velocity covariance and transport modeling in heterogeneous anisotropic porous formations 1. saturated flow, *Water Resources Research* **31**(1) (1995), 129–137.

[189] K. K. Sabelfeld, Coagulation of Aerosol Particles in Intermittent Turbulent Flows. Rarefied Gas Dynamics, vol. 2, pp. 77–84, in: *Proceedings of the 21st Int. Symposium.* Marseille, France, July 26–31, 1998.

[190] K. K. Sabelfeld, On approximate evaluation of Wiener path integrals by the Monte Carlo method, *J. Comp. Mathem. and Mathem. Physics.* **19**(1) (1979), 29–43 (in Russian).

[191] K. K. Sabelfeld. *Monte Carlo methods in boundary value problems*, Springer-Verlag, Heidelberg, New York, Berlin, 1991.

[192] K. K. Sabelfeld, Evaluation of elastic coefficients from the correlation and spectral tensors in respond to boundary random excitations. Proceed. Intern. Conference on Inverse and Ill-Posed pproblems of mathematical physics. dedicated to professor M. M. Lavrentiev on occation of his 75-th birthday. Novosibirsk, August 21–24, 2007.

[193] K. K. Sabelfeld, Expansion of random boundary excitations for some elliptic PDEs. Monte Carlo Methods and Applications. vol.13 (2007), N5–6, 403–451.

[194] K. K. Sabelfeld, Stokes flows under random boundary velocity excitations, *J. Stat. Physics* **133**(6) (2008), 1107–1136.

[195] K. K. Sabelfeld and A. A. Kolodko, Monte Carlo simulation of the coagulation processes governed by Smoluchowski equation with random coefficients, *Monte Carlo Methods and Applications* **3**(4) (1997), 275–311.

[196] K. K. Sabelfeld and D. Kolyukhin, Stochastic Eulerian model for the flow simulation in porous media, Monte Carlo Methods and Applications **9**(3) (2003), 271–290.

[197] K. K. Sabelfeld and O. A. Kurbanmuradov, Statistical modelling of diffusion in random velocity fields, *Sov. J. Numerical Analysis and Math. Modelling* **4**(1) (1989), 51–68.

[198] K. K. Sabelfeld and O. A. Kurbanmuradov, Numerical statistical model of classical incompressible isotropic turbulence, *Sov. J. Numer. Analysis and Math. Modelling* **5**(3) (1990), 251–263.

[199] K. K. Sabelfeld and O. A. Kurbanmuradov, Stochastic Lagrangian Models for Two-Particle Motion in Turbulent Flows, *Monte Carlo Methods and Appl.* **3** (1997), 53–72.

[200] K. K. Sabelfeld and O. A. Kurbanmuradov, One-particle stochastic Lagrangian model for turbulent dispersion in horizontally homogeneous turbulence, *Monte Carlo Methods and Applications* **4**(2) (1998), 127–140.

[201] K. K. Sabelfeld and O. A. Kurbanmuradov, Two-particle stochastic Eulerian-Lagrangian models of turbulent dispersion, *Mathematics and Computers in Simulation* **47**(2–5) (1998), 429–440.

[202] K. K. Sabelfeld, S. V. Rogasinsky, A. A. Kolodko, and A. I. Levykin, Stochastic algorithms for solving Smoluchowsky coagulation equation and applications to aerosol growth simulation *Monte Carlo Methods and Applications* **2**(1) (1996), 41–87.

[203] K. K. Sabelfeld and I. A. Shalimova, *Spherical Means for PDEs.* VSP, Utrecht, 1997.

[204] K. K. Sabelfeld and I. A. Shalimova, Random walk methods for static elasticity problems, *Monte Carlo Methods and Applications* **8**(2) (2002), 171–200.

[205] K. K. Sabelfeld and I. A. Shalimova, Elastic half-plane under random displacement excitations on the boundary, *J. Stat. Physics* **32**(6) (2008), 1071–1095.

[206] K. K. Sabelfeld, I. Shalimova, and A. Levykin, Random Walk on Fixed Spheres for Laplace and Lamé equations, *Monte Carlo Methods and Applications* **12**(1) (2006), 55–93.

[207] K. K. Sabelfeld, I. Shalimova, and A. Levykin, Stochastic simulation method for a 2D elasticity problem with random loads, *Probabilistic Engineering Mechanics* (2007), accepted for publication.

[208] K. K. Sabelfeld and N. A. Simonov, *Random Walks on Boundary for Solving PDEs.* VSP, Utrecht, 1994.

[209] K. K. Sabelfeld and V. I. Tatarskii, On the approximate evaluation of Wiener path integrals, *Dokl. Akad. Nauk SSSR* **243**(4) (1978), 905–908 (in Russian).

[210] P. Saffman and J. S. Turner, On the collision of drops in turbulent clouds, *J. Fluid Mech.* **1** (1956), 16–30.

[211] P. Saladin and V. Fiorotto, Solute transport in highly heterogeneous aquifers, *Water Resources Research* **34**(5) (1998), 949–961.

[212] F. Sanso and G. Venuti, White noise stochastic BVP's and Cimmino's theory. PL da Vinci – IV Hotine-Marussi Symposium on Mathematical Geodesy: Trento, Italy, 2001, pp. 5–20.

[213] B. L. Sawford, Lagrangian Monte Carlo simulation of the turbulent motion of a pair of particles, *Quart. J. R. Met. Soc.* **108** (1982), 207–213.

[214] B. L. Sawford, The effect of Gaussian particle-pair distribution function in the statistical theory of concentration fluctuations in homogeneous turbulence, *Quart. J. R. Met. Soc.* **109** (1983), 339–354.

[215] B. L. Sawford, Lagrangian Statistical Simulation of Concentration Mean and Fluctuation Fields, *J. Climate and Applied Meteorology* **24** (1985), 1152–1166.

[216] B. L. Sawford, Recent developments in the Lagrangian stochastic theory of turbulent dispersion, *Bound. Layer Met.* **62** (1993), 1–4, 197–215.

[217] B. L. Sawford and F. M. Guest, Uniqueness and universality of Lagrangian stochastic models of turbulent dispersion, in: *Proc. AMS 8th Symp. on Turbulence and Diffusion.* San Diego, 1988, pp. 96–99.

[218] B. L. Sawford and J. C. R. Hunt, Effects of turbulence structure, molecular diffusion and source size on scalar fluctuations in homogeneous turbulence, *J. Fluid Mech.* **165** (1986), 373–400.

[219] H. Schwarze, U. Jaekel, and H. Vereecken, Estimation of Macrodispersion by Different Approximation Methods for Flow and Transport in Randomly Heterogeneous Media, *Transport in Porous Media* **49**(2) (2001), 267–287.

[220] H. Sigurgeirsson and A. M. Stuart, A model for preferential concentration, *Phys. Fluids* **14**(12) (2002), 4352–4361.

[221] H. P. Schmid, Source areas for scalar and scalar fluxes, *Boundary-Layer Meteorology* **67** (1994), 293–318.

[222] L. Smith and R. A. Freeze, Stochastic Analysis of Steady State Groundwater Flow in a Bounded Domain. 1. One-Dimensional Simulation, *Water Resour. Res.* **15**(3) (1979a), 521–528.

[223] L. Smith and R. A. Freeze, Stochastic Analysis of Steady State Groundwater Flow in a Bounded Domain. 2. Two-Dimensional Simulation, *Water Resour. Res.* **15**(6) (1979b), 1543–1559.

[224] M. K. Schneider and A. S. Willsky, A Krylov Subspace Method for Covariance Approximation and Simulation of Random Processes and Fields. *Multidimensional Systems and Signal Processing* **14**(4) (2003), 295–318.

[225] H. P. Schuepp, M. Y. Leclerc, J. I. MacPherson, and R. L. Desjardins, Footprint prediction of scalar fluxes from analytical solutions of the diffusion equation, *Boundary-Layer Meteorology* **50** (1990), 355–373.

[226] M. Shinozuka, Simulation of multivariate and multidimensional random processes. *J. Acoust. Soc. Am.* **49** (1971), 357–368.

[227] O. F. Smidts. Analyse probabiliste du risque du stockage de déchets radioactifs par la méthode des arbres d'événements continus, PhD thesis, Université Libre de Bruxelles, Brussels, 1997.

[228] T. T. Soong, *Random Differential Equations in Science and Engineering*, Academic Press, New York, 1973.

[229] R. B. Sowers, Multidimensional reaction-diffusion equation with white-noise boundary perturbations, *The Annals of Probability* **22**(4) (1994), 2071–2121.

[230] P. Spanos and R. Ghanem, Stochastic finite element expansion for random media, *J. Engineering Mechanics, ASCE* **115** (1989), 1035–1053.

[231] P. D. Spanos, V. Ravi, and S. Rao, Random Field Representation in a Biorthogonal Wavelet Basis, *J. Engineering Mechanics* **127**(2) (2001), 194–205.

[232] R. B. Stull, *An Introduction to Boundary Layer Meteorology*, Kluwer Academic Publishers, Dordrecht, 1988.

[233] D. Talay, Simulation and numerical analysis of stochastic differential systems: a review. *Rapport de recherche INRIA*, 1990. p. 1313.

[234] D. Tartakovsky and D. Xiu, Stochastic analysis of transport in tubes with rough walls, *J. Comp. Physics* **217**(1) (2006), 248–259.

[235] G. I. Taylor, Diffusion by continuous movements, *Proc. London Math. Soc. Ser. 2* **20** (1921), 196–211.

[236] D. J. Thomson, On the relative dispersion of two particles in homogeneous stationary turbulence and the implications for the size of concentration fluctuations at large times, *Quart. J. R. Met. Soc.* **112** (1986), 890–894.

[237] D. J. Thomson, Criteria for the selection of stochastic models of particle trajectories in turbulent flows, *J. Fluid. Mech.* **180** (1987), 529–556.

[238] D. J. Thomson, A stochastic model for the motion of particle pairs in isotropic high-Reynolds-number turbulence, and its application to the problem of concentration variance, *J. Fluid. Mech.* **210** (1990), 113–153.

[239] D. J. Thomson, The separation of particle pairs in the eddy-damped quasinormal Markovian approximation, *Phys. Fluid. Mech.* **8**(2) (1996), 642–644.

[240] D. J. Thomson and B. J. Devenish, Particle pair in kinematic simulations, *J, Fluid Mechanics* **526** (2005), 277–302.

[241] A. F. B. Tompson, R. Ababou, and L. W. Gelhar, Implementation of the three-dimensional turning bands random field generator, *Water Resour. Res.* **25**(10) (1989), 2227–2243.

[242] A. F. B. Tompson and L. W. Gelhar, Numerical simulation of solute transport in tree-dimensional randomly heterogeneous porous media, *Water Resour. Res.* **26**(10) (1990), 2541–2562.

[243] H. Triebel, *Theory of function spaces*. Birkhäuser, Leipzig Basel Boston Stuttgart, 1983.

[244] C. Turfus and J. C. R. Hunt, A stochastic analysis of the displacements of fluid elements in inhomogeneous turbulence using Kraichnan's method of random modes, in: G. Comte-Bellot and J. Mathieu (eds.), *Advances in Turbulence*, pp. 191–203, Springer-Verlag, Berlin, 1987.

[245] S. A. Utev, On central limit theorem for set schemes of random variables with φ-mixed conditions, *Teor. Ver. i Prim.* **35** (1990), 110–117 (in Russian).

[246] R. Valentini et al., Respiration as the main determinant of carbon balance in European forests, *Nature* **404** (2000), 861–865.

[247] H. Van Dop, F. T. M. Nieuwstadt, and J. C. R. Hunt, Random walk models for particle displacements in inhomogeneous unsteady turbulent flows, *Phys. Fluids* **28** (1985), 1639–1653.

[248] N. G. Van Kampen, *Stochastic process in Physics and Chemistry*, North–Holland, Amsterdam, 1984.

[249] E. Van Marcke, *Random Fields: Analysis and Synthesis*, The MIT Press, Cambridge, MA, 1988.

[250] H. L. Van Trees,, *Detection, Estimation and Modulation Theory*. parts 1–3, Hohn Wiley and Sons, New York, 1968.

[251] H. Vereecken, G. Lindenmayr, O. Neuendorf, U. Döring, and R. Seidemann, TRACE: A mathematical model for reactive transport in 3D variably saturated porous media, Technical report, Forschungszentrum Jülich, ICG-4, 1994.

[252] V. S. Vladimirov, *Generalized functions in mathematical physics*, Nauka, Moscow, 1979 (in Russian).

[253] B. G. Wager and E. D. Nadegina, *Atmospheric boundary layer under horizonatal inhomogeneity*, Gidrometeoizdat, Leningrad, 1979 (Russian).

[254] M. M. R. Williams and S. K. Loyalka, *Aerosol Science.Theory and Practice*, Pergamon, New York, 1991.

[255] J. D. Wilson and T. K. Flesch, Trajectory curvature as a selection criterion for valid Lagrangian stochastic models, *Boundary-Layer Meteorol.* **84** (1997), 411–426.

[256] J. D. Wilson and B. L. Sawford, Revew of Lagrangian stochastic models for trajectories in the turbulent atmosphere. *Boundary Layer Met.* **78** (1996), 191–210.

[257] D. Xiu and J. Shen, An efficient spectral method for acoustic scattering from rough surfaces, *Communications in computational physics* **2**(1) (2007), 54–72.

[258] X. F. Xu. A multiscale stochastic finite element method on elliptic problems involving uncertainties, *Computer Methods in Applied Mechanics and Engineering* **196**(25–28) (2007), 2723–2736.

[259] A. M. Yaglom, *Correlation theory of stationary and related random functions. Vol. I: Basic results*. Springer-Verlag, Berlin, 1987.

[260] J. Yang, D. Zhang, and Z. Lu, Stochastic analysis of sturated-unsaturated flow in heterogeneous media by combining Karhunen–Loève expansion and perturbation method, *J. Hydrology* **294** (2004), 18–38.

[261] P. K. Yeung, Direct numerical simulation of two-particle relative diffusion in isotropic turbulence, *Phys. Fluids* **6** (1994), 3416–3428.

[262] P. K. Yeung and S. B. Pope, Lagrangian statistics from direct numerical simulation of isotropic turbulence, *J. Fluid. Mech.* **207** (1989), 531.

[263] D.-H. Yu, *Natural Boundary Integral method and its Applications*, Science Press, Kluwer, 2002

[264] B. A. Zeldin and P. D. Spanos, Random Field Representation and Synthesis Using Wavelet Basis, *ASME J. Applied Mechanics* **63**, (1996), 946–952.

[265] D. Zhang and Z. Lu, An efficient, high-order perturbation approach for flow in random porous media via Karhunen–Loève and polynomial expansions, *J. Comp. Physics* **194** (2004), 773–794.

Index

Bei Fragen zur Produktsicherheit wenden Sie sich bitte an:
If you have any questions regarding product safety,
please contact:

Walter de Gruyter GmbH
Genthiner Straße 13
10785 Berlin
productsafety@degruyterbrill.com